D1206307

Fine Needle Aspiration Cytology of Superficial Organs and Body Sites

Fine Needle Aspiration Cytology of Superficial Organs and Body Sites

Kim R. Geisinger
Professor and Director
Surgical Pathology and Cytopathology
Wake Forest University
School of Medicine
The Bowman Gray Campus
Winston-Salem, North Carolina

Jan F. Silverman
Professor and Chairman
Department of Pathology
Allegheny General Hospital
Pittsburgh, Pennsylvania

CHURCHILL LIVINGSTONE
A Division of Harcourt Brace & Company
New York • Edinburgh • London • Philadelphia • San Francisco

Churchill Livingstone
A Division of Harcourt Brace & Company

The Curtis Center
Independence Square West
Philadelphia, Pennsylvania 19106

Library of Congress Catalog card number is 99–71799

FINE NEEDLE ASPIRATION CYTOLOGY
OF SUPERFICIAL ORGANS AND BODY SITES

ISBN 0–443–07963–3

Printed in the United States of America

Last digit is the print number: 9 8 7 6 5 4 3 2 1

To our families . . .

Lori and Mary

Kristen, Brian, Mischell, Jeff, and Laura

Jessica and Melissa

and

To all others who have touched our lives personally and professionally.

Contributors

David J. Dabbs, MD
Professor and Director of Anatomic Pathology
Department of Pathology and Laboratory Medicine
Allegheny General Hospital
Pittsburgh, Pennsylvania

Andrew S. Field, FRCPA FIAC Dip Cytopath (RCPA)
Deputy Director and Senior Staff Specialist
Department of Anatomical Pathology
St. Vincent's Hospital
Sydney, New South Wales, Australia

Kim R. Geisinger, MD
Professor and Director of Surgical Pathology and Cytopathology
Department of Pathology
Wake Forest University School of Medicine
Winston-Salem, North Carolina

Scott E. Kilpatrick, MD
Assistant Professor
Department of Pathology and Laboratory Medicine
University of North Carolina
Chapel Hill, North Carolina

Celeste N. Powers, MD, PhD
Professor and Director of Cytopathology
Department of Surgical Pathology and Cytopathology
Medical College of Virginia
Virginia Commonwealth University
Richmond, Virginia

Robert O. Rainer, M.D.
Assistant Professor
Department of Pathology
Wake Forest University School of Medicine
Winston-Salem, North Carolina

Dominic S. Raso, M.D.
Pathologist
Pathology Consultants of Central Virginia, Inc.
Lynchburg, Virginia

Jan F. Silverman, MD
Professor and Chairman
Department of Pathology and Laboratory Medicine
Allegheny General Hospital
Pittsburgh, Pennsylvania

Michael W. Stanley, MD
Professor of Pathology
University of Minnesota
Chairman
Department of Pathology
Hennepin County Medical Center
Minneapolis, Minnesota

Preface

Superficial masses are very frequent clinical occurrences that may cause great concern for the patient as well as intrigue and bewilderment for the clinician. Fine needle aspiration biopsies of these inflammatory, reactive, and neoplastic lesions present vast cytomorphologic arrays that challenge even the most experienced pathologist. To optimize diagnostic yield, we firmly advocate that the combination of pattern recognition at low magnification and a detailed high-power examination of cytomorphologic attributes always be utilized. Accordingly, the fundamental goals of *Fine Needle Aspiration Cytology of Superficial Organs and Body Sites* include strong emphasis on the surgical pathologist's approach to aspiration biopsies of these mass lesions. Thus, a practical working knowledge of the underlying histopathologic bases for the cytomorphologic findings in Diff-Quik- and Papanicolaou-stained aspiration smears is essential.

Throughout the text, photomicrographs, and charts, we have attempted to portray our own style of daily practice. However, in endeavoring to provide a greater breadth of expertise to this book, we have enlisted authors who also embrace the same practice philosophy. As described in this text, specific diagnoses may often be rendered after integrating clinical, radiographic (especially in orthopaedic lesions), cytomorphologic, and, at times, ancillary diagnostic results. When finalizing our interpretations, we often utilize terminology identical to that used in conventional surgical pathology reports. However, in other instances, the aspiration specimen will only allow generation of a differential diagnosis that may be detailed as a comment in the cytology report. In general, we believe this diagnostic integration will allow the pathologist to minimize false-negative and avoid false-positive diagnoses. In contrast to deep-seated mass lesions, the pathologist is often the clinical aspirator of superficial masses. In many situations, this is preferable to specimens procured by relatively inexperienced clinicians. Furthermore, the pathologist's diagnoses are enhanced by the integration of the "gross" findings when performing the clinical examination and procedure. Regardless of who performs the aspirate, we adhere to the belief that an immediate interpretation of the aspirate is highly valuable. This "quick read" permits not only the determination of specimen adequacy but also may generate a preliminary diagnosis in many cases. In addition, it allows for the procurement of additional material for extra direct smears, cell blocks, immunocytochemistry, electron microscopy, molecular biological assays, cytogenetic examinations, and microbiological culture.

We believe that the experienced authors who have contributed in creating this text have brought their considerable expertise to this publication. Although our philosophy and approach to fine needle aspiration biopsies are quite similar, the variations that may be present reflect the specific practices from our various institutions. We hope that this text will resonate with your practice and the reader will discover this integrated diagnostic approach to be readily applicable in all types of pathology practice settings.

Contents

CHAPTER 1

Lymph Nodes

In both adults and children, lymphadenopathy can be secondary to myriad causes and represents a common clinical problem, the consequences of which range from negligible to very serious and fatal. In conjunction with a good clinical history and physical examination, fine needle aspiration (FNA) biopsy of these nodes frequently supplies sufficient information to allow the clinician to triage the patient and then observe, treat with antibiotics, initiate anticancer therapy, stage the malignant process, or obtain a lymph node biopsy with histologic examination.[1–10] Thus, FNA biopsy cytomorphology, combined as required with ancillary diagnostic procedures, advances clinical management in a rapid, cost-effective, and safe manner.

One important clinical question is whether the lymphadenopathy is localized or generalized. In general, a localized process favors a benign diagnosis, whereas malignancy (especially the non-Hodgkin's lymphomas) is more likely to present with generalized lymph node enlargement.[1] However, a localized rapidly enlarging node with tender, erythematous, and swollen surrounding tissues favors a reactive process. A lymph node that has noticeably increased in size over days to weeks also favors a benign etiology.

In our experience, a benign nonspecific or reactive hyperplasia of lymphoid tissue is one of the most common causes of lymphadenopathy. Infectious etiologies that may be recognized include pyogenic bacteria, mycobacteria,[11–15] fungi,[16] cat-scratch disease,[17,18] toxoplasmosis,[19–22] and infectious mononucleosis.[23,24] Other nonneoplastic conditions include sarcoidosis, Rosai-Dorfman disease,[25–27] and Kikuchi's disease.[28] Metastatic neoplasms, especially carcinomas and melanomas, are another common source of enlarged lymph nodes.[1–3,5,29–31] Finally, primary malignant lymphomas represent an additional important cause of lymphadenopathy. It is this last group that represents the most problematic aspect of FNA biopsy diagnosis.[32–36] Most investigators agree that metastatic malignancies and specific infections can be identified in aspiration biopsies in most cases. It is the distinction between reactive hyperplasias and the malignant lymphomas that presents the greatest challenge. Here, more than in any other situation, ancillary diagnostic tests are useful.

TECHNIQUE OF FNA BIOPSY OF LYMPH NODES

It is important that the differential diagnosis of lymphadenopathy be considered in the individual patient before the FNA. The experienced aspirator is in an excellent position to maximize the diagnostic efficiency of the procedure by following a set protocol and maintaining a constant awareness of the possibility of lymphoma or infection. In our institutions, pathologists and cytotechnologists attending aspirations of other sites (e.g., lung, mediastinum, abdomen) play a vital role in initiating the workup for a potential lymphoma or infectious disease.

Needle-stick injuries are a potential source of major concern for aspirators sampling lymph nodes in patients who are HIV-positive and those with possible infectious lymphadenopathy. Ideally, the aspirator should regard all aspirates as potentially infectious and develop a consistent technique designed to minimize risks for all aspirations. Sterile gloves are worn, and care is taken with handling the aspiration gun, needle, and syringes to avoid needle-stick injuries. For a right-handed operator, the syringe holder is held in the right hand and every movement to disengage or replace needles or syringes is begun by the left hand starting at the base of the gun, moving down the syringe to the needle. The left hand is never brought "back up" to the needle. Each sample is expelled onto slides and the needle is rinsed, then grasped and immediately disposed of in a sharps container; the syringe holder with needle *in situ* is never placed on the work area. Glasses (or protective perspex shields) and surgical face masks and gowns are worn when performing aspirates on hepatitis B- or C-positive and HIV-positive cases to minimize conjunctival and mucosal contamination. Forceful expelling of the material within the needle onto the glass slides and complicated techniques of selecting material for smearing from slides are strongly discouraged. Multiple air-dried and alcohol-fixed slides, however, are prepared using transferral techniques from one slide to another. FNA biopsies are best performed in designated procedural rooms.

In some circumstances, a nonaspirating technique using a 22- to 25-gauge needle without an attached syringe is used on the first approach to a lymph node aspirate. Once through the skin, this technique allows the aspirator to feel the needle penetrate the node capsule, and multiple passes through the node can be obtained for each skin puncture. No local anesthetic is used. The syringe holder technique (needle attached to syringe) is used on some occasions to gain a larger, if potentially blood-stained, sample for culture, flow cytometry, or cell block preparation. The exact technique used depends in part on the preference and experience of the aspirator, as well as the clinical situation. The aspirator can gain an initial impression at the time of smearing the material expelled from the needle as to the nature of the lesion. Lymphomatous lymph nodes tend to feel soft and buttery under the needle, and the expelled material is grayish-pink and smears smoothly. Hodgkin's disease lymph nodes and those infected by *Mycobacterium* tend to be firm to the needle and gritty with a low cell yield on smearing, so aspiration is required. Metastatic carcinomas tend to be firm, although metastatic squamous cell carcinomas commonly yield copious necrotic debris and keratin, and adenocarcinomas may produce slippery mucus and debris. Suppurative lymphadenitis is soft and tender to the needle and often yields obvious pus, which may have an offensive smell. These variations provide a "wake-up call" to the operator to ensure adequate safety precautions and to obtain material for culture, special stains, flow cytometry, and a cell block. This may require extra passes.

In all FNA biopsies of lymph nodes, air-dried Romanowsky-stained and alcohol-fixed Papanicolaou-stained smears are prepared. We strongly advocate the use of both stains in every case. Whenever there is any suspicion of an infection, air-dried smears for methenamine silver and auramine stains and alcohol-fixed smears for Gram's and mucicarmine stains are prepared to facilitate the diagnosis of fungal, mycobacterial, bacterial, and cryptococcal infections.

If an infection is suspected, material must be put aside for bacteriologic workup. Skin disinfection with alcohol swabs is the only skin preparation used. A separate pass can be performed, or alternatively, after expelling material for smears, the needle and syringe from each pass can be rinsed in a small amount (2–3 ml) of sterile saline and placed in a sterile container for bacteriologic culture. Additional passes or needle and syringe washes can be collected in Hank's balanced salt solution with 10–20% fetal calf serum for preparation of a cell block using serum and thrombin and the centrifuged material. The cell block allows routine paraffin block-based immunocytochemistry to be applied to FNA material to diagnose metastatic tumors and malignant lymphomas. Routine immunoperoxidase studies on cell blocks in cases of suspected lymphoma include CD3, CD20, CD79a, CD45, kappa and lambda light chains, and at times CD68 studies. Histochemical stains for bacteria, mycobacteria, and fungi can also be used. Gram's stain of both direct smears and cell blocks can be useful, but in many cases culture of FNA material provides a bacteriologic assessment including antibiotic sensitivities, which is of far greater benefit to the patient.

Material is routinely placed in a suitable medium for flow cytometry studies if there is any possibility of lymphoma. Direct immunohistochemical staining of cytospin smears is another technique for the diagnosis of lymphomas.[34–37] After FNA diagnosis of a lymphoma in our institutions, excisional biopsy for definitive classification of the lymphoma is the preferred management in some patients, including immunocompromised patients, but such an approach may differ in other centers. Factors such as localization of the node (e.g., retroperitoneum) and health status of the patient may influence whether a histologic examination will be done in some cases.

SAMPLE COLLECTION

Ideally, a lymph node FNA biopsy should be assessed on site to ensure adequate cellularity and to consider the possibility of a lymphoid proliferation. If an aspirate is suspicious for lymphoma, an additional sample is drawn for flow cytometry and put in a buffered salt solution. Additional material should also be obtained to make a cell block in case immunostaining is required for nonhematopoietic antigens.

GENERAL CYTOMORPHOLOGIC ATTRIBUTES

As with aspiration biopsies of all body sites, a systematic assessment of the specimen is essential to achieve the correct diagnosis. This includes evaluation of the overall smear cellularity, the pattern of arrangement of cells, the predominant cell type, and the nature of any material in the smear background.

Primary lymphoid proliferations, both benign and malignant, produce moderate to highly cellular smears with marked dispersal of cells. An exception to this dispersed, dissociative picture is the presence of lymphohistiocytic aggregates, which are derived from germinal centers and consist of irregular syncytial clusters of variably sized lymphocytes, histiocytes, and probable dendritic reticulum cells.[1,38] Most metastatic neoplasms, however, include tissue fragments as evidence of true intercellular cohesion in the smears; one common exception is melanoma. A consistent background component of aspirates of both benign and malignant lymphoid proliferations is the lymphoglandular body, which represents pale, irregular fragments of lymphoid cytoplasm that are sheared away from the cell during the aspiration procedure.[1] Other background elements may include mucus, foreign material, and necrotic debris. Determining the predominant cell type is useful in distinguishing benign from malignant lymphoid lesions: in most benign lymphoid abnormalities, small mature-appearing lymphocytes dominate.

ANCILLARY TECHNIQUES

Flow Cytometry

Flow cytometric analysis is a valuable diagnostic aid when assessing hematologic malignancies.[39] The use of flow cytometry for cell surface phenotyping complements cytomorphologic assessment when evaluating the involvement of lymph nodes for non-Hodgkin's lymphoma (particularly in cases of partial involvement or follicular [nodular] patterns and in distinguishing various subsets of B-cell non-Hodgkin's lymphomas).

Flow cytometry has been used to study human hematopoietic neoplasms since the early 1970s; with technological advances in both the microcomputer and laser industries, flow cytometry entered many clinical laboratories in the middle to late 1980s. The technology is used to study cell surface markers, DNA content, and proliferative activity. Along with bone marrow and peripheral blood, FNA biopsies readily yield material for flow cytometric analysis, adding objective immunologic data to permit classification of a particular lymphoma in the new Revised European and American Lymphoma scheme.[40]

Using monoclonal antibody staining, flow cytometry is a more rapid technique than molecular diagnostic modalities such as gene rearrangement studies. Many monoclonal antibodies directed toward lymphoid antigens are available, especially because FNA yields with fresh tissue and the repertoire of antibodies is not limited by fixation and antigen retrieval steps. Flow cytometry is a semiautomated analytic process that can be used to assess a large number of cells and detect a small neoplastic population, yielding high sensitivity and specificity. These advantages generally outweigh the disadvantages of the high cost of equipment and antibody reagents, and the lack of correlation of the morphologic and immunologic phenotype achieved with immunocytochemistry. Flow cytometry is a useful adjunct in the diagnosis of B-cell non-Hodgkin's lymphomas,[41] but it has less utility in T-cell non-Hodgkin's lymphomas and little use in Hodgkin's disease.[42]

Molecular Techniques

Material obtained by FNA biopsy can be used in assays for gene rearrangements using DNA or RNA probes. Southern blot analysis, dot blot, polymerase chain reaction, and *in situ* hybridization can be used.[43–48]

Clonal rearrangements of genes for heavy and light immunoglobulin chains and T-cell receptors are very supportive of the diagnosis of non-Hodgkin's lymphomas, as are oncogene rearrangements, such as the bcl-2 gene in follicular lympohomas.[44] A clonal population can be demonstrated even though neoplastic lymphoid cells are mixed with reactive cells.

Specific chromosomal translocations can be sought in aspirated material to confirm a diagnosis of Burkitt's lymphoma or small cleaved cell lymphoma or to type leukemias.[49]

Nucleic acid from microorganisms such as herpes simplex, cytomegalovirus, and *Pneumocystis carinii*

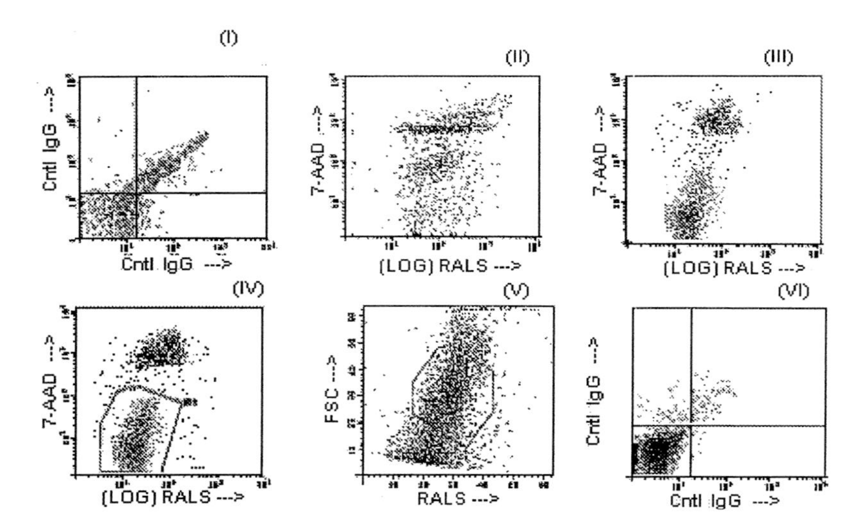

FIGURE 1–1. Use of a viability marker, 7-actinomycin (7-AAD). The compound is an important adjunct to the analysis for lymph node FNAs because it eliminates an unwanted artifact. The ungated, baseline control is seen in Panel I. In this cytogram, the 45° angle commonly produced by debris and necrotic cells is noted. This is an unwanted artifact that hinders analysis. Panel II reveals the pattern seen when extensive nonviable cells are present. Most of the events retain the fluorescent 7-AAD. Panel III shows a population of viable cells that were capable of removing the 7-AAD from their cytoplasm. This population can be gated to include only the live cells for analysis, as noted in Panel IV. A logical NOT gate will exclude debris and clumps, as noted in Panel V. These two gates must be combined because debris, which also causes nonspecific binding, will not take up 7-AAD, and it needs to be excluded from the analysis population. Panel VI shows the results. The control population includes viable cells, and the background monoclonal binding is confined to the first quadrant.

can be identified in aspirated material by *in situ* hybridization.

GENERAL APPROACH FOR IMMUNOPHENOTYPING LYMPH NODES

Currently, there is no standard panel for analyzing lymphoid neoplasms by flow cytometry, but it should include markers for B cells (surface immunoglobulin, CD19, CD20, CD79a), T cells (CD3, CD4, CD8), and nonlymphoid cells (CD14, CD45). Our complete lymphoid panel consists of CD19 vs. CD5, CD10 vs. CD33, CD 20 vs. Kappa, CD20 vs. Lambda, CD 2 vs. HLA-DR, CD7 vs. CD3, and CD4 vs. CD8. In many cases we also include a CD45 vs. CD14, as well as using 7-actinomycin D to gate out dead and dying cells in all tubes (Fig. 1–1).[50] Assessing the immunologic status of cells can include the presence or absence of monoclonality, antigen or light scatter aberrancy, population excesses, and reactive changes.[51–62] The CD19 vs. CD5 study is a good example. It is used to gauge the number of normal B cells and T cells, because CD19 is a pan-B marker and CD5 is associated with T cells. In low-grade non-Hodgkin's lymphomas, there may be dual expression of CD19 and CD5. Looking at the intensity of antigen expression on the surface is helpful in subclassifying the type of lymphoma. Thus, the antigen aberrancy helps to confirm the presence of a lymphoma as well as to subclassify it.

FNA biopsies produce a variable yield of cells. As a general rule, at least 100,000 cells are needed for flow cytometry. This number permits a limited panel of antibodies, such as CD5 vs. CD19, CD20 vs. kappa, and CD20 vs. lambda. This provides a B-cell sum and T-cell sum. Usually this panel can detect the presence of a B-cell non-Hodgkin's lymphoma, but it does not further classify the lymphoma.

NORMAL IMMUNOPHENOTYPE

A good understanding of the normal complement of lymphocytes in a lymph node is essential for interpretation of flow cytometry results. Normal lymph nodes contain areas where circulating T cells predominate, but during an immune response these can become sites of B-cell proliferation (Fig. 1–2). In addition to the predominant lymphoid cells, a lymph node will also contain histiocytes and connective tissue elements, which we typically do not test for using monoclonal antibodies. However, these cells, especially the histiocytes, can produce false-positive results through nonspecific binding of the monoclonal antibodies. In a resting node, T cells will predominate (60% T cells and 40% B cells); this T-cell compartment is made up of two major types of cells. Helper cells are CD4-positive and express a full complement of other T-cell antigens. Suppressor T cells are characterized by the presence of CD8. Both of these cell types also express a full complement of other T-cell antigens. These other antigens include CD2, CD3, CD5, and CD7. Normal T cells will not express both CD4 and CD8 at the same time. In a normal node, the number of CD4-positive cells is usually twice the number of CD8-positive cells. There are no firm rules regarding the percentage of CD4-positive cells vs. CD8-positive cells when assess-

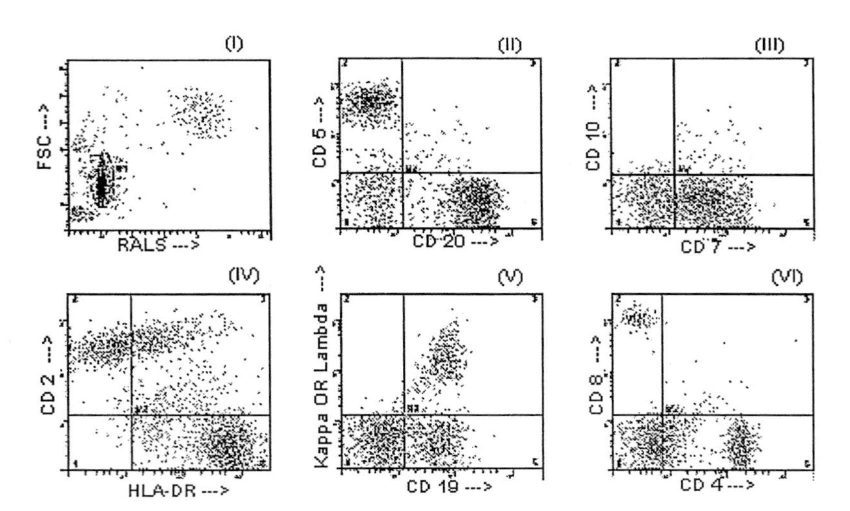

FIGURE 1–2. A normal lymph node. Panel I reveals the typical light scatter properties. The overwhelming majority of the cells have low forward and right-angle scatter. Panel II shows the mixture of B cells (CD20) and T cells (CD3). There is usually a 60/40 split. A population of CD10 (follicular center cells) is not usually seen in a resting lymph node as seen in Panel III. CD7 is a T-cell marker, and the percentage of positive cells should be close to the CD5-positive percentage. Panel IV shows a population of reactive T cells (CD2 +; HLA-DR +) and a population of B cells positive for HLA-DR. The B-cell population can also be separated into two distinct populations by kappa and lambda, as noted in Panel V. CD19 is used to bring out the B cells, and the kappa-positive population should equal the lambda-negative percent positive population, and vice versa. Panel VI shows the T-cell compartment. There should be two populations of T cells, CD4+ and CD8+. There should be twice as many CD4+ cells, and the sum of the two populations should approximate a pan-T marker.

ing a lymph node aspirate. T-cell lymphomas are typically subset-restricted and express either CD4 or CD8. In reactive processes one may observe a predominance of either suppresser or helper T cells, but as a general rule the CD4/CD8 ratios are higher in lymph nodes than in peripheral blood. This can skew the ratio quite considerably, but it is a normal finding. Immature thymic phenotypes characterized by CD1, CD4, and CD8 coexpression, in addition to the normal complement of T-cell antigens, are rarely seen.[59]

The ratio of T cells to B cells in a normal node is highly variable, but it should be close to one. B cells are typically characterized by the presence of CD19, CD20, CD79a, and kappa or lambda surface immunoglobulin. B-cell differentiation is classified by the relation to the germinal center, subdividing them into three groups of cells. The prefollicular cells are characterized by either the lack of surface immunoglobin or the presence of IgM or IgD antibodies on the cell surface. The follicular center cell is characterized by a switch to IgG surface immunoglobulin and surface CD10, whereas the postfollicular cell has IgG on its surface. Anatomically, the B-cell zones of a lymph node can be divided into a follicular zone, a mantle zone, and a marginal zone. The follicular zone contains follicular center cells and is thus characterized by the presence of CD10 and IgG surface immunoglobulin. The mantle zone contains memory B lymphocytes that home to their node of origin and manifest coexpression of CD5. The marginal zone contains circulatory lymphocytes that express CD11c. In a normal node, there are many diverse populations of B cells, and no one clone is expressed more than any other clone. As a result of this diversity, the percentage of kappa-bearing B cells is roughly 1.5 times the number of lambda-bearing B cells but varies between 0.8 and 2.3. This ratio is an important one, as it is the primary marker for monoclonality of B cells.

REACTIVE AND NEOPLASTIC IMMUNOPHENOTYPES

Benign reactive lymphoid cells mark predominantly as CD3 and CD4 cells, whereas neoplastic lymphoid cells demonstrate light chain restriction (more commonly kappa) and monoclonal surface immunoglobulin (Slg).

Low-grade B-cell lymphomas have B-cell antigens (CD19, CD20, CD79a), light chain restriction, and Slg expression.[64,65] Follicular center cell lymphomas often demonstrate CD10 and prominent Slg, but lack CD5.[66,67] Small lymphocytic lymphomas characteristically possess CD5 and weak Slg and are CDl0-negative.[64,65,68] Monocytoid B-cell lymphomas are usually CD11C-positive and negative for CD5 and CD10.[69] Mantle cell lymphomas express CD5 and Leu 8 with restricted Slg.[63,70] They are also negative for CD23, which is usually positive in small lymphocytic lymphoma.

High-grade lymphomas such as large cell or immunoblastic types have a mature B-cell phenotype (CD79a) but may not express Slg, CD19, or CD20.[65,66] Burkitt's lymphomas have a similar mature B-cell typing and often show cytoplasmic immunoglobulin (Clg) or Slg, usually IgM. Multiple myeloma and plasmacytomas have strong Clg but lack CD45 and B-cell surface antigens other than CD79a.

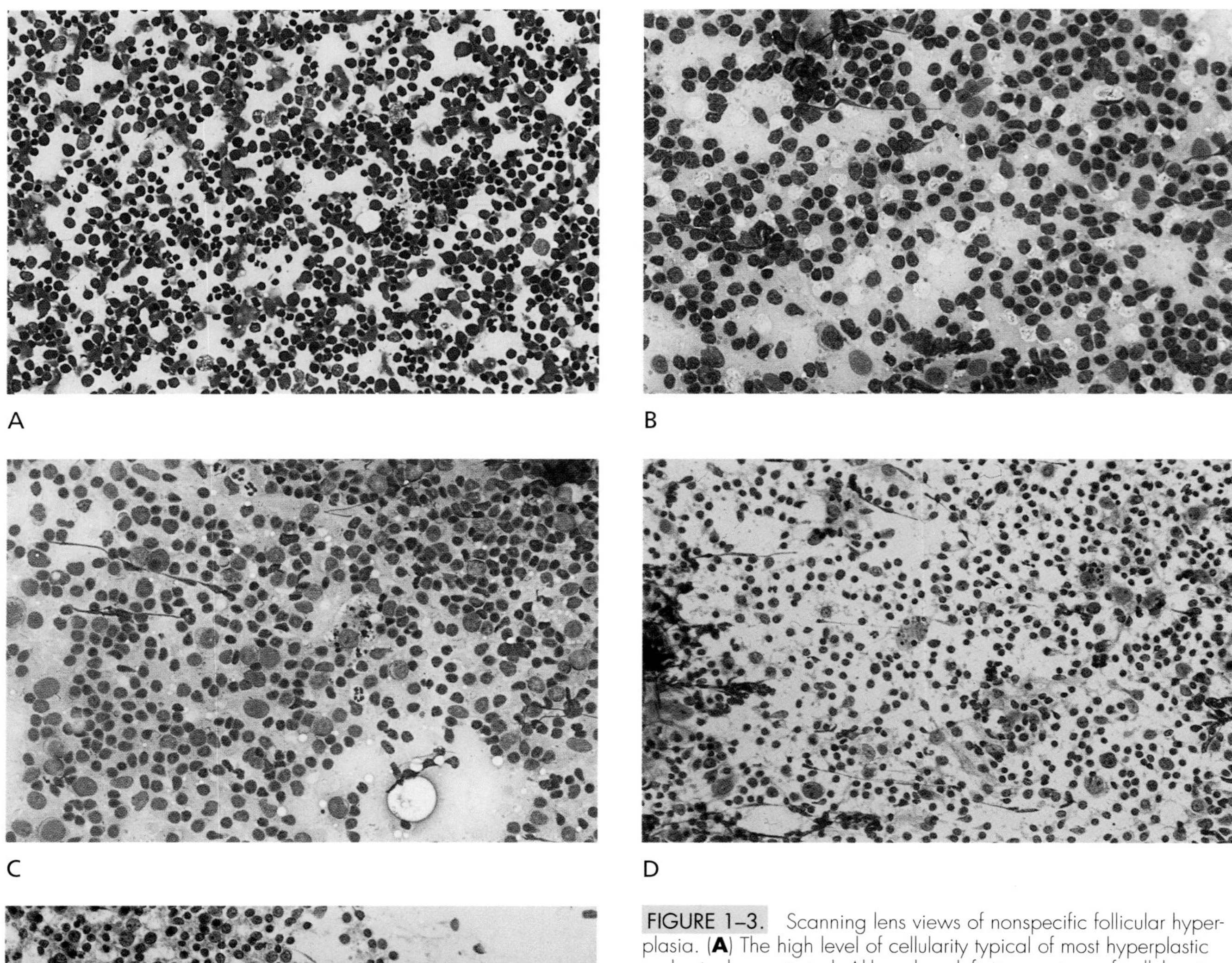

FIGURE 1–3. Scanning lens views of nonspecific follicular hyperplasia. (**A**) The high level of cellularity typical of most hyperplastic nodes is demonstrated. Although a definite spectrum of cellular sizes is readily apparent, small mature-appearing lymphoid cells predominate numerically. (**B**) The predominance of small lymphocytes is readily apparent in this aspirate. However, intermediate and large cells are also present. The lack of cohesion among the aspirated lymphocytes is also easily appreciated. (**C**) This hyperplastic aspirate includes many of the features typical of this benign proliferation, including high cellularity, a dissociated cellular pattern, marked heterogeneity of the lymphocyte size, a predominance of small lymphocytes, and a solitary tingible-body macrophage. (**D**) This is the Papanicolaou-stained material of the same aspirate in **C**. The tingible-body macrophages are characterized by abundant cytoplasm containing karyorrhectic debris. Again, the predominance of small lymphocytes is easily recognized. (**E**) The nucleus of the tingible-body macrophage in this aspirate is almost obscured by the phagocytized nuclear debris. Again, note the high cellularity and the lack of cohesion among the aspirated lymphoid cells. (**A–C,** Romanowsky stain; **D** and **E,** Papanicolaou stain)

Peripheral T-cell lymphomas are a heterogeneous group but express one or more pan-T cell markers (CD2 and CD3) and often have an abnormal predominance of CD4 helper cells.[60,71–73] In other instances, CD8 suppressor cells may be overexpressed, CD4 and CD8 may be coexpressed, an immature T-cell antigen (e.g., CD1) may be expressed, and CD25 or CD38 activation antigens may be demonstrated.

Lymphoblastic lymphomas have an immature (thymic) T-cell phenotype with CD1 and terminal deoxynucleotidyl transferase, as well as CD4 and CD8 coexpression.[74–76] To distinguish these lymphomas in the mediastinum from thymomas, which may have similar lymphoid markers, correlation with cytomorphologic, demographic, and/or molecular diagnostic features is required.

SPECIFIC ENTITIES

The following presents an integrated approach to the FNA workup and diagnosis of a variety of reactive and neoplastic conditions involving lymph nodes. Cytomorphologic features, immunophenotypic pattern, and other pertinent ancillary studies will be presented for each lesion.

Nonspecific Reactive Lymphoid Hyperplasia

In many instances, the underlying cause of reactive lymphoid hyperplasia is never determined. Many antigens, including a number of viruses, may induce benign proliferative responses among various components of the lymphoid tissue within lymph nodes. Clinically, the most common picture is that of a solitary lymph node up to 3 to 4 cm in the head and neck region. The most common histologic pattern is that of follicular hyperplasia, in which well-delineated germinal centers are prominent and are composed of a heterogeneous population of lymphocytes and tingible-body macrophages. Less often, interfollicular expansion by plasmacytoid cells, immunoblasts, and histiocytes is the cause of the nodal enlargement. Aspiration biopsy, in general, does not affect the histologic interpretation of subsequently excised nodes.[77]

The crucial cytomorphologic feature of reactive hyperplasia is the recognition of a heterogeneous population of lymphoid cells.[38] A mixture of small lymphocytes and large lymphoid cells including immunoblasts is present with a complete spectrum of intermediate morphologic forms (Figs. 1–3 to 1–5). However, the small mature-appearing lymphocyte predominates numerically. Tingible-body macrophages with intracytoplasmic apoptotic debris are not pathognomonic of a hyperplastic process (Figs. 1–6 and 1–7); they may be seen in high-grade malignant lymphomas, such as Burkitt's lymphoma. However, large numbers of tingible-body macrophages in the presence of a full spectrum of lymphoid cells with small lymphocytes predominating strongly favors a benign diagnosis.[6] Lymphohistiocytic aggregates of dendritic reticulum cells and large and small cleaved and noncleaved cells have their origin in germinal centers (Figs. 1–8 to 1–11). Capillaries may be prominent in smears (Fig. 1–12).

In air-dried Romanowsky-stained smears, artifacts due to crush and slow air-drying may make it difficult to distinguish small mature lymphocytes from lymphoblasts and small cleaved cells. Thus, the smears should be examined for better-preserved and better-stained areas. Alcohol-fixed Papanicolaou-stained smears clearly show the lymphocyte cell membrane and highlight membrane or chromatin irregularities of small lymphoid cells.

Cytomorphologic Features of Nonspecific Reactive Hyperplasia

- Generally high cellularity
- Mixture of dispersed lymphoid cells with a full spectrum of sizes and differentiation, in which small lymphocytes predominate
- Plasma cells, plasmacytoid cells, and immunoblasts may be evident
- Lymphohistiocytic aggregates
- Tingible-body macrophages

Specific causes of lymphadenopathy should be carefully excluded by examination with special stains and by looking for markers such as caseous granular material, suggesting tuberculosis. The differential diagnosis of the polymorphic cellular smear includes Hodgkin's disease, small lymphocytic lymphoma, and lymph nodes only partially involved by other neoplasms, including the non-Hodgkin's lymphomas. Reed-Sternberg (R-S) cells and eosinophils should be diligently sought, especially along the edges of the smear.

Reactive hyperplasias produce variable flow cytometric results (see Fig. 1–2). A monoclonal population is absent, but subclones that greatly skew the kappa/lambda ratio may be present. There is also a mixture of T and B cells, with roughly 50% of each type of cell. Within the T-cell compartment, the number of helper CD4 cells is slightly increased. A mixture of small and large cells can be seen by light scatter properties. All cells have a normal mature immunophenotype, and no antigen aberrancy is present, except in florid follicular hyperplasias, where the larger cells have increased CD20 expression and CD10

(Text continues on page 10.)

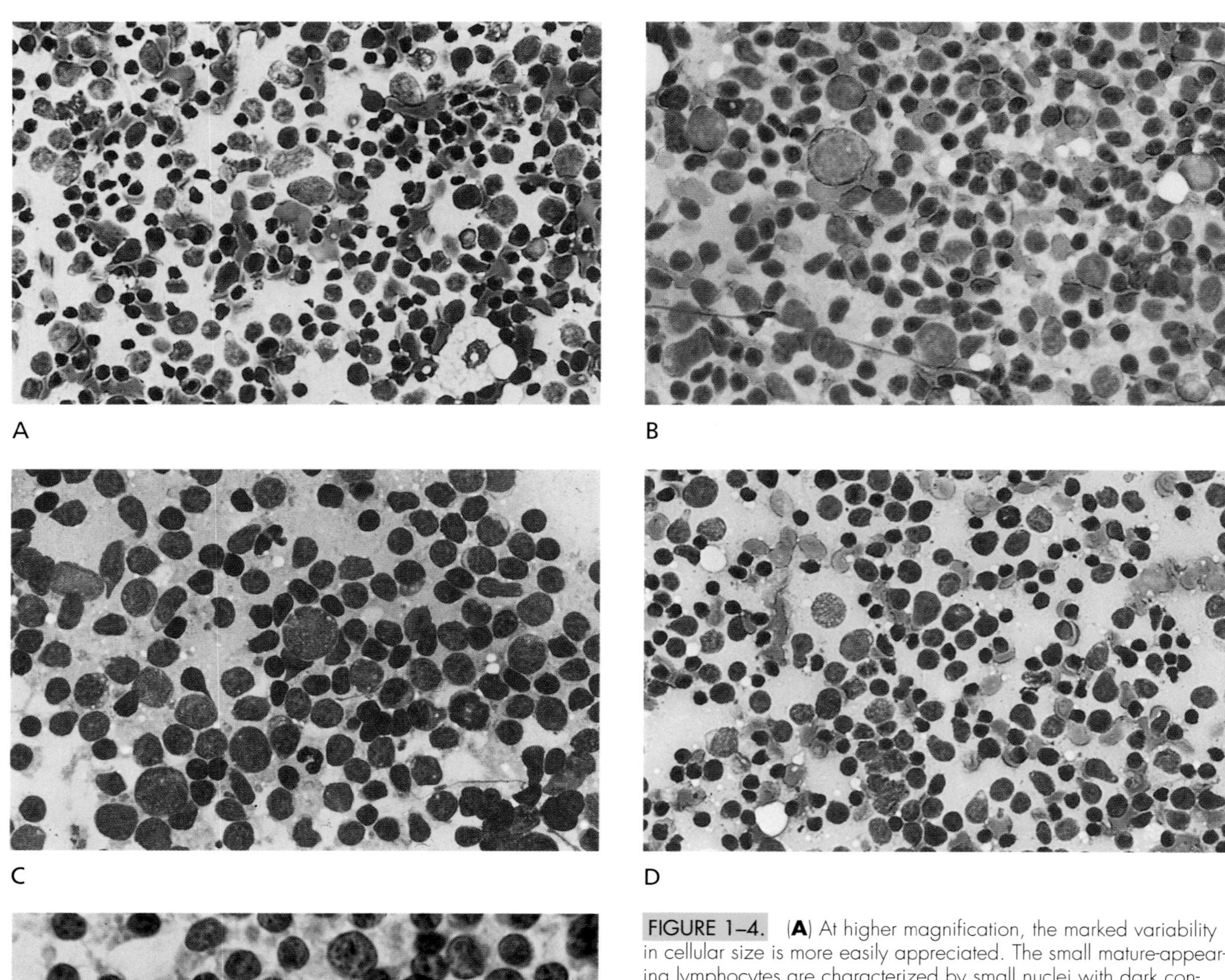

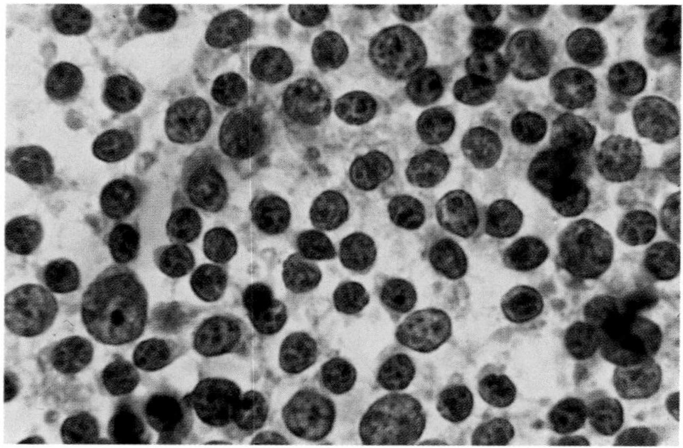

FIGURE 1–4. (**A**) At higher magnification, the marked variability in cellular size is more easily appreciated. The small mature-appearing lymphocytes are characterized by small nuclei with dark condensed chromatin, inconspicuous nucleoli, and very high nuclear-to-cytoplasmic (N/C) ratios. The largest lymphoid cells have more finely reticulated chromatin, one to several small but distinct nucleoli, and at times a visible rim of cytoplasm. Note the spectrum of intermediate or transitional lymphoid cells. Also present is a single benign histiocyte with abundant clear cytoplasm. (**B**) The largest lymphoid cells have solitary round nuclei with finely reticulated chromatin, one to multiple prominent nucleoli, and basophilic cytoplasm. The dissociative cellular pattern is apparent. (**C**) This hyperplastic aspirate manifests large lymphoid cells including immunoblasts, plasma cells, and probable eosinophils. Again, note the distinct spectrum of cellular sizes with the predominance of smaller lymphoid cells. (**D**) An uncommon component of aspirates of most patients with nonspecific hyperplastic nodes is the plasma cell. It is characterized by a solitary round nucleus that is eccentrically positioned within a moderate volume of basophilic cytoplasm. Often, a perinuclear clear zone is recognized. (**E**) Small lymphocytes predominate in this hyperplastic smear. They are characterized by dark condensed and coarse chromatin, delicate but distinct nuclear membranes, and very high N/C ratios. (**A–D,** Romanowsky stain; **E,** Papanicolaou stain)

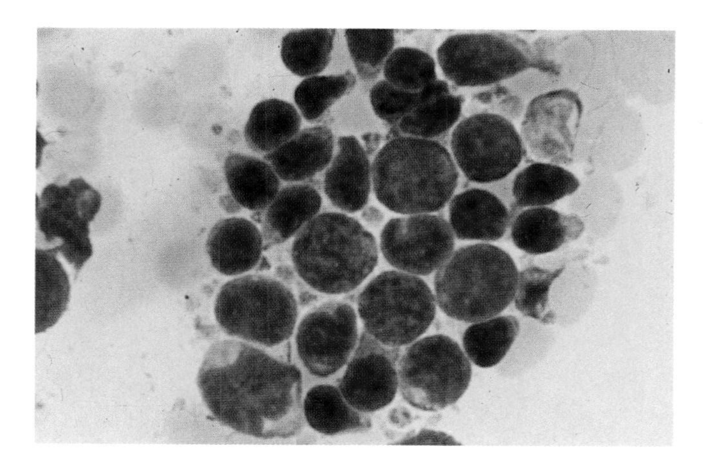

FIGURE 1–5. In this field of a hyperplastic specimen, intermediate-sized lymphoid cells with cleaved nuclei are easily appreciated. Although the lymphocytes are closely packed, true intercellular cohesion is lacking. Note the lymphoglandular bodies. (Romanowsky stain)

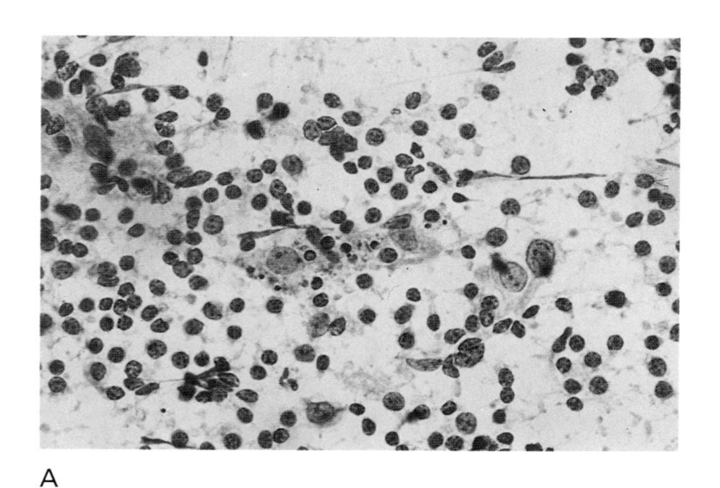

A

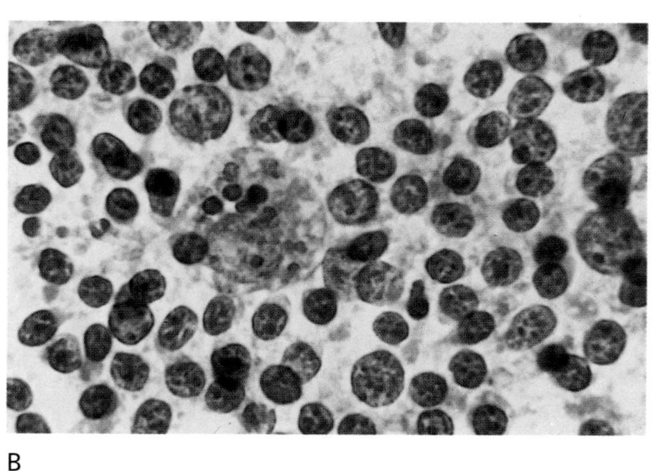

B

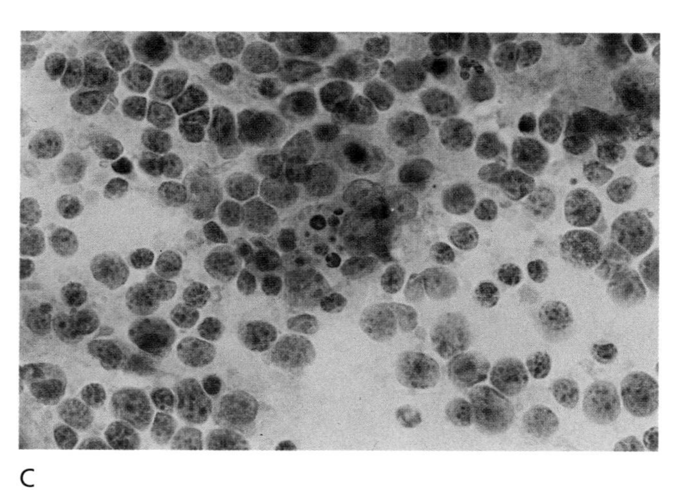

C

FIGURE 1–6. **(A–C)** Tingible-body macrophages are characterized by benign-appearing nuclei that vary from round to indented (kidney bean) in contour. They have delicate membranes and very finely granular pale chromatin. Minute nucleoli may be apparent. The voluminous pale cytoplasm contains varying amounts of phagocytized apoptotic debris. In most instances, the tingible-body macrophages are markedly outnumbered by the lymphoid cells. (Papanicolaou stain)

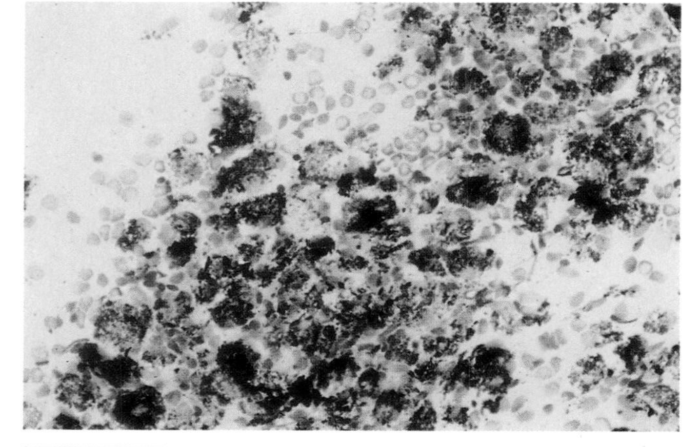

FIGURE 1–7. Dermatopathic lymphadenopathy. The cytoplasm of many histiocytes contains dark granular melanin pigment. (Romanowsky stain)

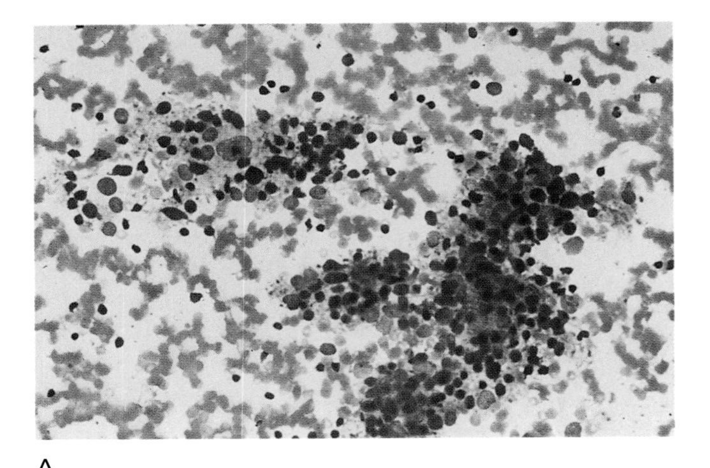

A

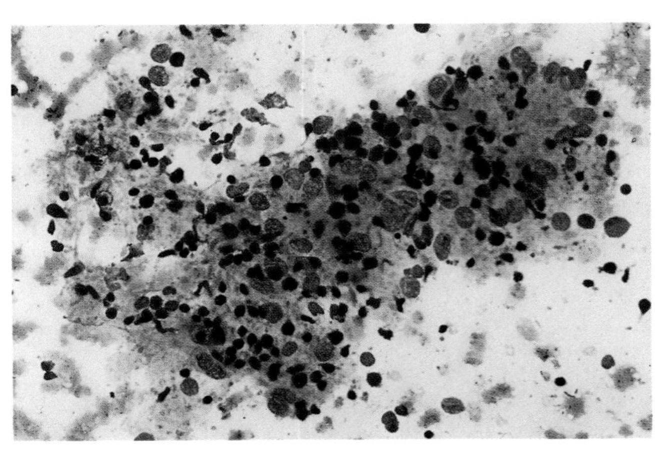

B

FIGURE 1–8. (**A, B**) Lymphohistiocytic aggregates are fragments of aspirated germinal center cells. These aggregates vary in both size and shape and contain admixtures of benign macrophages and lymphocytes. The mottled light and dark staining in these aggregates is characteristic. (Romanowsky stain)

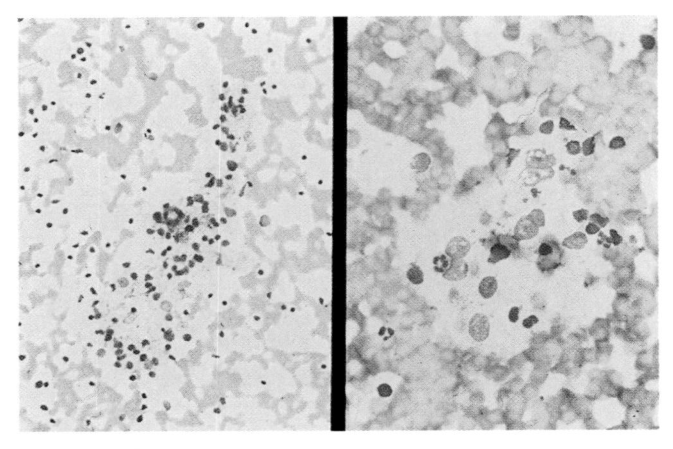

FIGURE 1–9. The staining intensity of the lymphohistiocytic aggregates depends on the relative proportions of lymphocytes and histiocytes. As in this case, when lymphocytes are relatively sparse, the abundant pale histiocytic cytoplasm stands out. (Romanowsky stain)

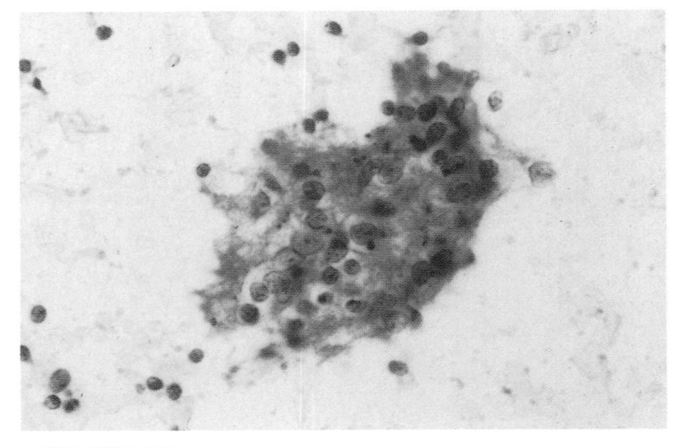

FIGURE 1–11. Frequently, a large proportion of the perimeter of the aggregates is occupied by histiocytic cytoplasm. This is in contrast to aggregates of many metastatic carcinomas, in which the malignant nuclei occupy a large proportion of the cluster's border. (Papanicolaou stain)

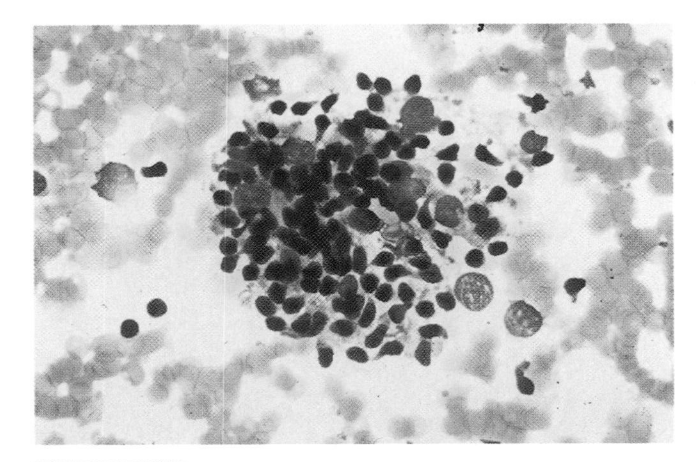

FIGURE 1–10. This lymphohistiocytic aggregate is from the other end of the spectrum: it is composed mostly of small mature-appearing lymphocytes, conferring a dark-blue staining reaction. (Romanowsky stain)

may not be present. The flow cytograms are usually very clean, and nonspecific binding is rarely seen. Granulocytic markers are negative, and histiocytic markers may or may not be present.

Immunophenotype of Reactive Hyperplasia

- Variable findings due to mix of B and T cells
- CD10 may be positive
- Increased expression of CD20
- Both kappa- and lambda-positive cells
- Reactive T cells (CD3 and CD4 predominate)

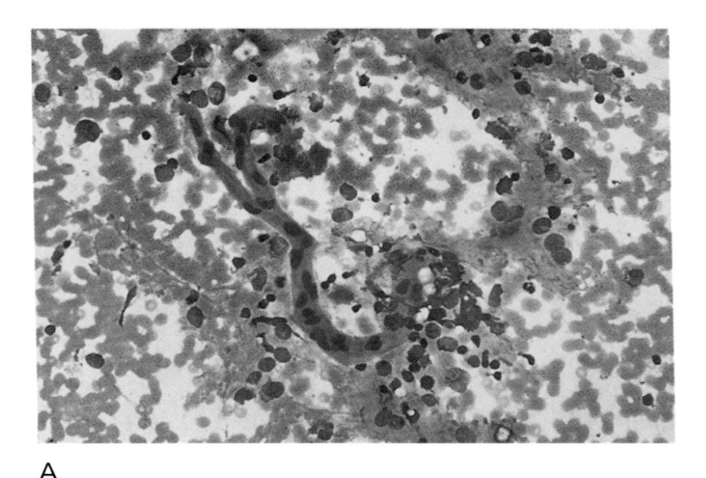

A

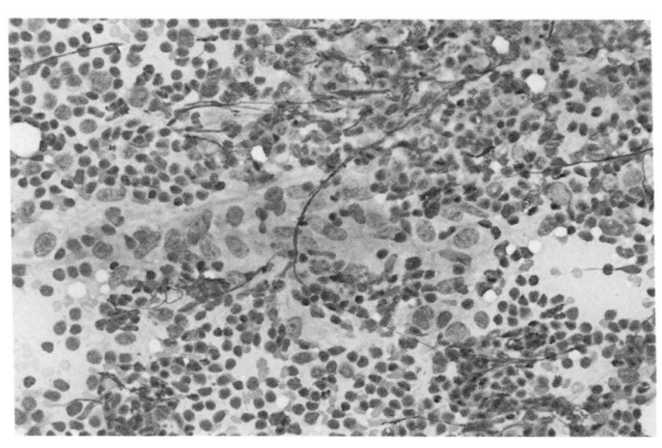

B

FIGURE 1–12. (**A**) Well-formed capillaries are seen with some frequency in aspirated hyperplastic nodes. They are characterized by a tubular configuration and small endothelial cell nuclei. (**B**) At times, the endothelial cells may have more abundant cytoplasm and larger nuclei with delicate cytoplasm. The tubular configuration in central lumen may not be apparent. (Romanowsky stain)

Infectious Mononucleosis

Infectious mononucleosis is usually a self-limited infection caused by the Epstein-Barr virus. Most often, adolescents and young adults are affected with fever, pharyngitis, splenomegaly, and a peripheral atypical lymphocytosis. The lymphadenopathy is usually cervical in distribution but may be generalized. In patients with a typical clinical picture, lymph node aspiration biopsy is unlikely. However, an unusual presentation or prolonged lymphadenopathy may lead to FNA biopsy.

FNA biopsy smears are markedly cellular, with a heterogeneous lymphoid response resembling follicular hyperplasia but with increased numbers of plasmacytoid cells and immunoblasts.[23,24] The latter are large cells with solitary nuclei with smooth nuclear membranes, fine chromatin, and a large, often central nucleolus (Fig. 1–13). The cytoplasm is eccentric, usually basophilic, and abundant. We believe that FNA biopsy cannot make a specific diagnosis of infectious mononucleosis but rather can suggest this diagnosis to be confirmed by serology. The differential diagnosis includes other florid nonspecific reactive changes, follicular center cell and large cell lymphomas, and Hodgkin's disease. The predominant cell remains a small lymphocyte rather than a small, intermediate, or cleaved large lymphoid cell of lymphoma, whereas R-S cells lack the deep-blue eccentric cytoplasm of the occasional binucleated immunoblast of infectious mononucleosis. Although the Epstein-Barr virus uses the CD21 receptor to infect B cells, many of the symptoms and the hypercellularity encountered with infectious mononucleosis are consequences of proliferating suppressor T cells. These cells proliferate to squelch the proliferating immortalized B-cell clone, producing a characteristic predominance of CD8-positive cells with a normal complement of other T-cell antigens.[78, 79] These cells will not coexpress CD4. Monotypic B cells are not identified, and B cells are rare.

Cytomorphologic Features of Infectious Mononucleosis

- High cellularity, with small lymphocytes predominating
- Polymorphous lymphoid population with prominent immunoblasts and plasmacytoid cells
- Large atypical lymphocytes with huge nuclei and abundant eccentric cytoplasm

Immunophenotype of Infectious Mononucleosis

- Abundance of CD8-positive lymphocytes
- Normal complement of other T-cell antigens
- Rare B cells

Suppurative Lymphadenitis

Pyogenic infection of lymph nodes with secondary enlargement is more frequently seen in children than adults. Commonly, lymph nodes in the head and neck region are involved, presumably draining a regional bacterial infection. The nodes may be painful, tender, and associated with fever.

Aspiration smears are usually highly cellular and dominated by large numbers of intact and degenerated neutrophils (Fig. 1–14). Bacteria may be visualized, especially with the Romanowsky stains. Fungi may also be seen. A separate needle pass or the sterile saline washings of the needle and syringe should be placed in a sterile container and promptly delivered to

(*Text continues on page 13.*)

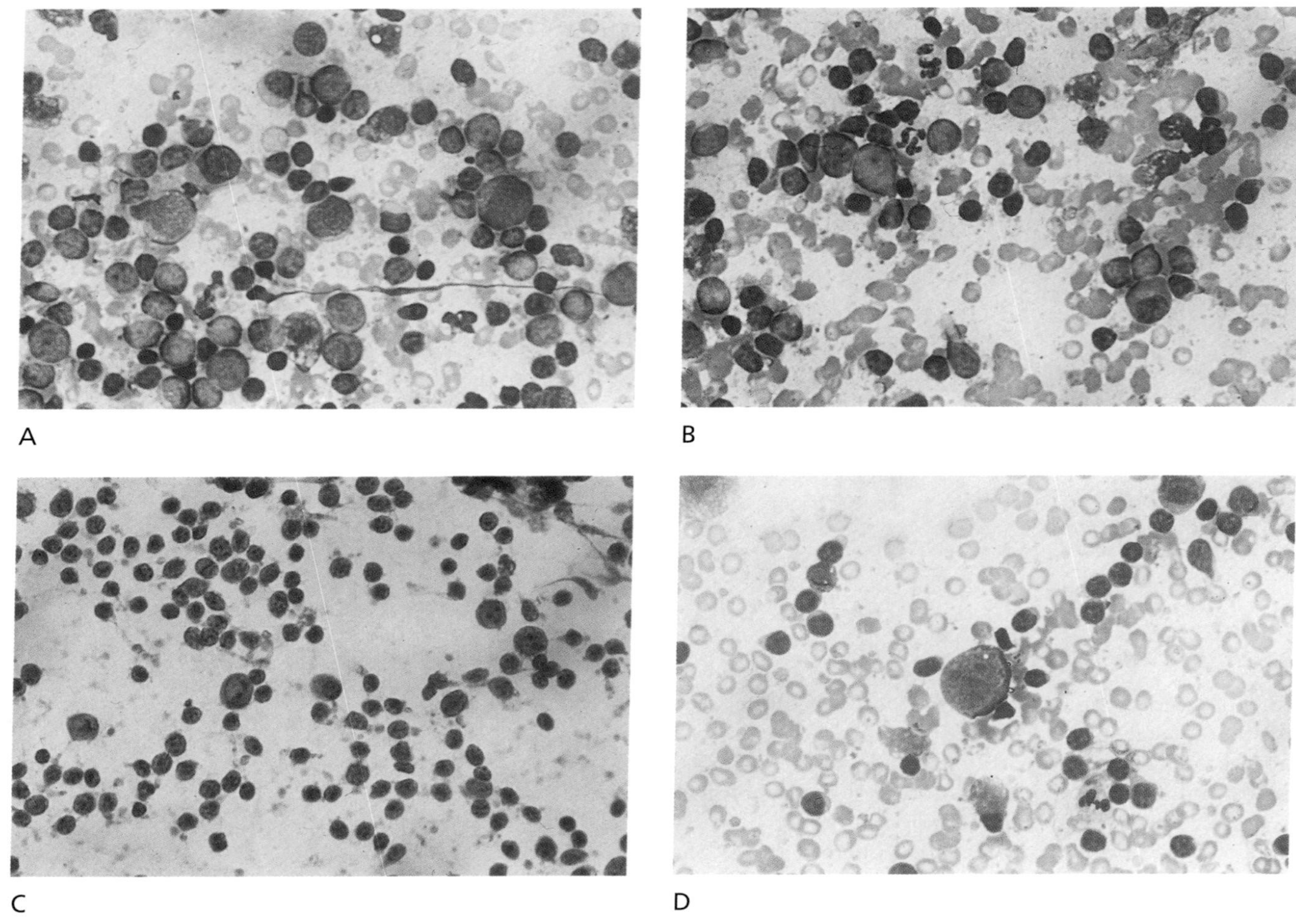

FIGURE 1–13. Infectious mononucleosis. (**A–C**) Although small mature-appearing lymphocytes are present, a relatively high proportion of lymphoid cells are quite large, including immunoblasts. They are characterized by round nuclei with very finely reticulated chromatin, one to multiple prominent nucleoli, and often a visible rim of basophilic cytoplasm. Plasmacytoid lymphoid cells are also apparent. The fine pale chromatin of the immunoblasts contrasts sharply with the dark condensed chromatin of the small lymphocytes. Lymphoglandular bodies are evident. (**D**) The intense cytoplasmic basophilia of this immunoblast is characteristic of infectious mononucleosis. (**A, B, D,** Romanowsky stain; **C,** Papanicolaou stain)

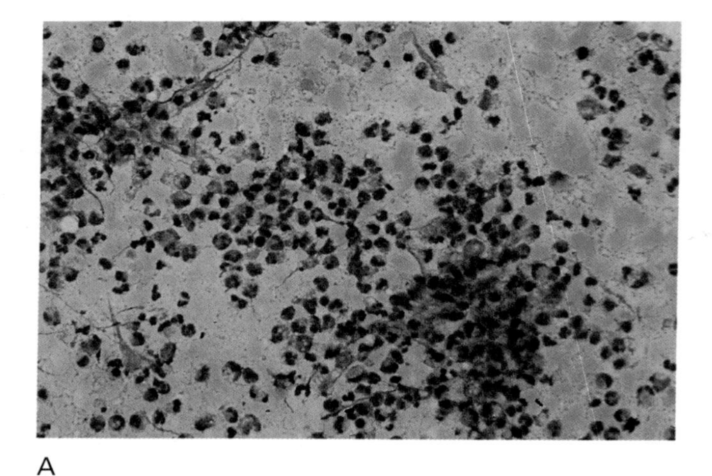

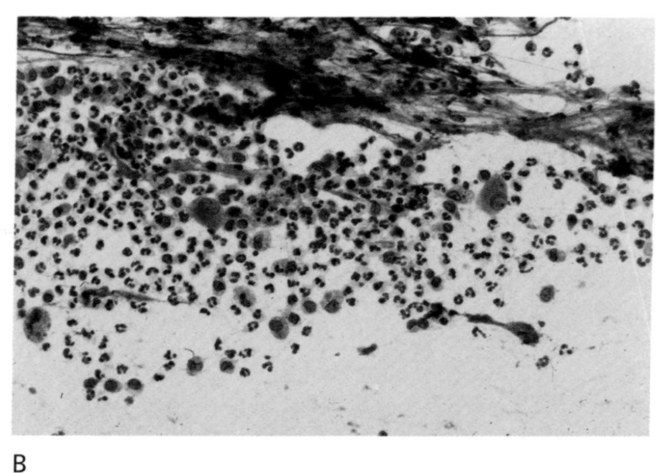

FIGURE 1–14. (**A, B**) Suppurative lymphadenitis. Neutrophilic leukocytes are the predominant inflammatory cells and may appear clumped in aggregates. Many manifest varying degrees of nuclear degeneration. Lymphocytes make up a small proportion of the total cellular population. (**A,** Romanowsky stain; **B,** Papanicolaou stain)

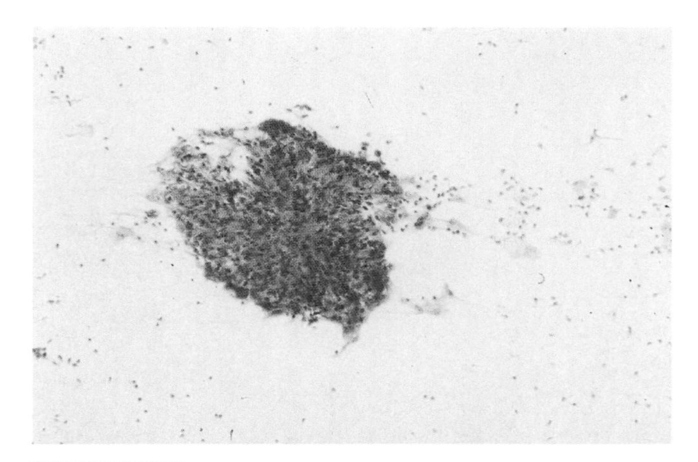

FIGURE 1–15. Granulomatous lymphadenopathy in a patient with sarcoidosis. Smear cellularity is not nearly as high as that characteristic of follicular hyperplasia. Granulomas are characterized by cohesive aggregates of epithelioid cells and lymphocytes. Because of the relatively low N/C ratios, cytoplasm appears abundant. (Papanicolaou stain)

the microbiology lab for culture and sensitivity studies.

Cytomorphologic Features of Suppurative Lymphadenitis

- Variable cellularity, often high
- Predominant cell is neutrophil, both intact and degenerated
- Neutrophils are both dispersed and loosely aggregated
- Lymphocytes and macrophages
- Bacteria may be seen, both within phagocytes and extracellularly

Flow cytometric evaluation is not needed; if performed, the findings are nonspecific. When using T- and B-cell antigens alone, many of the cells fail to stain. However, these cells are CD45-positive, indicating they are hematopoietic in origin. Myeloid antigens such as CD13 can demonstrate the presence of neutrophils. Staining with 7-actinomycin D often reveals numerous dead cells.[50] Necrosis also produces a substantial degree of nonspecific binding by the monoclonal antibodies.

Granulomatous Lymphadenitis

Granulomatous lymphadenitis is a form of chronic inflammation frequently seen in histologic and aspiration specimens of lymph nodes (Figs. 1–15 to 1–19). Common causes include sarcoidosis, foreign body reactions, and infections. Classically, infectious agents include mycobacteria and fungi. Sarcoid is a diagnosis of exclusion, whereas FNA biopsy may reveal the etiologic agent in many other diseases.

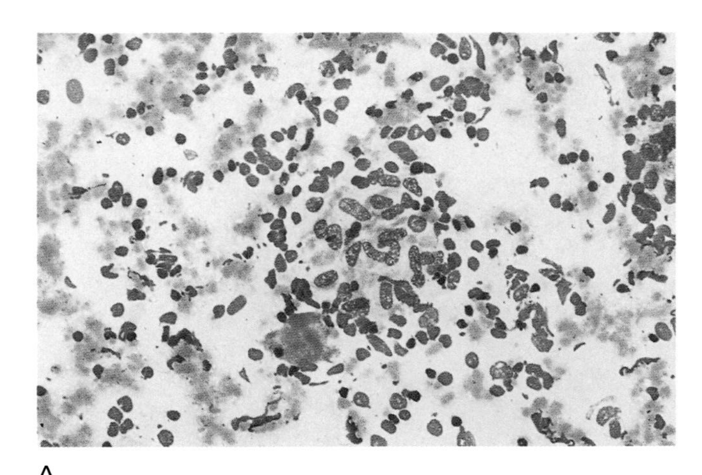

A

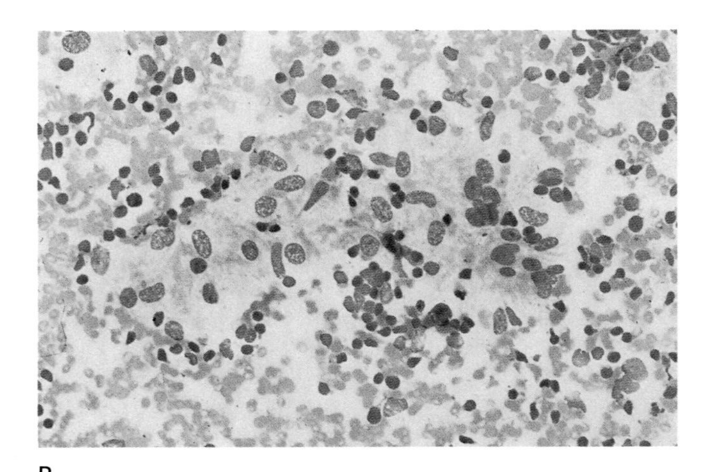

B

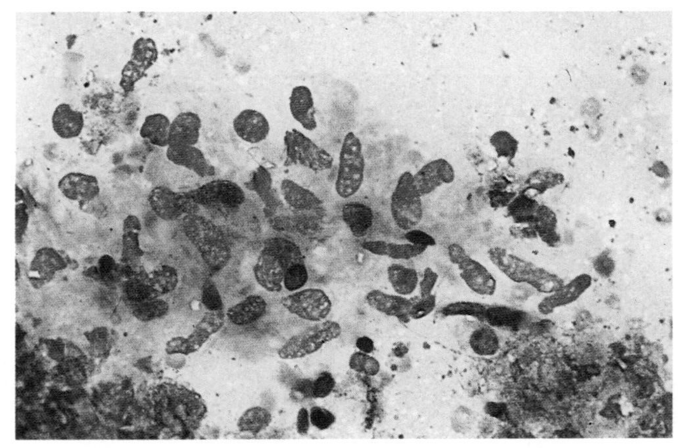

C

FIGURE 1–16. **(A–C)** Sarcoidosis. Granulomas had a syncytial appearance because of indistinct intercellular borders. In addition to the pale abundant cytoplasm, epithelioid cells are characterized by solitary elongated nuclei. The latter often have a characteristic bend or indentation in the center of their long axis. Nucleoli are small and inconspicuous. (Romanowsky stain)

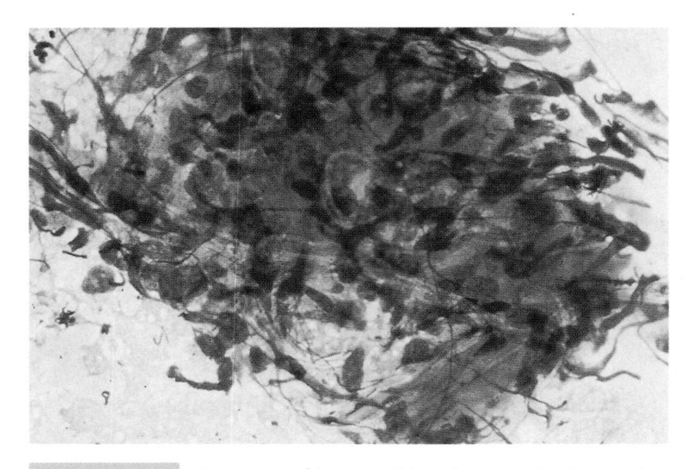

FIGURE 1–17. Fragment of large cell lymphoma with dense sclerotic connective tissue. Fragments of aspirated lymphoma associated with dense collagen may simulate a granuloma in that the malignant lymphoid nuclei may be compressed and elongated by the sclerosis. To avoid this diagnostic pitfall, other portions of this smear need to be examined; they typically demonstrate the dissociated pattern of malignant lymphoma. (Romanowsky stain)

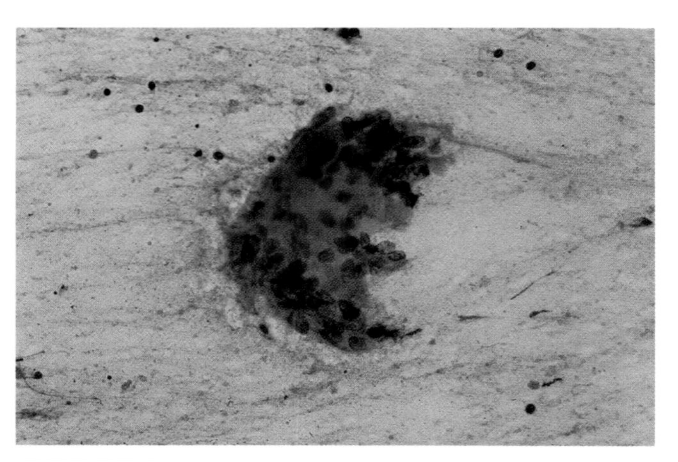

FIGURE 1–18. A characteristic component in many instances of granulomatous lymphadenopathy is the multinucleated inflammatory giant cell. It is characterized by abundant cytoplasm that contains few to many rather uniform nuclei typical of histiocytes. (Papanicolaou stain)

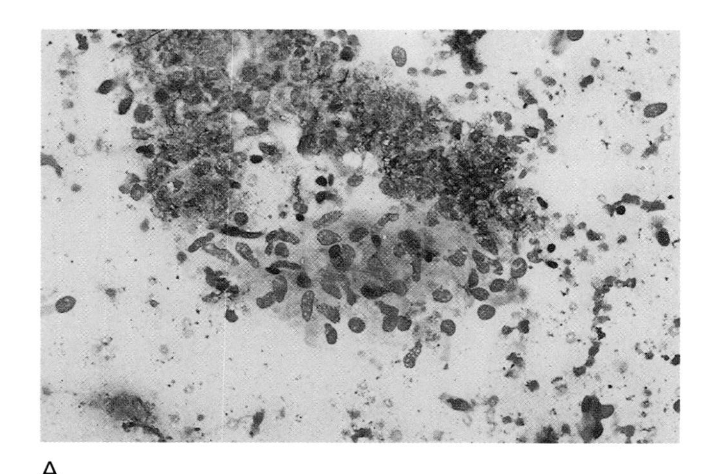

A

B

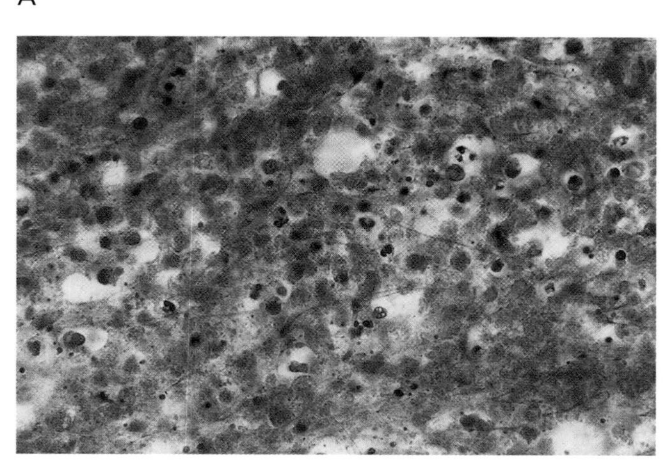

C

FIGURE 1–19. (**A–C**) Tuberculosis. In addition to well-formed epithelioid cell granulomas, the smears typically possess caseous necrotic material. The latter is typically closely associated with the granulomas. It is characterized by granular to amorphous debris in which cellular outlines occasionally are evident. (**A,** Romanowsky stain; **B** and **C,** Papanicolaou stain)

In developed countries, mycobacterial infections had become an uncommon clinical event. Today the AIDS epidemic and migration have caused a resurgence. In some parts of the world, such as India, tuberculosis has remained a serious and common problem. Before AIDS, clinical infections by atypical mycobacteria were relatively uncommon. However, atypical mycobacteria have become a major cause of infections in immunosuppressed persons, especially those with AIDS. The classic histopathologic change in tuberculosis consists of granulomatous inflammation associated with caseous necrosis, but in markedly immunocompromised patients, a well-developed granulomatous response may not occur.

The typical feature of tuberculous lymphadenitis is epithelioid cell granulomas associated with necrosis (Figs. 1–19 and 1–20). Epithelioid cells are modified histiocytes with a moderate amount of pale cytoplasm[11,13] and a solitary nucleus with a characteristic elongated, bent, or centrally indented shape, the so-called footprint, snowshoe, sand shoe, or boomerang nucleus (see Fig. 1–16). Chromatin is very finely granular and pale-stained; nucleoli are small and generally inconspicuous. Syncytial arrangement of these cells results in granulomas, which are present in a background of necrotic, relatively agranular, homogeneous debris. Multinucleated giant cells of the Langhans type with aggregated nuclei and copious, at times finely vacuolated cytoplasm are present in variable numbers (see Fig. 1–18). Their nuclei are often arranged peripherally in a horseshoe rimming of the cytoplasm. Recently, Pandit et al.[11] have described the "eosinophilic structure" as a characteristic component of smears from nodes involved by tuberculous lymphadenitis. With the hematoxylin and eosin stain, these consist of sharply delineated masses of eosinophilic material containing no recognizable cells, surrounded by a clear zone or halo. They proposed that these represented further degeneration within granulomas.[11] However, they are not specific for tuberculosis.[12]

Although sarcoidosis is a diagnosis of exclusion, aspiration biopsies demonstrating prominent granulomatous inflammation without necrosis in the proper clinical setting assist in confirming the presence of sarcoidal inflammation in enlarged lymph nodes. For example, transbronchial needle aspirates can diagnose sarcoidosis in hilar nodes.

Reflecting the diffuse nature of the granulomatous process within the sampled node, the overall smear cellularity tends to be low, as there are relatively few lymphocytes (see Figs. 1–15 and 1–16). The smears are dominated by cohesive aggregates of epithelioid histiocytes, which may also be present as dispersed cells. Multinucleated inflammatory giant cells (see Fig. 1–18), fibroblasts, and mostly small lymphocytes may also be present. There should be no evidence of necrotic debris in the smear background. Negative acid-fast, auramine, and methenamine silver-stained direct smears and cell blocks are mandatory in making the diagnosis. Aspirated material should be culture-negative.

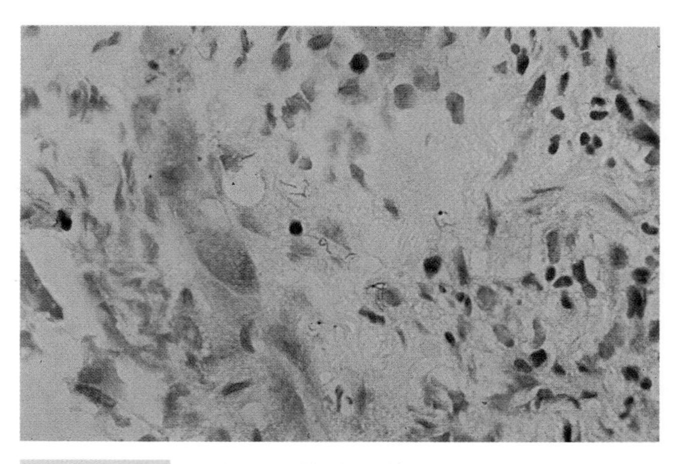

FIGURE 1–20. With a modified acid-fast stain, the positive mycobacteria may be easily demonstrated. At times they are numerous, but in other instances careful scrutiny is essential to document their presence.

There is no specific immunophenotypic pattern for sarcoid. Low cellular yield usually precludes flow analysis.

Cytomorphologic Features of Granulomatous Lymphadenitis

- Low to moderate cellularity
- Epithelioid cells singly or aggregated in granulomas
- Epithelioid cells: solitary elongated, bent nuclei with bland chromatin and small nucleoli and moderately pale cytoplasm
- Small lymphocytes
- Multinucleated inflammatory (Langhans') giant cells may be present
- Neutrophils may be present
- Necrotic debris may be prominent

Such changes are often not present in patients who are immunosuppressed (Fig. 1–21). Rather, smears may be dominated by dispersed macrophages with abundant cytoplasm, which in the Romanowsky stains may have a reticulated cross-hatched appearance due to numerous rodlike, nonstaining phagocytized bacilli.[14,15] These "negative images" of unstained bacilli may also be recognized extracellularly.[14,15] Their presence can be confirmed by an acid-fast or auramine stain directly on the aspirated material. Usu-

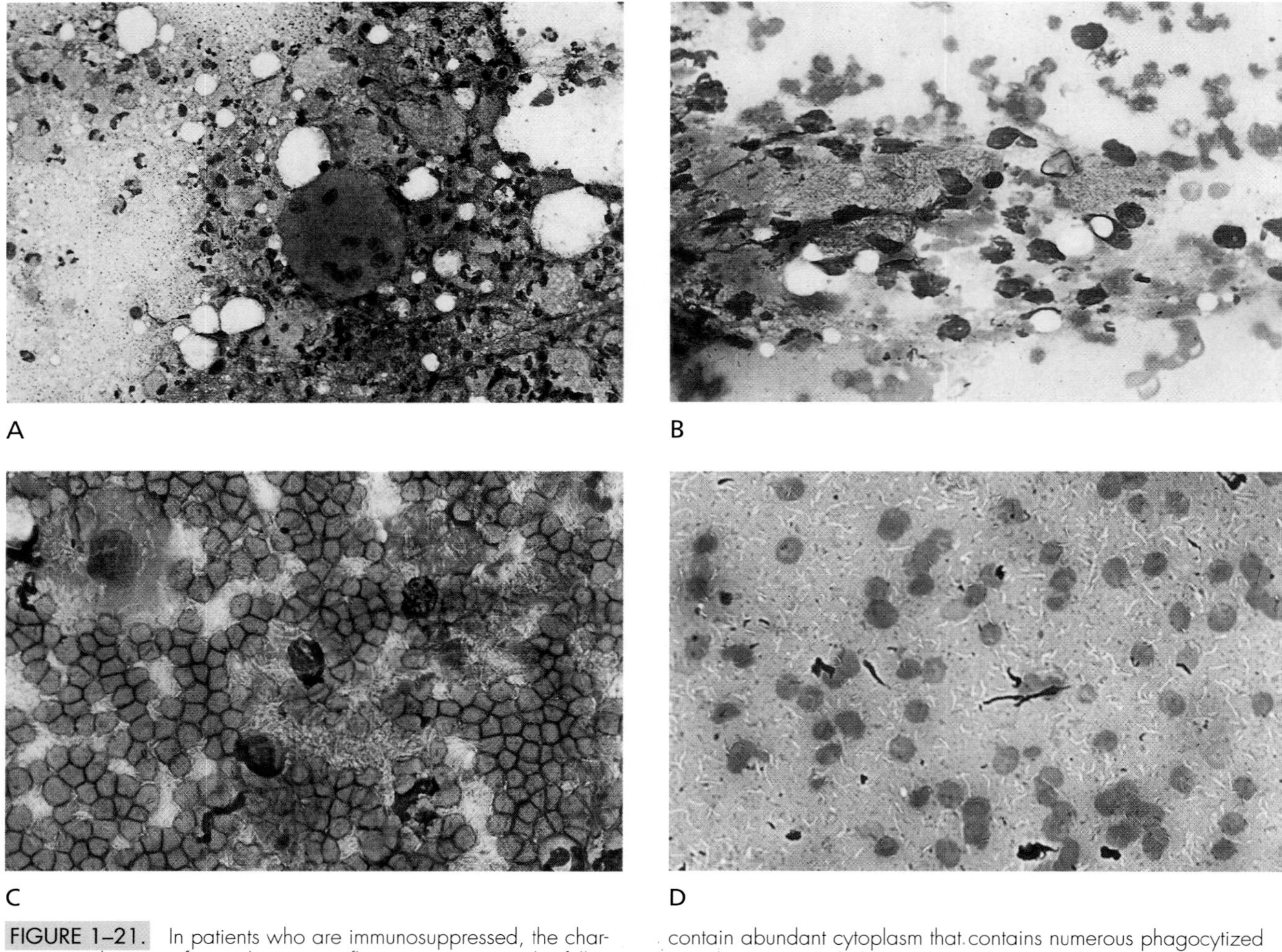

FIGURE 1–21. In patients who are immunosuppressed, the characteristic changes of granulomatous inflammation may not be fully developed. (**A**) A solitary multinucleated inflammatory giant cell is mixed with lymphocytes and neutrophils in this patient who was immunosuppressed secondary to AIDS. (**B, C**) Well-formed epithelioid cell granulomas may not be evident. Rather, modified histiocytes contain abundant cytoplasm that contains numerous phagocytized mycobacteria. With the Romanowsky stains, they appear as clear rod-shaped structures. (**D**) These negative images may also be evident in the acellular smear background. (Romanowsky stain)

ally, these are atypical mycobacteria. However, one recent study did not find reduced granulomatous inflammation in HIV-positive persons with tuberculosis.[80] In addition, it is mandatory to submit material for culture and sensitivity studies.

Other infectious agents may be seen in aspirated cellular material. Although special stains highlight the organisms, they may be visible in routine preparations as well (Figs. 1–22 to 1–24).

Vaccination with BCG often results in lymphadenopathy, especially of regional draining lymph nodes. Although the cytomorphology overlaps with tuberculosis, neutrophils and necrotic debris occur much more often with BCG disease, whereas well-formed granulomas are less frequently seen.[81]

Generally, flow cytometry is of little help in these infections. There may be an increase in helper T cells. Additionally, macrophages can be detected by using second-line antibodies such as CD14 or CD15. HLA-DR can be strongly expressed on the epithelioid histiocytes, and these cells will also express CD4, albeit more weakly than normal helper T cells.

Toxoplasmosis

Toxoplasmosis, caused by infection with the protozoan *Toxoplasma gondii,* is often an asymptomatic process, but it may produce a wide spectrum of disease depending on several factors, most importantly the route of transmission and the patient's immune status. The most common clinical manifestation in immunocompetent persons with an acute infection is lymphadenopathy. Although any lymph nodes may be involved, the cervical chains, especially posterior nodes, are most frequently affected.

FNA biopsy yields cellular smears that have lym-

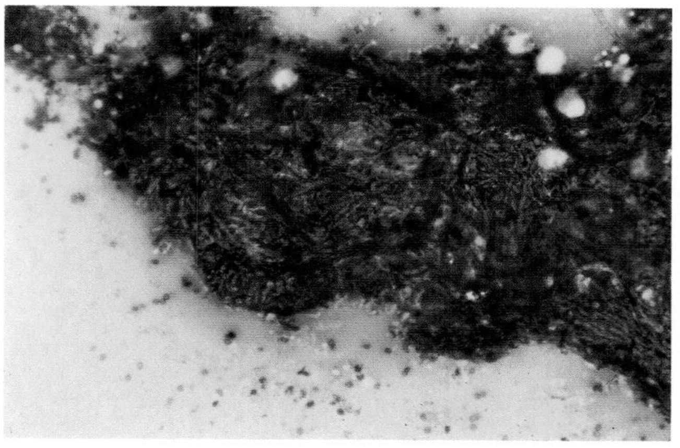

A

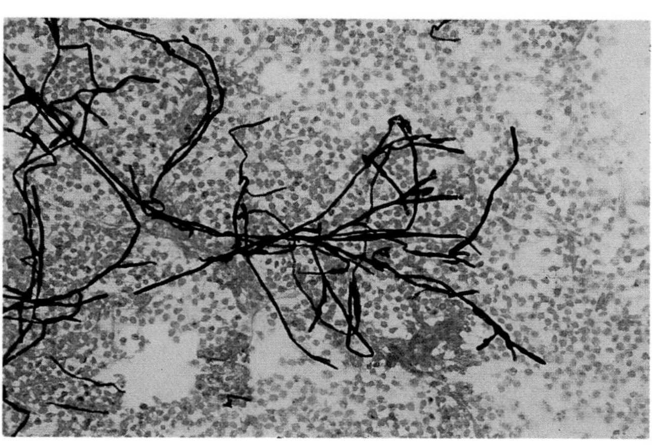

B

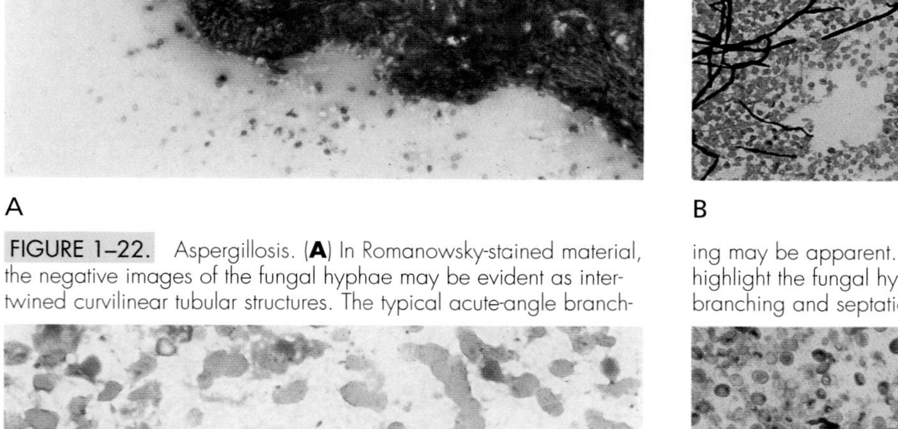

FIGURE 1–22. Aspergillosis. (**A**) In Romanowsky-stained material, the negative images of the fungal hyphae may be evident as intertwined curvilinear tubular structures. The typical acute-angle branching may be apparent. (**B**) Special stains such as methenamine silver highlight the fungal hyphae. They typically show acute-angle branching and septation.

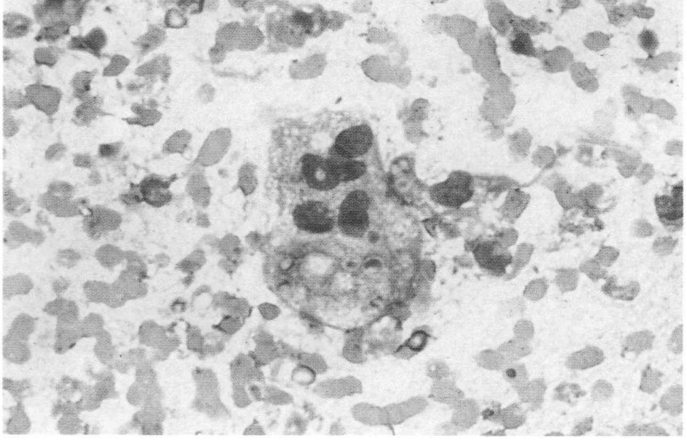

A

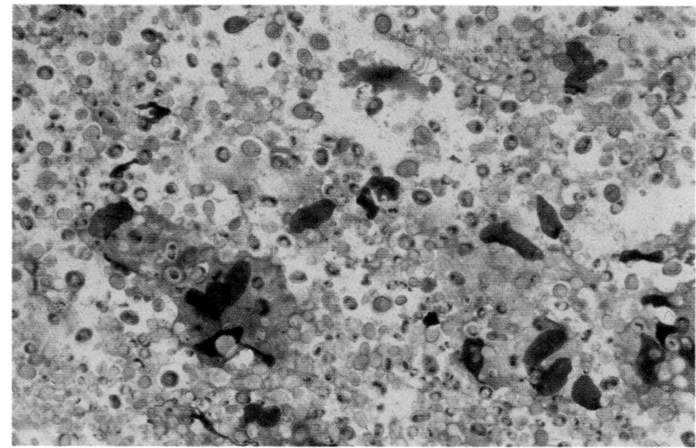

B

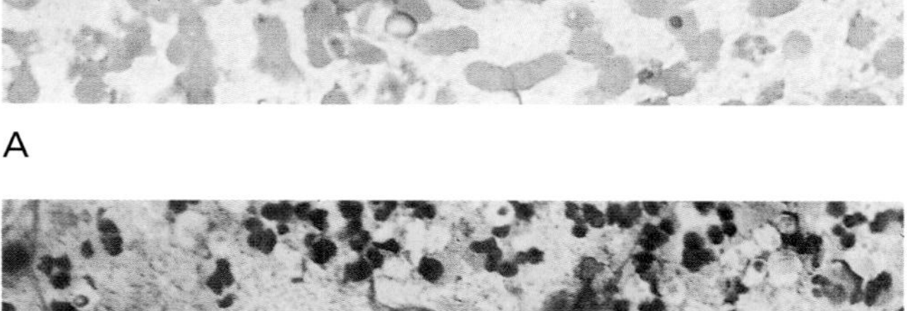

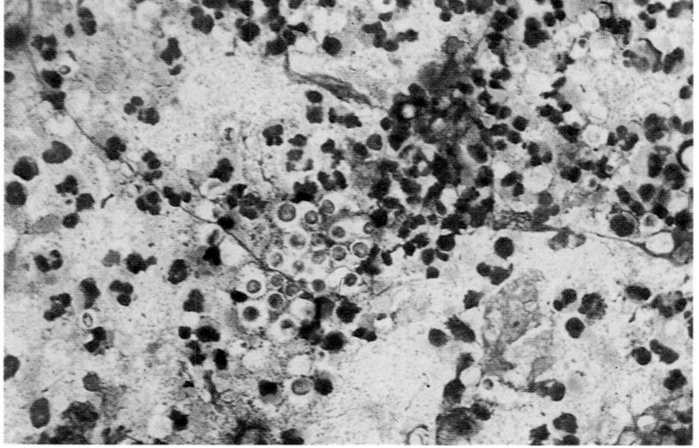

C

FIGURE 1–23. (**A–C**) Cryptococcoses. When cryptococci infect lymph nodes, a granulomatous reaction is often not well developed or apparent at all. (**A**) The yeasts of the cryptococci are present within the cytoplasm of a multinucleated inflammatory giant cell. They are characterized by rounded contours, variably staining internal structures, and visible clear halos. (**B, C**) In many cases, the organisms are free of phagocytes. In this specimen, they manifest very thick clear capsules with a basophilic staining of the central body. A budding of daughter yeast cells is apparent. (Romanowsky stain)

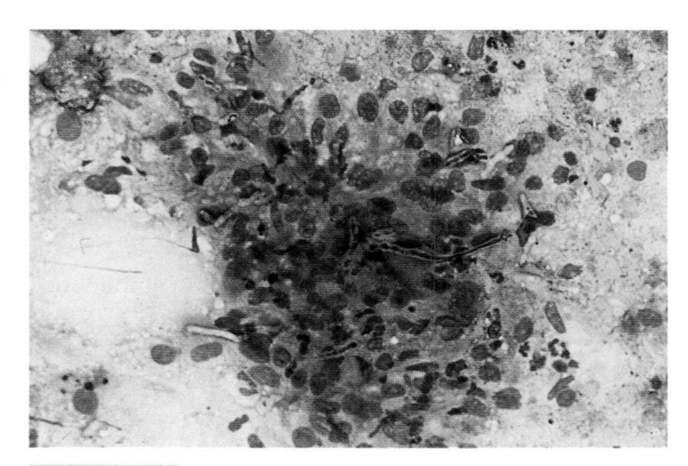

FIGURE 1–24. *Scytalidinum* infection. Fungal hyphae are admixed with epithelioid histiocytes and lymphocytes. (Romanowsky stain)

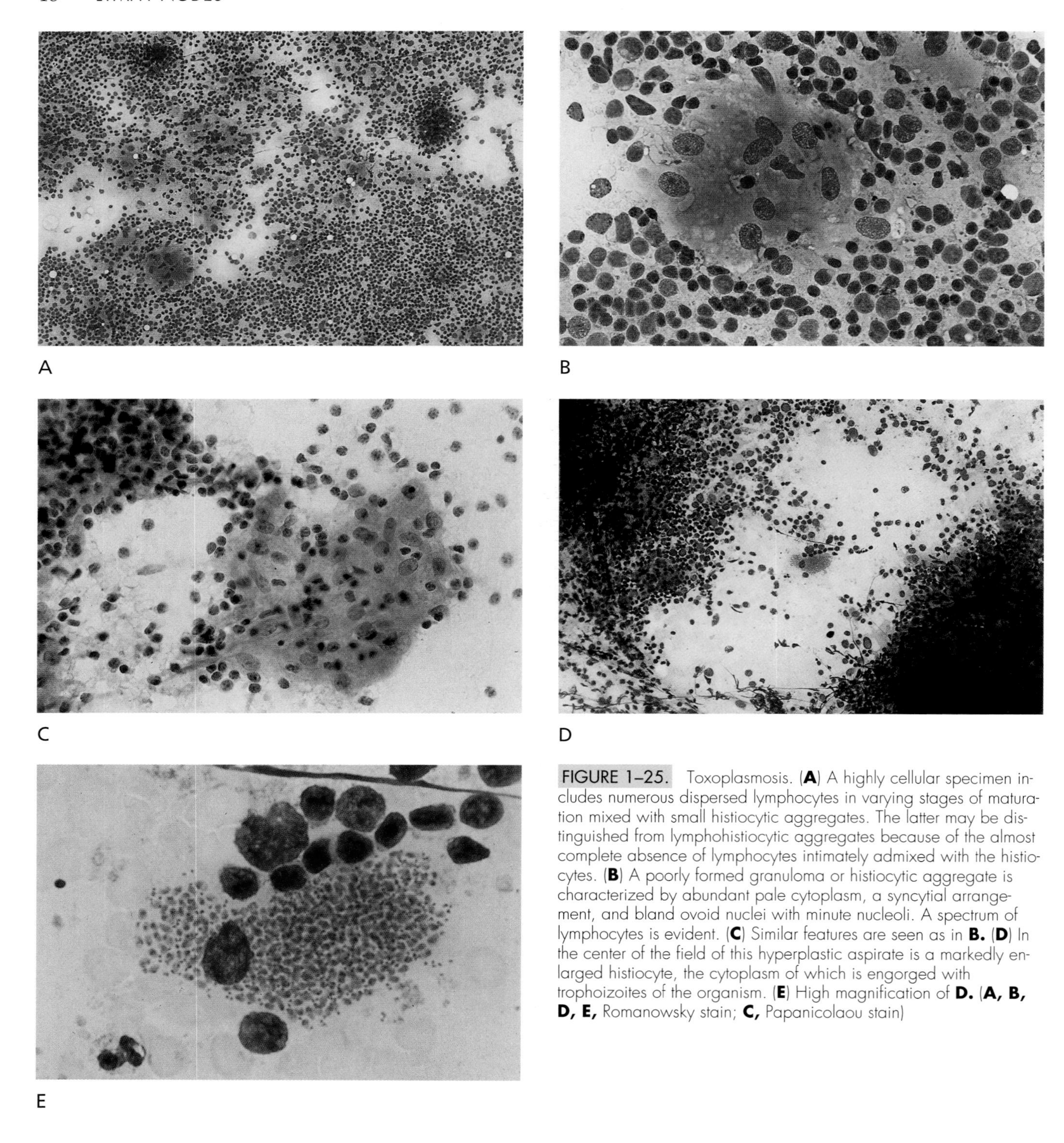

FIGURE 1–25. Toxoplasmosis. (**A**) A highly cellular specimen includes numerous dispersed lymphocytes in varying stages of maturation mixed with small histiocytic aggregates. The latter may be distinguished from lymphohistiocytic aggregates because of the almost complete absence of lymphocytes intimately admixed with the histiocytes. (**B**) A poorly formed granuloma or histiocytic aggregate is characterized by abundant pale cytoplasm, a syncytial arrangement, and bland ovoid nuclei with minute nucleoli. A spectrum of lymphocytes is evident. (**C**) Similar features are seen as in **B.** (**D**) In the center of the field of this hyperplastic aspirate is a markedly enlarged histiocyte, the cytoplasm of which is engorged with trophoizoites of the organism. (**E**) High magnification of **D.** (**A, B, D, E,** Romanowsky stain; **C,** Papanicolaou stain)

phoid cells in all stages of maturation, tingible-body macrophages, lymphohistiocytic aggregates, single histiocytes, and the characteristic small clusters of histiocytic cells (Fig. 1–25). The cytology reflects the typical histopathology of follicular hyperplasia, with small epithelioid cell clusters apposed to the germinal centers.[19–22] However, the monocytoid B cell, which prominently expand sinuses in tissue sections, are more difficult to recognize in smears. Orell et al.[82] have briefly described them as having large oval nuclei with pale-stained chromatin. The presence of the actual organism is exceedingly rare.

Cytomorphologic Features of Toxoplasmosis

- High cellularity
- Polymorphous lymphoid population resembling follicular hyperplasia
- Small clusters of histiocytes
- Dispersed histiocytes

Whenever the cytomorphology is suggestive of toxoplasmosis, the diagnosis must be confirmed by the use of appropriate serologic tests.

Cat-Scratch Disease

Cat-scratch disease is a self-limited benign inflammation of lymph nodes, usually cervical, that is usually associated with exposure to cats. It is due to infection with *Bartonella henselae* or in a small number of cases *Alipia felis*. Many patients have a rash and fever as well. Although it may occur at any age, most patients are children and young adults. Histologically, there is a characteristic mixture of stellate microabscesses and associated granulomatous inflammation.

Aspiration smears show a polymorphous cell population of variably sized lymphocytes, lymphohistiocytic aggregates, tingible-body macrophages, neutrophils, and granulomas.[17,18] Suppurative granulomas are the most distinctive feature and consist of aggregates of epithelioid cells surrounding and infiltrated by neutrophils (Figs. 1–26 and 1–27). Dispersed histiocytes with ingested neutrophils can be prominent. Granular cellular debris is present in the background. If granulomas are not present, a combination of neutrophils and individually dispersed macrophages in a background of reactive lymphoid cells suggests the diagnosis. Donnelly et al.[17] demonstrated the causative pleomorphic microbes in the smears using a modified Steiner stain in most of their specimens.[17] A classic Warthin-Starry stain can also be used. In the absence of demonstrable organisms, a combination of clinical history, serology, and cytomorphology establishes the diagnosis of cat-scratch disease. However, the differential diagnosis of the neutrophilic infiltrate includes the causes of suppurative lymphadenitis; if suppuration is inconspicuous, the smears resemble a nonspecific reactive node.

Cytomorphologic Features of Cat-Scratch Disease

- Very polymorphous mixture of lymphoid cells, tingible-body macrophages, and histiocytes
- Suppurative granulomas
- Numerous neutrophils
- Granular background debris

Flow cytometric analysis is of little use when assessing this entity. It has features in common with suppu-

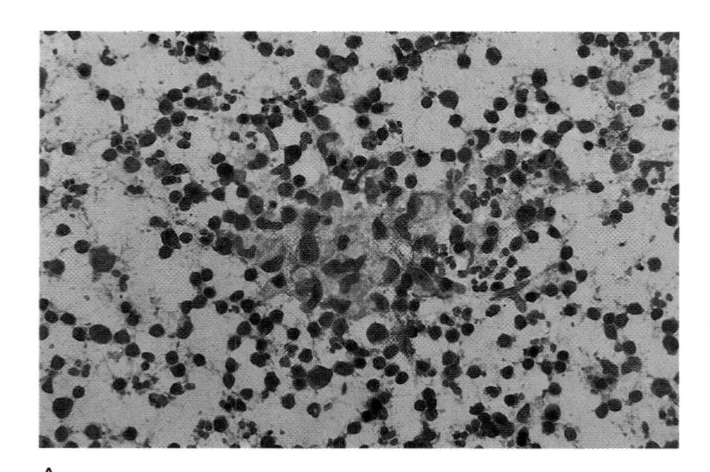

A

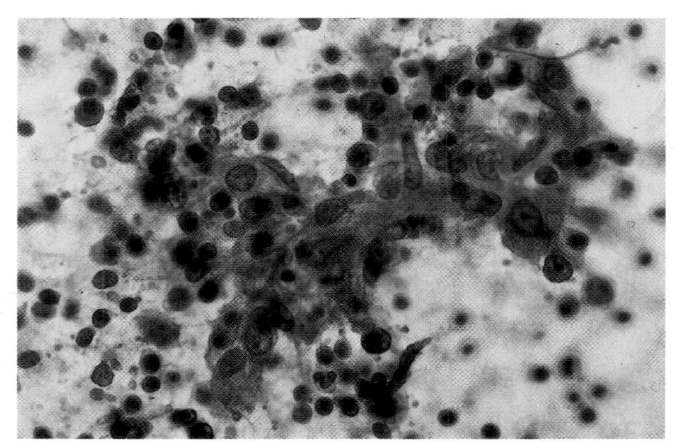

B

C

FIGURE 1–26. Cat-scratch disease. (**A, B**) Suppurative granulomas are composed of an admixture of epithelioid cells, histiocytes, neutrophils, and lymphocytes. The epithelioid cells have elongated nuclear contours with delicate membranes, fine pale chromatin, and inconspicuous nucleoli. (**C**) This specimen includes scattered histiocytes, neutrophils, lymphocytes, and necrotic debris. (**A** and **C,** Romanowsky stain; **B,** Papanicolaou stain)

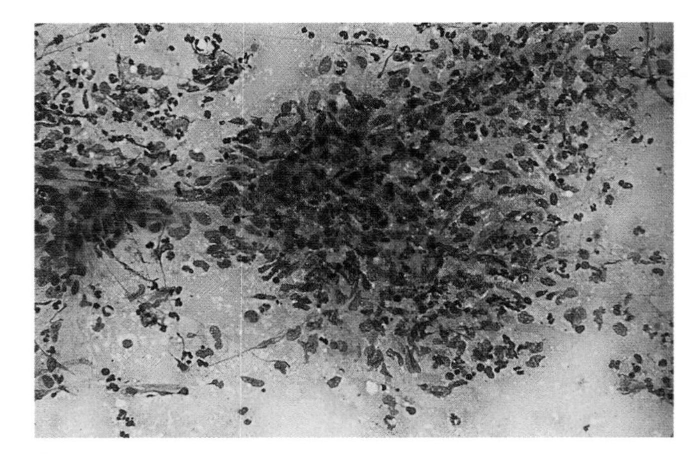

A

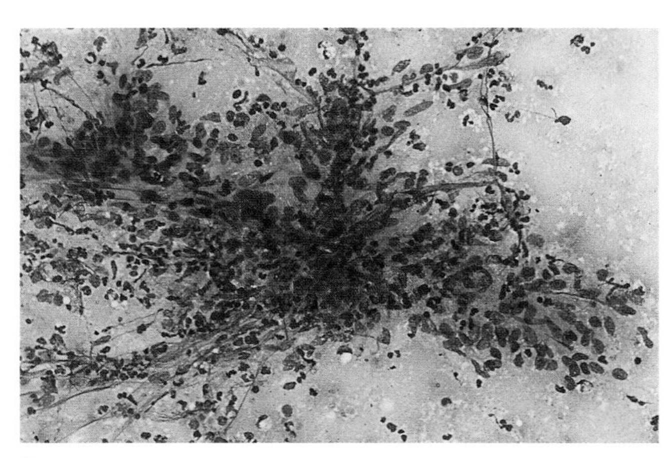

B

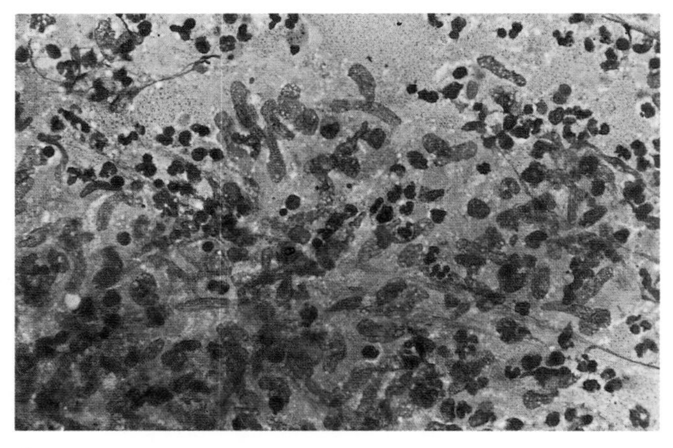

C

FIGURE 1–27. Suppurative granulomatous inflammation is manifested by large granulomas admixed with segmented leukocytes. The granulomas are characterized by loose syncytial arrangements of epithelioid cells and other modified histiocytes. The epithelioid cells have elongated nuclear contours, often with a bending or bowing of the long axis of the nucleus. (Romanowsky stain)

rative lymphadenitis, from which it cannot be separated.

FNA OF LYMPH NODES IN IMMUNOCOMPROMISED PATIENTS

Recently there has been an increase in the number of immunocompromised patients, including persons undergoing treatment for solid and hematologic malignancies; bone marrow, heart, lung, and liver transplant patients; and HIV-positive persons. The diagnostic usefulness of FNA in these patients with lymphadenopathy is well established.[80,83–87] The method is rapid, inexpensive, and well tolerated and provides excellent material for cytologic and bacteriologic assessment. The general aim of the FNA in immunocompromised patients is to exclude an infectious agent (mycobacterium avium-intracellulare [MAI], tuberculosis, fungi, specific suppurative bacteria), Kaposi's sarcoma, and lymphoma. The role of culture of FNA material cannot be overstated. It is an essential part of the procedure in a group of patients whose inflammatory reactions to common agents may be both deficient and atypical, and in whom unusual agents occur. We use a standard aspiration technique, with special attention given to making extra slides for auramine, acid-fast, and methenamine silver stains, as well as putting aside material for bacteriologic culture, flow cytometry, and cell block. In HIV-positive persons, transplant recipients, and patients with a hematologic cancer, a suppurative lymphadenitis, with its hallmarks of a necrotic background, large numbers of degenerated and viable neutrophils, and a minor component of macrophages and mixed lymphoid population, should suggest either a bacterial infection (such as *Staphylococcus, Streptococcus, Serratia, Pseudomonas, Klebsiella*) or a fungal infection (disseminated *Aspergillus, Cryptococcus, Candida,* or *Scytalidium dimidiatum*).

A granulomatous lymphadenitis, with its characteristic multinucleated giant cells, plentiful epithelioid histiocytes, small lymphocytes, and granulomatous tissue fragments, should raise the traditional differential diagnosis of *Mycobacterium tuberculosis* or sarcoidosis.[80,85,87] In HIV-positive patients, the only finding may be large numbers of plump histiocytes with cross-hatched cytoplasm representing "negative image" bacilli, which are also found in the background Romanowsky-stained serum (see Fig. 1–21).[14,15] *Mycobacterium avium intracellulare* can easily be seen in the macrophages and in the background on modified Ziehl-Neelsen and auramine stains. If neutrophils are prominent and the patient is not HIV-positive, cat-scratch fever (neck or axillary nodes) and lymphogranuloma venereum (inguinal lymph nodes) should be considered. Toxoplasmosis should also be considered when a follicular hyperplasia pattern with a polymorphous lymphoid population in which small lymphocytes predominate is seen. Tingible-body macrophages and occasional small histiocytic clusters and even occasional toxoplasmal cysts assist in the diagnosis.

The histologic stages of HIV lymphadenopathy recognized in surgical pathology are not readily diagnosable on FNA.[88] However, some lymph node FNA biopsies in HIV-positive patients produce hypocellular smears with predominant plasmacytoid lymphocytes and plasma cells with a lesser number of lymphocytes, consistent with HIV infection-related involution of the lymph node (Fig. 1–28).

The cellular yield from lymph nodes containing Kaposi's sarcoma is often disappointingly low, with the aspirate often producing mainly blood and serum with only scattered irregular tissue fragments consisting of spindle cells (Fig. 1–29). These cells are irregularly arranged in tissue fragments that may resemble granulomas and have enlarged, mildly irregular, and hyperchromatic nuclei.[89] Metachromatic stroma can be seen between cells in the Romanowsky stain, and the nuclei lack the "sand shoe" indentations of epithelioid histiocytes. Hemosiderin may occasionally be seen. In occasional cases, Kaposi's sarcoma produces highly cellular smears. If neutrophils are admixed with spindle cells and histiocytes, then bacillary angiomatosis should be considered and a Warthin-Starry stain used.

Most lymphomas seen in HIV-positive patients are intermediate to high-grade B-cell neoplasms, often presenting in soft tissues, the central nervous system, and the gastrointestinal tract as well as lymph nodes.[90–92] Burkitt-like small noncleaved lymphoma accounts for one third of the lymphomas; it often has plasmacytoid features and cytologically overlaps with immunoblastic lymphoma.[91] The other two thirds of lymphomas in HIV-infected patients are large cell, immunoblastic, and anaplastic large cell neoplasms, with a common association with Epstein-Barr virus.[92] Often, classification on morphology alone is difficult because of the overlapping cell types. Flow cytometry is valuable in both categorization and distinction from benign proliferations. The diagnosis of Hodgkin's disease can be suggested by FNA biopsy in HIV-infected patients, but distinction from viral lymphadenitis relies on excisional biopsy.[93]

Similarly, FNA biopsy of lymph nodes in transplant patients can provide a rapid diagnosis of the monoclonal proliferations identical to immunoblastic lymphoma, multiple myeloma, large cell lymphoma, or Burkitt-like lymphoma.[94]

However, distinguishing plasmacytic hyperplasias and polymorphic lymphoproliferations, with their mixed populations of small and large lymphocytes, in the spectrum of posttransplant lymphoproliferative disorders requires flow cytometry and probably histopathologic assessment.[95,96] Many of these proliferations regress when immunosuppression is decreased, whereas others progress with a high mortality rate, even with chemotherapy. Thus, a full diagnostic evaluation is required.[95] T-cell lymphomas are uncommon in HIV-positive patients and in posttransplant patients.[97]

Rosai-Dorfman Disease (Sinus Histiocytosis with Massive Lymphadenopathy)

This idiopathic disease characteristically involves the cervical neck nodes of children and young adults, particularly African-Americans. Characteristically, the lymphadenopathy is bilateral, painless, and often quite impressive in size. Histologically, lymph nodes are altered by prominent distention of the sinuses by benign histiocytes that have phagocytized erythro-

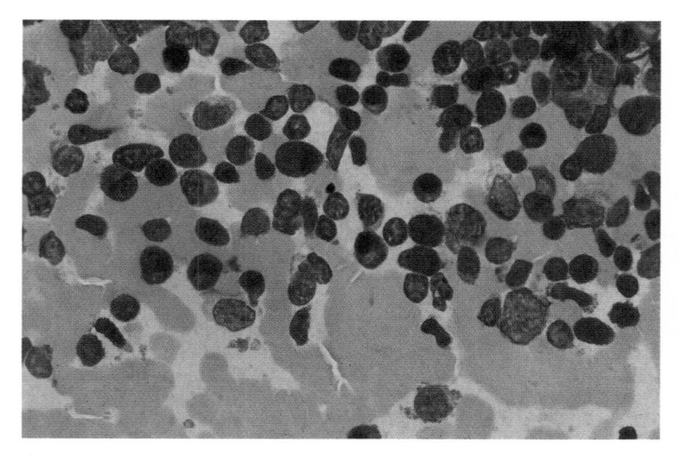

A

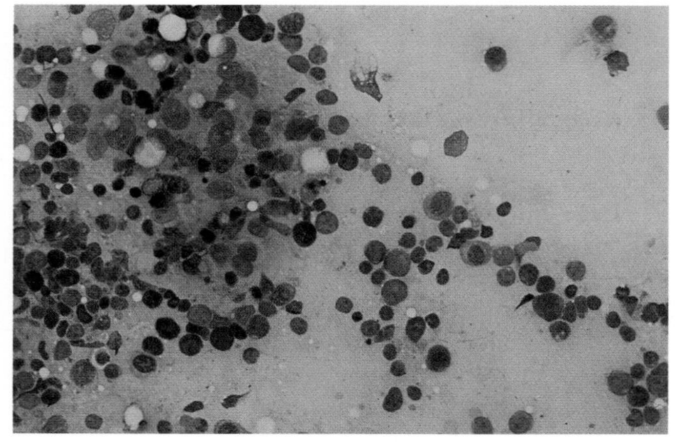

B

FIGURE 1–28. Castleman's disease. (**A**) An aspirate of a node in an HIV-positive person. A relatively large proportion of the aspirated cells are plasma cells and plasmacytoid lymphocytes. Smaller and larger lymphoid cells are also apparent. (**B**) Plasmacytoid lymphocytes are joined by small lymphocytes and an aggregate of histiocytes. (Romanowsky stain)

A

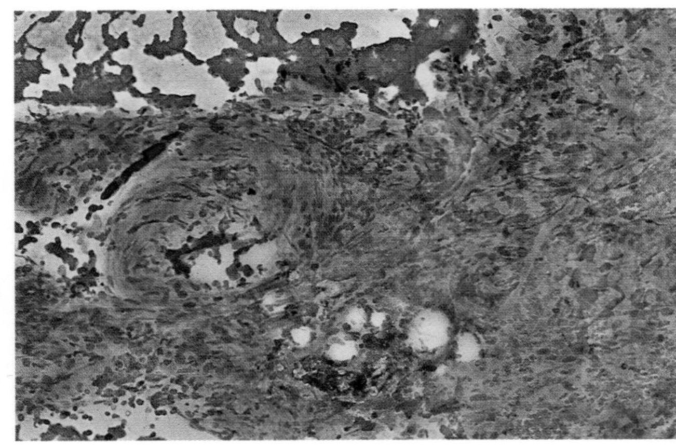

B

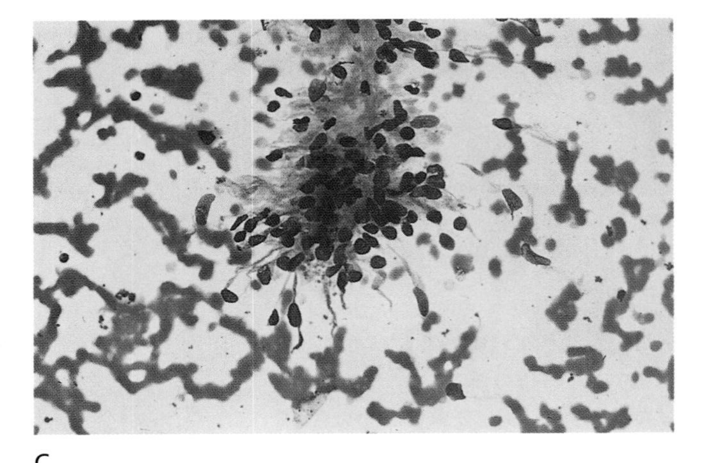

C

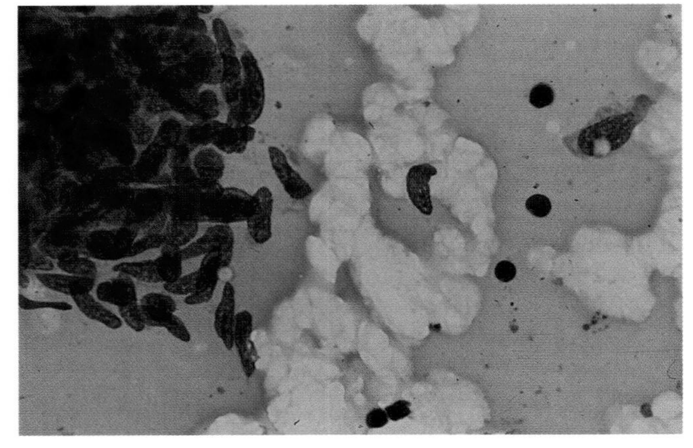

D

FIGURE 1–29. Kaposi's sarcoma. (**A**) A large fragment of aspirated sarcoma is characterized by a haphazard arrangement of tumor cells. The latter have solitary ovoid dark nuclei and indistinct cytoplasm. The edges of the fragment are irregular and frayed. Individually dispersed neoplastic cells are also apparent surrounding this fragment. (**B**) This aspirate of Kaposi's sarcoma from another patient demonstrates the vascular nature of the neoplasm, with small slitlike and rounded vascular spaces within the neoplastic mass. Again, the neoplastic cells are characterized by blunt ovoid nuclei and indistinct cellular borders. (**C**) Individual neoplastic cells may demonstrate long, thin, tapered cytoplasmic tails that often appear bipolar. In general, nucleoli are not well developed. (**D**) Although the morphology of Kaposi's sarcoma is not specific, in the proper clinical setting the morphology of these neoplastic cells is certainly in keeping with such an interpretation. Again, note the haphazard arrangement of the elongated nuclei within the aggregate of malignant cells. Individually dispersed tumor cells are also evident. (Romanowsky stain)

cytes and/or lymphocytes. Follicles are reduced in extent.

Cytologic smears may be moderately to highly cellular and may include both small and, to a lesser extent, large lymphoid cells. The diagnostic feature is the large number of phagocytic histiocytes that contain viable-appearing lymphocytes and red blood cells in their pale cytoplasm.[25–27] To diagnose or suggest Rosai-Dorfman disease in an FNA biopsy, these cells need to be seen in large numbers (Fig. 1–30). Although these cells are immunocytochemically positive for S-100 protein, they lack the nuclear grooves and complex indented nuclei of the proliferating cells seen in Langerhans' histiocytosis.

Usually this entity cannot be diagnosed by flow cytometry. There may be a huge population of CD45-negative cells, which suggests a nonlymphoid malignancy, but secondary markers demonstrate bright staining with CD38, indicating that these cells are macrophages.

Kikuchi's Lymphadenitis

Histiocytic necrotizing (or Kikuchi's) lymphadenitis is a very uncommon lymph node disorder that most often affects the cervical lymph nodes of young adult women, particularly those of Japanese or other Asian descent. It is a benign self-limited process of unknown etiology. Histopathologically, lymph nodes show several karyorrhectic areas in the paracortex composed of phagocytic and nonphagocytic histiocytes, immunoblasts, plasmacytoid monocytes, karyorrhectic debris, and an almost complete absence of neutrophilic leukocytes.

Tsang and Chan[28] published what we believe to be the largest series of aspiration biopsies of Kikuchi's

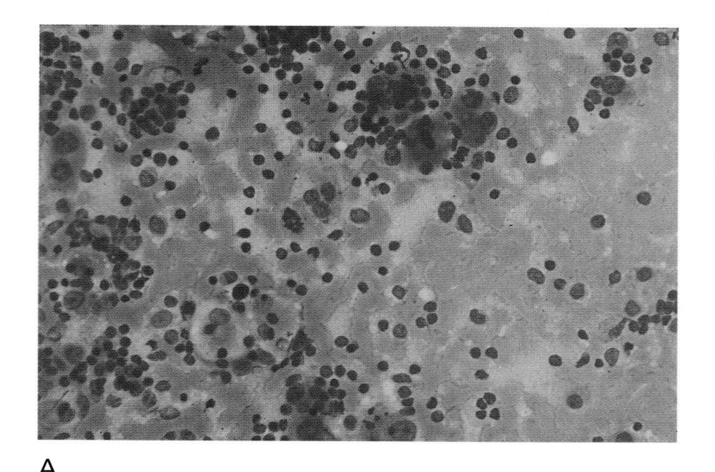

A

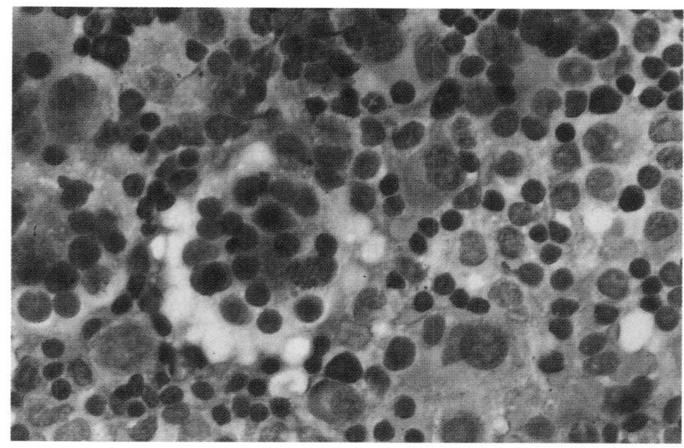

B

FIGURE 1–30. Rosai-Dorfman disease. (**A**) Small and large lymphocytes are mixed with dispersed histiocytes. The cytoplasm of many of the histiocytes contains one too many engulfed lymphocytes. (**B**) The nucleus of a histiocyte is almost obscured by numerous phagocytized lymphocytes. A few scattered plasmacytoid cells are also evident. (Romanowsky stain)

lymphadenitis. According to these authors, there are two characteristic cell types seen in the smears. The phagocytic histiocyte, the most distinctive form, is a large cell with round contours and a crescent-shaped nucleus, the convex side of which appears to fuse with the cell membrane (Fig. 1–31). These nuclei manifest irregular, twisted contours and inconspicuous nucleoli. Their abundant cytoplasm is pale and contains phagocytized eosinophilic and basophilic debris. Similar-appearing debris is also present extracellularly and at times is extensive. The other typical cell type is the plasmacytoid monocyte, characterized by an eccentrically positioned round nucleus and basophilic cytoplasm without a perinuclear hof. Immunoblasts are always present in the smears, at times in large numbers, while neutrophils are minimal or totally absent.

Necrosis can also be seen in infarcted nodes and nodes involved by metastatic carcinoma (usually squamous) and high-grade lymphomas. However, in these cases neutrophils are usually present. Systemic lupus erythematosus yields a polymorphic lymphoid infiltrate in which plasmacytoid cells are prominent with karyorrhectic debris, tingible-body macrophages, neutrophils, and the pathognomonic but rare hematoxylin bodies, which represent mauve cytoplasmic inclusions in inflammatory cells.

OVERVIEW OF MALIGNANT LYMPHOMAS

As well stated by Pontifex and Haley,[32] the use of FNA biopsies to diagnose lymphoid malignancies has been controversial. An ever-declining proportion of clinical oncologists and histopathologists regard this application with a high level of suspicion and even disdain. However, the weight of evidence in the literature and in practice shows that most malignant lymphomas can be diagnosed primarily and treated on the basis of FNA biopsy cytomorphology, at times supplemented with ancillary diagnostic procedures. We believe that with the judicious use of ancillary tests, and recognizing certain limitations, most lymphomas, both Hodgkin's disease and the non-Hodgkin's tumors, can be diagnosed and subclassified. The major differential diagnosis in most cases is nonspecific reactive hyperplasia.

The most important ancillary diagnostic tests involve the detection of specific lymphoid cell surface markers.[37,65] In our laboratories, we rely on flow cytometric determination of these antigens in aspirated cellular material, although immunocytochemical studies can also provide useful information and offer the advantage of direct correlation of cytomorphology and antigen expression. The number of antigens evaluated depends, to a large extent, on the number of as-

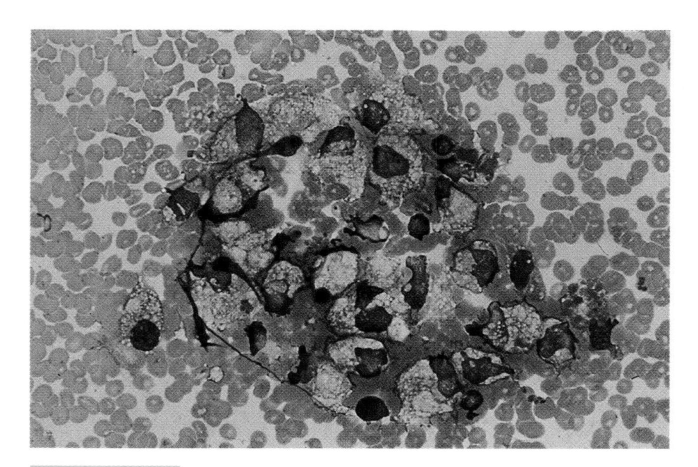

FIGURE 1–31. Kikuchi's lymphadenitis. This aspirate is characterized by loosely arrayed histiocytes with abundant foamy cytoplasm and benign-appearing nuclei. Some of the latter have a sickled contour and appear to be apposed against the cell membrane. A few small lymphocytes are also scattered within the aggregate. Note the absence of neutrophils. (Romanowsky stain)

pirated cells available. Most important is the detection of kappa and lambda light chain restriction in monoclonal B-cell lymphoid populations (Fig. 1–32). Other desirable markers include CD19, CD20, CD79a, CD5, CD10, CD3, and CD7. The first three are good markers of B-cell lymphomas. CD10 is typically present in follicular center cell lymphomas. Normally a T-cell antigen, CD5 is often aberrantly coexpressed in small cell types of B-cell lymphomas. In many T-cell lymphomas, the neoplastic elements are positive for CD7 and negative for CD3, demonstrating an aberrant expression of T-cell antigens. A myeloid cell marker may be useful for the diagnosis of granulocytic sarcoma. Some institutions determine the proliferative activity of the lymphoma cells either by immunocytochemistry or flow cytometry to facilitate grading of the malignant lymphoma.[98] We have little experience with this specific application.

The use of molecular techniques to determine rearrangements of the genes for T-cell receptors or immunoglobulin chains has a role in this setting,[45,46] but cell surface marker analysis usually provides sufficient information.

Flow cytometric analysis is of limited use in the diagnosis of nonhematopoietic neoplasms. Perhaps the most revealing finding is the lack of staining with any of the monoclonal antibodies (Fig. 1–33). The most significant negative marker is CD45 (known as leukocyte common antigen), which is present on most hematopoietic cells. It is not found on erythroid precursors, plasma cells, or epithelial neoplasms. Thus, a sample with good viability that has a population of CD45-negative cells should be considered highly suspicious for a metastatic neoplasm. Additionally, one can look for alterations of the light scatter properties produced by epithelial tumor cells, which are usually

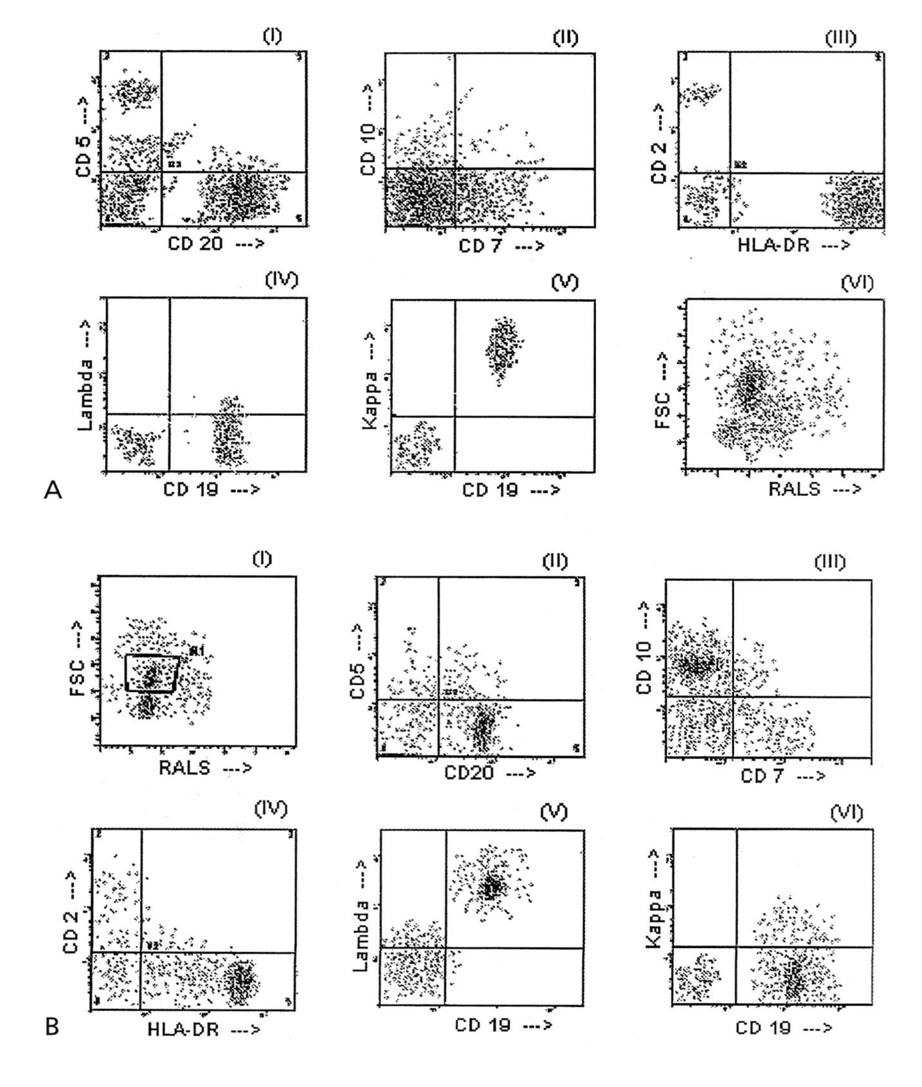

FIGURE 1–32. Follicular Center Cell Lymphomas, Large Cell. (**A**) These graphs depict some finding one might encounter in a large cell lymphoma. Panel I shows the T and B cell markers, and the increased numbers of B Cells demonstrated by the CD20. Although Follicular Center Cell lymphomas tend to coexpress CD10, when they undergo large cell transformation, they can lose this antigen as shown in Panel II. Typical for B Cells HLA-DR is expressed, and this antigen can show increased intensity in B Cell lymphomas as seen in Panel III. Monoclonality is depicted in Panel IV and Panel V. The intensity of immunoglobulin is often increased 2–3 logs compared to the control. Panel IV and Panel V. The intensity of immunoglobulin is often increased 2–3 logs compared to the control. Panel VI shows the increased size of the cells as seen by the relatively high FSC. (**B**) Two populations of cells are seen by light scatter in Panel I. The following cytograms show the characteristics of the larger cells. The vast majority of Cells are B cells as shown by the CD20 expression in Panel II. Panel III shows the expression of CD10 by the cells. Panel IV confirms the B cell nature of this process through the expression of HLA-DR. The monoclonality of these cells is shown in Panel V and Panel VI. The smaller cells showed an identical immunophenotype. The presence of a smaller cell population and a larger cell population would make this a Mixed small cleaved, and large cell lymphoma.

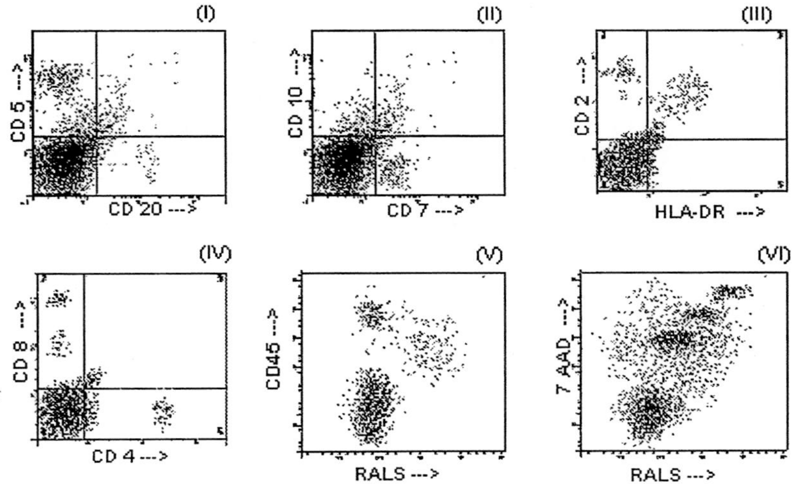

FIGURE 1–33. How a population of metastatic cells might appear. Panels I–IV show no marking with the usual T- and B-cell markers. A population of background lymphocytes may be present. Panel V shows that the vast majority of cells also fail to mark with leukocyte common antigen (CD45). It is also important to perform a viability assay such as 7-actinomycin (Panel VI), because necrotic debris can also show a similar pattern.

larger than lymphoid cells because they contain more cytoplasm.

HODGKIN'S DISEASE

Accounting for roughly 40% of lymphoid neoplasms, Hodgkin's disease principally affects older adolescents or young adults, with a second smaller incidence peak in middle-aged persons. Most frequently, supraclavicular and cervical lymph nodes are involved, but other sites, including mediastinal masses and retroperitoneal nodes, may also be involved.

Smear cellularity tends to be relatively low, reflecting the prominent fibrosis of nodular sclerosing Hodgkin's disease, the most common form of this disease.[99–103] The smears contain a polymorphous lymphoid cell population in which small mature lymphocytes predominate numerically and are mixed with eosinophils, plasma cells, and often histiocytes (Fig. 1–34). The diagnosis relies on the presence of R-S cells and their mononuclear variants (Figs. 1–35 to 1–39). R-S cells have considerable cytoplasm, pale blue with the Romanowsky stain, with two or more large complex or polylobated nuclei. The nuclei have irregular, focally thickened membranes with irregular, often coarse chromatin and large variably shaped blue nucleoli. In alcohol-fixed Papanicolaou-stained smears, the malignant nature of the nuclei is readily apparent. Mononuclear variants possess similar nuclear features (see Fig. 1–36). Although the R-S cells may be relatively sparse in the smears, they usually stand out because of their large size and are often located at the margin of the smear because of the fluid dynamics of smearing. One potential pitfall is the mistaken identification of benign immunoblasts as mononuclear variants. Immunoblasts usually are smaller, have perfectly smooth nuclear and nucleolar contours, and have more basophilic cytoplasm. In any smear in which small lymphocytes predominate, R-S cells must be looked for to exclude Hodgkin's disease. Their nuclei are often eccentrically positioned. We have found the application of CD30 (Leu-M_1) or CD15 (Ber-H_2) staining to be useful in confirming the light microscopic impression (Fig. 1–40).

Cytomorphologic Features of Hodgkin's Disease

- Low to moderate cellularity
- Polymorphous background of lymphocytes, eosinophils, plasma cells, and histiocytes
- Predominance of small lymphocytes
- Classic R-S and mononuclear variants in low numbers

Currently there is not a reliable method of detecting these neoplasms by flow cytometry.[54] There are several reasons. First, R-S cells are large and fragile, and many are lost in processing. Second, Hodgkin's disease has a cellular background of mixed benign cells, including numerous small lymphocytes, histiocytes, eosinophils, fibroblasts, granulocytes, and plasma cells. Some attempts have been made to enrich sam-

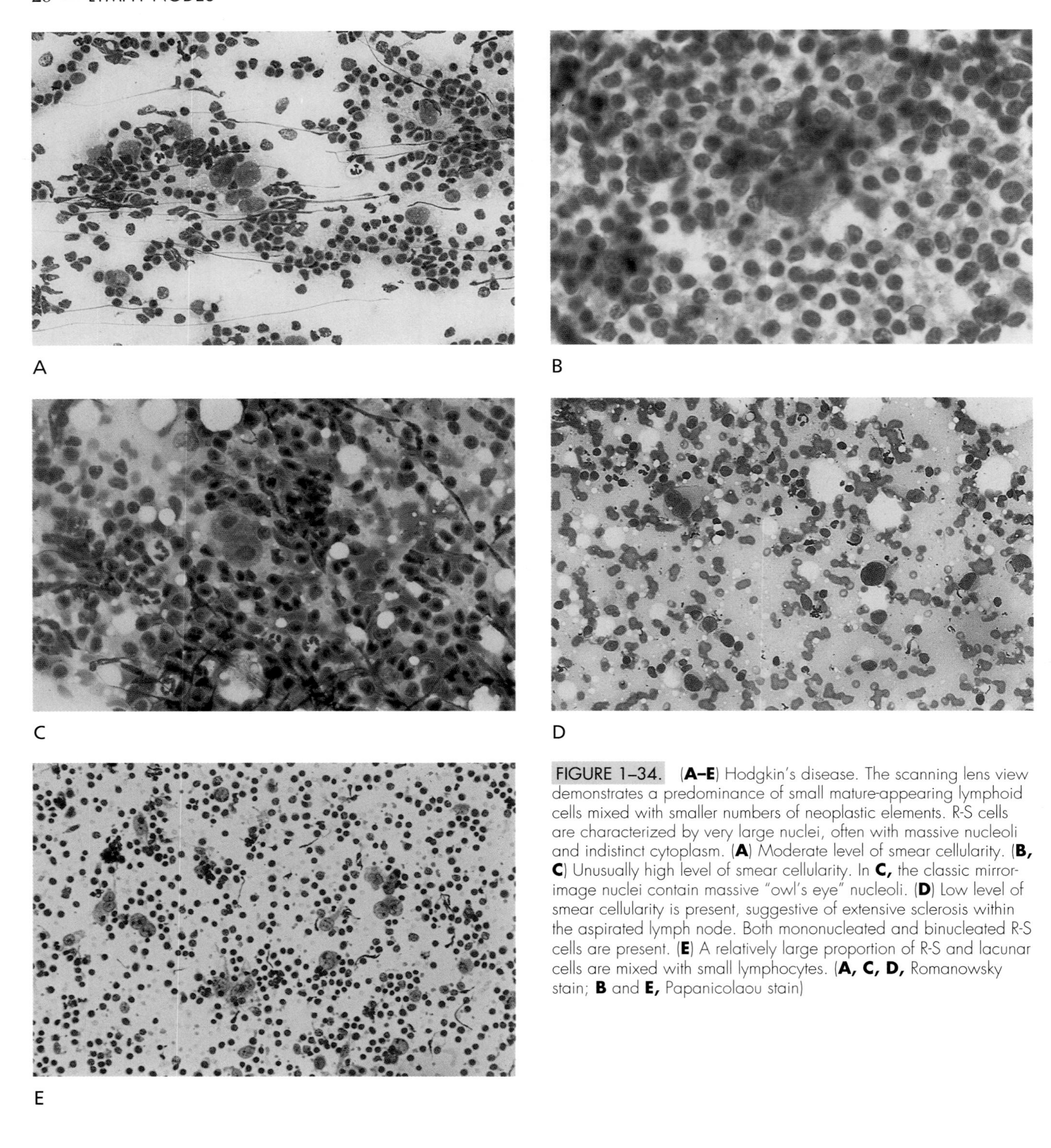

FIGURE 1–34. (**A–E**) Hodgkin's disease. The scanning lens view demonstrates a predominance of small mature-appearing lymphoid cells mixed with smaller numbers of neoplastic elements. R-S cells are characterized by very large nuclei, often with massive nucleoli and indistinct cytoplasm. (**A**) Moderate level of smear cellularity. (**B, C**) Unusually high level of smear cellularity. In **C,** the classic mirror-image nuclei contain massive "owl's eye" nucleoli. (**D**) Low level of smear cellularity is present, suggestive of extensive sclerosis within the aspirated lymph node. Both mononucleated and binucleated R-S cells are present. (**E**) A relatively large proportion of R-S and lacunar cells are mixed with small lymphocytes. (**A, C, D,** Romanowsky stain; **B** and **E,** Papanicolaou stain)

ples for R-S cells, but these efforts have not resulted in a useful product. Third, R-S cells are coated by helper T lymphocytes, which form a corona around the R-S cells, making them inaccessible to staining by monoclonal antibodies and increasing the percentage of CD-4-positive cells. Thus, a greatly elevated CD4/CD8 ratio is suggestive but not diagnostic of Hodgkin's disease. Immunocytochemistry may be especially valuable in distinguishing Hodgkin's disease from T-cell–rich B-cell lymphomas.[104]

(*Text continues on page 29.*)

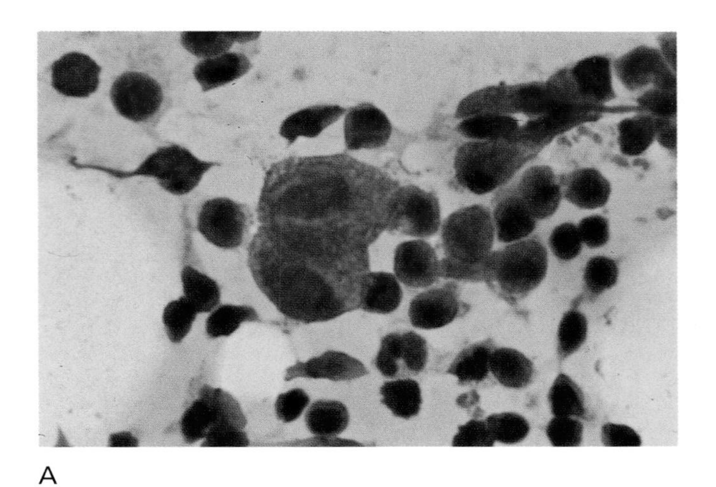

A

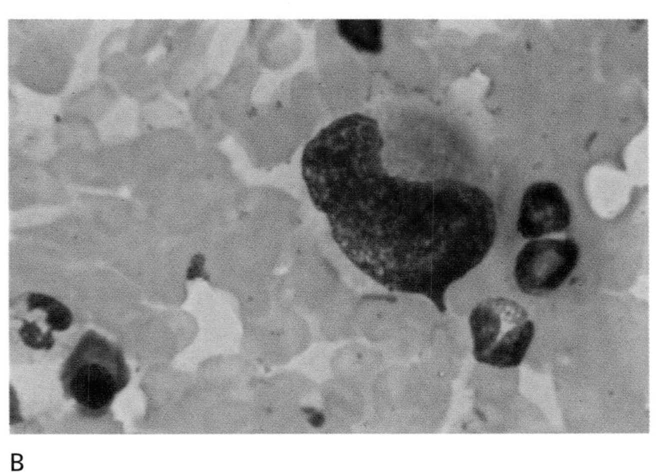

B

FIGURE 1–35. Classic R-S cells have two massive nuclei that closely resemble each other. They have finely reticulated chromatin and very large nucleoli. Cytoplasm may or may not be distinct. Frequently, the two nuclei slightly overlap. (Romanowsky stain)

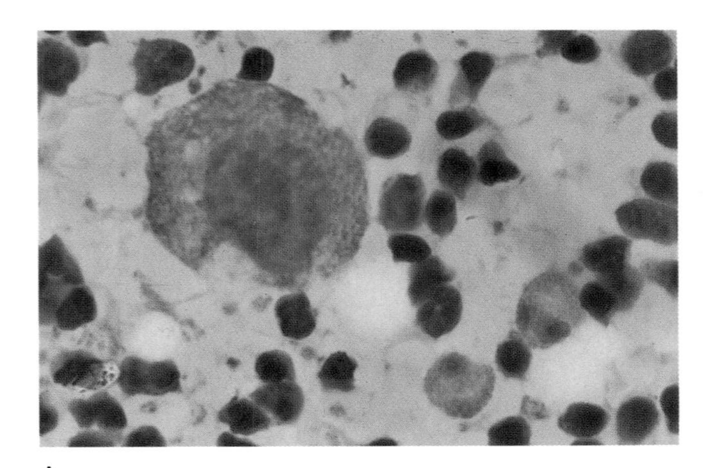

A

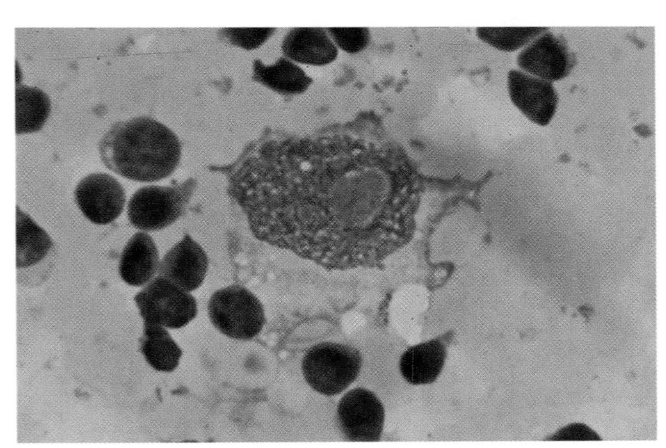

B

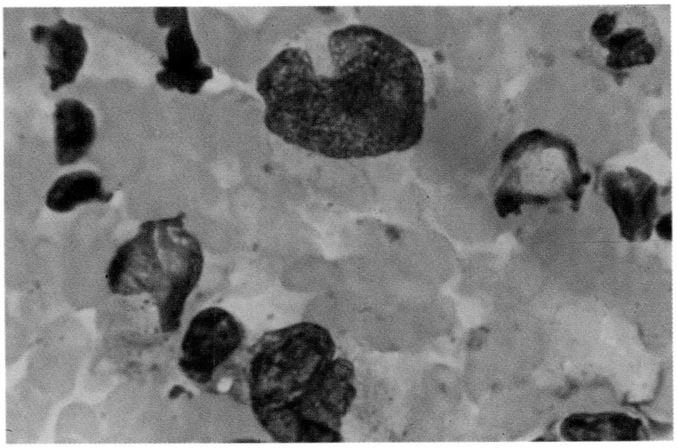

C

FIGURE 1–36. **(A–C)** Mononucleated R-S cell variants also have immature chromatin and usually a solitary huge nucleolus. In these examples, faintly basophilic cytoplasm is evident; minute vacuoles are also apparent. (Romanowsky stain)

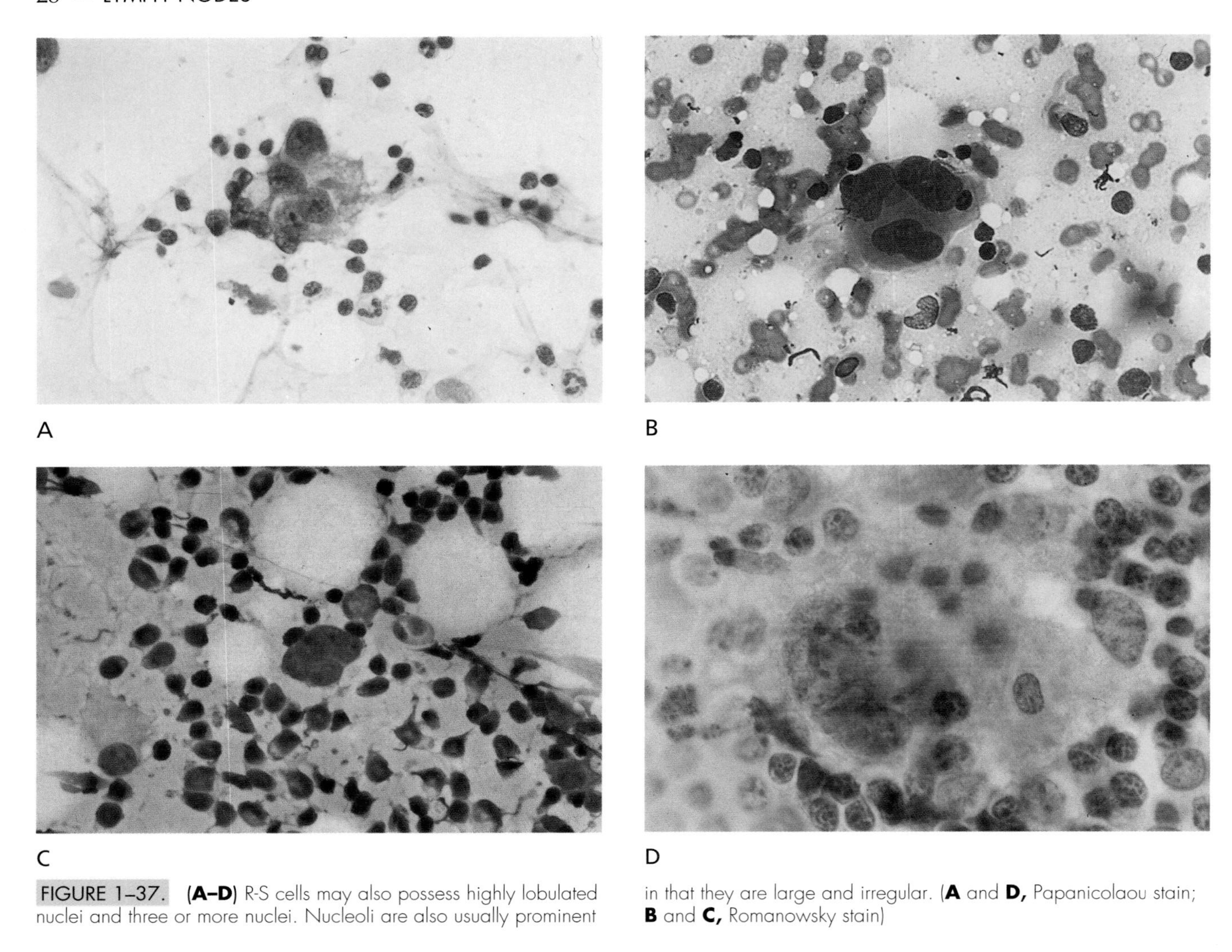

FIGURE 1–37. (**A–D**) R-S cells may also possess highly lobulated nuclei and three or more nuclei. Nucleoli are also usually prominent in that they are large and irregular. (**A** and **D,** Papanicolaou stain; **B** and **C,** Romanowsky stain)

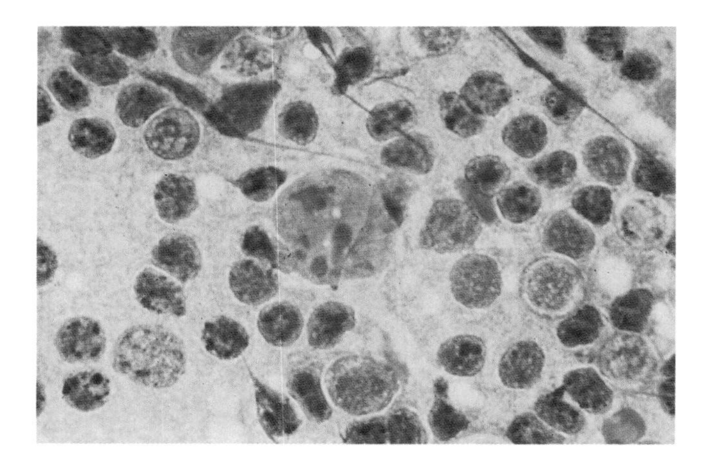

FIGURE 1–38. The contours of the nuclei and the nucleoli of R-S cells are often highly irregular. Because of the large size of these cells, they usually are readily apparent, even when present in low numbers. Note the inconspicuous cytoplasm. (Romanowsky stain)

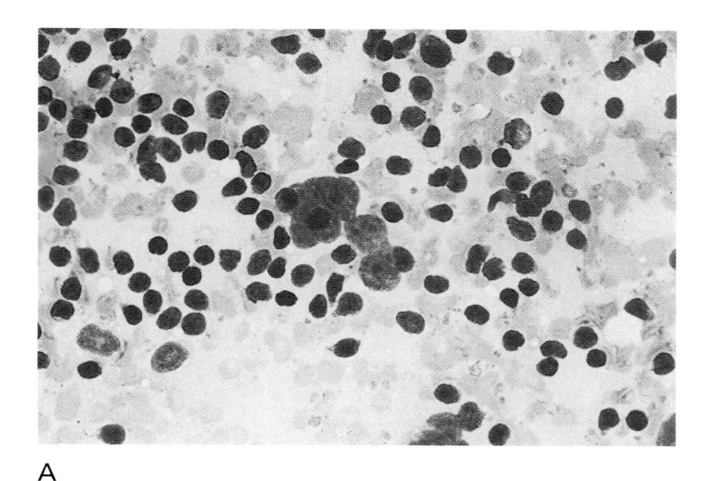

A

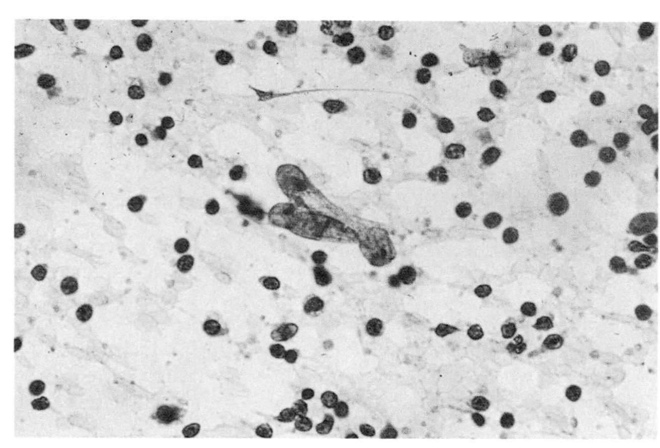

B

FIGURE 1–39. (**A, B**) Lacunar cells are characterized by highly lobulated nuclei with dark condensed chromatin and no evidence of attached cytoplasm. They are characteristic of aspirates of the nodular sclerosing variant of Hodgkin's disease. (**A,** Romanowsky stain; **B,** Papanicolaou stain)

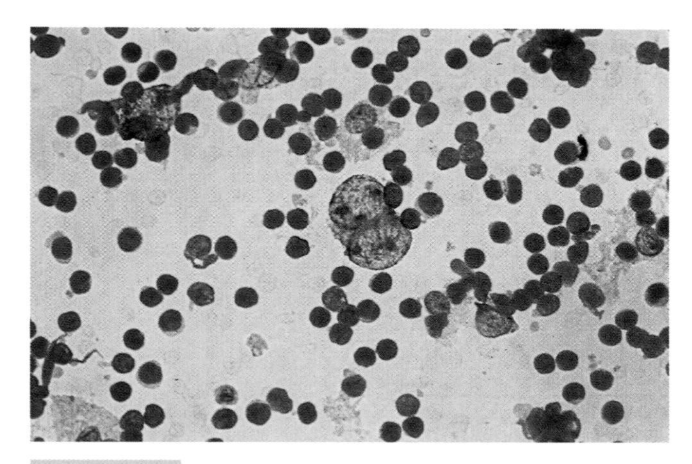

FIGURE 1–40. R-S cells are positive for Leu-M_1 in their cytoplasm. Note that the small lymphocytes are nonreactive.

Non-Hodgkin's Lymphoma

In the United States, the Working Formulation is currently the gold standard used to classify lymphoid neoplasms.[105,106] This classification scheme uses the morphologic appearance to classify lymphomas as of low, intermediate, or high grade, but does not take into account the biology or the pathogenesis of the neoplasm (Table 1–1). A more recent classification scheme, known as the REAL scheme, uses specific morphologic, immunologic, and genetic features or combinations of such attributes.[40] For instance, many mantle zone lymphomas (MZLs) express the bcl-1 oncogene.[107] This protein is expressed as a novel fusion protein resulting from a translocation involving chromosomes 11 and 14. This is a specific entity that has a specific marker, and is recognized by the REAL classification.[40] Flow cytometry can be used to assess the immunologic phenotype of a neoplasm for classification under the REAL classification.

The fundamental approach to lymphoma diagnosis in FNA biopsy relies on the morphologic assessment of cellularity, cell type, and special features, followed by ancillary studies, including flow cytometry and immunocytochemistry. Flow cytometry is generally recommended in all cases because cytomorphology alone cannot diagnose all the subcategories of non-Hodgkin's lymphomas. For example, high-grade lymphomas such as large cell and small noncleaved cell types are readily diagnosed on light microscopy, but follicular center cell lymphomas with a mixed cell population are extremely difficult to differentiate from reactive lymph nodes.

The presence or absence of monoclonality is crucial in diagnosing lymphoid malignancies. Although it has been shown that some monoclonal proliferations behave like dysplastic processes, it is generally believed that monoclonal proliferations of lymphoid tissue represent malignancy. For B cells, monoclonality can readily be documented by the presence or absence of kappa or lambda surface immunoglobulin restriction. Normal cells are admixed with neoplastic cells, so there may be times when the percentage of kappa-positive or lambda-positive B cells is less than 100% in a B-cell non-Hodgkin's lymphoma. Although T-cell malignancies are also monoclonal, the T-cell receptor is far more complex, and the presence or absence of a range of antigens must be assessed.

TABLE 1–1.

R.E.A.L. Classification	Kiel Classification	Working Formulation
B-Cell Neoplasms		
Low-Grade Non-Hodgkin's Lymphomas		
Chronic lymphocytic	Chronic lymphocytic	A. Small lymphocytic consistent with chronic lymphocytic lymphoma
Prolymphocytic	Prolymphocytic	
Hairy cell leukemia	Hairy cell leukemia	
Plasmacytoid B-cell	Lymphoplasmacytoid	A. Small lymphocytic, plasmacytoid
Lymphoplasmacytic lymphoma	Lymphoplasmacytic	
Plasmacytoma/myeloma	Plasmacytic	
Mantle cell lymphoma	Centrocytic	
Monocytoid and marginal zone	Nodal marginal zone lymphomas	
Extranodal marginal zone B-cell lymphomas (MALT lymphoma)		
Follicular, grade I	Follicular	B. Follicular, predominately small cleaved cell
Follicular, grade II	Follicular	C. Follicular, mixed small cleaved and large cell
Diffuse	Diffuse	E. Diffuse small cleaved cell F. Diffuse mixed small and large cell
High-Grade Non-Hodgkin's Lymphomas		
Follicular, grade III	Follicular	D. Follicular large cell
Diffuse large B-cell lymphoma	Diffuse	G. Diffuse, large cell
Diffuse large B-cell lymphoma	Immunoblastic	H. Large cell immunoblastic
Primary mediastinal B-cell lymphoma		
Mantle cell lymphoma, blastoid variant	Centocytoid	
Burkitt's lymphoma	Burkitt's lymphoma	J. Small noncleaved Burkitt's and non-Burkitt's
Precursor B-cell lymphoma/leukemia	Lymphoblastic	L. Lymphoblastic
T-Cell Neoplasms		
Low-Grade Non-Hodgkin's Lymphomas		
Chronic lymphocytic	Chronic lymphocytic	A. Small lymphocytic consistent with chronic lymphocytic lymphoma
Large granular lymphocytic leukemia		
Prolymphocytic leukemia	Prolymphocytic leukemia	
Mycosis fungoides/Sezary syndrome	Small cell cerebriform	
Peripheral T-cell lymphoma, lymphoepithelial type	Lymphoepithelioid (Lennert's) lymphoma	
Angioimmunoblastic T-cell lymphoma	Angioimmunoblastic	
Peripheral T-cell lymphoma, unspecified	T-zone lymphoma	
Peripheral T-cell, medium-sized cell	Pleomorphic, small cell	A. Small lymphocytic consistent with chronic lymphocytic lymphoma
Intestinal T-cell lymphoma		
Adult T-cell lymphoma/leukemia		
High-Grade Non-Hodgkins Lymphomas		
Peripheral T-cell lymphoma, mixed medium and large cell	Pleomorphic, medium-sized and large cell	
Peripheral T-cell lymphoma, large cell	T-cell immunoblastic	
Anaplastic large cell lymphoma, T and null cell type	T-cell large cell anaplastic (Ki-1+)	
Precursor T-cell lymphoma, T-cell lymphoblastic lymphoma/leukemia	T-cell lymphoblastic	L. Lymphoblastic

Large Cell Lymphoma

The most common lymphoma diagnosed in aspiration biopsies is non-Hodgkin's large cell lymphoma. This neoplasm affects both adults and children and is relatively common in AIDS patients and transplant recipients. In the Working Formulation, they fall into intermediate and high grades.[105,106]

Aspiration smears are highly cellular (Fig. 1–41) and characterized by a monomorphic population of atypical noncohesive large lymphoid cells.[1,32,47,65,108] In

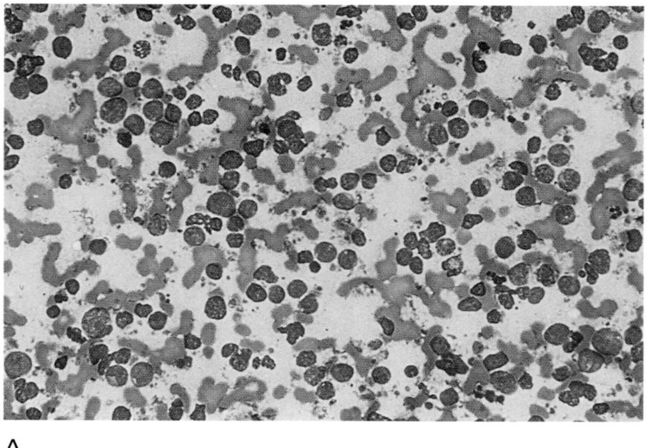

A

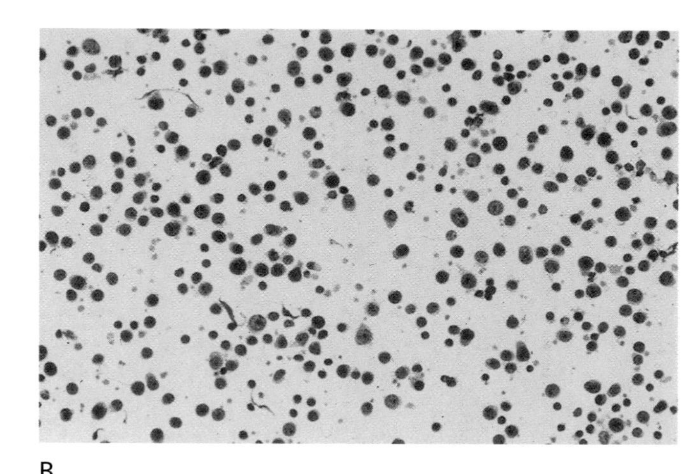

B

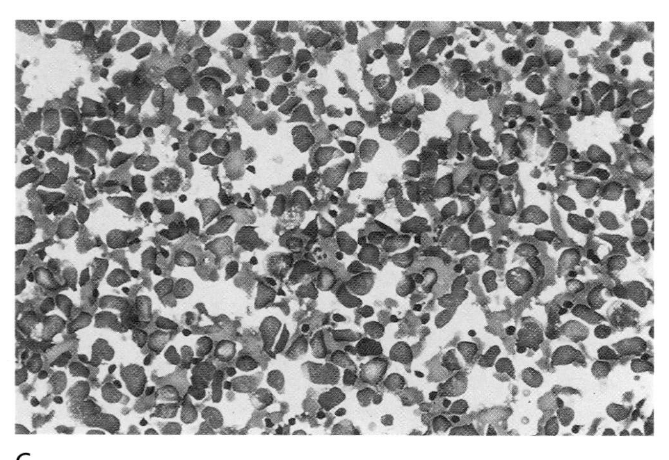

C

FIGURE 1–41. (**A–C**) In the scanning lens view, aspiration biopsies of large cell lymphoma manifest a very high level of smear cellularity. Most often, monomorphic populations of large malignant lymphoid cells are dissociated without evidence of intercellular cohesion. **A** and **B** are from the same patient. (**A** and **C,** Romanowsky stain; **B,** Papanicolaou stain)

fact, it is this monomorphism of large atypical lymphoid cells that readily permits the diagnosis of this malignancy (Figs. 1–41 to 1–44). Nuclei are large and irregular in shape, with membranes having thick and thin regions, variably clumped and cleared chromatin, and one to several prominent nucleoli. Cytoplasmic basophilia may be intense and vacuolization may be prominent, usually in the form of minute vacuoles (see Fig. 1–42). The cells may resemble large centrocytes (cleaved) or centroblasts (noncleaved) immunoblasts with single nucleoli in large nuclei with eccentric cytoplasm, or may have large, even polylobated hyperchromatic nuclei. The nuclear features, although appreciable in Romanowsky-stained smears, are more clearly seen in Papanicolaou-stained smears; air-dried smears allow better assessment of cytoplasmic features. Tingible-body macrophages are usually scant (Fig. 1–45). Necrosis is seen occasionally.

The differential diagnosis includes anaplastic carcinomas and malignant melanomas, both of which can have a prominent dissociated cell population in smears. The presence of true cohesion and tissue fragments in carcinomas, the presence of pigment granules and prominent nuclear pseudoinclusions in melanoma, and the nuclear and cytoplasmic features of lymphomas assist in the diagnosis. The use of flow cytometry and immunocytochemistry almost always confirms the diagnosis.

Cytomorphologic Features of Large Cell Lymphoma

- High cellularity; numerous lymphoglandular bodies
- Dispersed uniform to variably sized large lymphoid cells
- Large pleomorphic nuclei with variable contours
- Prominent nucleoli
- Variable nuclear-to-cytoplasmic (N/C) ratios

(*Text continues on page 34.*)

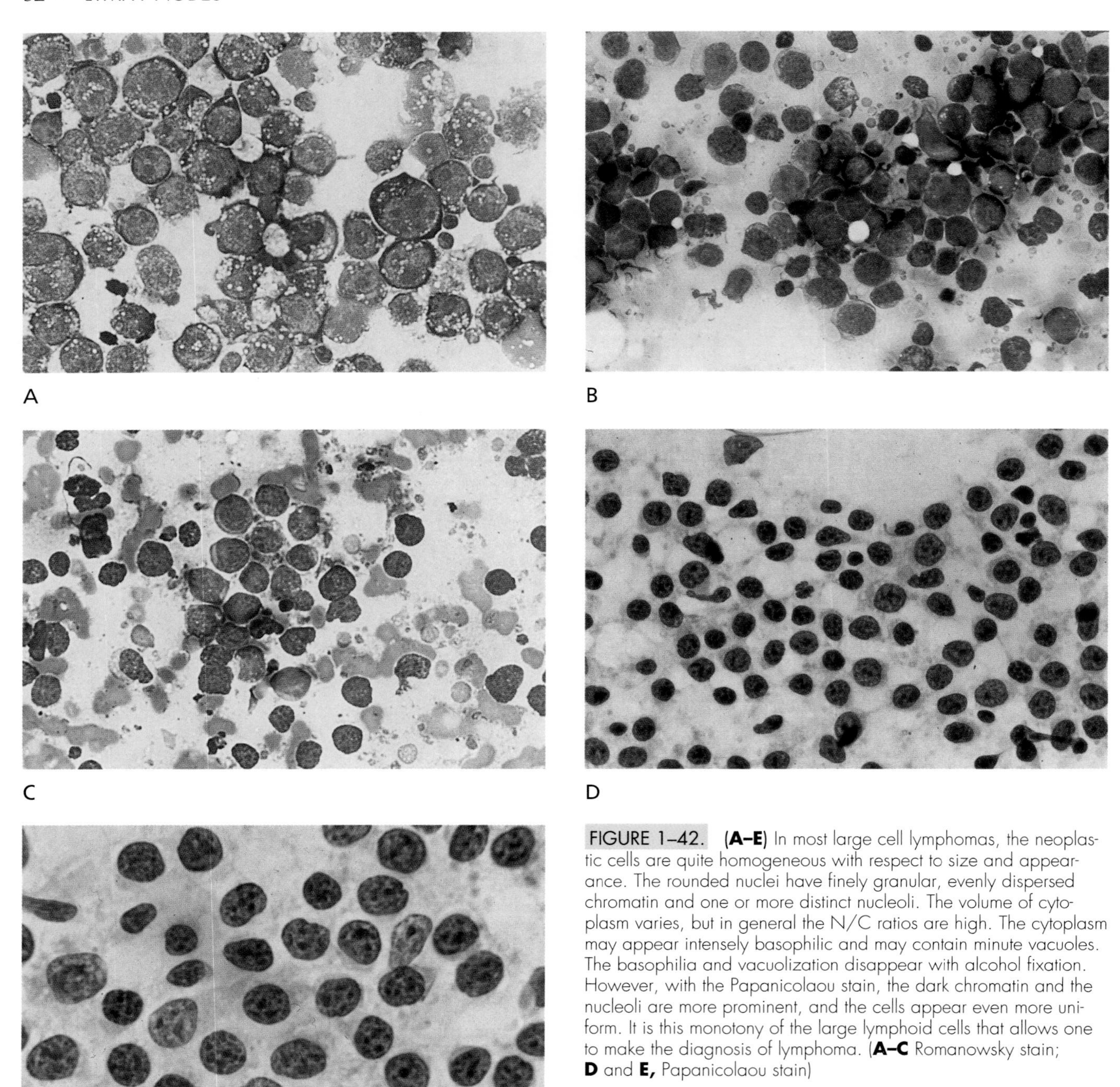

FIGURE 1–42. **(A–E)** In most large cell lymphomas, the neoplastic cells are quite homogeneous with respect to size and appearance. The rounded nuclei have finely granular, evenly dispersed chromatin and one or more distinct nucleoli. The volume of cytoplasm varies, but in general the N/C ratios are high. The cytoplasm may appear intensely basophilic and may contain minute vacuoles. The basophilia and vacuolization disappear with alcohol fixation. However, with the Papanicolaou stain, the dark chromatin and the nucleoli are more prominent, and the cells appear even more uniform. It is this monotony of the large lymphoid cells that allows one to make the diagnosis of lymphoma. (**A–C** Romanowsky stain; **D** and **E,** Papanicolaou stain)

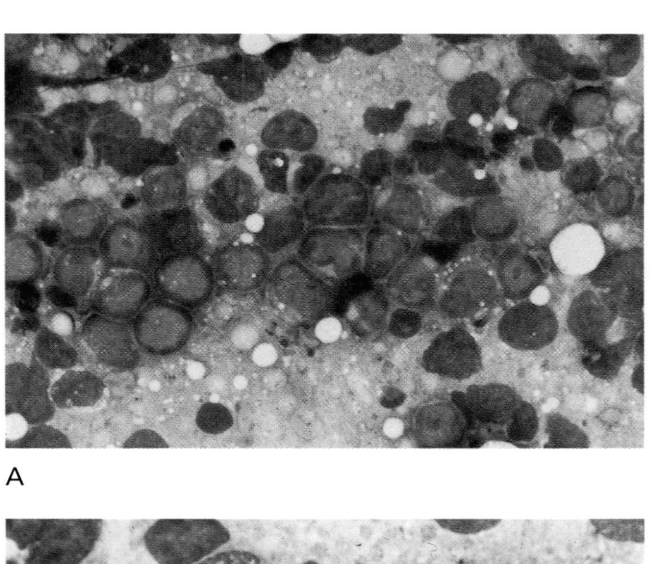

A

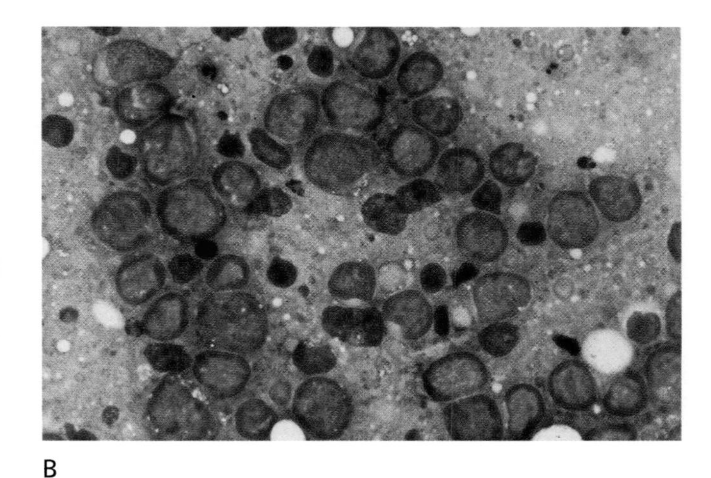

B

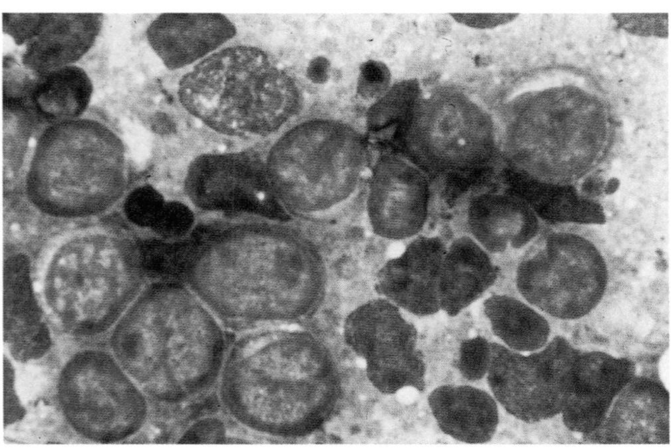

C

FIGURE 1–43. (**A–C**) Large cell lymphoma. In some neoplasms, many of the malignant nuclei show prominent irregularities in contour, including indentations and cleavage lines. Nucleoli tend to be well developed. (Romanowsky stain)

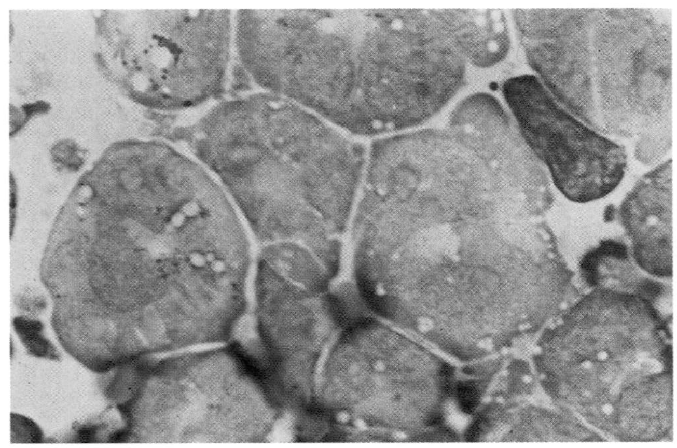

FIGURE 1–44. In some large cell lymphomas, striking pleomorphism is evident. For example, the tumor cells here are characterized by huge nuclei with highly irregular shapes, including indentations, folds, and cleavage planes. In addition, each nucleus houses multiple massive nucleoli. The quantity and quality of cytoplasm vary among the neoplastic cells. Some have vacuoles, whereas in others, red-staining lysosomes are apparent. (Romanowsky stain)

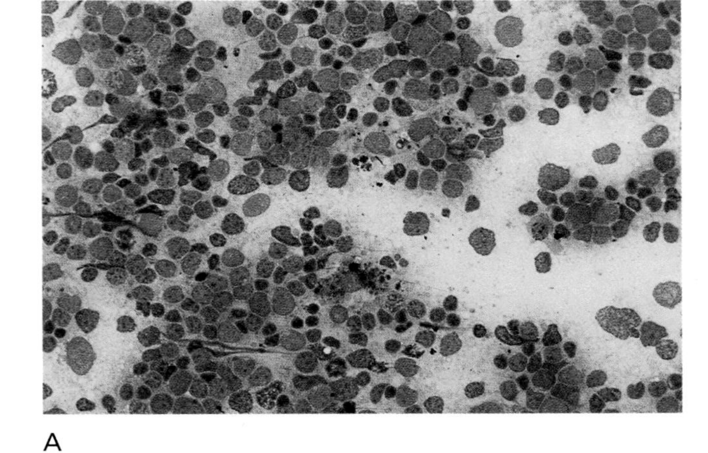

A

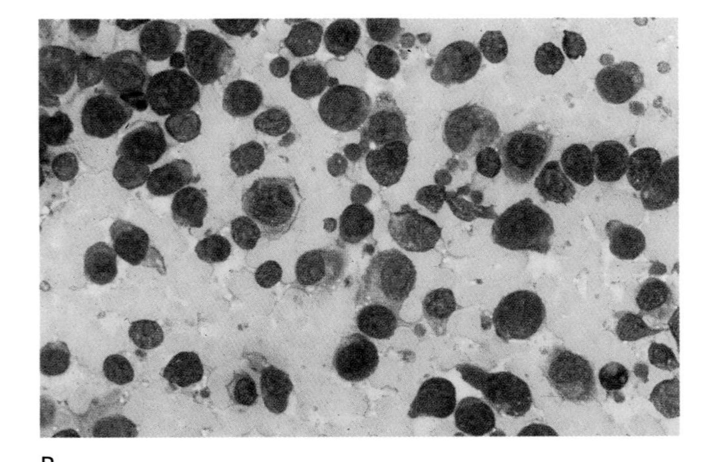

B

FIGURE 1–45. (**A, B**) High-grade lymphomas in HIV-positive patients may have a large cell appearance, including immunoblasts. Note the tingible-body macrophages in the smear background; they are a marker of high cell turnover rates. (Romanowsky stain)

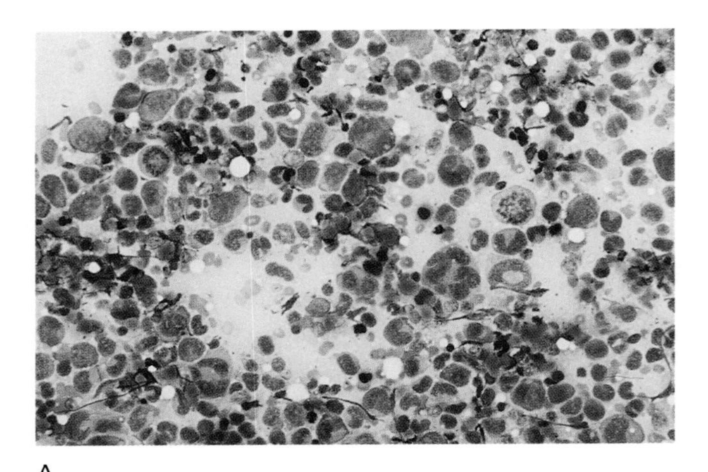

A

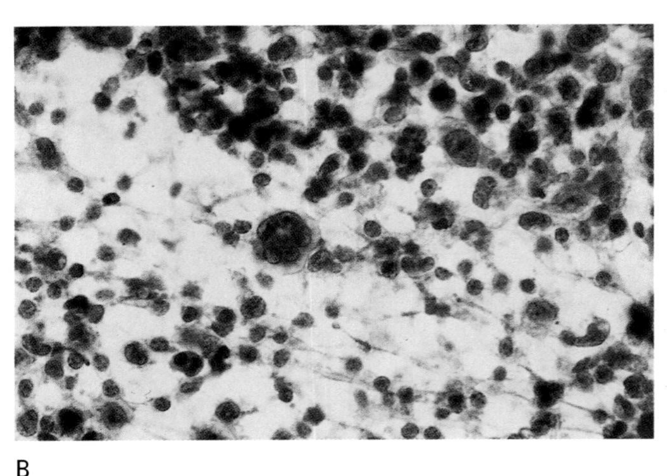

B

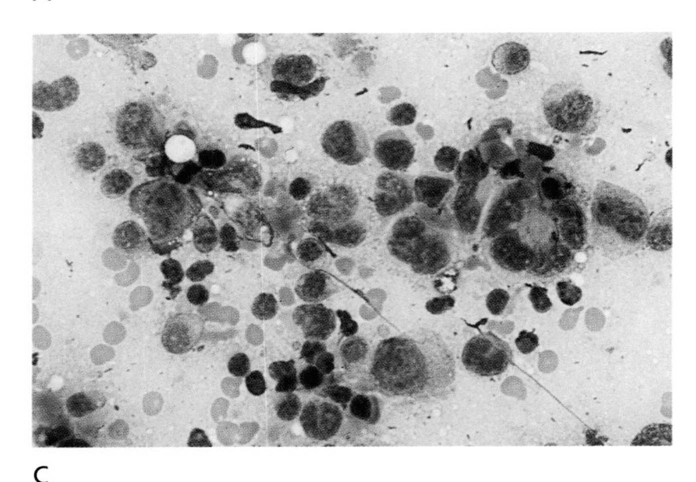

C

FIGURE 1–46. **(A–C)** Anaplastic Ki-1-positive lymphomas are characterized by high smear cellularity and striking pleomorphism. The malignant lymphoid cells vary remarkably in size and shape. Many of the tumor cells resemble those of a large cell non-Hodgkin's lymphoma, whereas others are small lymphocytes with variably shaped nuclei and lymphoblasts. In addition, one can characteristically find multinucleated tumor giant cells. Many of these have a wreathlike arrangement of nuclei around the periphery of the cell. A lack of cohesion among the aspirated neoplastic cells is usually apparent. (**A** and **C,** Romanowsky stain; **B,** Papanicolaou stain)

A recently recognized variant is the anaplastic large cell lymphoma, the definition of which is still in evolution.[109] Usually, the neoplastic cells are immunocytochemically positive for CD30 (Ki-1). The malignant cells manifest remarkable pleomorphism (Fig. 1–46). Many resemble those of a large cell lymphoma, whereas others are smaller, resembling small cleaved cells and blasts. Still others are monstrous and often multinucleated; some may simulate R-S cells. Nucleoli may or may not be prominent. In histologic preparations of lymph nodes, the tumor cells may resemble metastatic carcinoma in their pattern of involvement. However, in aspiration smears, a dissociated pattern is apparent, and pleomorphism is usually well developed.[110–115]

Small Cleaved Cell Lymphoma

This malignancy is essentially restricted to the adult population. Histologically, the lymph node architecture is effaced by either a nodular or diffuse prolifera-

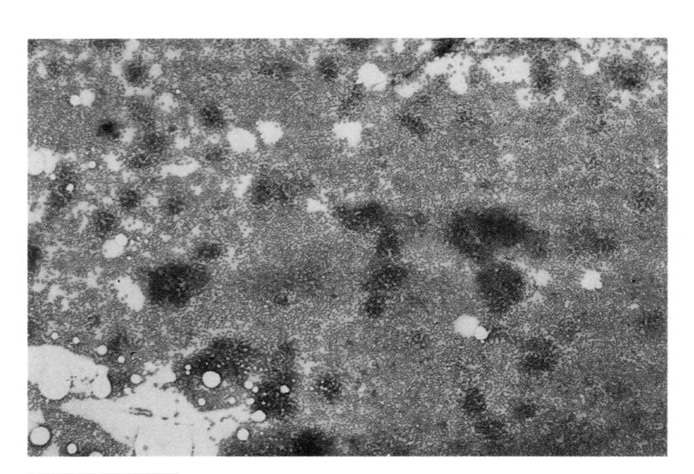

FIGURE 1–47. Aspirates of small cleaved cell lymphoma typically yield highly cellular smears with a dissociative pattern of malignant lymphocytes. Histologically, the tumors may have a follicular, diffuse, or mixed pattern. In aspiration smears, the neoplastic follicles may present as dense aggregates of cells surrounded by dispersed neoplastic cells. The histology of this aspirated lymph node was a mixed follicular and diffuse arrangement. (Romanowsky stain)

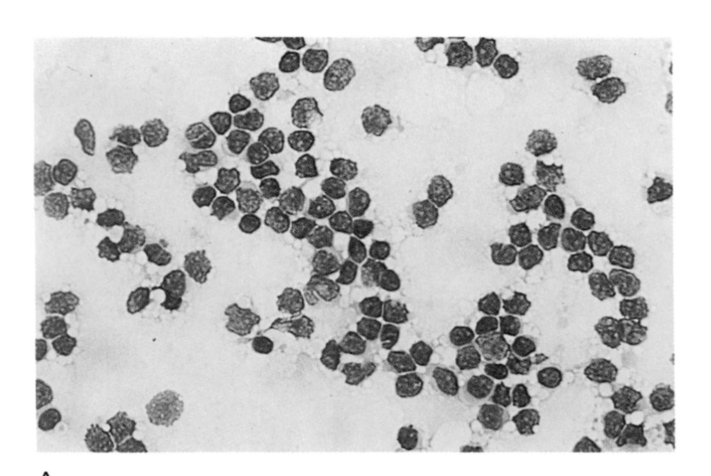

A

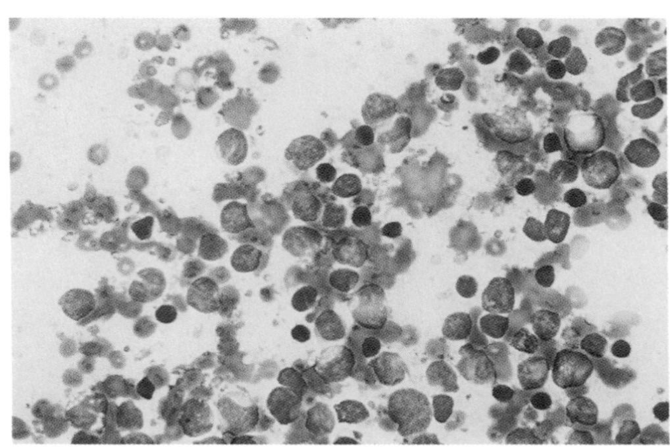

B

FIGURE 1–48. In small cleaved cell lymphoma, the neoplastic elements are relatively small and homogeneous in appearance. They are characterized by irregular nuclear shapes, including notches, indentations, and cleavage planes. The chromatin is coarsely granular and darkly stained; nucleoli are generally inconspicuous. The N/C ratios are very high. (Romanowsky stain)

tion of small atypical lymphoid cells. In the Working Formulation, nodular and diffuse patterns are classified as of low and intermediate grades with prognostic differences, respectively. This distinction requires histologic examination of the node.[105,106]

Smears are generally highly cellular and show a homogeneous population of dispersed small mononucleated lymphoid cells.[1,65] The tumor cells have diameters up to double that of small mature lymphocytes, high N/C ratios, and thickened and irregularly contoured nuclei featuring clefts, deeper indentations, and linear creases (Figs. 1–47 and 1–48). Areas of parachromatin clearing and small nucleoli may be evident. It has been suggested that aggregates of these cells in the smears may point to a nodular architecture of the lymphoma.[36] However, as stated above, histology is generally required for this distinction.

Cytomorphologic Features of Small Cleaved Cell Lymphoma

- High cellularity with monotonous small to intermediate dispersed lymphoid cells
- Irregular nuclei with notches, clefts, and creases and moderately clumped chromatin with parachromatin clearing
- High N/C ratios

The immunophenotype of this malignancy was discussed earlier.[64]

Mixed Small Cleaved and Large Cell Lymphoma

As with small cleaved cell lymphoma, this follicular center cell neoplasm occurs almost exclusively in adults. In the Working Formulation, this is an intermediate-grade lymphoma and consists of an admixture of small and large atypical lymphoid cells, with a predominance of the former.[64] According to Sneige et al.,[65,116] when large tumor cells represent 5–15% of the neoplastic elements, the specimen should be classified as a mixed lymphoma (Figs. 1–49 to 1–51).

The major problem in the diagnosis is distinguishing this heterogeneous mix of malignant lymphoid cells from a benign reactive hyperplastic node, and immunophenotyping is essential (Fig. 1–52). FNA biopsy material from reactive nodes with follicular hyperplasia can show various regions on the slides in which small cleaved cells and larger lymphoid cells can be quite prominent, to the exclusion of small

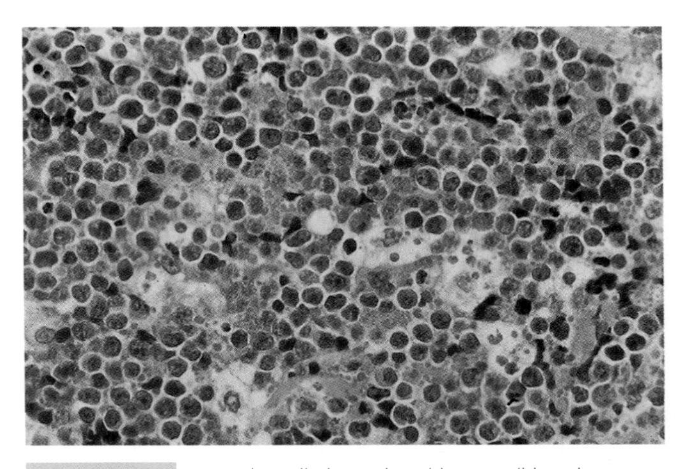

FIGURE 1–49. Mixed small cleaved and large cell lymphoma. High smear cellularity demonstrates predominance of small lymphoid cells with irregularly shaped nuclei and very high N/C ratios. A minority of the neoplastic cells are larger with more vesicular chromatin and distinct nucleoli; in addition, scanty rims of cytoplasm are visible. Lymphoglandular bodies are prominent. (Romanowsky stain)

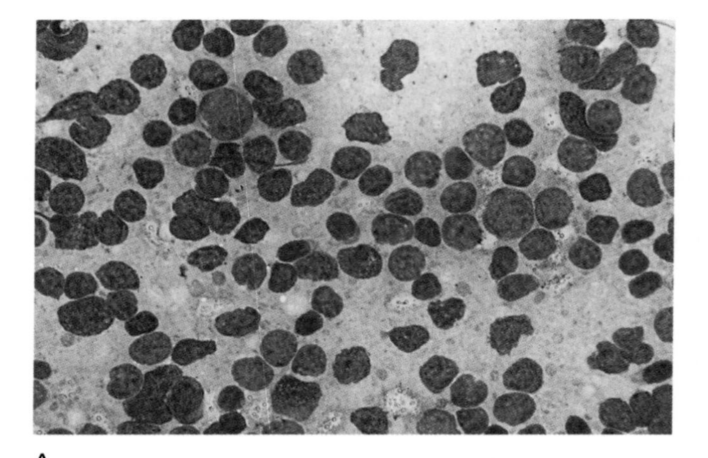

A

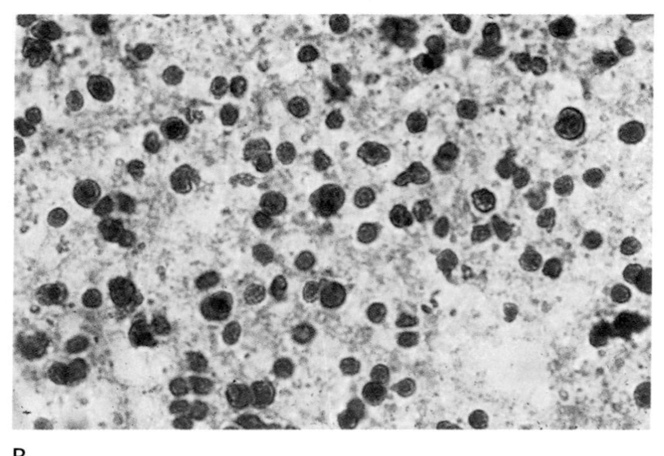

B

FIGURE 1–50. (**A, B**) Mixed small cleaved and large cell lymphoma. The coarsely granular chromatin in the small malignant lymphocytes is apparent. (**A,** Romanowsky stain; **B,** Papanicolaou stain)

lymphocytes. These areas, however, representing smeared germinal centers, generally have tingible-body macrophages, and by scanning the cellular smear, foci in which small lymphocytes predominate can be seen.

Follicular center cell lymphomas account for the vast majority of non-Hodgkin's lymphomas and are well categorized in the Working Formulation. Many are associated with a chromosomal translocation that gives rise to the bcl-2 oncogene.[44] Immunologically, these lymphomas express pan-B antigens with surface immunoglobulin expression. Many also coexpress CD10[39,41] and demonstrate increased CD20 expression when compared with normal B cells.[39] These cells lack most other antigens, such as CD5. It is thought that these neoplasms represent a progression from follicular, predominantly small cleaved cell lymphoma to diffuse, predominately large cell lymphomas. CD10 is less likely to be coexpressed and CD20 less apt to be abnormal as these lymphomas migrate toward the more aggressive end of the spectrum.[66] The aggressive lymphomas are also much more likely to have overt necrosis and dying cells, which can be picked up in a viability study.

Follicular center cell lymphomas are generally straightforward to detect by flow cytometry. Monotypic surface immunoglobulin can usually be detected. However, the type of heavy chain is variable, and a potential pitfall lies in the fact that a small percentage of these tumors express pan-B cell antigens but lack detectable surface immunoglobulin. To make matters worse, some florid follicular hyperplasias may also lack detectable surface immunoglobulin. When we encounter this situation, we typically do not claim that the flow cytometric findings are diagnostic for a lymphoma, but rather that they are suspicious for one.

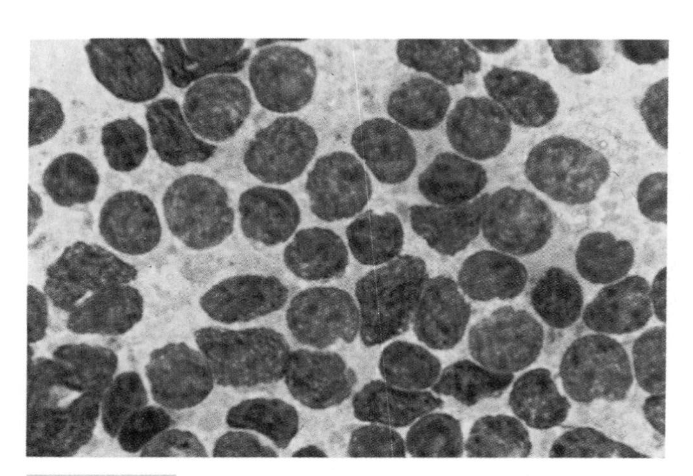

FIGURE 1–51. Mixed small cleaved and large cell lymphoma. Slight irregularities in the nuclear contours are apparent in the small malignant lymphocytes, which have extremely high N/C ratios. (Romanowsky stain)

Immunophenotyping of Follicular Center Cell Lymphomas

- Pan-B cell markers (CD19, CD20, CD79a) expressed
- CD10 may be expressed
- Variably increased CD20 expression
- Monoclonal SIg

Small Lymphocytic Lymphoma

This category of low-grade lymphomas represents a heterogeneous group; the greatest difficulty is the inability to separate them from benign conditions on morphologic grounds alone. Aspiration smears are

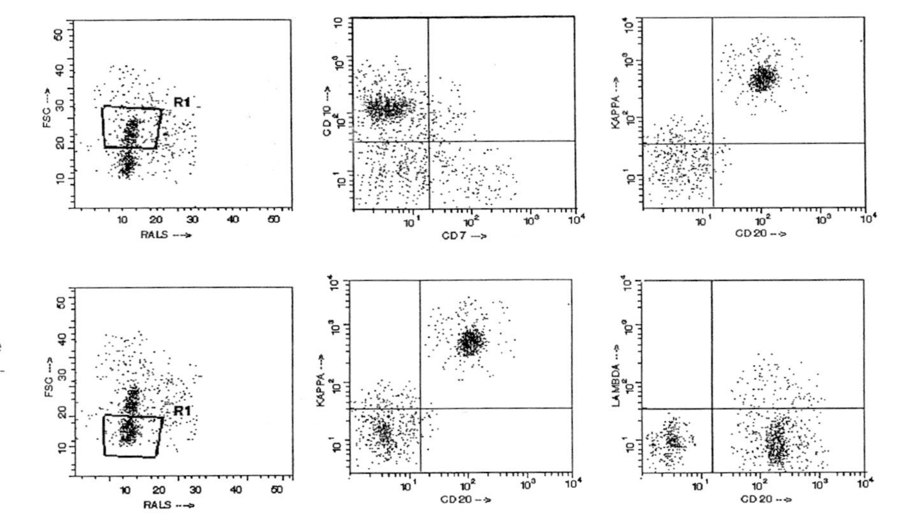

FIGURE 1–52. Mixed Small Cleaved and Large Cell Lymphoma (Follicular Grade II). These cytograms depict a mixed lymphoma. The light scatter properties reveal two distinct populations of cells. The upper set of graphs depicts gating of the large cells while the lower graphs show the small cells. The cells do co-express CD10 and monotypic Kappa. The large and small cells have an identical immunophenotype.

highly cellular and composed of numerous dispersed small lymphocytes with very high N/C ratios, clumped chromatin, and very inconspicuous nucleoli (Fig. 1–53). Leong and Stevens[6] have warned about the similarity with benign inactive lymph nodes, which yield a similar predominant small lymphocyte pattern. The demonstration of light chain restriction is essential for the aspiration biopsy diagnosis of this lymphoma. Coexpression of CD20 and CD5 is expected as well. These common neoplasms affect middle-aged and older populations, progress slowly, and typically involve multiple lymphoid sites. They roughly correspond to chronic lymphocytic leukemia and are mainly B-cell neoplasms.

Chronic lymphocytic leukemia and its nodal counterpart, small lymphocytic lymphoma (SLL), characteristically coexpress pan-B cell antigens, CD19 and CD20, as well as CD5, and can be detected by the presence of monotypic kappa or lambda surface immunoglobin (Fig. 1–54). However, immunologic findings are highly heterogeneous with respect to the density of antigen expression on the cell surface.[64] One constant feature is that the expression of CD5 is always weaker than normal T cells present in the sample. The CD20 is often weakly expressed and is not detectable in some cases. The CD19 is usually well expressed and is almost always brighter than the CD20. Surface immunoglobulin is also weakly expressed, and it may be difficult to detect a shift in the intensity of either kappa or lambda.

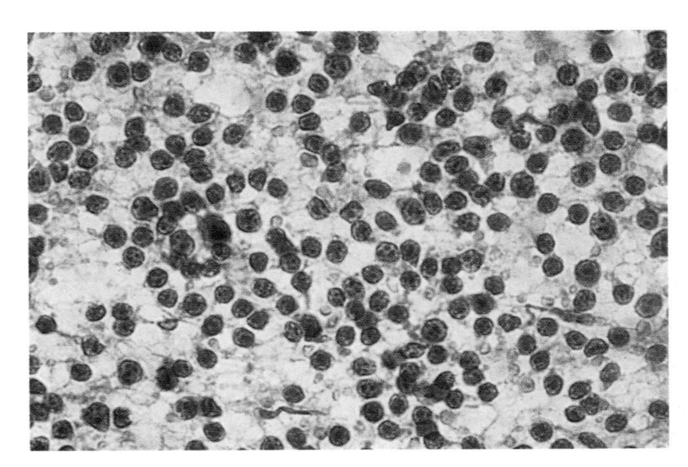

FIGURE 1–53. In aspiration smears, small lymphocytic lymphomas are characterized by highly cellular specimens with a striking uniformity of small lymphoid cells. They are characterized by distinct nuclear membranes, coarsely granular darkly stained chromatin, and very high N/C ratios. The presence of rare larger lymphoid cells does not detract from this diagnosis. Lymphoglandular bodies are evident. (Papanicolaou stain)

SLL may undergo several transformations, notably Richter's and prolymphocytic. Richter's transformation is an aggressive change to a high-grade lymphoma with large and almost blastlike cells. There is no definitive means of detecting this process by flow cytometry because these cells have an immunophenotype identical to the parent SLL.[68] Prolymphocytic transformation has no specific immunologic feature to distinguish it from SLL, but it may coexpress CD11c and have increased expression of CD20 and surface immunoglobulin expression compared to SLL.

A rare entity to keep in the differential diagnosis is the so-called T-cell variant of SLL, which looks and clinically acts like SLL. However, it is a peripheral T-cell neoplasm, not a B-cell monoclonal proliferation. Although not well defined, immunologically it is characterized by expression of either CD4 or CD8, and the cells show an aberrant immunophenotype. For flow cytometry, it is best to use a combination of CD5 or CD7 in conjunction with CD3. Most tumor cells form an abnormal cluster when compared to the normal T cells.

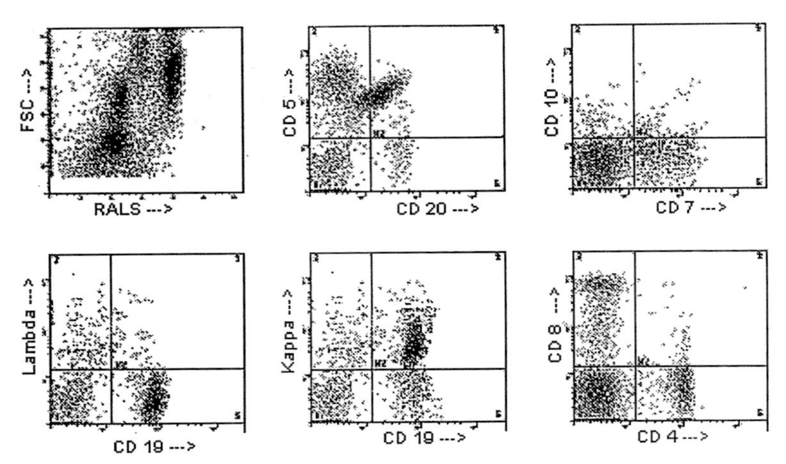

FIGURE 1–54. Typical findings for chronic lymphocytic leukemia (CLL). Panel I shows the light scatter properties. A population of small cells can be seen falling below the normal lymphocytes on the forward scatter (FSC) axis. One of the hallmarks for this disease is shown by the co-expression of CD5 and CD20 in Panel II. CLL always has a weaker CD5 expression than normal T cells, and the CD20 intensity is usually weak. Panel III shows the lack of CD10 in CLL, and Panels IV and V show the monotypic expression of surface immunoglobulin. Panel VI shows the T-cell compartment with the mixture of CD4- and CD8-positive cells.

Mantle Zone Lymphoma

Although SLL and chronic lymphocytic leukemia and its variants have in common the coexpression of CD5 and pan-B markers, this finding is not specific, and MZLs also have this immunologic fingerprint.[63,70] These lymphomas, also known as centrocytic lymphomas, have a more aggressive course and are difficult to maintain in remission. MZL is composed of small lymphocytes with variably irregular nuclear outlines, some resembling small cleaved cells and others prolymphocytes.[70] SLL represents a spectrum that has some overlap with MZL, but with rare exceptions most cases of MZL can be separated from SLL by following a few simple guidelines when assessing the flow cytometric data. Generally, MZL has much brighter expression of CD20 and surface immunoglobulin. SLL is a neoplasm of prefollicular center cells and thus tends to have IgM or IgD on its surface; MZL, a lymphoma of postfollicular center cells, tends to have IgG on its surface. Another discriminator is the use of CD23. This protein appears to be expressed on SLL cells but is not found on MZL cells (Fig. 1–55).

Small Noncleaved Cell Lymphoma

This neoplasm, categorized as a high-grade lymphoma in the Working Formulation, characteristically occurs in its nonendemic form as an intra-abdominal tumor involving lymph nodes with infiltration of visceral organs. This lymphoma represents a large proportion of all lymphomas arising in patients with AIDS[84] and is also fairly common in transplant recipients. Aspiration smears are extremely cellular, with intermediate-sized lymphoid cells with round nuclei, high to moderate N/C ratios, moderately to coarsely clumped chromatin, and one or more prominent small nucleoli (Figs. 1–56 and 1–57).[117–119] Although typi-

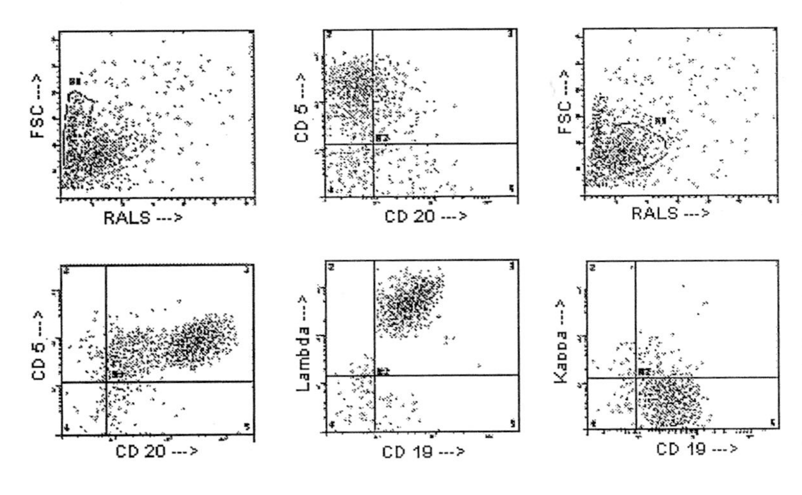

FIGURE 1–55. Mantle zone lymphoma; contrast with the CLL findings. The light scatter properties are depicted in Panel I. Two populations of cells are seen in this cytogram. The first population is shown in Panel II. These cells are shown to be T cells in Panel II by the expression of CD5 and the lack of CD20. Panel III shows a gate to assess the population of cells with increased right-angle light scatter. Like CLL, these cells coexpress CD5 and CD20, as shown in Panel IV. However, the intensity of CD20 is often much brighter than typically seen in CLL. Panels V and VI also show the monoclonality of this process, but once again the immunoglobulin intensity is typically higher in a mantle zone lymphoma.

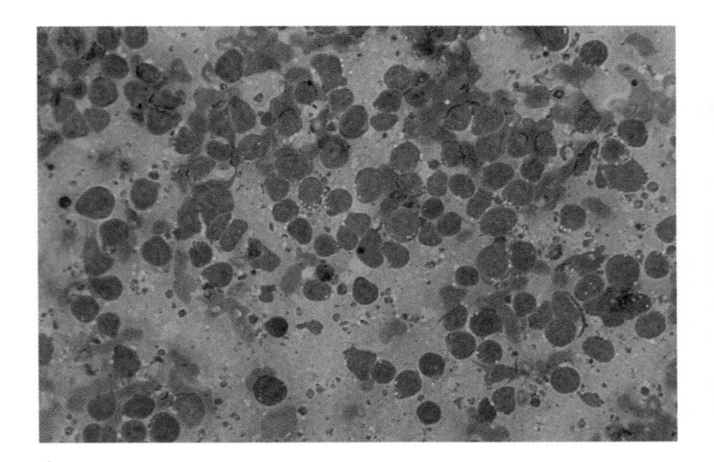

A

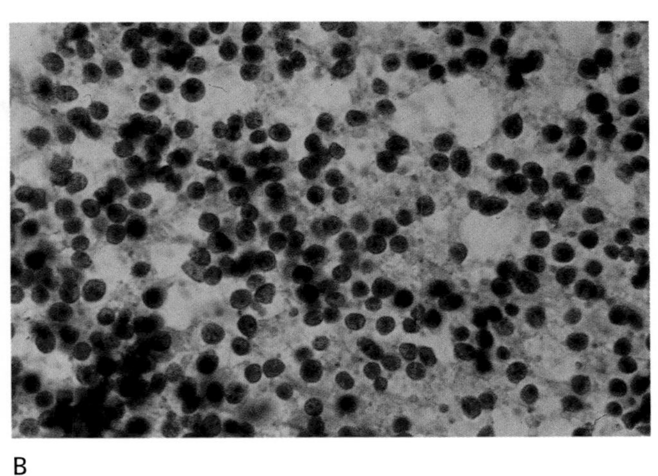

B

FIGURE 1–56. (**A, B**) Aspiration smears of small noncleaved cell lymphoma are extremely cellular with a dissociated pattern of uniform-appearing blasts. They have round nuclei with moderately granular chromatin, multiple small nucleoli, and rims of basophilic cytoplasm. The latter may contain lipid-positive vacuoles. Lymphoglandular bodies are present. (**A,** Romanowsky stain; **B,** Papanicolaou stain)

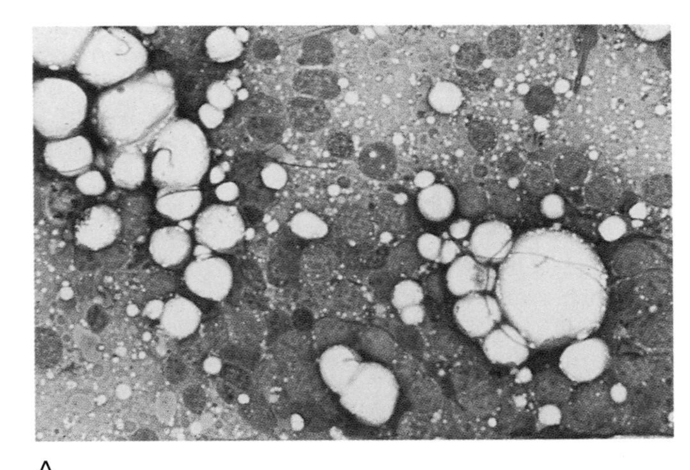

A

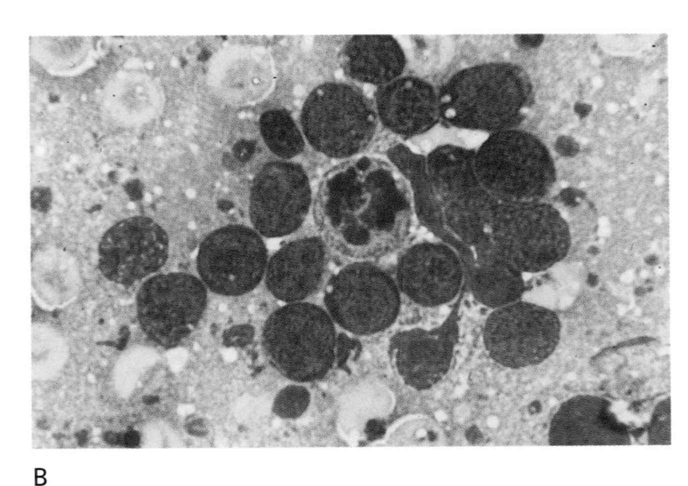

B

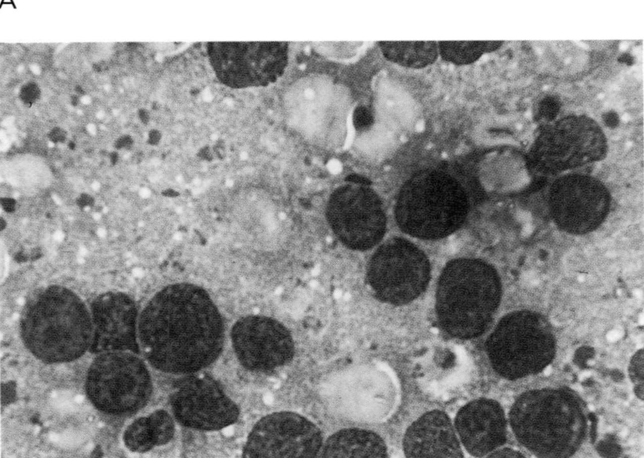

C

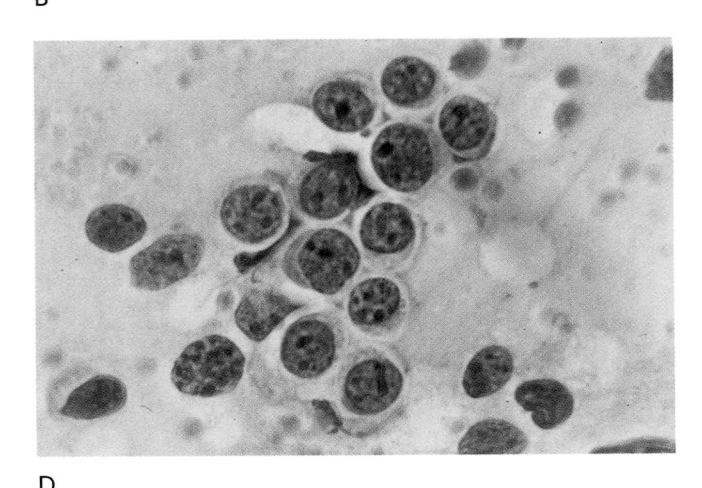

D

FIGURE 1–57. (**A–D**) In small noncleaved cell lymphoma, the neoplastic cells are quite homogeneous. Nuclei often appear perfectly round, with clumped chromatin and one to multiple small but distinct nucleoli. Cytoplasmic basophilia and vacuolization may be prominent. Lymphoglandular bodies are often well developed. In **D,** note the mitotic figure. (**A–C,** Romanowsky stain; **D,** Papanicolaou stain) cally perfectly round, some nuclei may be polylobated or irregular. Multiple small nucleoli are typical of the Burkitt's variant, whereas solitary and larger nucleoli are expected in the non-Burkitt's type. The deeply basophilic cytoplasm is eccentric, varies from scanty to moderate in volume in air-dried Romanowsky smears, and contains small but prominent lipid vacuoles in some neoplastic cells. As one might expect from the "starry-sky" histologic pattern, tingible-body macrophages are a prominent accompaniment. Plentiful mitotic figures and evidence of apoptosis are evident.

Cytomorphologic Features of Small Noncleaved Cell Lymphoma

- High cellularity with dispersed intermediate-sized uniform lymphoid cells
- Nuclei are mainly round with coarse chromatin and one to several obvious small nucleoli
- Cytoplasm is moderate and basophilic and may contain small lipid vacuoles
- Tingible-body macrophages are numerous

Burkitt's lymphoma arises when the *C-myc* oncogene is continuously expressed. It is one of the fastest-growing tumors known. It has a characteristic phenotype, coexpressing pan-B antigens and surface immunoglobin.[120] Often IgM lambda is detectable. Like follicular center cells, CD10 is present. When assessing these neoplasms, a proliferative fraction may be helpful because almost all the cells should be cycling and thus positive for Ki-67.

Immunophenotype in Small Noncleaved Cell Lymphomas

- Expression of surface immunoglobulin
- Expression of a full complement of B-cell antigens
- Coexpression of CD10
- May coexpress CD34

Lymphoblastic Lymphoma

This malignancy typically involves children, particularly teenage boys. In many cases, peripheral lymphadenopathy is associated with a mediastinal mass, which may actually call clinical attention to the disease. As with the Burkitt's type of lymphoma, lymphoblastic lymphoma represents a high-grade neoplasm. Most are T-cell neoplasms.

Aspiration smears are usually extremely cellular, with a disassociated homogeneous pattern of small uniform blastic cells.[75,76] The cells are 1.5 to 2 times the diameter of small mature lymphocytes, have extremely high N/C ratios, and manifest variable nuclear contours (Figs. 1–58 and 1–59). Although the nuclei may appear smooth and round, closer inspection reveals that most are irregular, with convolutions and indentations. The chromatin is very finely granular, almost powdery in consistency, and nucleoli are generally inconspicuous. Tingible-body macrophages are scant.

Cytomorphologic Features of Lymphoblastic Lymphoma

- High cellularity with lymphoglandular bodies
- Dispersed uniform blasts
- Nuclei are variable, often convoluted, with fine chromatin and inconspicuous nucleoli
- Very high N/C ratios

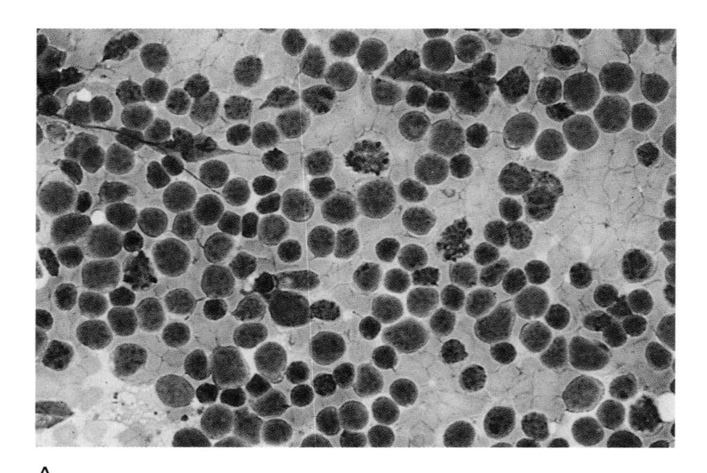
A

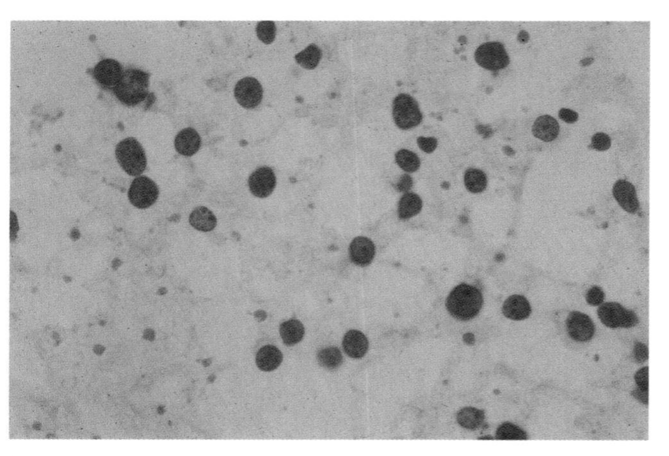
B

FIGURE 1–58. **(A,B)** Lymphoblastic lymphoma. The neoplastic cells are small and uniform. They have delicate irregular nuclear membranes, very finely granular (powdery) chromatin, and inconspicuous nucleoli. In addition, the N/C ratios are extremely high, as cytoplasm is usually not apparent. (**A,** Romanowsky stain; **B,** Papincolaou stain)

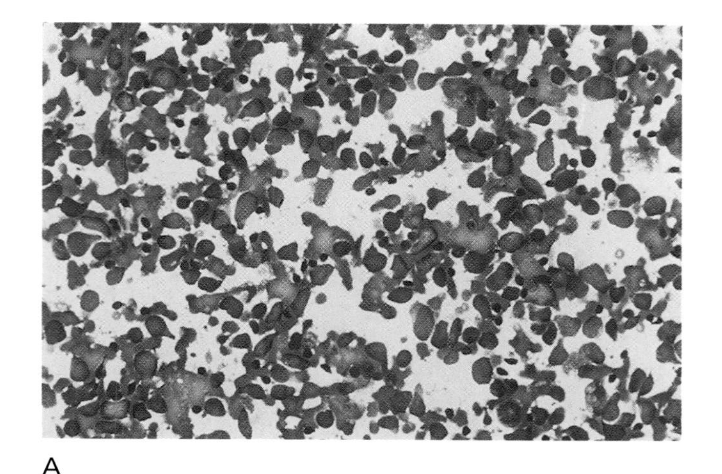

A

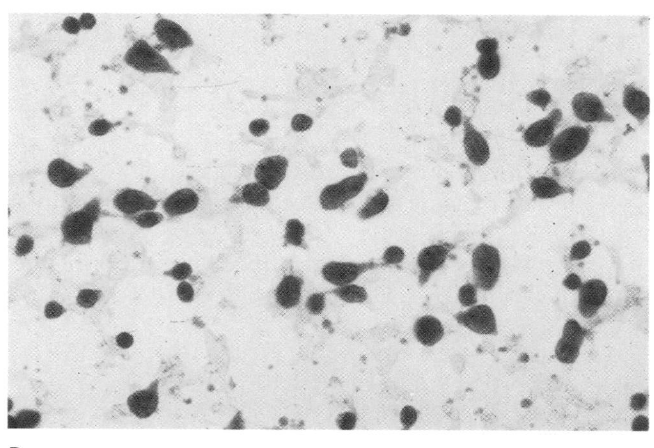

B

FIGURE 1–59. Lymphoblastic lymphoma. This is an unusual morphologic variant of lymphoblastic lymphoma in an aspirate of mediastinum lymph nodes in a 16-year-old boy. The smear cellularity is high, with relatively uniform cells. Although the nucleoli features are typical of lymphoblastic lymphoma, these cells have more abundant cytoplasm than is often seen. In this case, many of these cells have long, tapered cytoplasmic tails. (**A,** Romanowsky stain; **B,** Papanicolaou stain)

These aggressive lymphomas vary greatly on both morphologic and immunologic grounds.[74] Most are T cell in origin; the rest are B cell. Many represent tissue involvement by acute lymphoblastic leukemia. These cells demonstrate a thymic immunophenotype expressing CD1, TdT, CD7, and CD5, and may or may not coexpress CD3, which is a late antigen. Additionally, these cells may coexpress CD4 and CD8 on the same call, or alternatively may lack both antigens. We have seen one case that morphologically and clinically behaved like a lymphoblastic lymphoma, but demonstrated subset restriction by expressing CD4 and a full complement of T-cell antigens. These cells lacked CD8. Lymphoblastic lymphomas may also demonstrate B-cell differentiation and then are identical to acute lymphoblastic leukemias immunologically (Fig. 1–60). These cells express CD19 and CD10 but lack surface immunoglobulin and CD20. Cases of acute lymphoblastic lymphoma that show surface immunoglobin and CD20 are typically classified as having a Burkitt's immunophenotype.

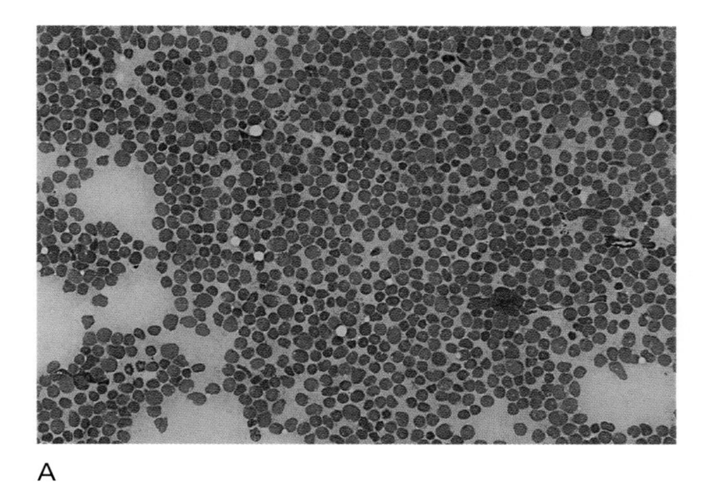

A

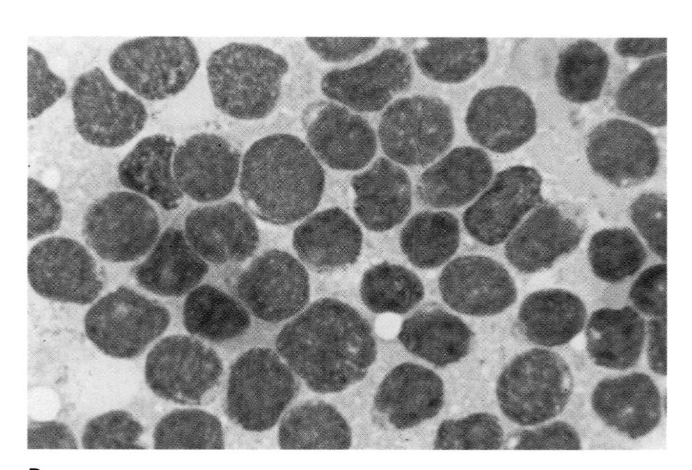

B

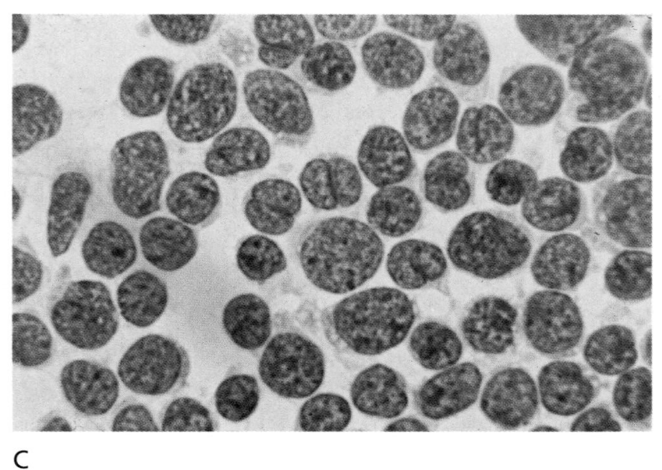

C

FIGURE 1–60. (**A–C**) Acute lymphoblastic leukemia in a nodal aspirate. (**A**) High smear cellularity and homogeneity of the malignant blasts are apparent. (**B, C**) Higher magnification reveals delicate uniform chromatin, minute nucleoli, and high N/C ratios. (**A** and **B,** Romanowsky stain; **C,** Papanicolaou stain)

Immunophenotype of Lymphoblastic Lymphoma

- Mostly T cell
- Coexpression of CD4 and CD8
- May be missing some T-cell antigens, especially CD3
- Rare B-cell type shows pre-B cell immunophenotype with CD19, CD10, and CD34 but no CD20 or SIg

METASTATIC NEOPLASMS

The use of FNA biopsy of both superficial and deeply situated enlarged lymph nodes to diagnose metastatic tumors is widely practiced. Several authors have published their experiences in this setting.[2,3,5,29–31] Metastatic carcinoma is the most common diagnosis rendered on aspirated nodes in our institutions. This procedure provides rapid results, is minimally invasive in patients who are often quite sick or even terminal, and is inexpensive.

Recently, Steel et al.[5] reported their results in aspiration biopsies of lymph nodes in more than 1100 patients. More than 80% of the malignant cytologic diagnoses were metastatic carcinoma or melanoma. The site most likely to contain cancer (85%) was the supraclavicular region. Their diagnostic accuracy was 96% in carcinomas and 100% in melanomas; their rare false-negative specimens were thought to be due to sampling errors. Of the 215 benign enlarged nodes, there were 3 false-positive and 7 suspicious diagnoses rendered. The suspicious diagnoses represented examples of granulomatous inflammation and an epidermal inclusion cyst. All 3 false-positive specimens were benign lymphoid hyperplasia mistaken for non-Hodgkin's lymphomas.

Also recently, Cervin et al.[30] retrospectively analyzed 152 FNA biopsies of supraclavicular lymph nodes (Virchow's node). Two thirds of the patients had a malignant diagnosis. The left supraclavicular nodes were slightly more often involved by neoplasm than were the right nodes. Malignant neoplasms arising in the breast, lungs, and head and neck regions showed no difference in pattern of spread to either the right or left sides. However, 84% of all metastatic pelvic cancers and all primary abdominal neoplasms metastasized to the left supraclavicular nodes. Ten percent of the cancers represented lymphoreticular malignancies.

In FNA biopsies, most metastases morphologically resemble their primary neoplasms. For example, in aspirates of neck nodes, metastatic papillary thyroid cancer yields cells with the characteristic ovoid nuclei with fine chromatin, cytoplasmic pseudoinclusions, and longitudinal grooves. In addition, papillary arrangements may be well represented. It is not the purpose of this chapter, therefore, to describe the cytomorphology of most nodal metastases; rather, only a few specific situations will be discussed.

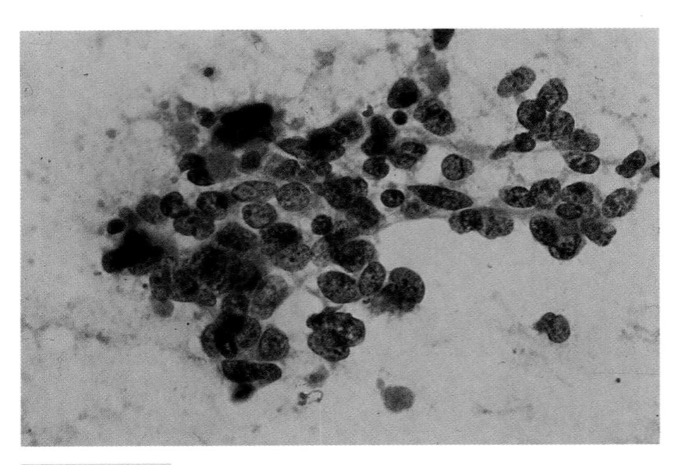

FIGURE 1–61. Metastatic small cell anaplastic carcinoma is characterized by small malignant cells with high N/C ratios. The nuclei have irregular contours, coarsely granular and darkly stained chromatin, and inconspicuous nucleoli. Some of the nuclei may resemble carrots. Intercellular cohesion is preserved and nuclear molding is prominent. (Papanicolaou stain)

One important consideration is the presence of a small cell malignancy in FNA biopsies of lymph nodes. To optimize the interpretation, careful scrutiny of the clinical data is essential. It is important to know the patient's age, the history of a previously diagnosed cancer, the site of the aspirated node, and a history of tobacco usage. In all age groups, non-Hodgkin's lymphomas need to be considered. In children, the most common nonlymphoreticular malignancy to metastasize to lymph nodes is neuroblastoma. The small tumor cells will manifest varying degrees of intercellular cohesion, with the formation of small spherical aggregates, broad sheets, and pseudorosettes.[121] In addition, larger neoplastic cells representing maturing neuroblasts and ganglion cells may be present. Characteristically, the smear background contains a filamentous material that corresponds to neuropil.

In our experience, the most common nonlymphoreticular small cell malignancy in adults is metastatic small cell undifferentiated (oat cell) carcinoma of bronchogenic origin. Nodal aspirates contain obviously malignant cells, both singly and clustered (Fig. 1–61). The latter usually manifest nuclear molding, in which adjacent nuclei are compressed against one another. The carcinoma cells have angulated nu-

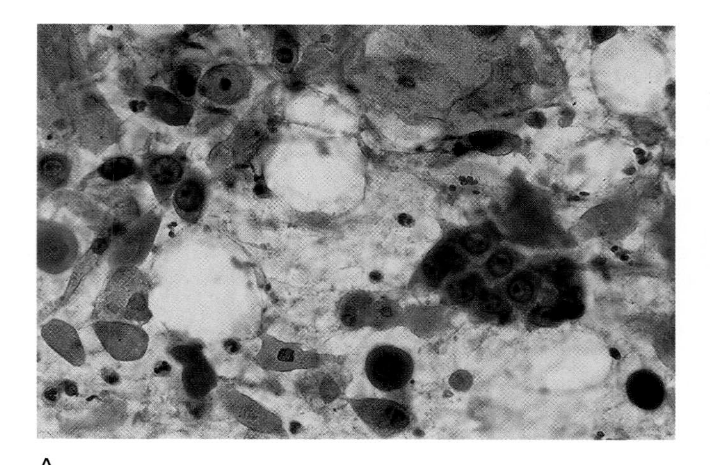

A

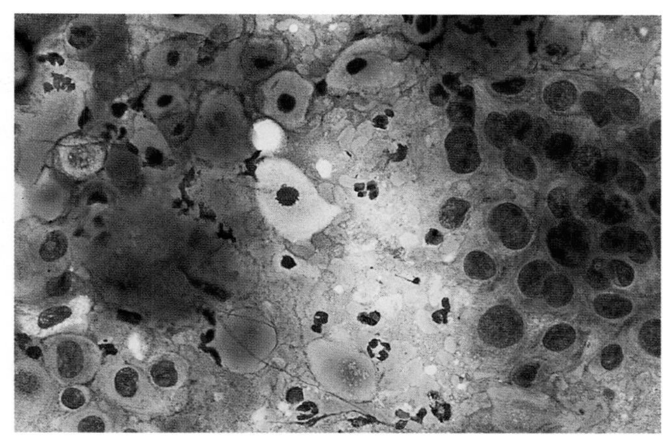

B

FIGURE 1–62. (**A, B**) In metastatic squamous cell carcinoma, the neoplastic cells may manifest dense orangeophilic cytoplasm and irregular nuclear shapes. In addition, a pyknotic-like chromatin pattern may be apparent. N/C ratios are highly variable. (**A,** Papanicolaou stain; **B,** Romanowsky stain)

clear contours, darkly stained coarse chromatin, very high N/C ratios, and often minute nucleoli. Crushed nuclear debris and strands of smeared chromatin are usually prominent. Characteristically, small cell carcinomas are positive for synaptophysin and cytokeratin and negative for CD45.

In aspiration biopsies of metastatic squamous cell carcinoma that contain tumor cells with obviously malignant nuclei, the diagnosis is almost always straightforward (Figs. 1–62 and 1–63). However, metastases of squamous cell carcinoma in lymph nodes may undergo extensive necrosis and keratinization and be associated with neutrophils or a granulomatous reaction. This is especially common in nodes in the head and neck region. Consequently, a viable tumor may form only a thin rim near the periphery of the node. Aspiration smears may consist predominately of enucleate squamous cells presenting as polygonal masses of cytoplasm; with the Papanicolaou stain, they often have an orangeophilic appearance, whereas they have a sky blue color with the Romanowsky stains. Centrally, empty holes may mark the grave site of the malignant nuclei. In this setting, the differential diagnosis includes benign congenital and acquired squamous-lined cysts, including branchial cleft cysts.[122]

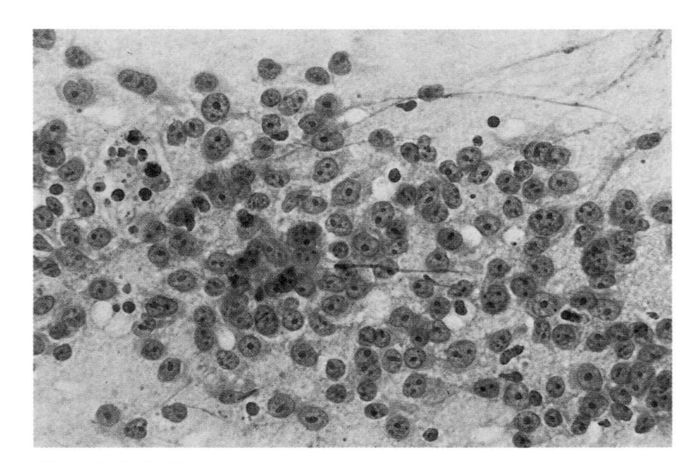

FIGURE 1–63. Nasopharyngeal carcinomas are essentially poorly differentiated squamous cell neoplasms with prominent nucleoli and very high N/C ratios. They may be difficult to distinguish from large cell lymphoma, although intercellular cohesion is usually apparent and lymphoglandular bodies are not prominent among the neoplastic cells. Note the tingible-body macrophage, a sign of rapid cell turnover. (Papanicolaou stain)

Melanomas often enter the differential diagnosis of dispersed cell patterns in node aspirates. Melanomas may feature bizarre nuclear pleomorphism, plentiful nuclear pseudoinclusions, and cytoplasmic pigment (Fig. 1–64). However, hemorrhage may produce pigment in tumor cells of other types of malignancy. Generally, the distinction from large cell lymphoma and anaplastic carcinoma can be made by demonstrating S-100 protein and HMB-45 in melanoma cells immunocytochemically.

Considerable information can be provided to the oncologist as to the probable primary site. Metastases from renal, intestinal, gonadal, prostatic, thyroid, breast, pulmonary (small cell), and cutaneous (melanoma) primaries all have typical morphologic features that suggest their origin (Fig. 1–65). In addition, cell blocks with immunocytochemical stains can be invaluable in confirming or making the diagnosis. Antibodies directed against cytokeratins, vimentin, S-100 protein, HMB-45, prostatic specific antigen, thyroglobulin, synaptophysin, carcinoembryonic antigen,

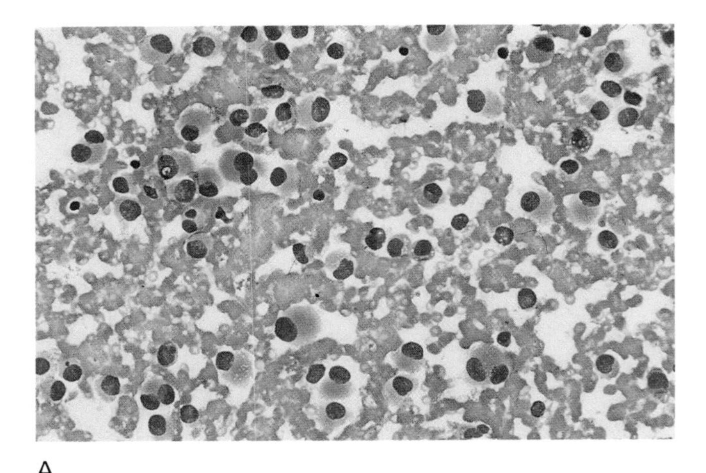

A

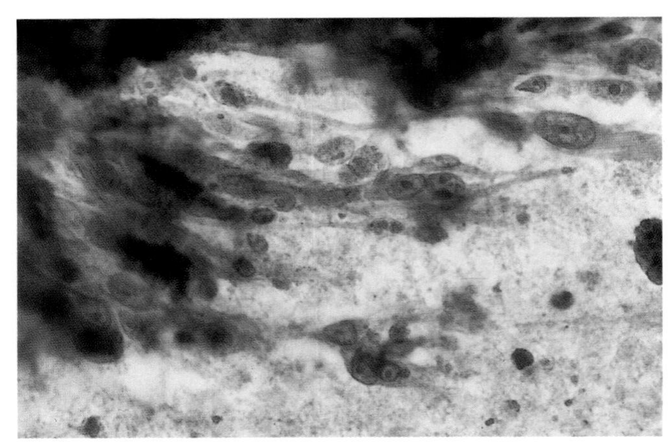

B

FIGURE 1–64. (**A, B**) Metastatic melanomas usually yield highly cellular smears with a dissociated pattern. (**A**) Characteristic dispersion of the cells is apparent. They have polygonal contours and one or two eccentric nuclei. Binucleation is characteristic. Note the lack of cohesion among the aspirated tumor cells. (**B**) Intense cytoplasmic pigmentation is apparent, a telltale sign of the derivation of a metastatic neoplasm in a lymph node aspiration specimen. Some of the neoplastic cells in this example have elongated or spindled contours. (**A,** Romanowsky stain; **B,** Papanicolaou stain)

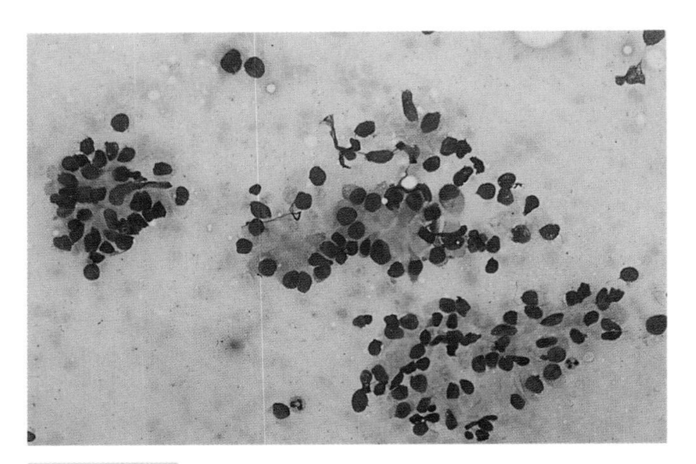

FIGURE 1–65. Metastatic adenocarcinoma of prostatic origin. The neoplastic cells are often arranged in small acinar units characterized by peripherally oriented round nuclei with nucleoli and central cytoplasm. The latter is delicate. (Romanowsky stain)

gross cystic fluid protein-15, and steroid hormone receptors can all be contributory.

REFERENCES

1. Frable WJ, Kardos TF. Fine needle aspiration biopsy. Applications in the diagnosis of lymphoproliferative diseases. Am J Surg Pathol 12(suppl 1):62–72, 1988.
2. Cardillo MR. Fine-needle aspiration cytology of superficial lymph nodes. Diagn Cytopathol 5:166–173, 1989.
3. Gupta A, Nayar M, Chandra M. Reliability and limitations of fine needle aspiration cytology of lymphadenopathies. An analysis of 1,261 cases. Acta Cytol 35:777–783, 1991.
4. Perkins SL, Segal GH, Kjoldsberg CR. Work-up of lymphadenopathy in children. Sem Diagn Pathol 12:284–287, 1995.
5. Steel BL, Schwartz MR, Ramzy I. Fine needle aspiration biopsy in the diagnosis of lymphadenopathy in 1,103 patients: role, limitations and analysis of diagnostic pitfalls. Acta Cytol 39:76–81, 1995.
6. Leong ASY, Stevens M. Fine-needle aspiration biopsy for the diagnosis of lymphoma: a perspective. Diagn Cytopathol 15:352–357, 1996.
7. Carter T, Feldman P, Innes D, Frierson H, Frigg A. The role of FNA cytology in the diagnosis of lymphoma. Acta Cytol 32:848–853, 1988.
8. Pilotti S, DiPalma S, Alasio L, Baroli C, Rilke F. Diagnostic assessment of enlarged superficial lymph nodes by fine needle aspiration. Acta Cytol 37:853–866, 1993.
9. Daskalopoulou D, Harhalakis N, Maouni N, et al. Institution fine needle aspiration cytology of non-Hodgkin's lymphomas. A morphologic and immunophenotypic study. Acta Cytol 39:180–186, 1994.
10. Prasad RRA, Narasimhau R, Sankaran V, Veliath AJ. Fine-needle aspiration cytology in the diagnosis of superficial lymphadenopathy: an analysis of 2,418 cases. Diagn Cytopathol 15:382–386, 1996.
11. Pandit AA, Khilneni PH, Prayag AS. Tuberculous lymphadenitis: extended cytomorphologic features. Diagn Cytopathol 12:23–27, 1995.
12. Arora VK, Singh N, Bhatia A. Are eosinophilic structures really a diagnostic criterion for tuberculosis? Diagn Cytopathol 15:360, 1996.
13. Gupta AK, et al. Critical appraisal of FNAB in tuberculous lymphadenitis. Acta Cytol 36:391, 1992.
14. Jannotta FS, Sidawy MK. The recognition of mycobacterial infections by intraoperative cytology in patients with acquired immunodeficiency syndrome. Arch Pathol Lab Med 113:120, 1989.
15. Stanley MW, Horwitz CA, Burton LG, Weisser JA. Negative images of bacilli and mycobacterial infection: a study of FNA smears from lymph nodes in patients with AIDS. Diagn Cytopathol 6:118–121, 1990.
16. Alfonso F, Gallo L, Winkler B, Suhrland MJ. FNA cytology of peripheral lymph node cryptococcosis. A report of three cases. Acta Cytol 38:459–462, 1994.
17. Donnelly A, Hendricks G, Martens S, Strovers C, Wiemerslages, Thomas PA. Cytologic diagnosis of cat scratch disease (CSD) by fine-needle aspiration. Diagn Cytopathol 13:103, 1995.

18. Stastny JF, Wakely PE Jr, Frable WJ. Cytologic features of necrotizing granulomatous inflammation consistent with cat-scratch disease. Diagn Cytopathol 15:108–115, 1996.
19. Christ ML, Feltes-Kennedy M. Fine-needle aspiration of toxoplasmic lymphadenitis. Acta Cytol 26:425–428, 1982.
20. Argyle JC, Schumann GB, Kjeldserg CR, Anthens JW. Identification of *Toxoplasma* cyst by FNA. Am J Clin Pathol 80:256–258, 1983.
21. Jayaram N, Ramaprasad AV, Chethan M, Sujay AR. *Toxoplasma* lymphadenitis. Analysis of cytologic and histologic criteria and correlation with serologic tests. Acta Cytol 41:653–658, 1997.
22. Gupta RK. Fine needle aspiration cytodiagnosis of toxoplasmic lymphadenitis. Acta Cytol 41:1031–1034, 1997.
23. Kardos TF, Kornstein MJ, Frable WJ. Cytopathology and immunopathology of infectious mononucleosis. Acta Cytol 32:722, 1988.
24. Stanley MW, et al. Fine needle aspiration of lymph nodes in patients with acute infectious mononucleosis. Diagn Cytopathol 6:323, 1990.
25. Layfield LJ. Fine needle aspiration cytologic findings in a case of sinus histiocytosis with massive lymphadenopathy (Rosai-Dorfman syndrome). Acta Cytol 34:767–770, 1990.
26. Trautman BC, Stanley MW, Goding GS, Rosai J. Sinus histiocytosis with massive lymphadenopathy (Rosai-Dorfman disease): diagnosis by fine needle aspiration. Diagn Cytopathol 7:513–516, 1991.
27. Stastny JF, Wilkerson ML, Hamati HF, Kornstein MJ. Cytologic features of sinus histiocytosis with massive lymphadenopathy. A report of three cases. Acta Cytol 41:871–876, 1997.
28. Tsang WYW, Chan JKC. Fine-needle aspiration cytologic diagnosis of Kikuchi's lymphadenitis. A report of 27 cases. Am J Clin Pathol 102:454–458, 1994.
29. Cochand-Priollet B, et al. Retroperitoneal lymph node aspiration biopsy in staging of pelvic cancer: a cytological study of 220 consecutive cases. Diagn Cytopathol 3:102, 1987.
30. Cervin JR, Silverman JF, Loggie B, Geisinger KR. Virchow's node revisited: analysis with clinicopathologic correlation of 152 fine-needle aspiration biopsies of supraclavicular lymph nodes. Arch Pathol Lab Med 119:727, 1995.
31. Cafferty LL, et al. Fine needle aspiration diagnosis of intraabdominal and retroperitoneal lymphomas by a morphologic and immunocytochemical approach. Cancer 65:72, 1990.
32. Pontifex AH, Haley L. Fine-needle aspiration cytology in lymphomas and related disorders. Diagn Cytopathol 5:432, 1989.
33. Katz RL, Caraway NP. FNA lymphoproliferative disease: myths and legends. Diagn Cytopathol 12:99, 1995.
34. Tani EM, Christensson B, Porwit A, Skoog L. Immunocytochemical analysis and cytomorphologic diagnosis on fine needle aspirates of lymphoproliferative diseases. Acta Cytol 32:209, 1988.
35. Chorny J, Katz RL. Overlooking malignant lymphoma on FNA: a diagnostic pitfall. Pathol Case Rev 1:92–95, 1996.
36. Kornstein MJ, Wakely PG Jr, Kardos TF, Frable WJ. Dendritic reticulum cells and immunophenotype in aspiration biopsies of lymph nodes. Value in subclassification of non-Hodgkin's lymphomas. Am J Clin Pathol 94:165–169, 1990.
37. Martin SE, et al. Immunologic methods in cytology: definitive diagnosis of non-Hodgkin's lymphomas using immunologic markers for T cells and B cells. Am J Clin Pathol 82:666–673, 1984.
38. Stani J. Cytologic diagnosis of reactive lymphadenopathy in fine needle aspiration biopsy specimens. Acta Cytol 31:8, 1987.
39. Braylan RC, Benson NA. Flow cytometric analysis of lymphomas [Review]. Arch Pathol Lab Med 113:627–633, 1989.
40. Harris NL, et al. A revised European-American classification of lymphoid neoplasms: a proposal from the International Lymphoma Study Group. Blood 84(5):1361–1392, 1994.
41. Borowitz MJ, et al. Monoclonal antibody phenotyping of B-cell non-Hodgkin's lymphomas: the Southeastern Cancer Study Group experience. Am J Pathol 121(3):514–521, 1985.
42. Borowitz MJ, et al. The phenotypic diversity of peripheral T-cell lymphomas: the Southeastern Cancer Study Group experience. Hum Pathol 17(6):567–574, 1986.
43. Rai KR, Sawitsky A. A review of the prognostic role of cytogenetic, phenotypic, morphologic, and immune function characteristics in chronic lymphocytic leukemia. Blood Cells 12:327–338, 1987.
44. Capaccioli S, et al. A bcl-2/IgH antisense transcript deregulates bcl-2 gene expression in human follicular lymphoma t(14;18) cell lines. Oncogene 13:105–115, 1996.
45. Lubnski J, et al. Molecular genetic analysis in the diagnosis of lymphoma in fine needle aspiration biopsies. I. Lymphomas versus benign lymphoproliferative disorders. Anal Quant Cytol Histol 10:391–398, 1988.
46. Katz RL, et al. The role of gene rearrangements for antigen receptors in the diagnosis of lymphoma obtained by fine-needle aspiration. Am J Clin Pathol 96:479–490, 1991.
47. Sneige N, et al. Cytomorphologic, immunocytochemical, and nucleic acid flow cytometry of 50 lymph nodes by fine-needle aspiration: comparison with results obtained by subsequent excisional biopsy. Cancer 67:1003–1010, 1991.
48. Molot RJ, et al. Antigen expression and polymerase chain reaction amplification mantle cell lymphomas. Blood 83:1626–1631, 1994.
49. Nowell PC, Croce CM. Chromosome translocations and onocogenes in human lymphoid tissues. Am J Clin Pathol 94:229, 1990.
50. Schmid I, et al. Dead cell discrimination with 7-amino-actinomycin D in combination with dual color immunofluorescence in single laser flow cytometry. Cytometry 13(2):204–208, 1992.
51. Stelzer GT, et al. Detection of occult lymphoma cells in bone marrow aspirates by multi-dimensional flow cytometry. Prog Clin Biol Res 377:629–635, 1992.
52. Rainer RO, Hodges L, Seltzer GT. CD45 gating correlates with bone marrow differential. Cytometry 22(2):139–145, 1995.
53. Takano Y, et al. Clonal Ig-gene rearrangement in some cases of gastric RLH detected by PCR method. Pathol Res Pract 188(8):973–980, 1992.
54. Morgan KG, et al. Hodgkin's disease: a flow cytometric study. J Clin Pathol 41(4):365–369, 1988.
55. Hoffkes HG, et al. Multiparametric immunophenotyping of B cells in peripheral blood of healthy adults by flow cytometry. Clin Diagn Lab Immunol 3(1):30–36, 1996.
56. Ault KA. Detection of small numbers of monoclonal B lymphocytes in the blood of patients with lymphoma. N Engl J Med 300(25):1401–1405, 1979.
57. Shackney SE, et al. Dual parameter flow cytometry studies in human lymphomas. J Clin Invest 66(6):1281–1294, 1980.
58. Wong GY, et al. Analysis of cell surface light chain immunoglobulin expression by flow cytometry in normal controls: a new mathematical approach. J Histochem Cytochem 33(2):119–126, 1985.
59. Page DM, et al. Two signals are required for negative selection of CD4+CD8+thymocytes. J Immunol 151(4):1868–1880, 1993.
60. Horning SJ, et al. Clinical and phenotypic diversity of T cell lymphomas. Blood 67(6):1578–1582, 1986.
61. Kelsoe G. The germinal center: a crucible for lymphocyte selection [Review]. Sem Immunol 8(3):179–184, 1996.
62. Paloczi K. Clinical applications of phenotypic analysis [Review]. Immunol Today 13(8):A27–A28, 1992.
63. Strickler JG, et al. Intermediate lymphocytic lymphoma: an immunophenotypic study with comparison to small lymphocytic lymphoma and diffuse small cleaved cell lymphoma. Hum Pathol 19(5):550–554, 1988.
64. Cossman J, et al. Low-grade lymphomas. Expression of developmentally regulated B-cell antigens. Am J Pathol 115(1):117–124, 1984.
65. Sneige N, et al. Morphologic and immunocytochemical evalu-

ation of 220 fine needle aspirates of malignant lymphoma and lymphoid hyperplasia. Acta Cytol 34:311, 1990.

66. Rudders RA, Ahl ET Jr, DeLellis RA. Surface marker and histopathologic correlation with long-term survival in advanced large-cell non-Hodgkin's lymphoma. Cancer 47(6): 1329–1335, 1981.
67. Scott CS, et al. Membrane phenotypic studies in B cell lymphoproliferative disorders. J Clin Pathol 38(9):995–1001, 1985.
68. Cherepakhin V, et al. Common clonal origin of chronic lymphocytic leukemia and high-grade lymphoma or Richter's syndrome. Blood 82(10):3141–3147, 1993.
69. Sheibani K, et al. Monocytoid B cell lymphoma, clinicopathologic study of 21 cases of a unique type of low-grade lymphoma. Cancer 62:1531, 1988.
70. Banks PM, et al. Mantle cell lymphoma: a proposal for unification of morphologic, immunologic and molecular data. Am J Surg Pathol 16:637, 1992.
71. Weiss LM, et al. Morphologic and immunologic characterization of 50 peripheral T-cell lymphomas. Am J Pathol 118:316–324, 1985.
72. Winberg CD. Peripheral T cell lymphoma: morphologic and immunologic observations. Am J Clin Pathol 99:426, 1993.
73. Katz RL, et al. Fine needle aspiration cytology of peripheral T-cell lymphoma. Am J Clin Pathol 91:120, 1989.
74. Cossman J, et al. Diversity of immunological phenotypes of lymphoblastic lymphoma. Cancer Res 43(9):4486–4490, 1983.
75. Kardos TF, et al. Fine needle aspiration biopsy of lymphoblastic lymphoma and leukemia. A clinical, cytologic and immunologic study. Cancer 60:2448, 1987.
76. Jacobs J, et al. Fine needle aspiration of lymphoblastic lymphoma: a multiparameter diagnostic approach. Acta Cytol 36:887, 1992.
77. Behm FG, O'Dowd GJ, Frable WF. Fine needle aspiration effects on benign lymph node histology. Am J Clin Pathol 82:195–198, 1985.
78. Ebihara T, Sakai N, Koyama S. CD8+ T cell subsets of cytotoxic T lymphocytes induced by Epstein-Barr virus infection in infectious mononucleosis. Tohoku J Exp Med 162(3): 213–224, 1990.
79. Anagnostopoulus I, et al. Morphology, immunophenotype, and distribution of latently and/or productively Epstein-Barr virus-infected cells in acute infectious mononucleosis: implication for the interindividual infection route of Epstein-Barr virus. Blood 85(3):744–750, 1995.
80. Lapeurta P, Martin SE, Ellison E. Fine-needle aspiration of peripheral lymph nodes in patients with tuberculosis and HIV. Am J Clin Pathol 107:317–320, 1997.
81. Gupta K, Singh N, Bhatia A, Arora VK, Singh UR, Singh B. Cytomorphologic patterns in Calmette-Guerin bacillus lymphadenitis. Acta Cytol 41:348–350, 1997.
82. Orell SR, Sterrett GF, Walters M N-I, Whitaker D. Manual and atlas of fine needle aspiration cytology, 2d ed. (pp. 72–73). Edinburgh: Churchill Livingstone, 1992.
83. Strigle SM, Rarck MU, Cosgrove MM, Martin SE. A review of FNA cytological findings in HIV infection. Diagn Cytopathol 8:41–52, 1992.
84. Strigle SM, Martin SE, Levine AM, Rarick MU. The use of FNA cytology in the management of HIV-related non-Hodgkin's lymphoma and Hodgkin's disease. J Acquir Immune Defic Syndr 6:1329–1334, 1993.
85. Margin-Bates E, Tanner A, Suvarna SK, et al. Use of FNA cytology for investigating lymphadenopathy in HIV-positive patients. J Clin Pathol 46:564–566, 1993.
86. Shabb N, et al. Fine-needle aspiration evaluation of lymphoproliferative lesions in human immunodeficiency virus-positive patients. Cancer 67:1008, 1991.
87. Llatjos M, Roneu J, Clotet B, et al. A distinctive cytologic pattern for diagnosing tuberculosis lymphadenitis in AIDS. J Acquir Immune Defic Syndr 6:1335–1338, 1993.
88. Burns BF, Wood GS, Dorfman RF. The varied histopathology of lymphadenopathy in the homosexual male. Am J Surg Pathol 9:287, 1985.
89. Hales M, Bottles K, Miller T, et al. Diagnosis of Kaposi's sarcoma by fine-needle aspiration biopsy. Am J Clin Pathol 88:20, 1987.
90. Levine AM. Review article: acquired immunodeficiency syndrome-related lymphoma. Blood 80:8, 1992.
91. Knowles DM, et al. Lymphoid neoplasia associated with the acquired immunodeficiency syndrome. Ann Intern Med 108:744, 1998.
92. Shibata D, et al. Epstein-Barr virus–associated non-Hodgkin's lymphoma in patients infected with the human immunodeficiency virus. Blood 81:2012, 1993.
93. Tirelli U, et al. Hodgkin's disease and HIV infection: clinicopathologic and virologic features of 114 patients from the Italian Cooperative Group on AIDS and tumors. J Clin Oncol 13:1758, 1995.
94. Gattus P, Castelli MJ, Peng Y, Reddy VB. Posttransplant lymphoproliferative disorders: a fine needle aspiration biopsy study. Diagn Cytopathol 16:392–395, 1997.
95. Knowles DM, et al. Correlative morphologic and molecular genetic analysis demonstrates three distinct categories of posttransplantion lymphoproliferative disorders. Blood 85:552, 1995.
96. Kowal-Vern A, et al. Characterization of postcardiac transplantation lymphomas. Histology, immunophenotyping, immunohistochemistry and gene rearrangement. Arch Pathol Lab Med 120:41–48, 1996.
97. van Gorp J, et al. Posttransplant T cell lymphoma. Report of three cases and a review of the literature. Cancer 73:3064, 1994.
98. Skook L, et al. Growth fraction in non-Hodgkin's lymphomas and reactive lymphadenitis determined by Ki-67 monoclonal antibody in fine-needle aspirates. Diagn Cytopathol 12:234, 1995.
99. Friedman M, et al. Appraisal of aspiration cytology in management of Hodgkin's disease. Cancer 45:1653, 1980.
100. Kardos TF, et al. Hodgkin's disease: diagnosis by fine needle aspiration biopsy: analysis of cytologic criteria from a selected series. Am J Clin Pathol 86:286, 1986.
101. Moriarity A, et al. Cytologic criteria for subclassification of Hodgkin's disease using fine needle aspiration. Diagn Cytopathol 5:122, 1989.
102. Das DK, et al. Fine needle aspiration cytodiagnosis of Hodgkin's disease and its subtypes. I. Scope and limitations. Acta Cytol 34:329, 1990.
103. Fulciniti F, Vetrani A, Aeppa P, et al. Hodgkin's disease: diagnostic accuracy of fine needle aspiration, a report of 62 consecutive cases. Cytopathology 5:226–233, 1994.
104. Tani E, Johansson B, Skoog L. T-cell-rich B-cell-rich lymphoma: fine needle aspiration cytology and immunocytochemistry. Diagn Cytopathol 18:1–7, 1998.
105. Robb-Smith AH. U.S. National Cancer Institute working formulation of non-Hodgkin's lymphomas for clinical use. Lancet 2(8295):432–434, 1982.
106. Lieberman PH, et al. Evaluation of malignant lymphomas using three classifications and the working formulation. 482 cases with median follow-up of 11.9 years. Am J Med 81 (3):365–380, 1986.
107. Wojcik EM, et al. Diagnosis of mantle cell lymphoma on tissue acquired by fine needle aspiration in conjunction with immunocytochemistry and cytokinetic studies. Acta Cytol 39:909, 1995.
108. Symmans WF, et al. Transformation of follicular lymphoma. Acta Cytol 39:673, 1995.
109. Nakamura S, Shiota M, Nakagawa A, et al. Anaplastic large cell lymphoma: a distinct molecular entity. A reappraisal with special reference to $p80^{NPM/ALK}$ expression. Am J Surg Pathol 21:1420–1432, 1997.
110. Akhtar M, Ali MA, Haider A, Antonius J, Hainau B, Dayel FA. Fine-needle aspiration biopsy of Ki-1-positive anaplastic large-cell lymphoma. Diagn Cytopathol 8:242–247, 1992.

111. Zakowski MF, Feiner H, Finfer M, Thomas P, Wollner N, Filippa DA. Cytology of extranodal Ki-1 anaplastic large cell lymphoma. Diagn Cytopathol 14:155–160, 1996.
112. McCluggage WG, Anderson N, Herron B, Caughley L. Fine needle aspiration cytology, histology and immunohistochemistry of anaplastic large cell Ki-l-positive lymphoma. Acta Cytol 40:779–785, 1996.
113. Sgrignoli A, Abati A. Cytologic diagnosis of anaplastic large cell lymphoma. Acta Cytol 41:1048–1052, 1997.
114. Jayaram G, Rahman NA. Cytology of Ki-l-positive anaplastic large cell lymphoma. A report of two cases. Acta Cytol 41:1253–1260, 1997.
115. Bizjak-Schwarzbartl M. Large cell anaplastic Ki-1-positive non-Hodgkin's lymphoma vs. Hodgkin's disease in fine needle aspiration biopsy samples. Acta Cytol 41:351–356, 1997.
116. Young NA, Al-Saleem TI, Al-Saleem Z, Ehya H, Smith MR. The value of transformed lymphocyte count in subclassification of non-Hodgkin's lymphoma by fine-needle aspiration. Am J Clin Pathol 108:143–151, 1997.
117. Das DK, et al. Burkitt-type lymphoma. Diagnosis by fine needle aspiration cytology. Acta Cytol 31:1, 1987.
118. Stastny JF, et al. Fine-needle aspiration biopsy and imprint cytology of small non-cleaved cell (Burkitt's) lymphoma. Diagn Cytopathol 12:201, 1995.
119. Labrecque LG, Lampert I, Kazembe P, et al. Correlation between cytopathological results and an *in situ* hybridization on needle aspiration biopsies of suspected African Burkitt's lymphomas. Int J Cancer 59:591–596, 1994.
120. Garcia CF, Weiss LM, Warnke RA. Small noncleaved cell lymphoma: an immunophenotypic study of 18 cases and comparison with large cell lymphoma. Hum Pathol 17(5):454–461, 1986.
121. Geisinger KR, Silverman JF, Wakely PG Jr. Pediatric cytopathology (pp. 307–313). Chicago: ASCP Press, 1994.
122. Engzell U, Zajicek J. Aspiration biopsy of tumors of the neck. I. Aspiration biopsy and cytologic findings in 100 cases of congenital cysts. Acta Cytol 14:51–57, 1970.

APPENDIX

Technical Principles of Flow Cytometry

In its most simplistic form, a flow cytometer consists of three basic elements: sample delivery, data acquisition, and data analysis. Sample delivery can be broken down into two phases: preparation of the sample and the flow dynamics of the flow cytometer. We routinely use monoclonal antibodies in flow cytometry and measure the fluorescence using logarithmic scale. With a few notable exceptions, lymphoid antigens are typically expressed brightly, or in a fairly high concentration, on the cells. Thus, it is easy to discern a true-positive result. Monoclonal antibodies have also increased their specificity for a particular antigen, and many commercially available antibodies for flow cytometry are entering FDA trials as reagents. Currently, few antigens require indirect labeling with an indirect step for detection on the cell surface. New fluorochromes are being added to the pathologist's repertoire. This is allowing us to do more with fewer cells, and this is imperative for analyzing FNA biopsies.

Sample delivery also is influenced by the dynamics of a flow cytometry system. Flow cytometers typically come in two varieties: an open flow cell and a closed flow cell. The open systems are used for sorting, whereas closed systems greatly simplify alignment of the instrument and are typically used in the newer clinical bench-top flow cytometers. The flow rates of samples are important. FNA biopsies typically do not yield a high cell count; thus, there are not many cells to analyze. In highly cellular samples, too many cells may reach the interrogation point, causing a blur.

Most of the preceding aspects of the sample delivery are out of the laboratory worker's direct control.

However, there is one important measure to enhance the ability to make a diagnosis. Cell viability is very important for obtaining accurate results by flow cytometry. If numerous dead cells are present, nonspecific binding of our specific monoclonal antibodies will occur. Some lymphomas are composed of fragile cells. During aspiration, many lymphoma cells lose their cytoplasm and thus their antigens. An antibiotic, 7-actinomycin D (7-AAD), is lipid-soluble and can diffuse into a cell. However, it is also a compound that can be actively pumped out of a cell by the MDR protein. Further, it absorbs the light-blue 488-nanometer laser light of a flow cytometer while admitting its energy transfer in the deep-red spectrum.[50] Thus, this compound can be used to distinguish live from dead cells while still allowing for detection in the green, orange, and near-red spectra. Cells that adsorb 7-AAD and cannot remove it are dead. Cells that can actively remove 7-AAD are alive, and it is these cells that should be analyzed. This simple incubation step readily enhances cell populations that can be distinguished by flow cytometry.

THE INTERROGATION POINT

The interrogation point of a flow cytometer is the point at which the monochromatic laser light strikes a single cell suspended in a fluid stream. The laser light scatters the light depending on the size of the cell. This is termed forward-angle light scatter. The bigger the cell, the bigger its shadow. In addition, light will scatter in all directions, and flow cytometers collect quantitative data at 90° to the path of the energy source. This is related to the complexity of the cell and is termed right-angle light scatter. Unlike myeloid cells, lymphoid cells typically do not contain granules, and this parameter is less important diagnostically.

The light scatter properties of a population of cells can be used to separate subsets of cells.[29] We stain cells with monoclonal antibodies conjugated with fluorescent dyes that highlight the proteins on the cell membrane specific for the various stages of maturation and function of lymphocytes. Additionally, neoplastic B cells usually express monotypic kappa or lambda light immunoglobin chains; this is a strong but not absolute indicator of malignancy.[41] Although one single cell could be representative of anything, 80% of a population of 20,000 cells all expressing the same set of cell surface proteins is a significant finding in the normally heterogeneous population of lymphoid cells.

The presence or absence of different proteins on a lymphoid cell is vital for the accurate determination of its immunologic differentiation. A major consideration is which antibodies should be used in a particular situation, accentuated with specimens obtained by FNA biopsy, because a limited number of cells are available for analysis. Three-color flow cytometry using FITC, phycoerythrin, and PerCP and Cy5 conjugates allows up to three unique proteins, such as CD19 and kappa and lambda light chains, to be assessed on every cell in a population.

DATA ANALYSIS

The functionally and immunologically distinct subpopulations of cells in an FNA biopsy of lymphoid tissue can be dissected from one another using a technique known as gating on the digitized data of the six parameters acquired at the interrogation point. Each cell can be assessed using combinations of multiple concurrent parameters through views known as color dot plots, where a population of cells can be assigned a color that is propagated through all views of the data. Hemodilution by peripheral blood of the aspirate leads to neutrophils and peripheral blood lymphocytes being counted in the events used to assess the specimen. The flow cytometer can be configured to stop counting based on the number of cells in the lymphocyte gate.

Data analysis of lymphoid populations can be greatly enhanced by applying several simple techniques. Flow cytometry is very useful for diagnosing B-cell neoplasms and most reactive conditions. However, it is limited when trying to diagnose T-cell malignancies, and it is of very limited use when assessing a lymph node involved with Hodgkin's disease.[54] Thus, methods that fully analyze B-cell populations are of greatest use. These include acquiring more events in the B-cell tubes and gating around the B cells identified by a specific monoclonal antibody that alleviates a great deal of nonspecific staining with kappa- and lambda-specific antibodies. For instance, we use CD20 with kappa and CD20 with lambda in separate tubes. We set an acquisition gate for 10,000 CD20-positive events. When analyzing the data, we typically use a histogram displaying forward-angle light scatter versus CD20. A gate is used in this histogram that isolates the CD20-positive B cells. We then assess the percentage of kappa- or lambda-positive cells within this gated population. The presence of both kappa- and lambda-bearing B cells usually indicates a benign or reactive condition. A single peak indicates monoclonality.

The sensitivity of flow cytometry for detecting B-cell malignancies can be further enhanced by examining subpopulations within the B cells. Neoplasms generally vary from normal populations. Although most of our diagnoses rely on the morphologic appearance of a neoplasm, lymphoid malignancies can vary their immunologic pattern as well as their cell size. Using the forward-angle light scatter versus CD20 histogram, one can look at both the large cell and the small cell portions of the B-cell population. Additionally, one can study populations that vary with respect to their expression of CD20. Once a gate is set, individual cells can be assessed for the presence of kappa or lambda surface immunoglobin.

Although data analysis of lymphoid populations focuses on cellular populations to assess clonal or aberrant antigen expression, the percentages of B and T cells within a sample and the B-cell and T-cell sum for all antigens are helpful, especially for T-cell neoplasms. These may be detected only by the fact that the T cells present have lost one of their antigens. Most lymphomas conserve many of the antigens normally found on lymphocyte.

CHAPTER 2

Breast

Perhaps more than any other clinical procedure, fine-needle aspiration (FNA) biopsy of the breast has catapulted the pathologist to the forefront of managing patients with breast disease. Pathologists now often perform FNA biopsies of patients with palpable lesions or attend at other FNA biopsies of the breast, including image-guided procedures. Pathologists are also leaders in teaching our clinical colleagues to obtain an FNA biopsy sample correctly and to prepare the slides for proper examination. All of these endeavors benefit the patient with breast disease.

Simultaneously, there are new challenges for the pathologist who sees patients with breast disease. Patients are presenting with smaller lesions because of advances in screening mammography. As a result, we are entering an era that is morphologically challenging as we more often encounter premalignant proliferative lesions that raise the patient's risk for developing breast cancer. Determining the criteria for understanding and interpreting these lesions is an active field of investigation.

The pathologist's role begins with specimen collection. In addition to routine smears for morphologic interpretation, samples can also be divided for relevant prognostic immunocytochemical studies, which may include hormone receptors, oncogene and suppressor gene analysis, and proliferation markers. Indeed, collection and reporting of this information has become the recommended standard of care.[1]

TECHNIQUE

For patients with breast disease, the technique of FNA biopsy has revolutionized the way that diagnoses, management, and treatment are rendered. Using a standard 22- to 26-gauge, 1″ to 1.5″ needle, with or without suction, the operator can obtain a sample of a lesion, stain it, and interpret it within minutes. From an economic perspective, this is a most cost-effective maneuver;[2–4] from a scientific point of view, the results are fast, efficient, and extremely accurate in the hands of an experienced cytopathologist;[5,6] from a humanistic viewpoint, it may allay the patient's anxiety about the nature of the lesion and allows her to participate in treatment options appropriate for her unique clinical situation. In addition, FNA biopsy does not cause skin scarring, which is very advantageous for the patient who has benign cytologic findings. Given that up to 70% of all breast surgery procedures are performed for benign breast disease,[7] the importance of the role of FNA biopsy becomes immediately apparent.

By improving the decision-making process of the selection of patients for traditional biopsy of breast tumors, FNA biopsy has become an indispensable tool in the management of breast lesions.[8] In patients who have had definitive therapy for breast carcinoma, the FNA biopsy maintains its role as the first line for investigation of recurrent lesions in the region of the scar[9] and identification of metastases.

After the immobilization of the target lesion with one hand, the other hand may be used to puncture the lesion directly with a needle to obtain a sample without aspiration. This allows greater tactile sense for the operator, which is especially useful when targeting small lesions. Alternatively, the needle is part of a syringe that is attached to a syringe holder, allowing one-handed aspiration.

Direct air-dried and alcohol-fixed smears are most useful for routine morphology, although cytospins[10] and Thinlayer slides[11] may be valuable where clinicians submit samples in fixative directly to the laboratory.

COMPLICATIONS

Complications due to breast aspiration are uncommon; local hematoma formation is the most common problem. Puncture of the thoracic cavity causing pneumothorax[12–14] is avoided by angling the needle toward the lesion but parallel to the chest wall, especially in patients with very small breasts and deep-seated lesions. Spreading tumor cells ("seeding" of the needle tract by tumor cells) is extremely rare; it is more of a problem with large-core needle biopsies.[15,16] Taxin et al.,[17] in a study of 308 patients treated conservatively with local excisions, found no evidence that FNA biopsy had adverse outcomes in terms of local recurrence or survival.

Needle tracts in excisional surgical biopsies from women who have had an FNA biopsy can be a source of concern to the histopathologist because the needle can physically displace epithelial elements into the stroma or lymphovascular channels, resulting in an appearance that simulates invasive malignancy (Fig. 2–1).[18–21] Ductal carcinoma *in situ,* after needling, may simulate invasive cancer. However, to appreciate the biopsy changes, the key point for the histopathologist to recognize is the profound hemorrhage, granulation, fibrin, and/or fat necrosis within and near the areas of epithelial displacement.[19–21]

PROCEDURE INDICATIONS

The FNA biopsy procedure is indicated for any breast mass, palpable or nonpalpable,[22,23] but it is best to obtain a mammogram first so that the lesion can be accurately characterized without superimposing artifactual image distortion from the FNA biopsy.[1] Characterization of the lesion by palpation, mammography, and FNA biopsy is referred to as the "triple test"; this approach offers the most cost-effective method for accurately determining the nature of the lesion so that definitive therapy can be planned.[1,24] When all three components of the triple test indicate malignancy, definitive therapy may be performed.[1,25]

PROCEDURE RELIABILITY

As with any test where there are positive (cancer) and negative (no cancer) results, we must be aware of the pitfalls of the procedure. These include operator experience, the number and causes of false-negative diagnoses, the number and causes of false-positive diagnoses, and the clinical ramifications of the result.

False-negative results may begin with the inexperienced aspirator who misses the lesion and obtains only blood and fat. Unsatisfactory specimen rates have ranged from 1–68%, according to Layfield et al.[25] Generally, with increasing operator experience, the number of inadequate specimens obtained from patients who have cancer decreases.[25,26] Operator training and credentialing is important[1,25,27] and should in-

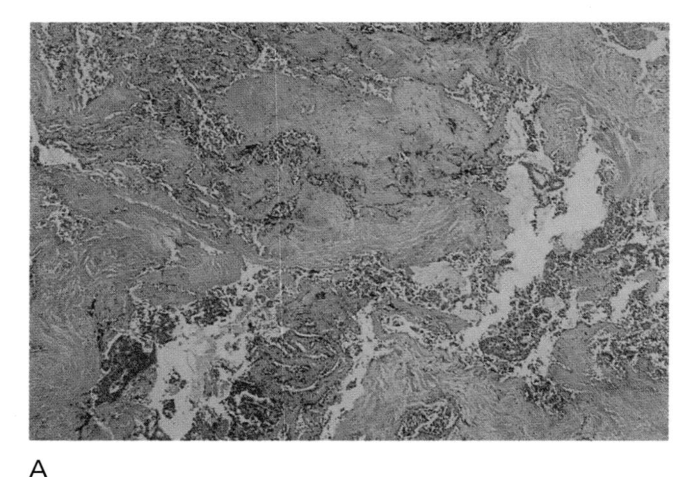
A

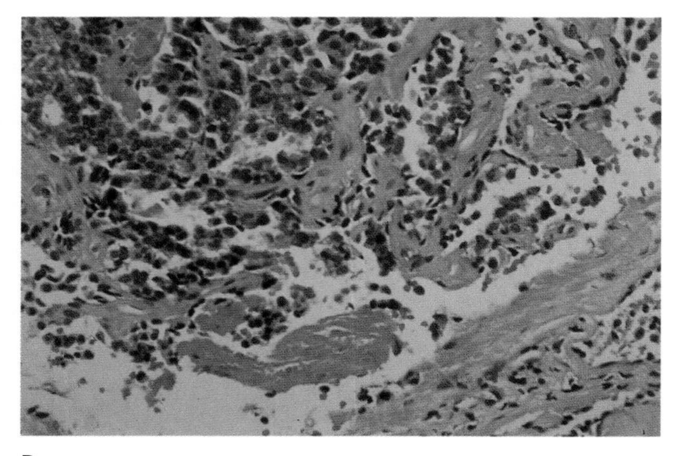
B

FIGURE 2–1. Tissue appearance after FNA biopsy. (**A**) Low-magnification appearance simulates invasive duct carcinoma. (**B**) Higher magnification demonstrates the needle tract associated with fibrin and epithelial misplacement. (Hematoxylin/eosin) clude performance of at least 30 breast FNA biopsies and supervision of the subsequent 20, with no more than a 20% insufficient sample rate, as recommended by the recent NIH-sponsored conference.[1] Many agree that an adequate breast FNA sample is "one that leads to the resolution of the patient's presenting problem."[1] The aspirator must be satisfied that the cytologic findings are compatible with the clinical impression of the gross lesion.[1] If the clinician is the aspirator, close immediate consultation with the pathologist (analogous to frozen-section reporting) is highly desirable. The pathologist must determine if the sample is interpretable, regardless of who the aspirator is. Recent studies supply guidelines for benign FNA biopsy findings based on tissue follow-up from several series: most studies found that the presence of at least six cell clusters, with at least five cells per cluster, yields the minimal rate of inadequate specimens, on the order of less than 2%.[24,28–30]

For cystic lesions in the breast, palpation of the lesional area after cyst fluid removal is mandatory so that an occult or intracystic tumor will not escape detection. For nontraumatic bloody or blood-tinged cyst contents, current guidelines[1] recommend cytologic examination; for greenish-gray "blue dome cyst" contents, cytologic examination is left to the operator's discretion.

Interpretive error of the smear is also a source of false-negative results, and this can lead to delay of an appropriate cancer diagnosis. Such interpretive errors may stem from the bland nature of the lesion—for instance, a low-grade cancer such as tubular carcinoma misinterpreted as benign,[31] or other relatively bland/paucicellular malignant lesions such as lobular, colloid, and papillary carcinomas and the small cell monomorphic carcinomas often seen in the elderly. Physical characteristics of the lesions may cause an increased number of false-negative results, such as small lesions that are not appropriately targeted.

Small lesions (<1 cm) may be the source of false-negative results (6–24% false-negatives).[25,32] Harvesting necrotic material from large lesions (>4 cm) is another source of error.[25,33] Desmoplastic lesions may not readily yield large numbers of cancer cells, as is often the case with invasive lobular carcinoma, and this also may be a source of a false-negative diagnosis.

False-negative diagnoses of breast carcinoma by FNA biopsy are reported to be in the range of 0–4.1%.[2,34–36] Large established centers report rates of less than 0.5%.[37–42] The specificity of the procedure is very high, approaching that of frozen sections, with reported rates of 96–100%.[25,43] Because of this very high degree of diagnostic accuracy in properly trained hands, definitive therapy, which may include surgery, chemotherapy, or radiation therapy, may proceed on the basis of an FNA diagnosis of malignancy.[44] To avoid false-positive diagnoses, the cytopathologist must be properly trained and experienced, must have an interpretation consistent with other components of the triple test, must closely follow the criteria set for the recognition of malignant cells, and must be aware of the pitfalls in making an erroneous diagnosis of cancer. For the patient with a positive triple test (all three components indicative of malignancy), post-FNA recommendations may include referral for definitive therapy, referral for definitive therapy after frozen-section confirmation, or referral for definitive therapy after frozen-section or permanent-section confirmation of the diagnosis, at the discretion of the attending physician.[1,45]

Excisional biopsy of the lesion is recommended for the situation of mixed or inconclusive triplets or for the presence of any unexplained cytologic atypia.[1]

CYTOPATHOLOGY REPORTING

All reporting must be done in the context of the triple test so that the operator is fully satisfied that the cytologic findings explain the patient's problem. Diagnostic reporting should be as specific as possible and should include prognostic information[1,46–50] that can be obtained from the cytopathologic examination. Reporting could include the general categories (benign, atypical/indeterminate, suspicious/probably malignant, malignant, unsatisfactory), with detailed explanations as to the microscopic findings that are diagnostically specific.[1,51] Malignant diagnoses should be as specific as possible, for the diagnosis may furnish a substantial element of prognostication.

Carcinomas of the special types—tubular, colloid, and papillary—should be specified, because these histologic types portend a good prognosis.[52] The 1990 NIH consensus conference on the management of early-stage breast cancer[53] cited the prognostic factors that should be included in every pathology report: tumor size, nuclear grade, histologic tumor type, lymph node status, steroid hormone receptor status, and cell proliferation rate. The triple test can supply much of this information, with the FNA biopsy supplying the histologic type, nuclear grade, estrogen/ progesterone receptors, and proliferation rate.

For the most common carcinomas, duct carcinomas of no special type, nuclear grading according to Fisher's modification of Black's nuclear grading system is appropriate. Nuclear grade is a powerful prognostic indicator,[1,53–57] correlates with other parameters of tumor aggressiveness,[58–60] and can be performed reproducibly on FNA biopsy specimens.[48–50,61–63] Just as the nuclear grade of breast carcinomas should be reported in the surgical pathology report,[53,64] so too should it be reported in the cytopathology report.[1] Nuclear grading on the FNA biopsy, a fundamental cytologic parameter, is easy to perform, in contrast to histologic grading, which is unreliable.[49,50]

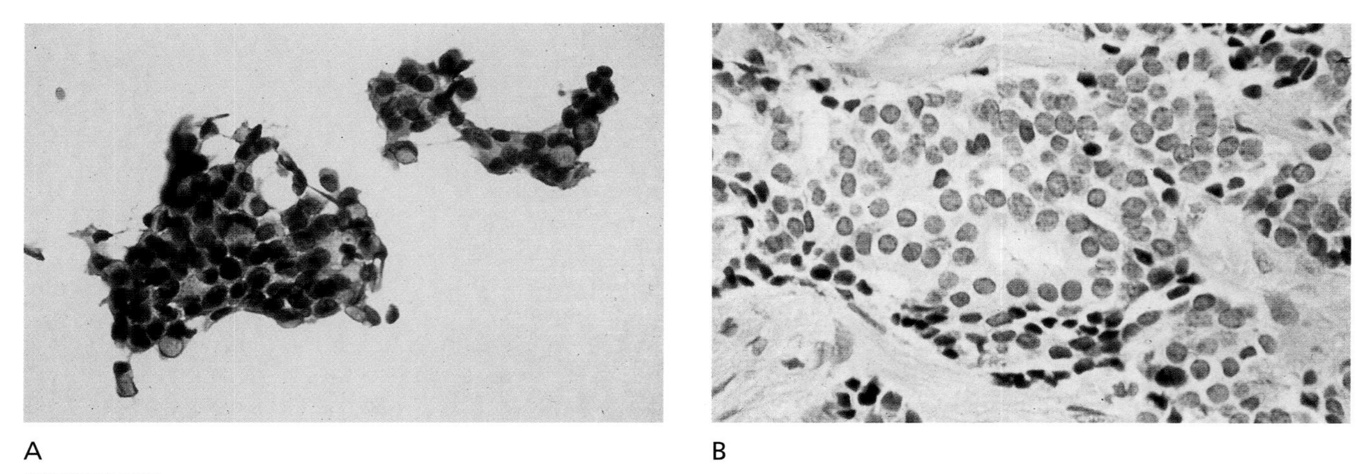

A B

FIGURE 2–2. Duct carcinoma, nuclear grade 1. (**A** and **B,** Papanicolaou stain and tissue with hematoxylin/eosin)

Nuclear grade 1 indicates a nucleus similar to normal duct epithelium. It has minimal enlargement, round, smooth nuclear membranes, uniform fine chromatin, and no nucleoli (Fig. 2–2).

Nuclear grade 2 is a nucleus that may be twice the size of nuclear grade 1, with smooth, uniform nuclear membranes and uniform chromatin; small nucleoli may be evident, and moderate anisonucleosis may be present (Fig. 2–3).

Nuclear grade 3 shows marked anisonucleosis with at least a threefold variation in diameter, marked hyperchromatism with angular nuclear contours, and

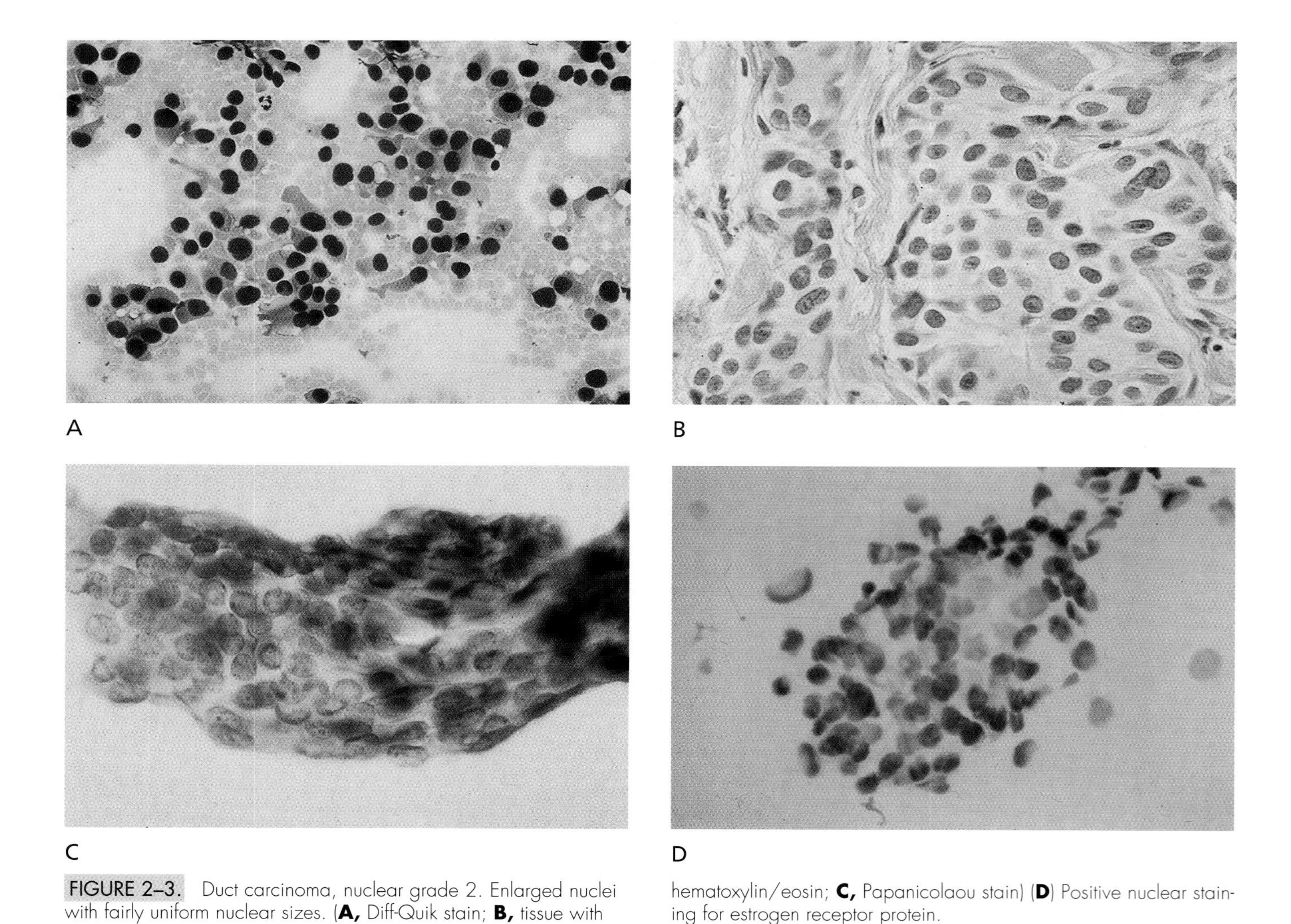

A B C D

FIGURE 2–3. Duct carcinoma, nuclear grade 2. Enlarged nuclei with fairly uniform nuclear sizes. (**A,** Diff-Quik stain; **B,** tissue with hematoxylin/eosin; **C,** Papanicolaou stain) (**D**) Positive nuclear staining for estrogen receptor protein.

sometimes prominent large nucleoli (Fig. 2–4).[48,49,52] Nuclear grading can be performed on both Papanicolaou- and Wright-stained smears.[49]

Prognostically, patients can be stratified into two main groups on the basis of nuclear grade, with "low nuclear grade" encompassing nuclear grades 1 and 2 and "high nuclear grade" for nuclear grade 3. This format, known as the simplified modified Black grade, may allow more practical and reproducible results among cytopathologists.[65,66]

Reporting of the results of ancillary studies performed on FNA biopsies of the breast is important, es-

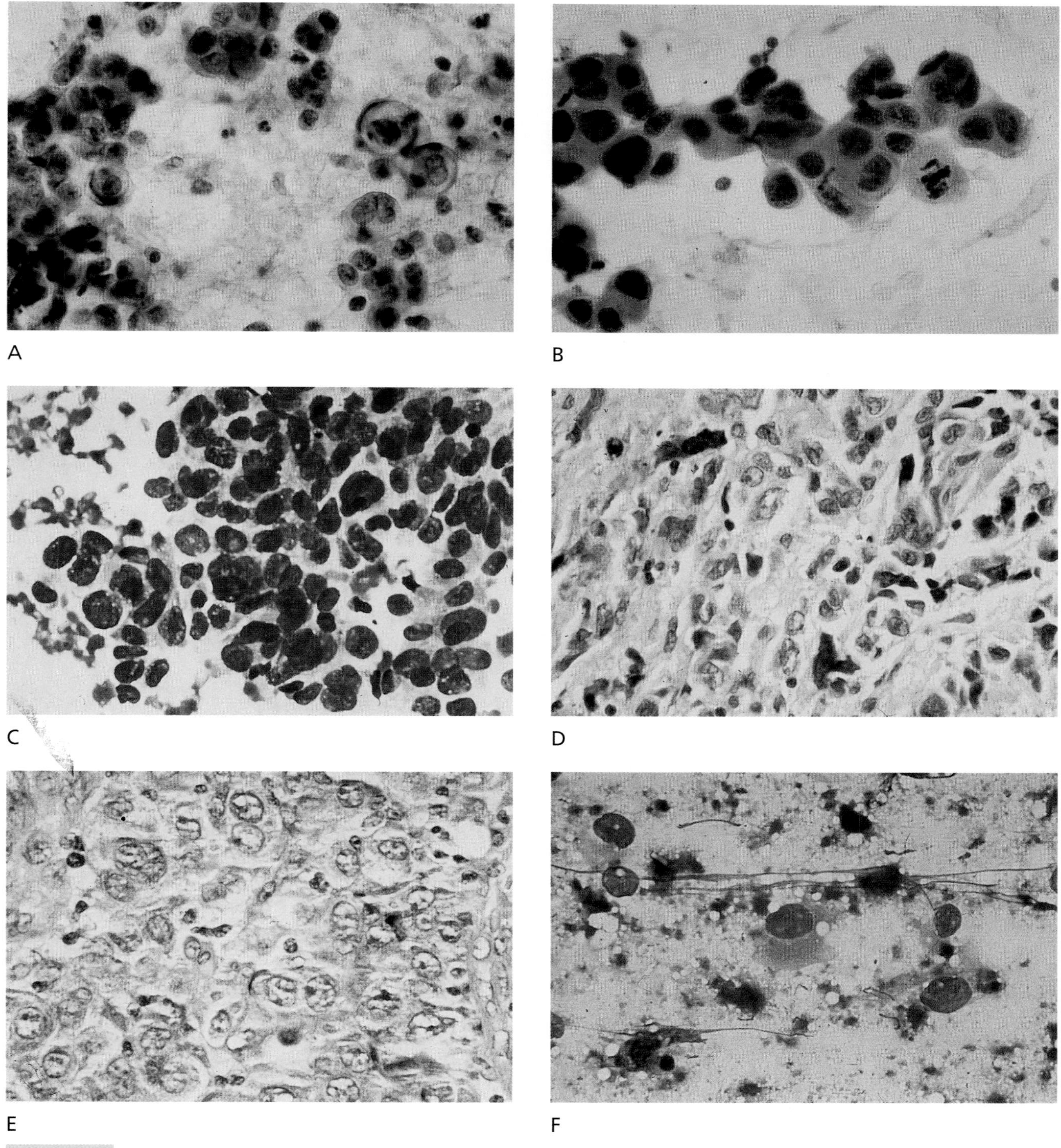

FIGURE 2–4. Duct carcinoma, nuclear grade 3, characteristically with pleomorphic hyperchromatic and angulated nuclei. (**A** and **B,** Papanicolaou stain; **C,** Diff-Quik stain; **D** and **E,** hematoxylin/eosin) (**F**) Intraductal comedocarcinoma, nuclear grade 3.

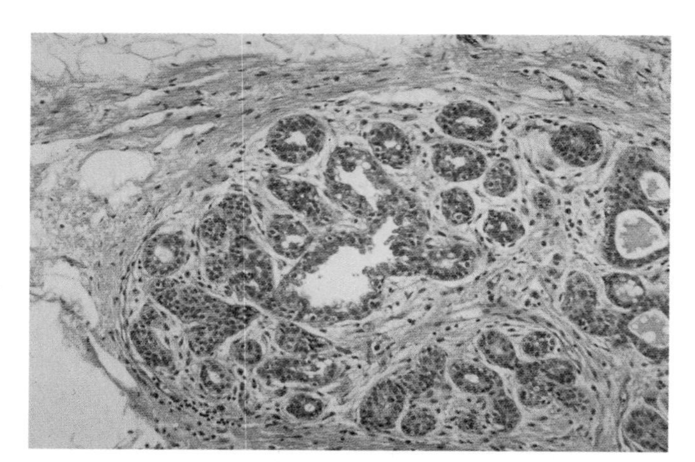

FIGURE 2–5. The normal breast lobule with acini contained within the loose connective tissue of the lobule. (Hematoxylin/eosin)

pecially for patients who may be treated with chemotherapy or radiotherapy before histologic examination.[67,68] Relevant ancillary studies such as immunoperoxidase stains for estrogen/progesterone receptors (see Fig. 2–3), oncogene amplification analysis, suppressor gene status, and Ki-67 proliferation rates can be performed on the sample.[69–82]

ANATOMY AND CYTOLOGY OF THE NORMAL BREAST

The normal breast consists of 12–20 lobules of glandular tissue, radially arranged about a duct system and supported by an abundant fibroadipose stroma. From the nipple, the large lactiferous ducts spread radially via medium-sized interlobular ducts that branch to become interlobular segmental ducts. It is from these segmental ducts that the intralobular ducts arise to penetrate a specialized, hormonally responsive, loose intralobular connective tissue stroma. Most of the bipolar naked stromal nuclei seen on FNA biopsies are from this stroma (Fig. 2–5). Within this loose stroma is the terminal-duct lobular unit (TDLU), the functional compartment of the breast. The lobule, the function of which is to prepare for lactation, responds cyclically to hormonal influences while maintaining a simple structure of acinar epithelial cells with interspersed myoepithelial cells, all resting on basement membrane. Whereas the bulk of the normal, nonlactating breast is composed of fat and connective tissue, the lactating breast is composed largely of actively secreting glandular acini (Fig. 2–6). An FNA biopsy of a nonlactating breast should thus contain a predominance of normal fat, fibrous tissue, and stromal cells with few ductal or acinar cells. In smears, glandular or ductal elements may be present as small glandular structures with a visible lumen, or ductal elements may present as familiar flat sheets with a honeycomb cellular arrangement (Fig. 2–7). Connective tissue cells in the FNA are single-spindled or rounded naked nuclei without discernible cytoplasm, spread uniformly throughout the smear (see Fig. 2–7).

Lactating breast, because of the tremendous increase in actively secreting glandular tissue, yields largely glandular tissue when aspirated by needle. These glandular units are quite different in appearance from the FNA contents of the nonlactating breast. The glands yield abundant large epithelial cells with large nuclei, prominent nucleoli, and heavily vacuolated cytoplasm (see Fig. 2–6).

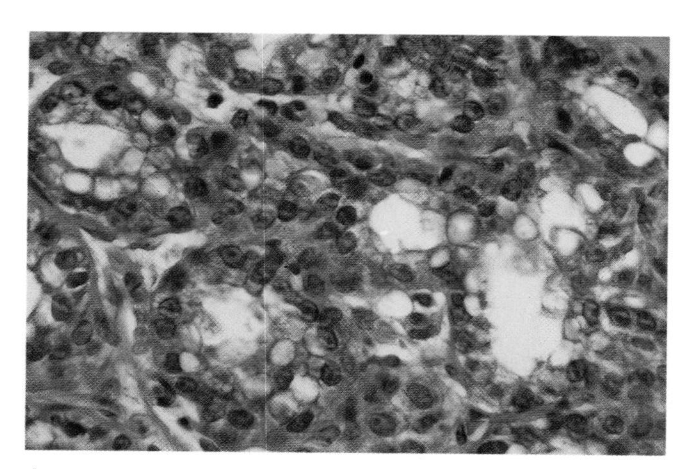

A

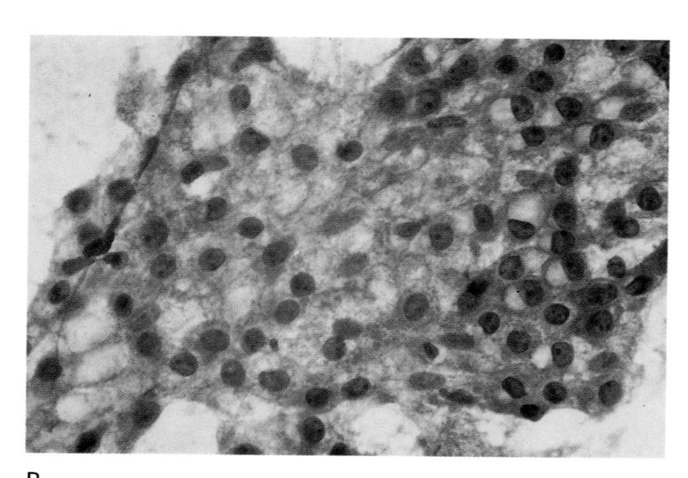

B

FIGURE 2–6. (**A**) Normal lobule with lactational secretory changes. Note the foamy vacuolated cytoplasm of the acini. (Hematoxylin/eosin) (**B**) Needle aspirate of lactational breast demonstrates sheets of acinar cells with enlarged nuclei, nucleoli, and abundant vacuolated cytoplasms.

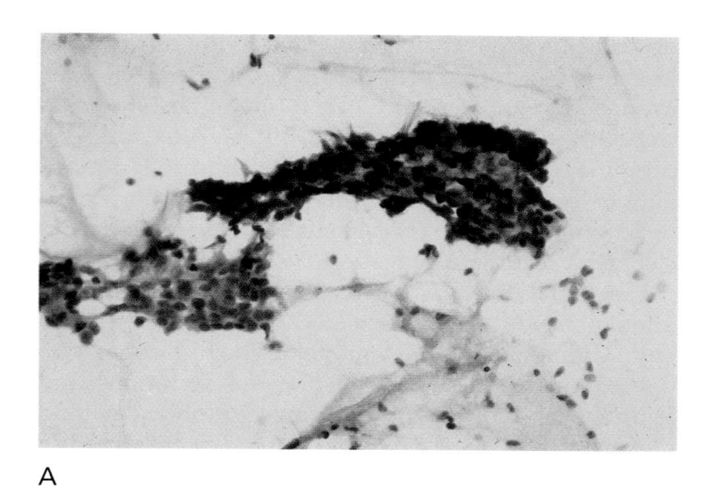

A

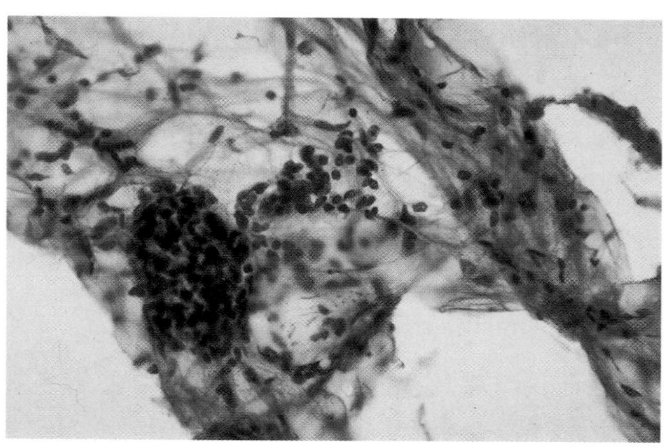

B

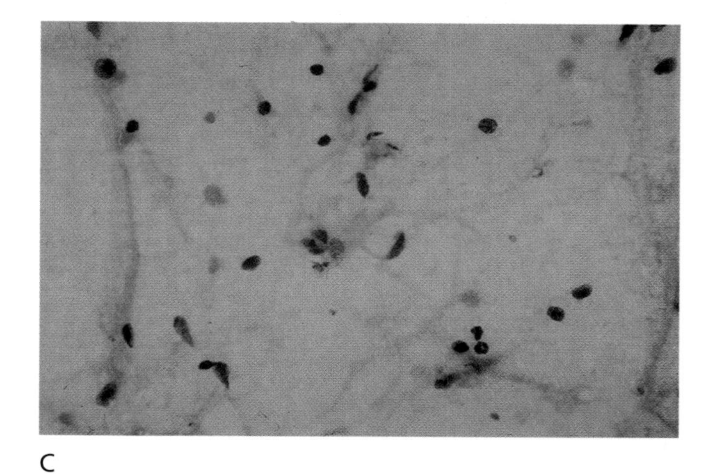

C

FIGURE 2–7. Normal breast. (**A**) Honeycomb sheets of few duct cells surrounded by naked stromal nuclei. (**B**) Few duct cells and abundant adipose with stromal cells in the background. (**C**) Scattered naked stromal nuclei. (Papanicolaou stain)

Atrophic breast tissue has undergone involutional change of the TDLU. Such alterations are readily seen in breast biopsies from most patients, regardless of age. Involutional changes result in replacement of the normal lobule and soft intralobular stroma with the dense collagenous stroma of the interlobular breast (Fig. 2–8). This is the so-called process of fibrosis of the breast that gives rise to the palpable irregularities and "graininess" in patients with fibrocystic change (FCC) of the breast.

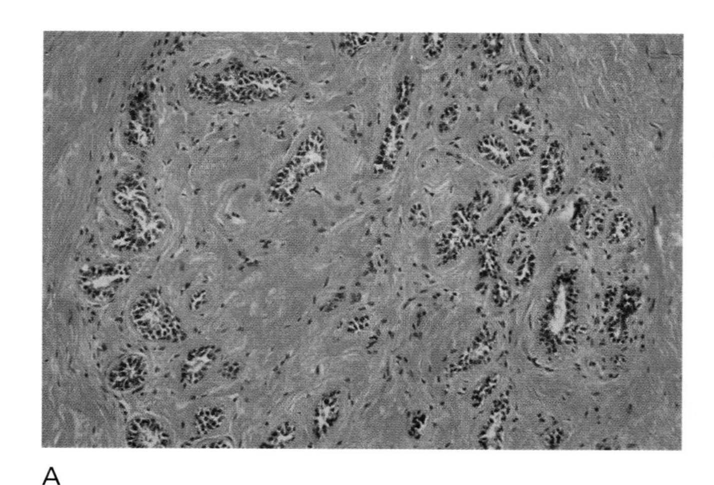

A

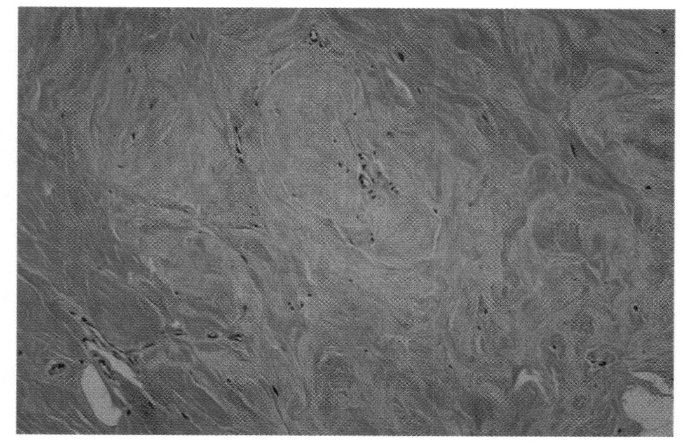

B

FIGURE 2–8. (**A**) Partial involutional "fibrosis" of the lobule. (**B**) Complete involutional change of a lobule with replacement by interlobular fibrous tissue. (Hematoxylin/eosin)

NONNEOPLASTIC INFLAMMATORY LESIONS OF THE BREAST

The clinical presentation of patients with inflammatory breast lesions most commonly includes local pain, tenderness, and swelling. Most of the inflammatory lesions develop during pregnancy. The vast majority are bacterial infections, and the tissue reaction is that of acute inflammation. Rarely, tuberculosis or fungal or viral infections are the etiology.[83–86]

Mastitis

FNA biopsy of mastitis with abscess grossly may yield yellowish or green fluid and microscopically demonstrates a plethora of neutrophils and a paucicellular background of epithelial elements (Fig. 2–9). Usually the epithelial elements are infiltrated by neutrophils and show the typical features of epithelial repair. The reparative epithelial elements consist of spidery or tapering flat sheets with maintenance of polarity. Cell size is enlarged with nuclear enlargement such that the nuclear to cytoplasmic (N/C) ratio is not increased. Nuclear membranes remain smoothly contoured, and nucleoli are often present, prominent and fairly uniform in size. Mitotic figures may be seen but are never atypical. When these cells degenerate, they become smaller, with pyknotic nuclei and a relatively increased N/C ratio.

Periductal mastitis or plasma cell mastitis most often presents in the fifth decade as localized dull pain. Cytologic findings parallel tissue histology, with dilated ducts filled with inspissated secretions and foam cells and surrounded by lymphocytes and plasma cells.[87] The foam cells, formerly considered to be of duct epithelium origin,[88] are macrophage-derived, demonstrating the immunophenotype of tissue histiocytes.[89]

Simple breast cysts (Fig. 2–10), a consequence of benign fibrocystic change, can on occasion become acutely inflamed. Patients present with severe pain, swelling, and redness of the overlying skin if the cyst is superficial. Such cysts when aspirated are turbid

A

B

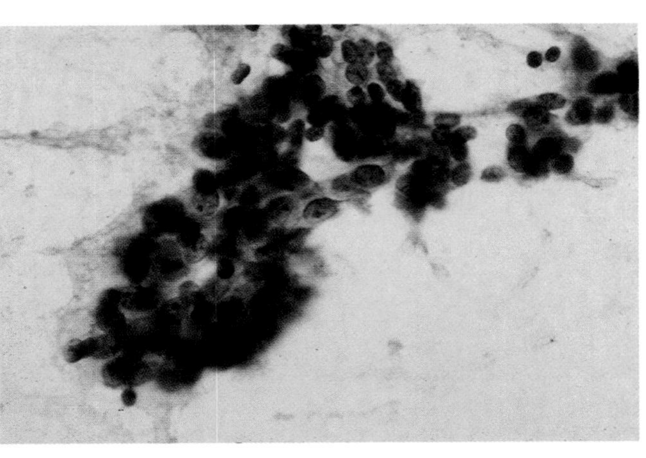

C

FIGURE 2–9. (**A**) Abundant polymorphonuclear leukocytes seen with mastitis. (**B** and **C**) Epithelial repair with tapering fragments of cells demonstrating uniform nuclear enlargement and nucleoli and percolated with neutrophils. (**A,** Diff-Quik; **B** and **C,** Papanicolaou stain)

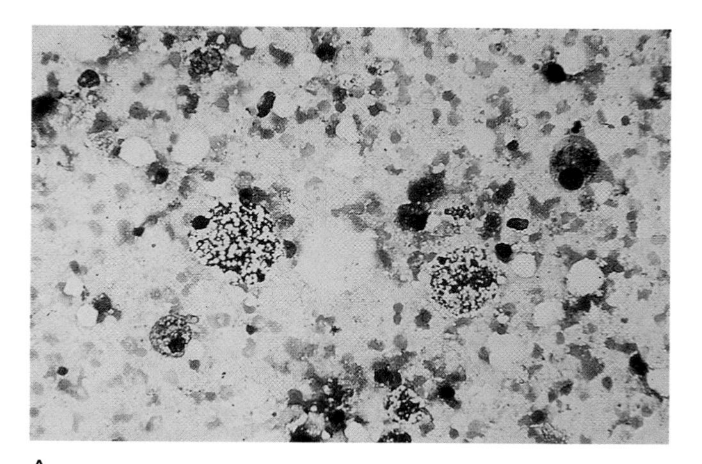

A

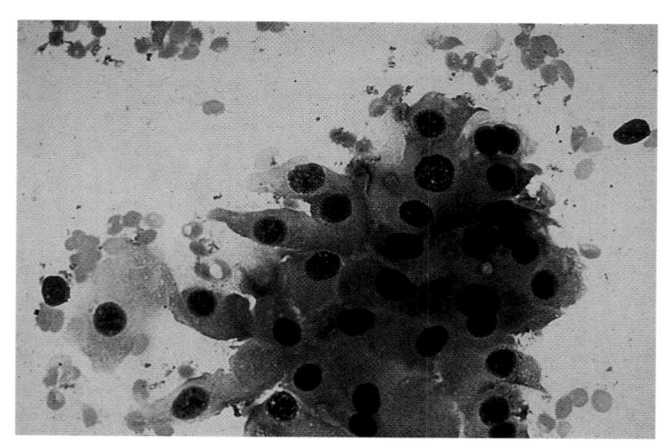

B

FIGURE 2–10. Contents of simple breast cysts include scattered ductal cells, degenerate foamy histiocytes, and apocrine metaplastic cells. (Diff-Quik)

and contain massive numbers of neutrophils in addition to foam cells and degenerated duct cells (Fig. 2–11). As with the contents of any breast cyst, except for perhaps clear contents, the aspirate should be examined cytologically.[1,90,91] Also, after aspiration of any cyst where some type of liquid is obtained, it is mandatory to repalpate the region of the cyst to make sure an underlying residual palpable abnormality is not overlooked.[1]

Subareolar Abscess

This inflammatory lesion probably begins in the lactiferous duct when squamous metaplasia occurs and results in duct obstruction by packing the lactiferous duct with orthokeratotic material. The obstructed duct then ruptures, spilling keratinaceous debris into the stroma. This incites a foreign-body giant cell reaction, chronic inflammation, and potentially sinus tract formation to the skin surface. The cystically dilated duct can become infected and chronically recur, requiring excision, antibiotic therapy, or both, for cure.[92,93] FNA cytology shows fragmented orthokeratotic or parakeratotic keratin, foreign body giant cells (Fig. 2–12), cholesterol clefts, acute and chronic inflammatory cells, squamous epithelial groups, enucleate squames, and various features of tissue repair.[93]

Although the subareolar abscess is a distinctive clinicopathologic entity with presentation of a painful red lesion, the cytologic findings of tissue repair[36,93–95] can be a pitfall for a diagnosis of malignancy. Granulation tissue repair is often cellular, with prominent, enlarged nuclei and conspicuous nucleoli. The tissue fragments typically have the readily recognized tapered or feathered edges, which is a key point to distinguish from malignancy.

Cytomorphologic Features of Subareolar Breast Abscess

- Squamous epithelium
- Anucleate squames, keratinaceous debris
- Foreign-body giant cells
- Acute and chronic inflammation
- Epithelial repair and inflammatory atypia (a diagnostic pitfall)
- Granulation tissue

Fat Necrosis

Fat necrosis of the breast is uncommon, but is most often the result of trauma, which sometimes goes unrecognized by the patient. Almost all patients give a history of trauma, prior surgical intervention, or radiation therapy.[87]

Histologically, fat necrosis demonstrates necrotic fat cells, foamy macrophages, inflammatory cell infiltrates, and fibroplasia, depending on the age of the lesion (Fig. 2–13). On FNA biopsy, one obtains fat with a bubbly appearance, blood, foam cells, and atypical cells. The cellular atypia may represent either adjacent, reactive epithelial cells or activated stromal cells. It is critical to recognize the atypical cells as benign in the context of the other morphologic features, because fat necrosis can mimic invasive cancer on mammography, and by its irregular outline and firmness can simulate cancer on palpation. The differential diagnosis includes carcinoma. However, the number of atypical cells in fat necrosis is usually very low. In a patient who has not had prior surgery, invasive carcinoma and fat necrosis are very uncommon.

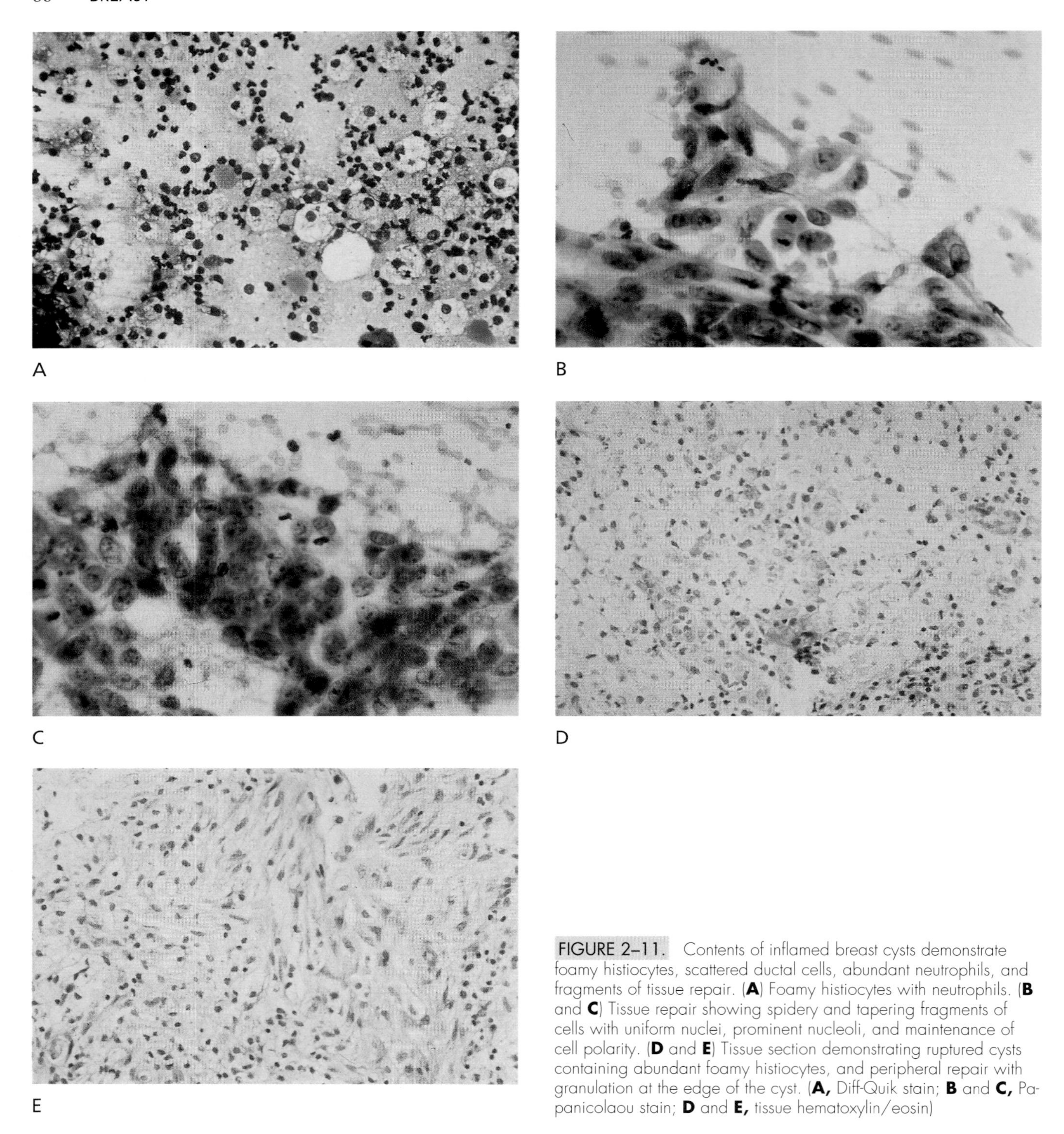

FIGURE 2–11. Contents of inflamed breast cysts demonstrate foamy histiocytes, scattered ductal cells, abundant neutrophils, and fragments of tissue repair. (**A**) Foamy histiocytes with neutrophils. (**B** and **C**) Tissue repair showing spidery and tapering fragments of cells with uniform nuclei, prominent nucleoli, and maintenance of cell polarity. (**D** and **E**) Tissue section demonstrating ruptured cysts containing abundant foamy histiocytes, and peripheral repair with granulation at the edge of the cyst. (**A,** Diff-Quik stain; **B** and **C,** Papanicolaou stain; **D** and **E,** tissue hematoxylin/eosin)

Fibrocystic Change

Formerly referred to as fibrocystic disease,[96] FCC accounts for the most common palpable abnormalities in the breast. It is a common source of anxiety that women experience about their health. The most common reason for surgical biopsy is to evaluate the nature of these ill-defined lumps.

FCC is seen mostly in the premenopausal years, predominately in women younger than 40. This disease of the breast lobule results in an unfolding of the lobule with concomitant cystic dilatation, with cysts reaching up to many centimeters in diameter. Although the exact pathogenesis remains elusive, Haagensen[84] suggested that periductal scarring that results in ductular obstruction causes retrograde cys-

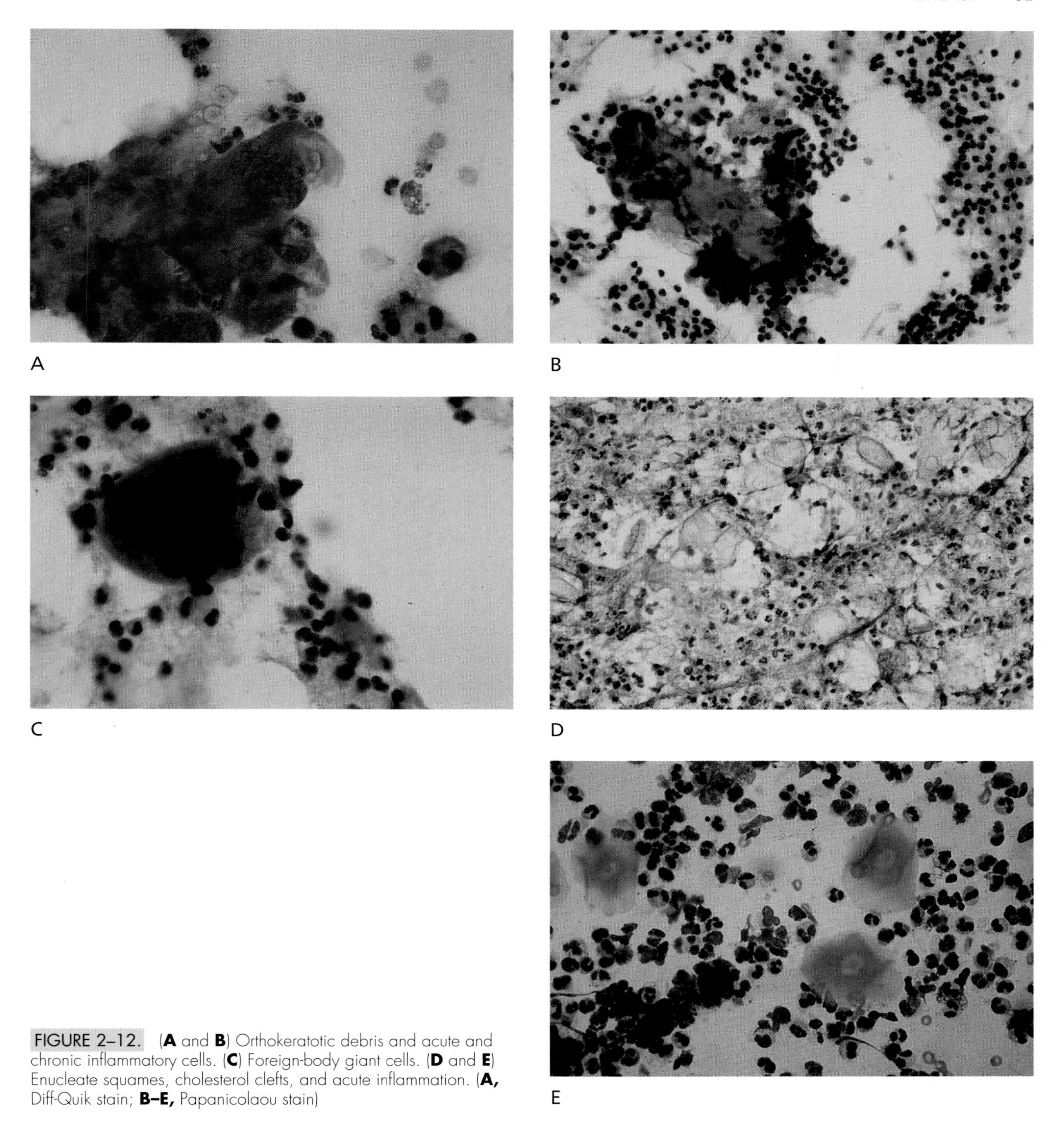

FIGURE 2–12. (**A** and **B**) Orthokeratotic debris and acute and chronic inflammatory cells. (**C**) Foreign-body giant cells. (**D** and **E**) Enucleate squames, cholesterol clefts, and acute inflammation. (**A,** Diff-Quik stain; **B–E,** Papanicolaou stain)

tic dilatation. Apocrine metaplasia is commonly seen in some dilated cysts, as well as in some normal lobules. The foam cells often seen in the cysts, in periductal locations and plastered along the walls of some cysts, are of macrophage rather than epithelial origin.[97] Stromal fibrosis, a salient feature of FCC, is derived from the extensive lobular involutional changes that occur, resulting in dense interlobular collagenous stroma (fibrosis) replacing the active loose stroma of the TDLU.

FNA biopsy of nonproliferative FCC yields small amounts of normal honeycomb arrangements of duct epithelium containing spindle nuclei of myoepithelial cells (see Fig. 2–7), foam cells, apocrine cells, bipolar stromal naked nuclei, and sometimes cyst fluid (see Fig. 2–10). Traditionally, the stromal naked nuclei

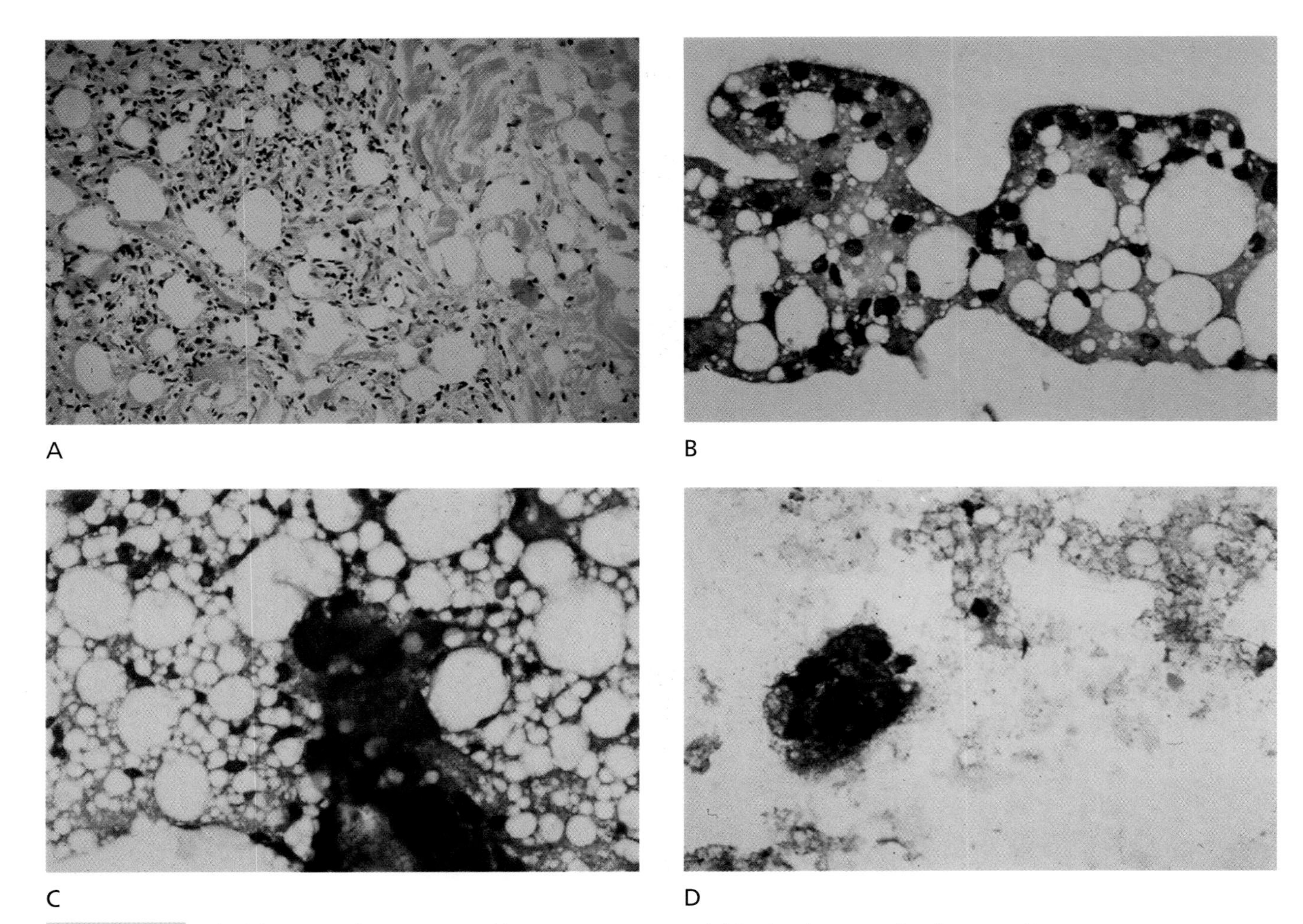

FIGURE 2–13. Fibroblastic granulation in area of fat necrosis. Bubbly fat with foamy macrophages and scattered atypical cells, a pitfall for the diagnosis of malignancy. (**A,** tissue hematoxylin/eosin; **B–D,** Diff-Quik stain)

have been considered to be myoepithelial cells, but in fact most likely they are stromal nuclei from the intralobular and interlobular connective tissue fibroblasts, because they do not react with antibodies specific for myoepithelial cells, such as smooth muscle myosin and calponin (Fig. 2–14).[98] Variable fibrous and adipose tissue fragments may be present.

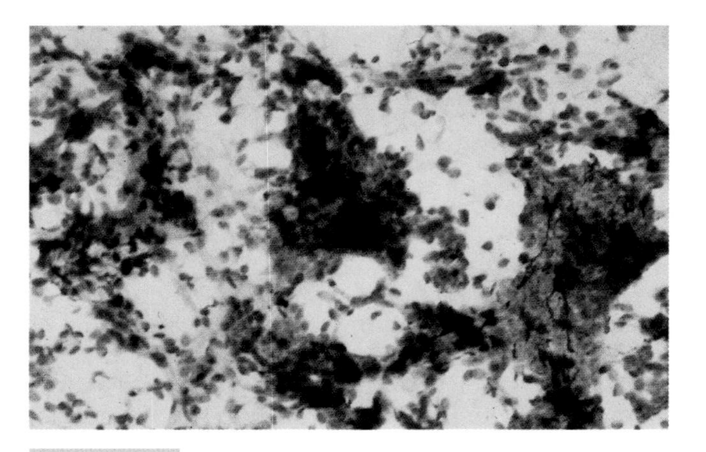

FIGURE 2–14. Myoepithelial cells in this fibroadenoma are decorated by calponin, a myoepithelial cell-specific antibody. There is no staining of bipolar nuclei. (Alcohol-fixed, previously Papanicolaou-stained)

Cytomorphology of Nonproliferative Fibrocystic Change

- Low epithelial cellularity
- Flat honeycomb epithelial sheets with maintenance of polarity, cohesion, and distinct cell borders
- Myoepithelial cells among epithelial cells
- Stromal naked nuclei
- Foam cells and apocrine cells
- Fat and fibrous stroma variable

Proliferative Fibrocystic Change, No Atypia

It is important to recognize proliferative cellular features in aspirates because of the elevated risk of breast cancer in this group of patients.[99,100] For moderate to florid hyperplasia without atypia, the risk increases twofold; it is elevated four- to fivefold in patients who have proliferation with atypia.[101–103] The high cellularity of the FNA biopsy of these lesions can be a potential diagnostic pitfall for carcinoma.[104]

Duct hyperplasia without atypia shows an increased cellularity of cohesive ductal groups with a honeycomb arrangement but with mild loss of polarity, manifested as some cellular and nuclear overlapping and mild anisonucleosis (Fig. 2–15). This corresponds to the cell streaming seen in histologic sections. The chromatin remains fine, and nuclear membranes are smooth, round to oval. Occasional small nucleoli are seen. Round or spindle myoepithelial nuclei are invariably present in epithelial groups. Foam cells, apocrine cells, and/or stromal naked nuclei can also be seen.

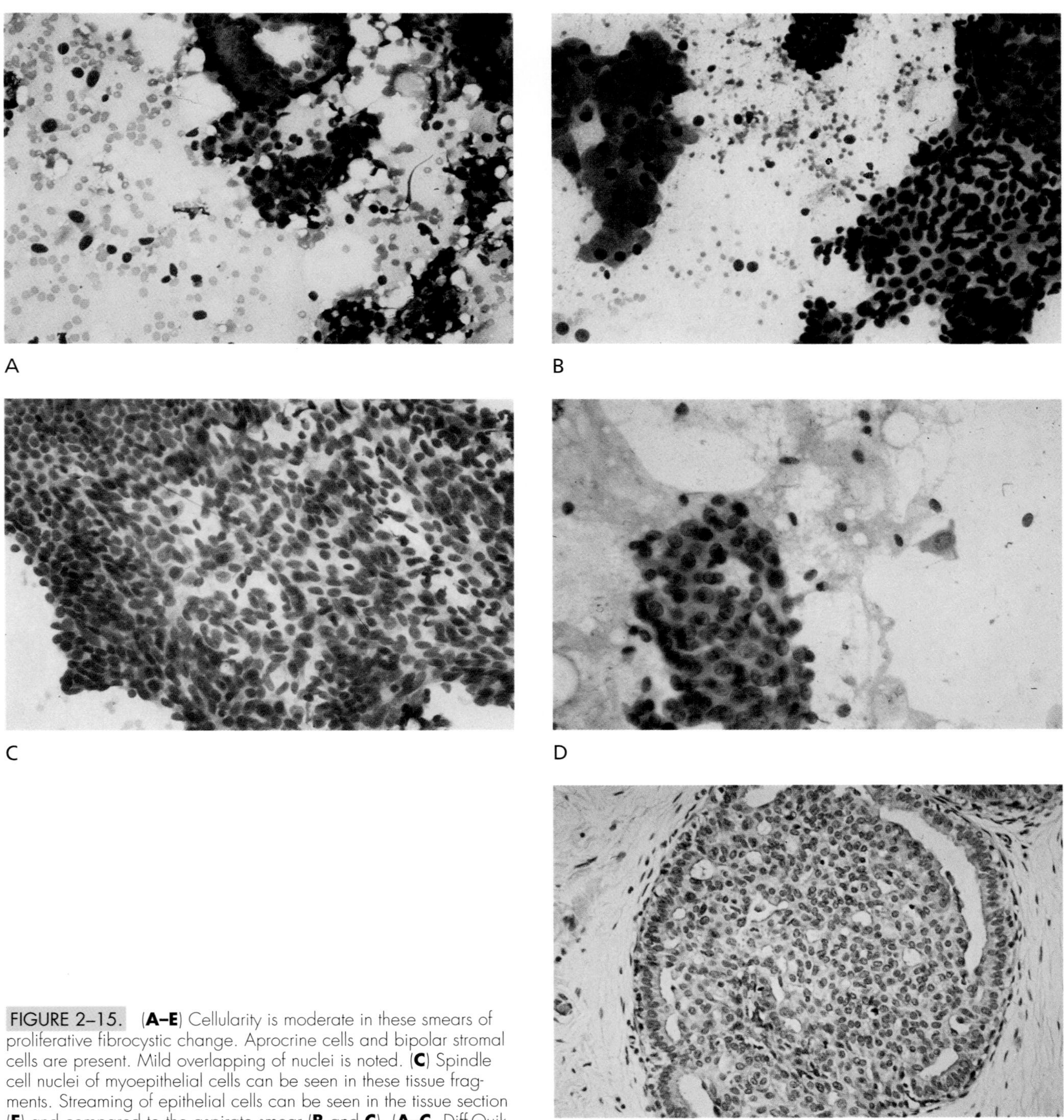

FIGURE 2–15. (**A–E**) Cellularity is moderate in these smears of proliferative fibrocystic change. Aprocrine cells and bipolar stromal cells are present. Mild overlapping of nuclei is noted. (**C**) Spindle cell nuclei of myoepithelial cells can be seen in these tissue fragments. Streaming of epithelial cells can be seen in the tissue section (**E**) and compared to the aspirate smear (**B** and **C**). (**A–C,** Diff-Quik stain; **D,** Papanicolaou stain; **E,** tissue hematoxylin/eosin stain)

Cytomorphologic Features of Proliferative Fibrocystic Change, No Atypia

- Moderate to high cellularity
- Cohesive epithelial sheets with mild loss of polarity
- Myoepithelial cells in epithelial groups
- Stromal naked nuclei
- Foam cells, apocrine cells

Proliferative Fibrocystic Change with Atypia

As mentioned earlier, one of the challenges of breast cytopathology is distinguishing among duct hyperplasia, atypical duct hyperplasia, and carcinoma. Sneige and Staerkel[105] and Masood et al.[106,107] presented criteria for recognizing atypical hyperplasia based on subsequent tissue examination.

Sneige and Staerkel used low-power patterns and high-power cellular detail to describe duct hyperplasia as epithelial groups mixed with myoepithelial cells and stromal cells in a complex or cribriform manner. Nuclei are bland and round to oval and override each other (Fig. 2–16). Few single cells may be present. Atypical hyperplasia has a monotonous population with some nuclear hyperchromasia. Moderate anisonucleosis is present, nucleoli are readily seen, and loss of polarity is readily evident as overlapping nuclei.[107–110] The presence of this degree of atypia in the FNA biopsy of breast should trigger an excisional biopsy because the cytologic features of atypical hyperplasia and noncomedo duct carcinoma *in situ* overlap.[105] Other investigators cannot reliably distinguish atypical hyperplasia from noncomedo carcinoma *in situ*.[106–109]

Cytomorphologic Features of Proliferative Fibrocystic Change with Atypia

- High epithelial cellularity
- Moderate cellular and nuclear overlapping
- Cell monotony
- Nucleoli common
- Myoepithelial nuclei in epithelial groups
- Stromal naked nuclei

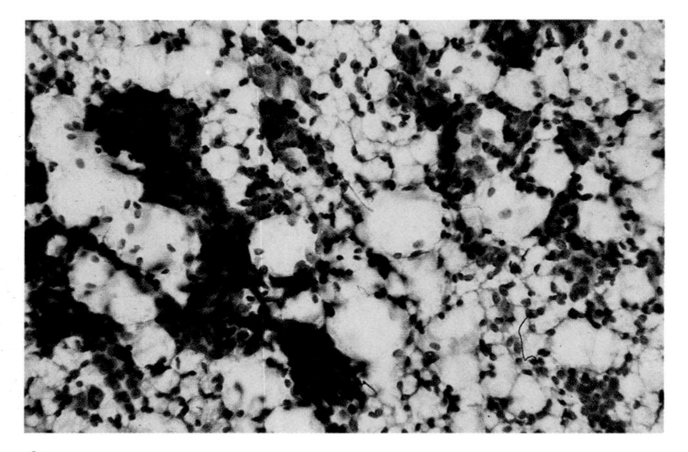

B

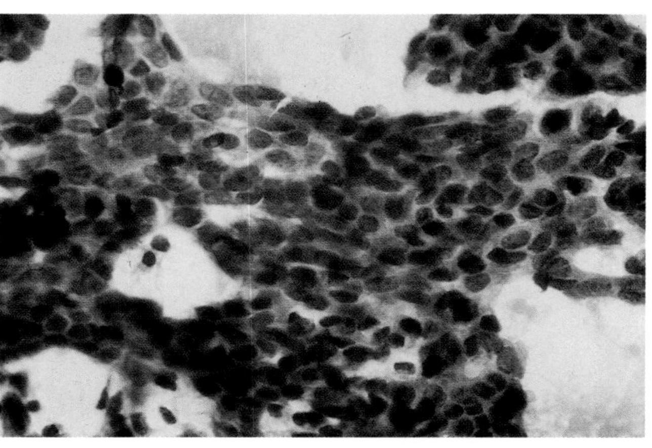

C

FIGURE 2–16. Proliferative fibrocystic change with atypia. (**A**) Hypercellular smears with stromal nuclei. (**B** and **C**) Complex architecture with epithelial bridging, loss of polarity, and cell monotony. (**A** and **C,** Papanicolaou stain; **B,** Diff-Quik stain)

BENIGN BREAST TUMORS

Fibroadenoma

In contrast to the ill-defined nature of the fibrocystic lesion, the fibroadenoma is a discrete, freely movable, and circumscribed tumor. This is a very helpful distinguishing clinical feature because the cytologic features may overlap with fibrocystic change.[111] Bottles et al.[112] performed a stepwise logistic regression analysis on a group of lesions and discovered that three-dimensional antler-like clusters of duct cells, stromal fragments, and marked cellularity were the most consistent cytologic features for the majority of fibroadenomas (Fig. 2–17). This can be an aid to distinguish this lesion from FCC.

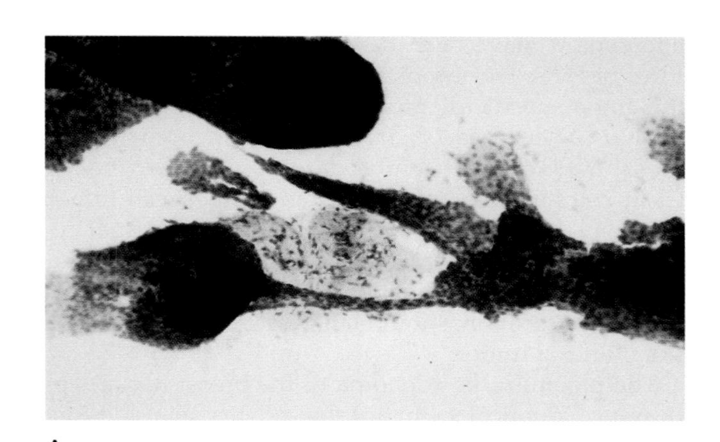

A

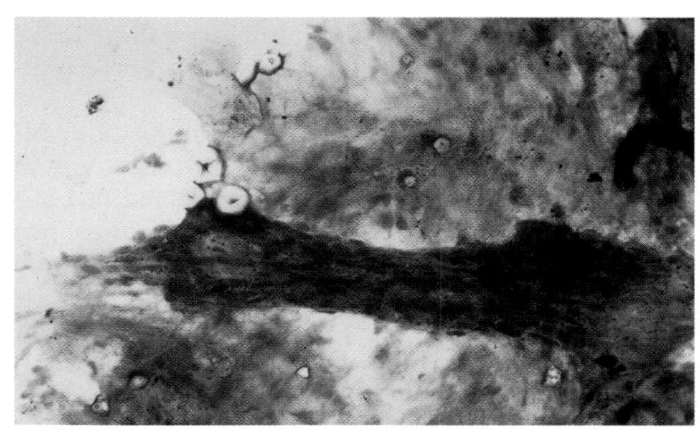

B

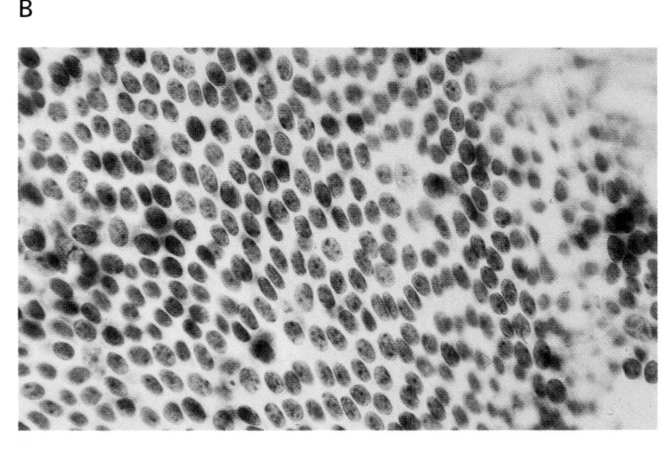

C

D

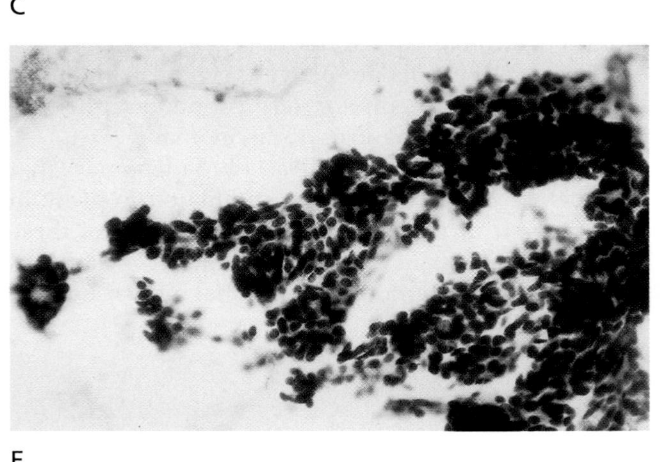

E

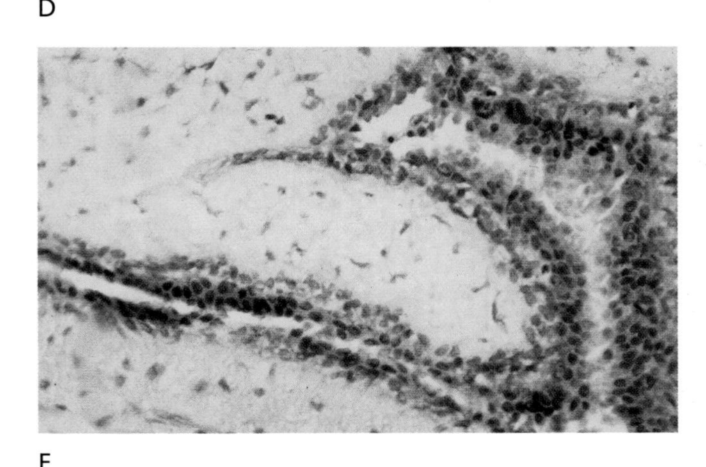

F

FIGURE 2–17. (**A**) Finger-like projections of cell groups in fibroadenoma. (**B** and **C**) Myxoid metachromatically stained stroma with variable background bipolar stromal cells. (**D**) Sheetlike arrangement of ductal cells and (**E**) antler-like fragments of duct epithelial cells. (**F**) Tissue section with the biphasic myxoid stromal and proliferative ductal elements. (**A,** Papanicolaou stain; **B** and **C,** Diff-Quik stain; **C** and **D,** Papanicolaou stain; **E,** tissues hematoxylin/eosin)

Cellularity may be so high that a diagnosis of carcinoma may be entertained.[113,114] In addition, atypia can be a feature of fibroadenomas, and it deserves careful scrutiny because of the pitfall for a false-positive diagnosis of carcinoma.[115]

Finger-like papillary projections of epithelial groups, abundant stromal naked nuclei, and myxoid stromal fragments are very common. In contrast, apocrine cells and foam cells are uncommon. Although the cytologic features themselves are not specific for fibroadenoma,[116] by combining the clinical and cytologic features one can arrive at a specific diagnosis of fibroadenoma.

The reasons for the atypia can be multifactorial, including the presence of preneoplastic changes within the fibroadenoma. Stanley et al.[115] concluded that adherence to the criteria for the diagnosis of malignancy was critical to avoid a false-positive diagnosis. Loose cellular cohesion and occasional single cells may be present, but antler-like clusters and naked stromal nuclei should mitigate against the diagnosis of carcinoma.

Cytomorphologic Features of Fibroadenoma

- High epithelial cellularity
- Antler-like and finger-like branching epithelial groups
- Epithelial atypia possible
- Abundant stromal naked nuclei
- Myxoid stromal fragments
- Foam cells and apocrine cells uncommon
- Pitfalls include high cellularity, atypia

SCLEROSING ADENOSIS AND ADENOSIS TUMOR

Sclerosing adenosis is a proliferative lesion often seen in the setting of FCC. Patients with adenosis have a slightly greater risk for the development of breast carcinomas.[101] Characterized by proliferating ductules/tubules associated with a densely sclerotic stroma, the gross and microscopic appearance can mimic carcinoma. The FNA biopsy findings are those of FCC, including uniform ductal cells, stromal nuclei, and stromal fragments (Fig. 2–18).

Sclerosing adenosis can achieve a palpable tumoral form, as described by Haagensen.[84] Because they may grossly mimic carcinoma, the hazard at frozen section is the failure to recognize the benign nature of the proliferating tubules. We examined the FNA from an adenosis tumor and observed benign cellular features analogous to those of proliferative FCC.[117]

Adenomas

True tubular adenomas of the breast are rare. There have been only two reports of its cytologic appearance,[118] describing features similar to fibroadenoma but lacking the myxoid stroma. Adenomas of the nipple (papillary adenomas) are benign florid epithelial proliferative lesions that appear essentially identical to proliferative FCC.[119–122] A variety of other adenomas, including eccrine spiradenoma[123] and ductal adenoma,[124] have been described with cytologic features of FCC. The importance of these lesions is not in making a specific diagnosis, but recognizing them as benign. Adenomas of skin adnexa can occur in the breast and can be confused with duct carcinoma. Clear cell hidradenoma occurs as a superficial tumor that may ulcerate the skin surface.[125] The unusual cytomorphologic appearance of polygonal clear and spindle cells, in combination with the superficial dermal-based location, should suggest the possibility of a skin adnexal tumor.

The pleomorphic adenoma of the breast has an appearance identical to that of the salivary gland lesion: abundant myxoid stroma with spindle myoepithelial cells and variable ductal cellularity.[126]

Lactating Adenoma

The lactating adenoma is a glandular proliferation with prominent, diffuse lactational changes. The latter consists of voluminous vacuolated cytoplasm because of the presence of secretory material. The secretory material is also abundant within the glandular lumens and ducts; when aspirated, the secretory product produces a dirty smear background. The specimen is very cellular, with a uniform population of vacuolated cells.[88,118,127–132] The cells are delicate, often with frayed cytoplasm, appearing as scattered stripped nuclei in the film of secretory product (Fig. 2–19). A background of stromal bipolar nuclei is generally not present. Because the cells are actively synthesizing lactational product, nucleoli are very prominent; this can be a pitfall for the diagnosis of carcinoma.[118]

Other breast lesions, such as fibroadenomas, that exist before pregnancy can be stimulated to grow in the pregnant or lactational state. Aspiration of these lesions in pregnancy shows epithelial lactational changes with an abundance of stromal bipolar nuclei.

Cytomorphologic Features of Lactating Adenoma

- High cellularity
- Vacuolated secretions in background
- Scattered stripped epithelial nuclei
- Epithelial cell cytoplasm frayed and vacuolated

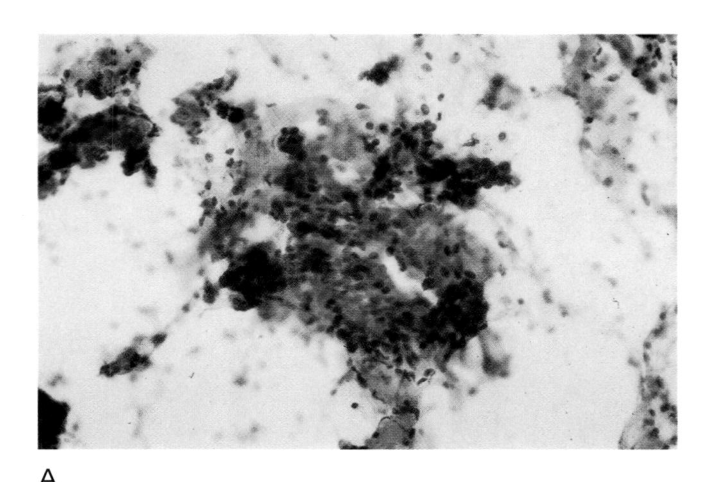

A

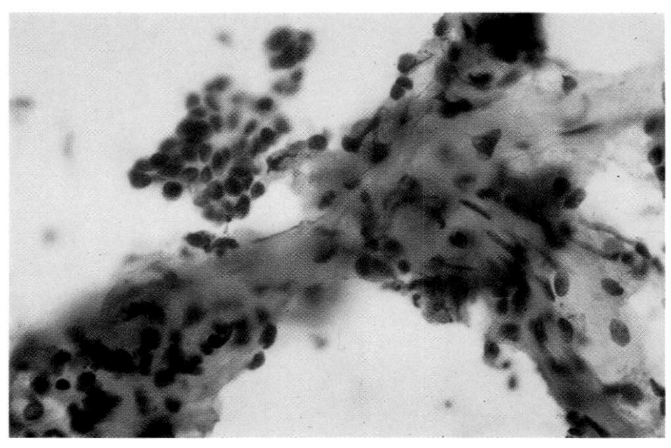

B

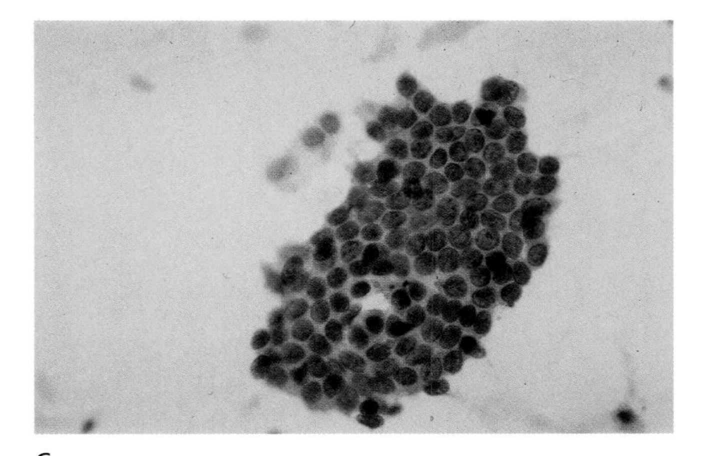

C

FIGURE 2–18. Ductal epithelial cells in sheetlike arrangement along with fragments of sclerotic fibrous tissue are characteristic of sclerosing adenosis. (Papanicolaou stain)

Papilloma of the Breast

Intraductal papilloma of the breast is a lesion seen most commonly in the fifth decade. It rarely gives rise to a palpable breast lesion. Papillary fragments of tight cohesive cell groups with depth of focus may be seen in aspirates (Fig. 2–20).[133] The diagnosis of a papillary lesion should always be confirmed on excisional biopsy because of the differential diagnosis of a papillary carcinoma.

Adenomyoepithelioma

Adenomyoepithelioma of the breast presents a confusing cytologic picture that can be a pitfall for the diagnosis of malignancy.[134,135] These tumors occur in patients in the fourth to sixth decades and exhibit a spectrum of biologic behavior from benign to malignant.[136,137] Aspiration cytology is hypercellular and shows dyscohesive cells and enlarged plasmacytoid cells with variably sized eccentric nuclei, inconspicuous nucleoli, and vacuolated or clear cytoplasm. Recognition of this unusual cytologic appearance and a cautious, conservative approach with biopsy recommendation is suggested.[134,135] The finding of myoepithelial cells with muscle actin stains by immunoperoxidase would be helpful in distinguishing adenomyoepithelioma from carcinoma.

Lipoma

Lipomas of the breast are rare, and unless one can prove that the aspiration needle is in the lesion, the finding of just adipose tissue on a smear would not be adequate to make the diagnosis of lipoma. Adipose tissue is a common finding on FNA biopsies of FCC and from normal areas of the breast. The diagnosis can be suggested when the lesion is circumscribed and freely moveable and the needle can be documented to be in the lesion, along with a supportive mammographic examination.

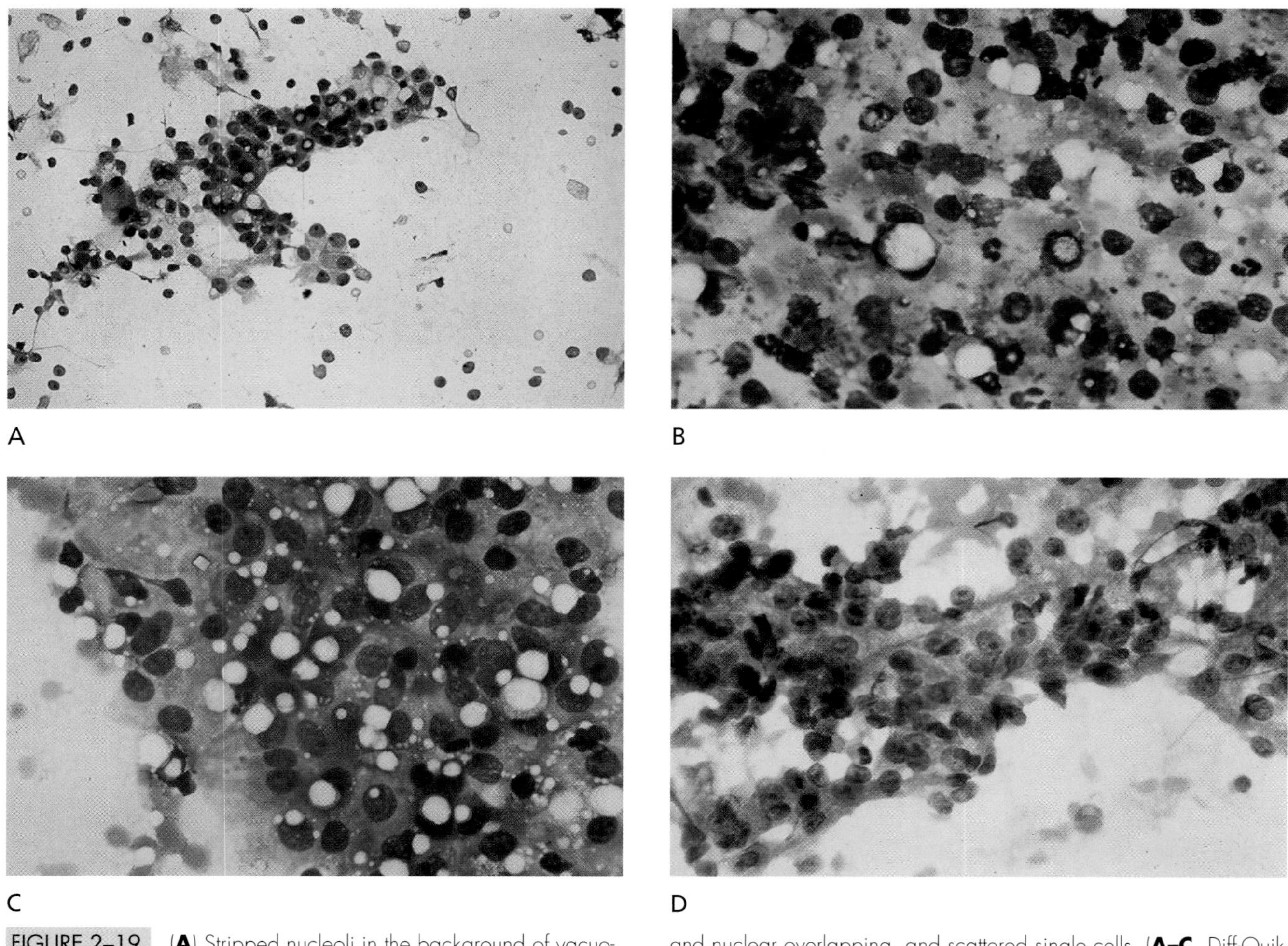

FIGURE 2–19. (**A**) Stripped nucleoli in the background of vacuolated material. (**B–D**) Vacuolated secretory cells characteristic of lactational change show pronounced cytoplasmic vacuolation, cellular and nuclear overlapping, and scattered single cells. (**A–C,** Diff-Quik stain; **D,** Papanicolaou stain)

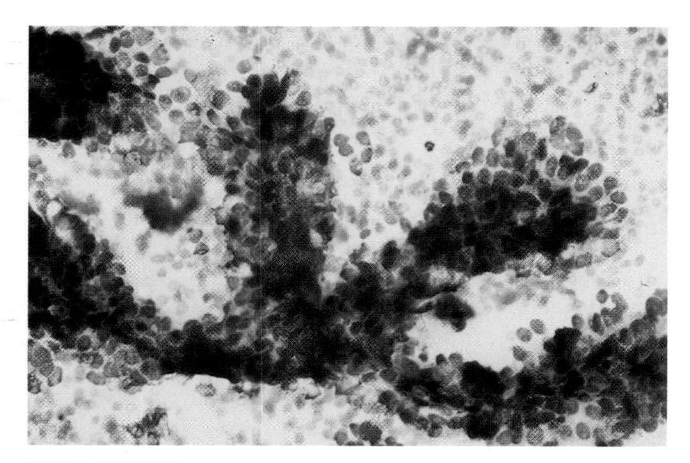

FIGURE 2–20. Papillary fragments seen with papillary carcinoma. Loss of cell polarity, papillary projections, and large nucleoli with high N/C ratios are characteristic. (Diff-Quik stain)

Granular Cell Tumor

These tumors occur in many different sites but are uncommon in the breast. The granular cell tumor can mimic invasive carcinoma mammographically, clinically, and grossly because it causes an intense desmoplastic stromal response.

FNA may reveal groups of cells with abundant granular cytoplasm and indistinct cell borders. Nuclei may show modest anisonucleosis with prominent nucleoli (Fig. 2–21). The granular cytoplasm can be fragile, and when stripped away the Romanowsky stains can appear suspicious. The Papanicolaou stains can be diagnostic, with eosinophilic staining of the granular cytoplasm seen to better advantage with this stain. The diagnosis can be confirmed with immunoperoxidase, showing S-100-positive and cytokeratin-negative cells.

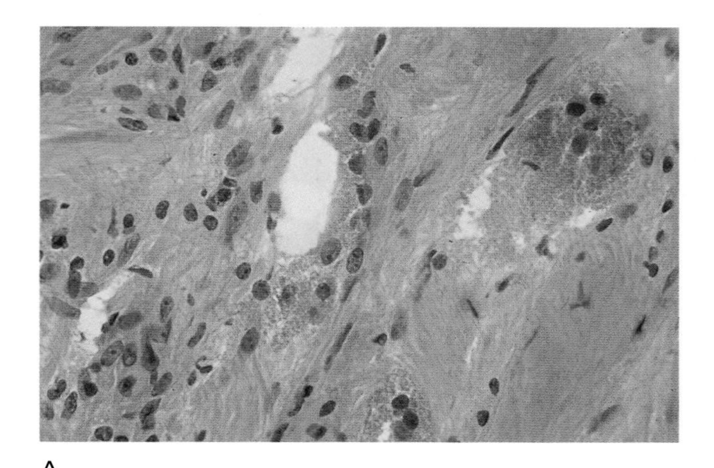

A

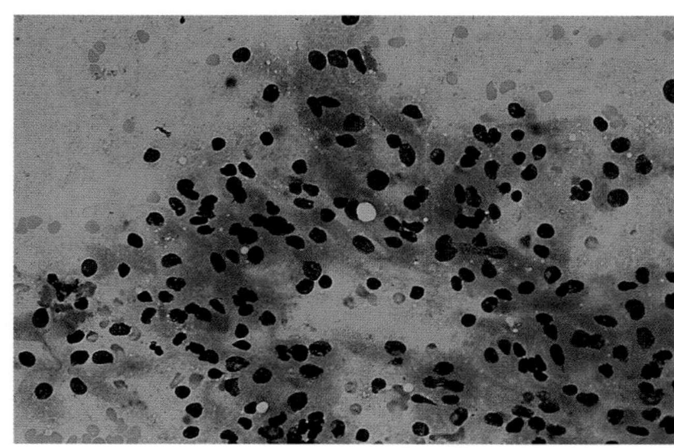

B

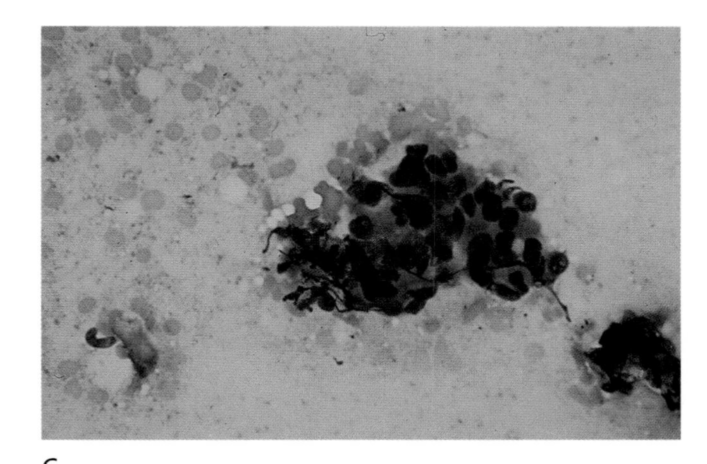

C

FIGURE 2–21. Granular cell tumor of the breast. (**A**) Characteristic large eosinophilic cells infiltrating sclerotic stroma. (**B,C**) Variable-sized cellular groups, some small and some larger, with indistinct cell cytoplasms simulating syncytial arrangements, with hyperchromatic nuclei and some nucleoli. (**A,** tissue hematoxylin/eosin; **B** and **C,** Diff-Quik stain)

The differential diagnosis can also include granular histiocytes: granular cell tumors and histiocytes are positive for CD68 (KP1), an antibody that reacts with cells containing abundant lysosomes.[138] Cytokeratin does not decorate tissue histiocytes.

Localized Amyloid Tumor of the Breast

Amyloid can cause breast masses that simulate malignancy.[139,140] In our series, one patient had bilateral amyloid tumors, but neither patient had evidence of systemic amyloidosis. Modified Wright-stained slides (Fig. 2–22) showed metachromatic magenta clumps of material; Papanicolaou-stained slides showed glassy eosinophilic cylindrical rods of amyloid. A few inflammatory cells can be in the background, as well as a giant cell reaction. When amyloid is appreciated in a smear, a presumptive diagnosis can be made and confirmed with Congo red stains performed on the cytology slides.

Gynecomastia

More often unilateral, gynecomastia presents as a disc-like swelling of the male breast in adolescents and the elderly. It may be idiopathic or secondary to a variety of medications. FNA biopsy is an accepted standard procedure for the evaluation of male breast lesions.[141]

FNA cytology reveals groups of ductal epithelial cells with a background of naked stromal nuclei (Fig. 2–23). Cellularity is usually low but can be quite high. Chromatin is bland, although nucleoli may be visible. This may manifest as "atypia," but the diagnosis of carcinoma should not be made without application of the full criteria for malignancy.[142,143]

Cytomorphologic Features of Gynecomastia

- Low to moderate cellularity
- Groups of ductal cells
- Nucleoli may be visible
- Stromal naked nuclei in background

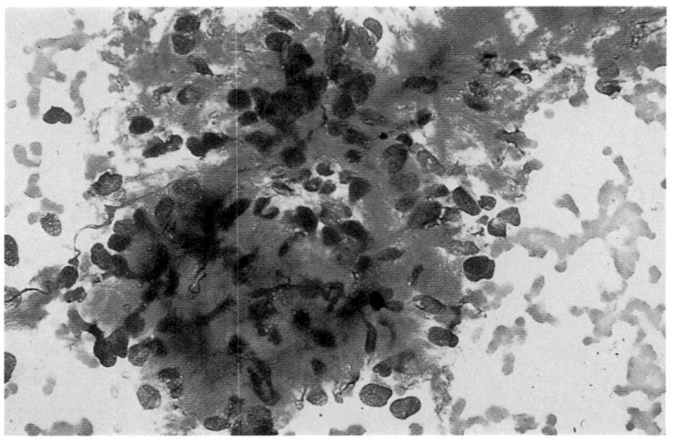
A

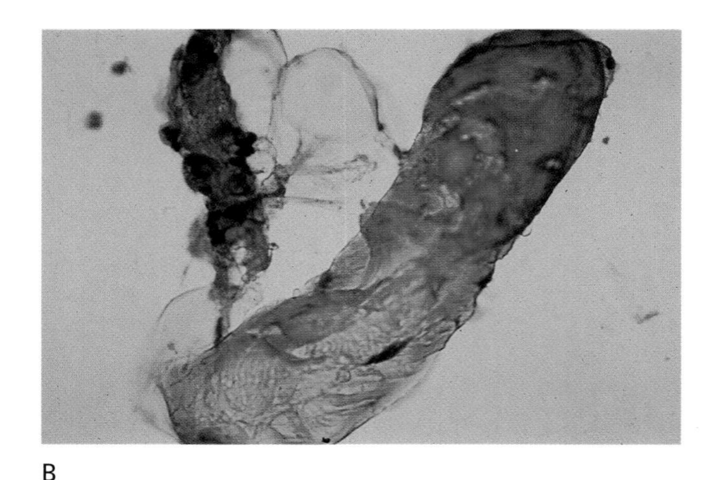
B

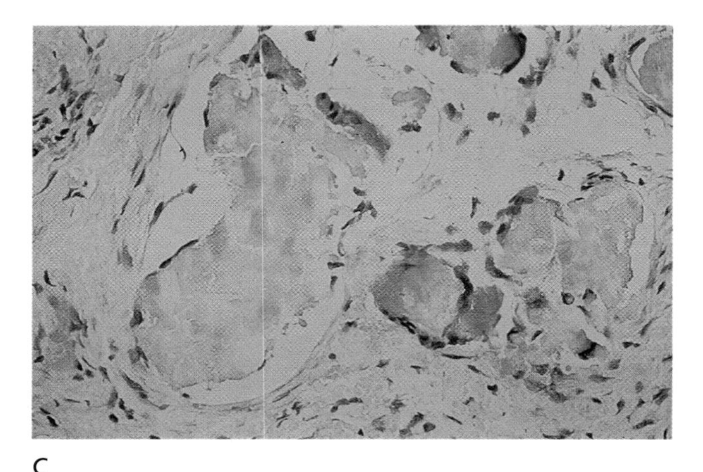
C

FIGURE 2–22. (**A**) Fragmented metachromatic material with adherent stromal cells. (**B**) Orange refractile tubular-shaped material representing amyloid. (**A,** Diff-Quik; **B,** Papanicolaou stain) (**C**) Tissue showing fragmented amorphous amyloid deposits with some peripheral giant cell reaction.

MALIGNANT BREAST TUMORS

Duct Carcinoma Not Otherwise Specified

Invasive duct carcinoma not otherwise specified (NOS) accounts for about 75% of all invasive carcinomas. Its peak incidence is in the fifth decade, but it accounts for 2% of all breast cancers under the age of 40.[144,145] Clinically invasive carcinomas often are stony hard to palpation, but palpation alone is insufficient to determine whether a tumor is malignant. The triple test of palpation, mammographic appearance, and cytologic features is the recommended approach when carcinoma is suspected.[1]

The tissue appearance of invasive carcinoma NOS demonstrates a wide spectrum, exemplified by the Scarf-Bloom-Richardson grading scheme, which includes the parameters of amount of tubule formation, nuclear grade, and mitotic index.[146] In addition, host stromal response is often desmoplastic, with varying degrees of an inflammatory host response and possibly tumor cell necrosis. In general, the yield of tumor cells on FNA is indirectly related to the amount of desmoplasia, a point especially important to take into account when using the nonaspiration sampling technique.[88]

The cytologic appearance of carcinoma NOS is somewhat variable, but there are many common features. Architecturally, cellularity should be high and display loosely cohesive syncytial groups of tumor cells. Generally there are many single cells, but cellular arrangements may vary from two or three cells to hundreds of cells per group. The syncytial arrangement precludes identification of cell borders, and nuclear overlapping is completely haphazard and prominent (Fig. 2–24). Cell size is greatly increased, although some carcinomas can consist of cells not much larger than normal duct cells. Nuclei are hyperchromatic with irregular nuclear membranes and are

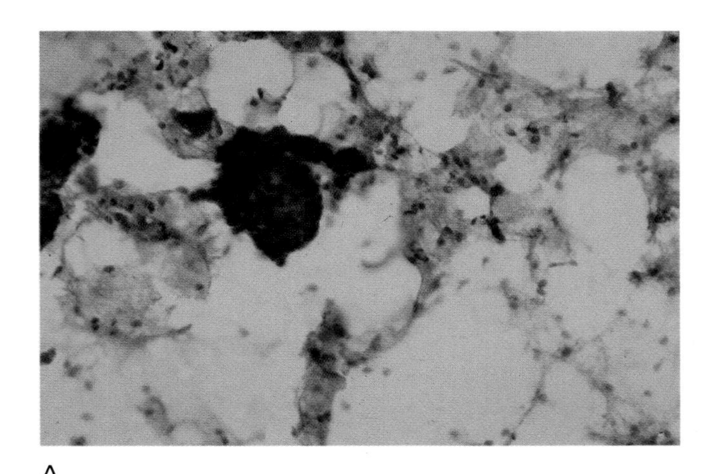
A

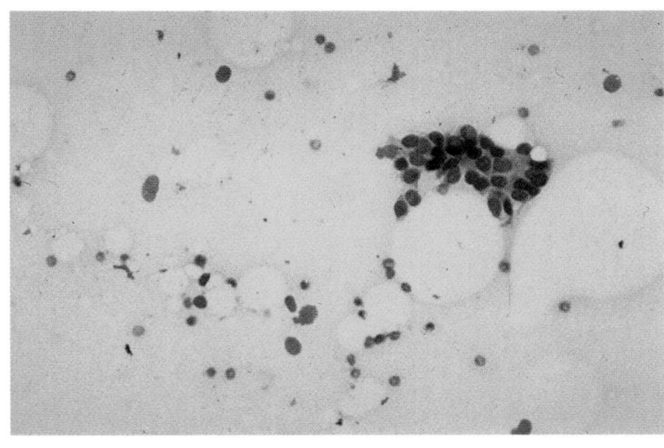
B

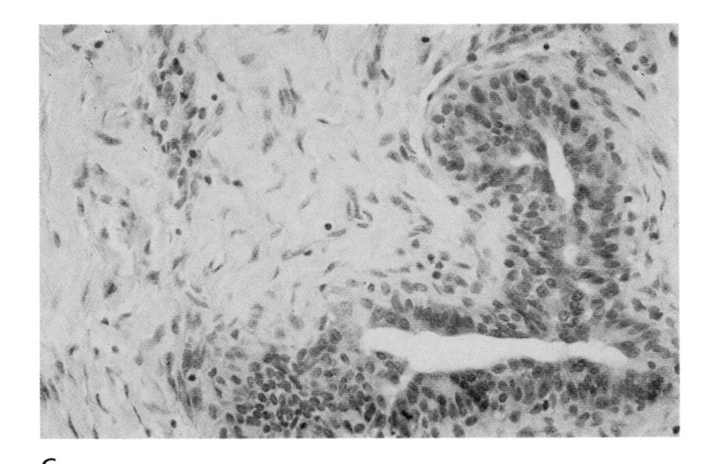
C

FIGURE 2–23. (**A,B**) Paucicellular specimen of gynecomastia showing groups of ductal cells and a background of bipolar stromal cells. (**C**) Duct epithelial hyperplasia and stromal hypercellularity. (**A,** Papanicolaou stain; **B,** Diff-Quik stain; **C,** tissue hematoxylin/eosin)

enlarged three to four times the size of red blood cells. Chromatin is coarse, clumped, or cleared, and nucleoli, when present, may be very large and often irregular in contour. These cytologic features display a spectrum that allows one to grade the nuclear appearance, as discussed previously. Cellular vacuolation with intracytoplasmic lumens or even signet ring cells may be present.[147] The background usually lacks the bipolar naked stromal nuclei seen in benign conditions, unless there is a mixture of benign and malignant cells in the smears. A tumor diathesis may be present. The recognition of the type of carcinoma is critical because of the prognostic implications.[1] Carcinomas NOS should be assigned a nuclear grade, as discussed previously.

Intraductal carcinoma, by definition, remains in the confines of the duct basement membrane and is recognized histologically as cribriform, solid, papillary, micropapillary, and comedo types.[145] Cytopathologists believe it is impossible to distinguish intraductal from invasive carcinoma on the FNA biopsy confidently, even though invasive lesions tend to show greater cellularity and less cohesive cellular groupings.[107,148–151] The cytologic findings of tumor cells (or any epithelial cellularity, for that matter) admixed with fat in the FNA biopsy is not a reliable sign of invasive malignancy because it is seen with high frequency with *in situ* carcinoma, invasive carcinoma, and even benign tumors.[152] Presumably, epithelial groups in intimate association with fat is a phenomenon associated with the aspiration procedure.

Cytomorphologic Features of Duct Carcinoma Not Otherwise Specified

- High cellularity
- Loosely cohesive and single atypical cells
- Syncytial group arrangements, loss of polarity
- No background stromal nuclei
- Tumor diathesis may be present

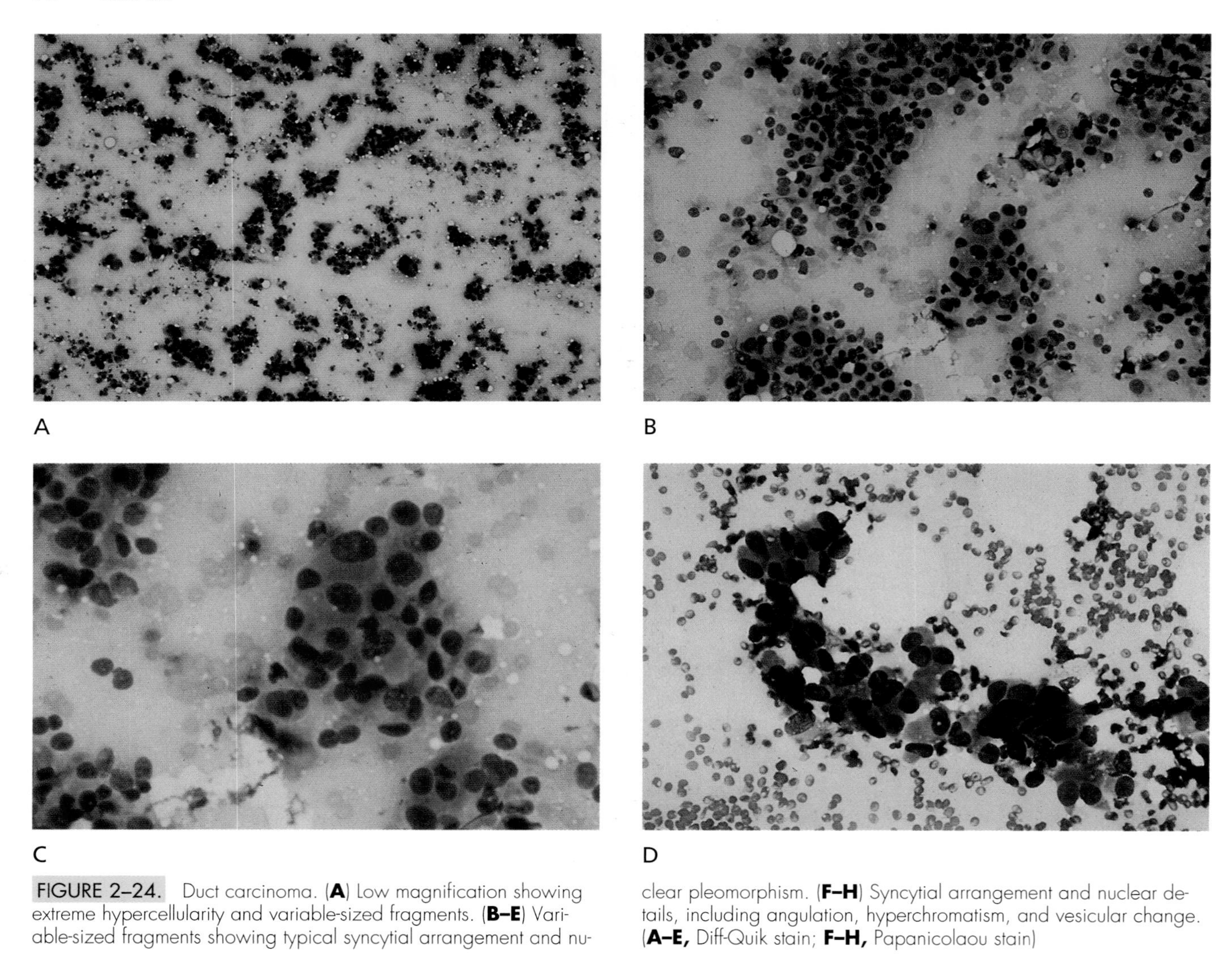

FIGURE 2–24. Duct carcinoma. (**A**) Low magnification showing extreme hypercellularity and variable-sized fragments. (**B–E**) Variable-sized fragments showing typical syncytial arrangement and nuclear pleomorphism. (**F–H**) Syncytial arrangement and nuclear details, including angulation, hyperchromatism, and vesicular change. (**A–E,** Diff-Quik stain; **F–H,** Papanicolaou stain)

DUCT CARCINOMAS OF SPECIAL TYPES

Mucinous (Colloid) Carcinoma

Mucinous or colloid carcinoma typically occurs in older women, usually in the sixth decade, and histologically consists of an abundant mucoid matrix within which are embedded islands of ductal cells in small groups. Cribriform/micropapillary carcinoma is the most commonly associated *in situ* lesion, and consequently the nuclear grade of colloid carcinoma is usually 1 or 2. By virtue of the abundant matrix, the matrix predominates on the FNA biopsy, staining metachromatically magenta on the modified Wright stain, with small glandular clusters of cells with relatively mild nuclear atypia (Fig. 2–25).[153–155] Nuclei may be vesicular with small nucleoli. Sometimes the paucity of cells can be a pitfall for a false-negative diagnosis, but this may be avoided by the finding of individual malignant cells. A false-positive pitfall is the mucocele-like tumors of the breast, but this lesion generally shows small sheets of bland duct cells without atypia in addition to the extracellular mucin.[156]

Cytomorphologic Features of Mucinous Carcinoma of the Breast

- Abundant mucinous matrix
- Scattered ductular aggregates
- Mild atypia with vesicular nuclei

Tubular Carcinoma

This special type of carcinoma carries an excellent prognosis, even with lymph node metastasis.[157] The histologic findings show infiltrating tubular structures

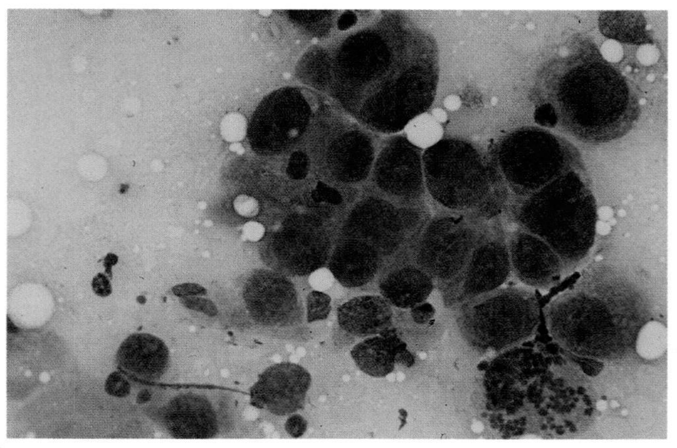

E

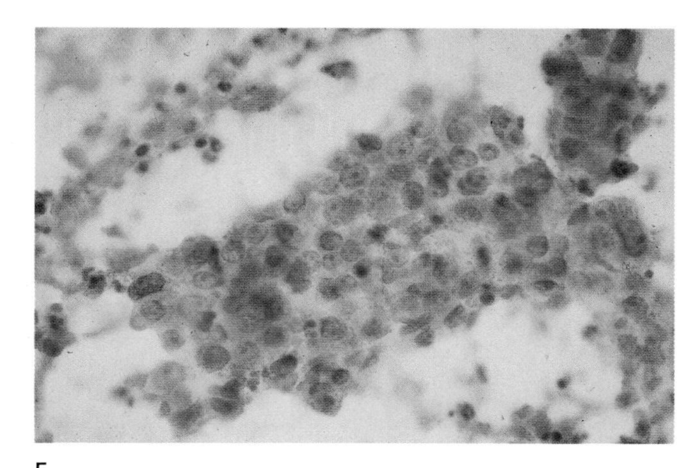

F

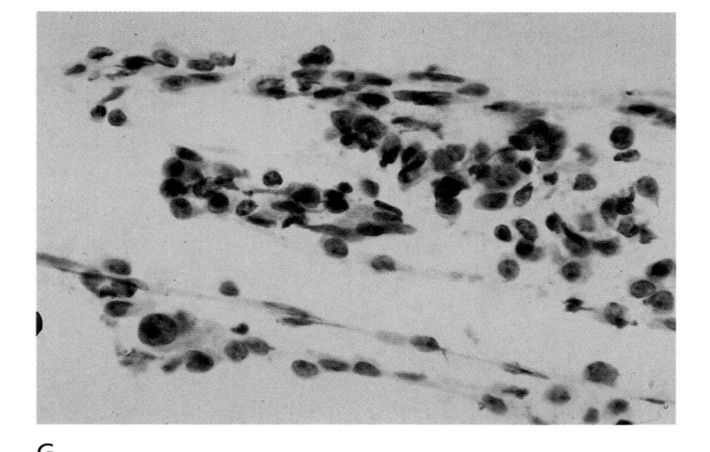

G

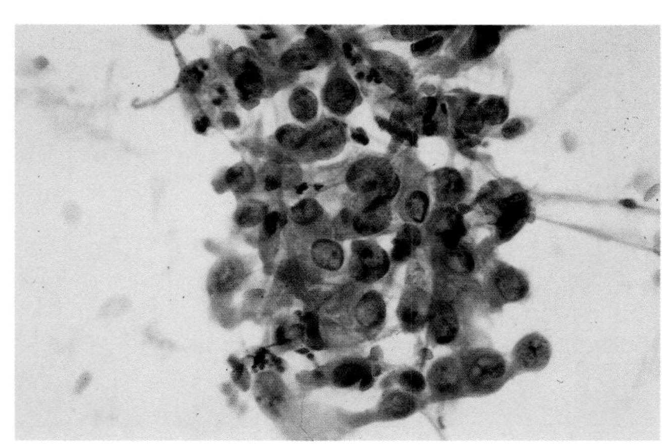

H

FIGURE 2–24. *Continued*

that are angulated or comma-shaped (Fig. 2–26), with small, regular, round, bland nuclei. This is the prototype of nuclear grade 1. These features tend to cause the underdiagnosis of this entity on FNA biopsy,[158–162] such that tissue biopsy is required for definitive diagnosis. Aspiration smears often show moderate cellularity with angulated (comma-like) tight epithelial groups and rarely tubular profiles. Presumably because of the small sizes of tubular carcinomas, stromal nuclei may be present in the smears as a result of the admixture of benign elements.

Cytomorphologic Features of Tubular Carcinoma

- Moderate cellularity
- Tubular profiles
- Angulated tight groups
- Minimal nuclear atypia
- Stromal bipolar nuclei sometimes

Papillary Carcinoma

This entity accounts for less than 1% of breast carcinomas in pure form. The aspiration smears typically show features of a low-grade carcinoma. Columnar cells are arranged in tight three-dimensional clusters or branching configurations in moderately cellular smears. Stromal cells are often present in the background because they are present in the supporting fibrovascular stalk. This combination of features may cause an underdiagnosis of carcinoma. A papillary neoplasm can be suggested, with recommendation for tissue confirmation.

Cytomorphologic of Papillary Carcinoma

- Moderate cellularity
- Three-dimensional papillary groups
- Columnar cells
- Naked nuclei background

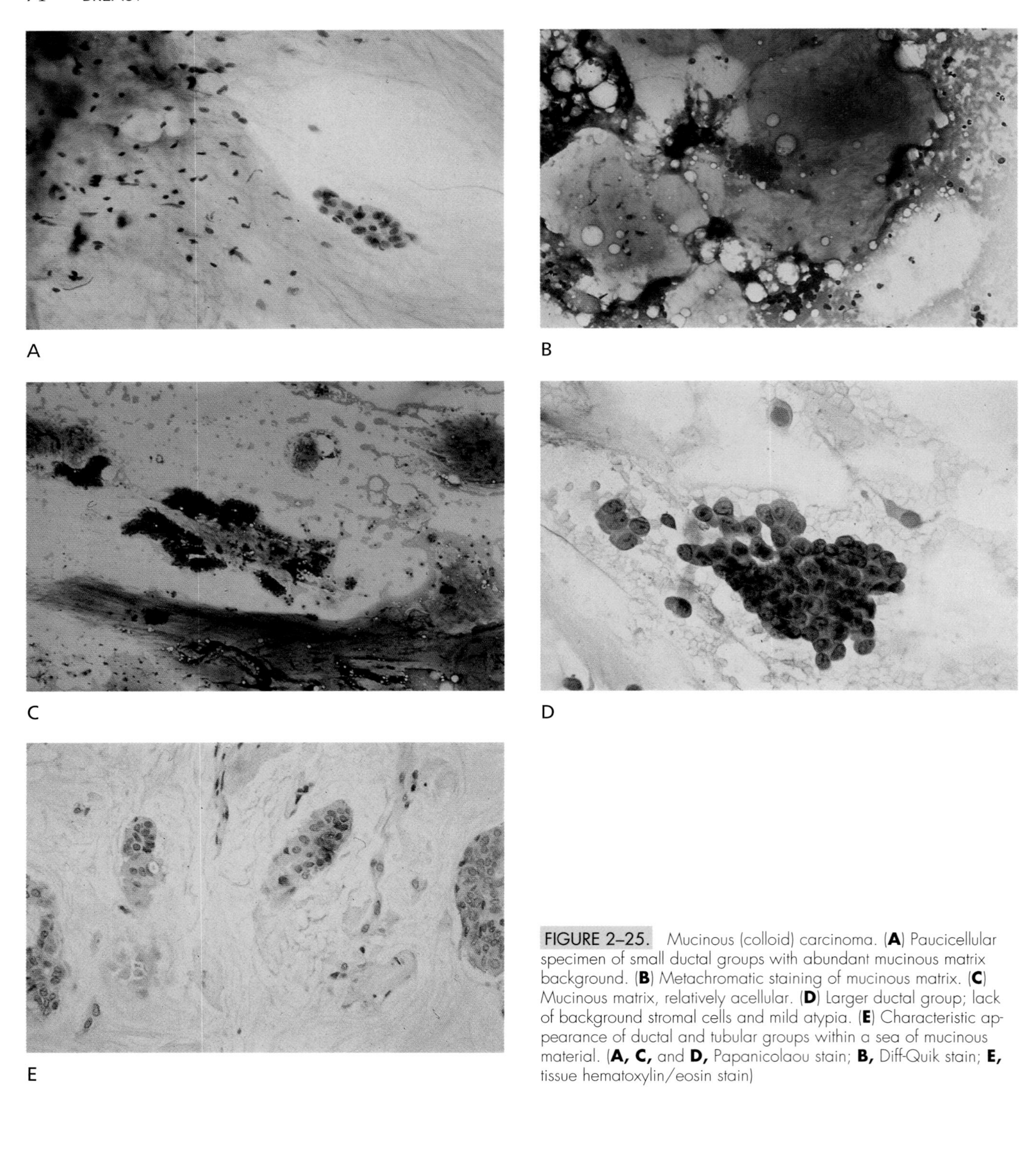

FIGURE 2–25. Mucinous (colloid) carcinoma. (**A**) Paucicellular specimen of small ductal groups with abundant mucinous matrix background. (**B**) Metachromatic staining of mucinous matrix. (**C**) Mucinous matrix, relatively acellular. (**D**) Larger ductal group; lack of background stromal cells and mild atypia. (**E**) Characteristic appearance of ductal and tubular groups within a sea of mucinous material. (**A, C,** and **D,** Papanicolaou stain; **B,** Diff-Quik stain; **E,** tissue hematoxylin/eosin stain)

Medullary Carcinoma

This was previously considered a special type of carcinoma with good prognosis, but recent studies have suggested a prognosis no different than duct carcinoma NOS.[163,164] Typically a lesion of the fifth and sixth decades, the tumor is histologically composed of syncytial sheets of cells with high nuclear grade, necrosis, and a florid lymphoplasmacellular infiltrate. These features are paralleled in the FNA biopsy, with syncytial groups of high-grade malignant cells, individual cells, necrotic diathesis, and many lymphoid cells percolating through the smear (Fig. 2–27). Bizarre pleomorphic cells may be present along with stripped nuclei. Fragments of desmoplastic fibrosis should not be present.

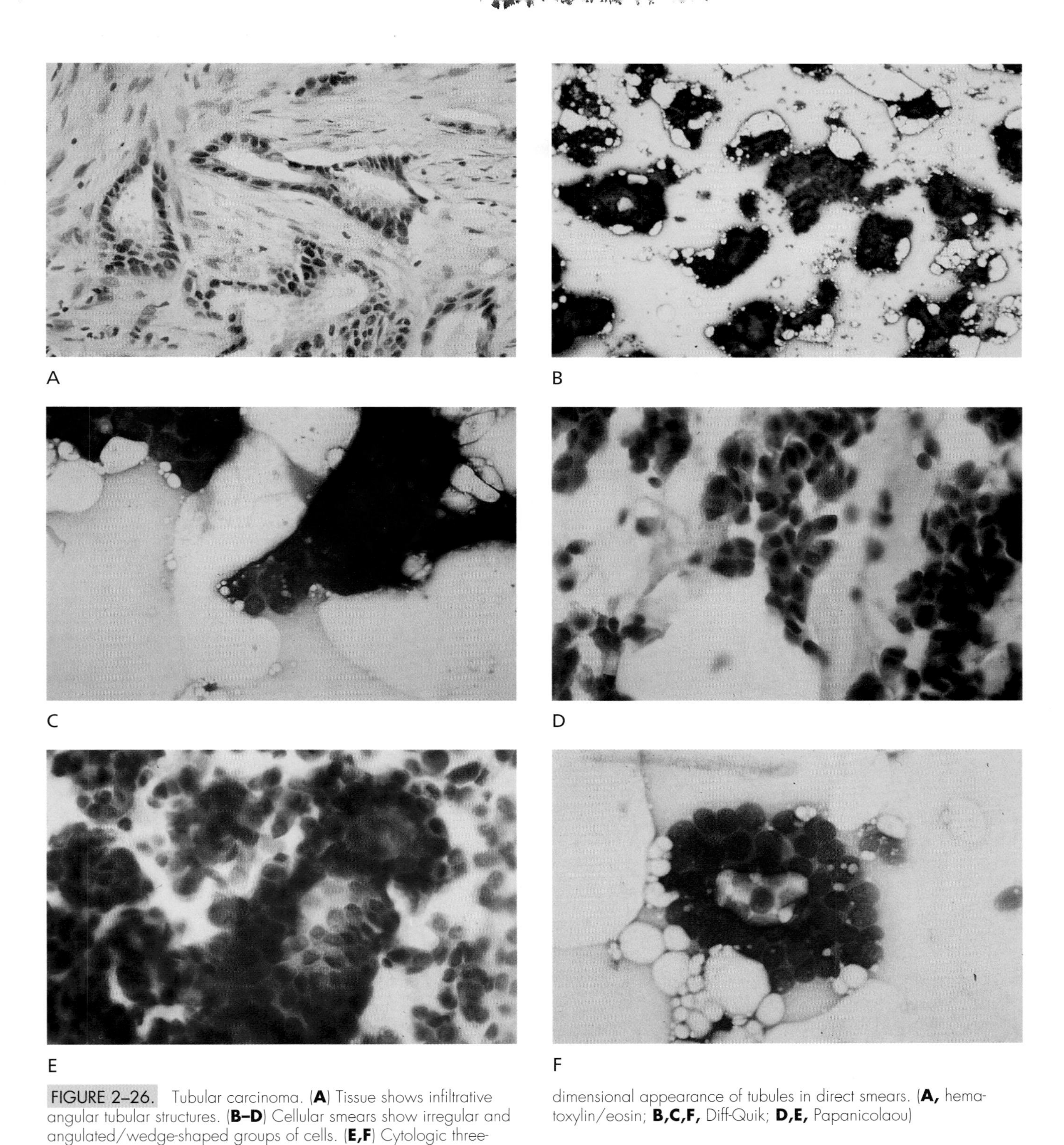

FIGURE 2–26. Tubular carcinoma. (**A**) Tissue shows infiltrative angular tubular structures. (**B–D**) Cellular smears show irregular and angulated/wedge-shaped groups of cells. (**E,F**) Cytologic three-dimensional appearance of tubules in direct smears. (**A,** hematoxylin/eosin; **B,C,F,** Diff-Quik; **D,E,** Papanicolaou)

Cytomorphologic Features of Medullary Carcinoma

- High cellularity
- Syncytial aggregates
- High nuclear grade
- Bizarre and pleomorphic cells
- Necrotic diathesis
- Lymphoplasmacellular infiltrate

Lobular Carcinoma of the Breast

Lobular carcinoma accounts for 5–10% of all breast carcinomas. It is more frequently bilateral or multicentric within the same breast compared to duct carcinomas.[165,166] Histologically, classic lobular carcinoma consists of small monomorphic cells with eccentric nuclei that are mildly hyperchromatic and infiltrate in a single linear filing pattern and targeted pattern

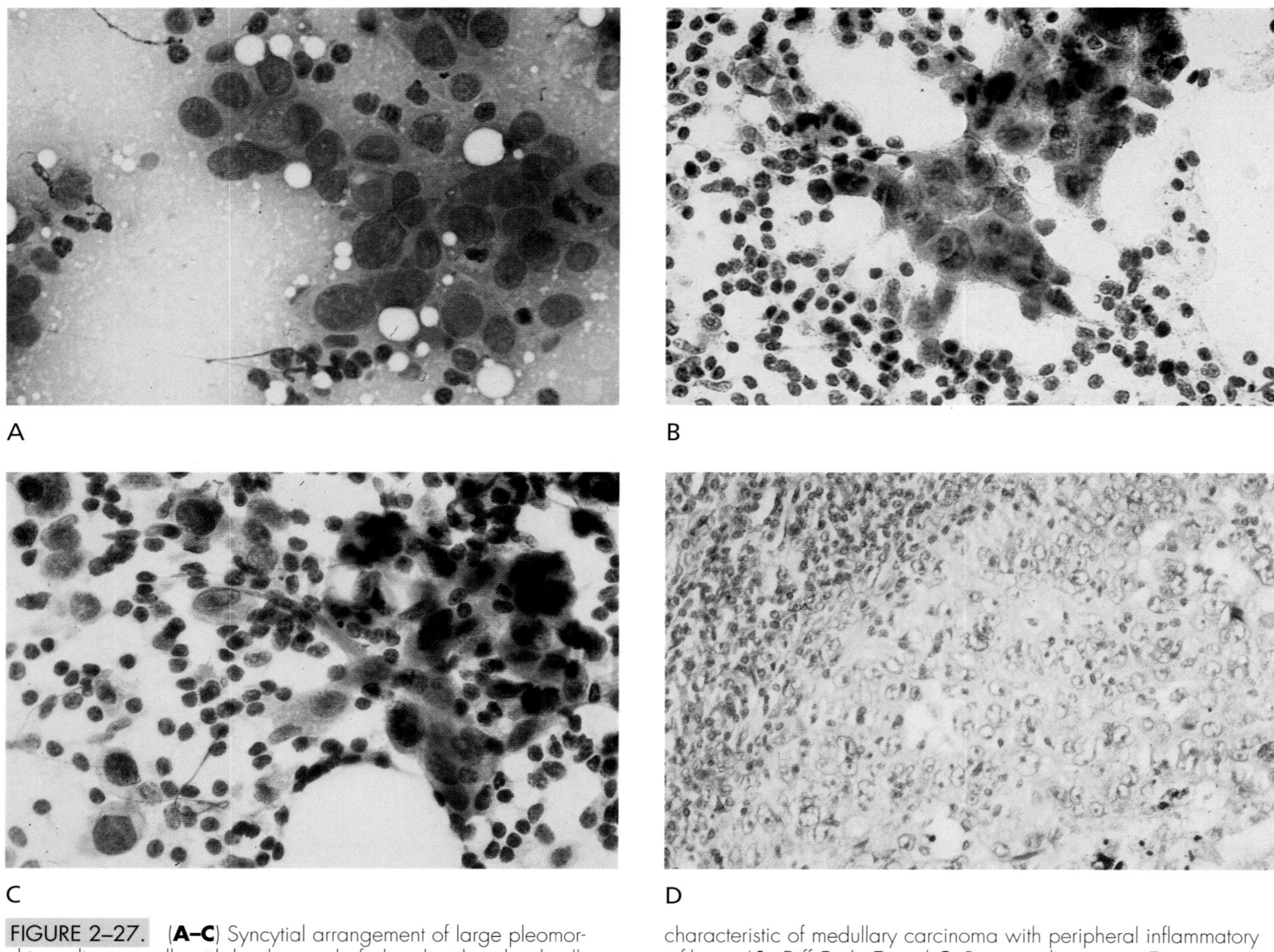

FIGURE 2–27. (**A–C**) Syncytial arrangement of large pleomorphic malignant cells with background of abundant lymphoid cells and percolating through the tumor cells. (**D**) Syncytial arrangement characteristic of medullary carcinoma with peripheral inflammatory infiltrate. (**A,** Diff-Quik; **B** and **C,** Papanicolaou stain; **D,** tissue hematoxylin/eosin stain)

around breast lobules. A pronounced fibrous desmoplastic response is present; for this reason, the cellularity of the FNA biopsy is commonly low. However, the linear arrangement of the cells can be preserved. Cohesiveness is poor and limited to the strands of cells in filing pattern, with many scattered single cells (Fig. 2–28). The N/C ratio is high in these small cells, which have finely granular chromatin. Signet ring cells may be observed, but their presence is not helpful in making the specific diagnosis of lobular carcinoma because they are often seen in duct carcinomas as well.[147] However, the presence of signet ring cells plus a bland population of tumor cells should suggest the diagnosis of lobular carcinoma. Greater than 10% signet ring cells may portend a poor prognosis in patients with lobular carcinoma.[167]

False-negative results are more common when it comes to the diagnosis of lobular carcinoma because of the hypocellular nature of the FNA biopsy specimens and the bland nuclei.[168] Nonclassic lobular carcinomas with solid, alveolar, mixed, or pleomorphic histologic patterns are more likely to be diagnosed as duct carcinomas because the cellularity is greater and the cells are larger for these variants.[168–170] Many cytologic features overlap with those of duct carcinoma.[171] As with duct carcinomas, distinguishing between *in situ* and infiltrating lobular carcinomas is impossible.[107,172]

Cytomorphologic Features of Lobular Carcinoma

- Low to moderate cellularity
- Short cords and individual cells
- Monotonous small cells
- Mild hyperchromasia, high N/C ratio
- Signet ring cells
- No stromal nuclei

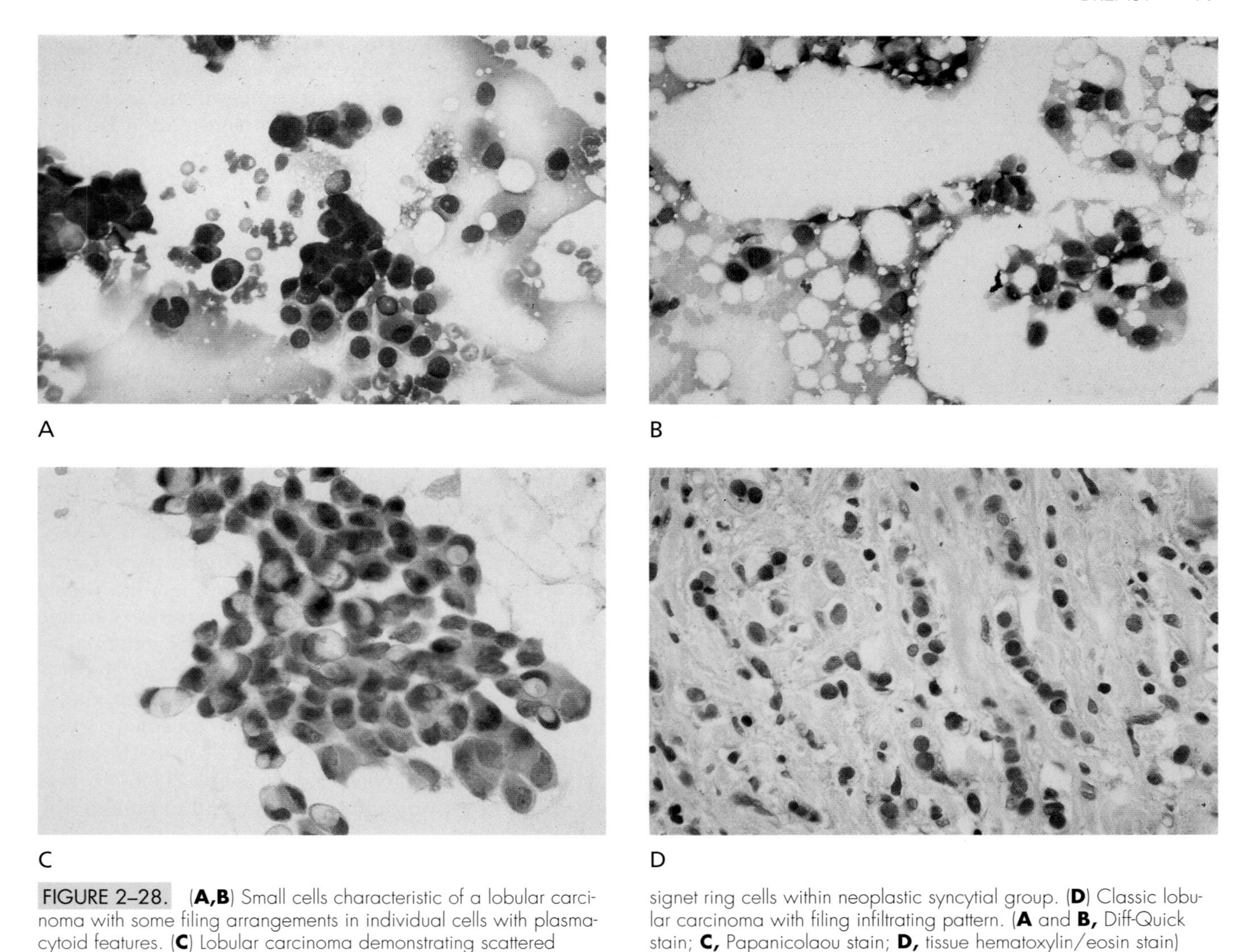

FIGURE 2–28. (**A,B**) Small cells characteristic of a lobular carcinoma with some filing arrangements in individual cells with plasmacytoid features. (**C**) Lobular carcinoma demonstrating scattered signet ring cells within neoplastic syncytial group. (**D**) Classic lobular carcinoma with filing infiltrating pattern. (**A** and **B,** Diff-Quick stain; **C,** Papanicolaou stain; **D,** tissue hematoxylin/eosin stain)

UNCOMMON TYPES OF BREAST CARCINOMA

Inflammatory Carcinoma

Advanced invasive duct carcinoma may involve most of the breast as part of its natural history if left untreated. Patients present with a huge, firm breast that may be focally ulcerated. The skin often shows *peau d'orange* effect because of lymphatic carcinomatosis of the skin. Blind aspirates into the breast may yield the cellular features of duct carcinoma NOS, but because of intense breast engorgement and edema, aspirates may not yield diagnostic cells. The FNA biopsy of inflammatory carcinoma often demonstrates a paucicellular specimen with malignant cells arranged in tight, three-dimensional clusters, as opposed to the frequent finding of individual cells in duct malignancies.[173] The clustering probably reflects the tumor cells obtained from lymphatic spaces.[173]

Apocrine and Secretory Duct Variants

Apocrine carcinoma is considered a variant of duct carcinoma. The cytologic features are those of apocrine cells, containing abundant granular eosinophilic cytoplasm, high N/C ratios, and hyperchromatic nuclei with prominent nucleoli.[174,175] Secretory carcinoma has some morphologic features that are similar to lipid-rich carcinoma and glycogen-rich clear cell carcinoma.[176–179] Apocrine carcinoma may consist of clear cells or mimic granular cell tumors with their coarse granular eosinophilic cytoplasm, referred to as a "myoblastomatoid" variant.[180] The diagnostic distinction can be made with cytokeratin-positive cells in apocrine carcinoma, and S-100-positive, cytokeratin-negative cells in the granular cell tumor.

Apocrine carcinoma can be distinguished from apocrine metaplasias based on marked cellularity, syncytial arrangement of cells, hyperchromasia and irregular nuclei, and necrotic diathesis.[181]

Intracystic Carcinoma of the Breast

As mentioned previously, when nontraumatic bloody fluid is obtained on FNA biopsy, it is imperative to repalpate to rule out an underlying mass. The intracystic carcinoma, though rare, can be the source of an occult carcinoma.[182] The cytologic appearance is that of a low-grade duct carcinoma, arranged in loosely cohesive groups and/or tight papillary clusters.[183]

Adenoid Cystic Carcinoma of the Breast

The classic cytologic appearance of adenoid cystic carcinoma on modified Wright stain consists of small basaloid cells with scant cytoplasm and small nucleoli adherent to clumps of metachromatic magenta-staining matrix, often forming three-dimensional spheres. Single basaloid cells are also present. Mucin-positive luminal material may be present.[184,185]

Squamous Cell Carcinoma of the Breast

This tumor is exceptionally rare as a primary breast tumor, and clinically a metastatic squamous carcinoma to the breast should be excluded. Malignant squamous cells may appear in the FNA biopsy for a variety of reasons, including metaplastic carcinoma, mixed squamous cell and duct carcinoma, and metastatic carcinoma from a distant site.[186–188] As in other anatomic sites, it may be difficult to diagnose the very well-differentiated squamous cell carcinoma because of overlapping cytologic features with epidermal cyst.[189]

METAPLASTIC CARCINOMA OF THE BREAST

Duct carcinomas of metaplastic type present a variety of morphologic types, including carcinoma with osseous, chondroid, spindle cell, and squamous differentiation.[190–195] These tumors account for less than 0.5% of breast tumors and occur in the fifth and sixth decades, similar to most variants of duct carcinoma. Metaplastic carcinomas that produce abundant matrix may have a better prognosis than those that produce pseudosarcomatous transformation.[190,196] The cytologic patterns of metaplastic carcinomas and primary breast sarcomas parallel the findings seen in aspirates of soft-tissue sarcomas.[197,198] Pleomorphic spindle cell patterns may be seen in malignant fibrous histiocytoma of the breast, complete with bizarre giant cells and osteoclast-like giant cells.[199,200] The bizarre giant cells in metaplastic carcinomas may show evidence of epithelial origin.[201] Immunocytochemistry may be a useful tool for determining the metaplastic nature of a carcinoma when sarcomatous transformation is present.[202,203]

METASTATIC MALIGNANCIES IN THE BREAST

Several excellent studies have reported the FNA cytology of metastatic malignancies to the breast,[204–209] with reported clinical occurrences in the range of 0.4–2.7%.[204] This figure includes patients who present with malignancy for the first time with metastasis to the breast.

The most frequent malignancies metastatic to the breast, in decreasing order, are melanoma, lymphoma, lung carcinoma, ovarian carcinoma, soft-tissue sarcomas, and gastrointestinal and genitourinary tumors.[210] Recognition of unusual cytomorphologic patterns of malignancies in the breast is essential to avoid unnecessary overtreatment, such as mastectomy.[211] Ancillary studies such as immunoperoxidase for estrogen/progesterone hormone receptors, cytokeratins, gross cystic disease fluid protein-15, leucocyte common antigen, S-100 protein, HMB-45, and electron microscopy can be most useful in making the correct diagnosis.

Recognition of Metastatic Breast Carcinoma

Breast carcinoma metastatic in distant sites is becoming more of a recognition problem because of the enhanced survival of patients with breast carcinoma. Similarly, the development of secondary malignancies in this patient population is more likely to occur. Lung adenocarcinomas, along with breast carcinomas, are the neoplasms with highest incidence in women. The need to identify the nature of lung lesions in patients with breast carcinoma becomes evident because of the markedly different treatments and prognosis for patients with primary lung adenocarcinoma compared to patients with metastatic breast carcinoma in the lung. Recent studies by Fabian and Dabbs[212] found considerable overlap in the morphology of primary lung adenocarcinoma and metastatic breast carcinoma in the lung. Immunocytochemical studies that may be applied to needle biopsies include CEA-D14 antibody, GCDFP-15 antibody, and ER/PR antibodies. Lung carcinomas are invariably intensely positive with CEA-D14, whereas metastatic breast carcinomas are largely negative and are variably positive for GCDFP-15 and ER/PR receptors.

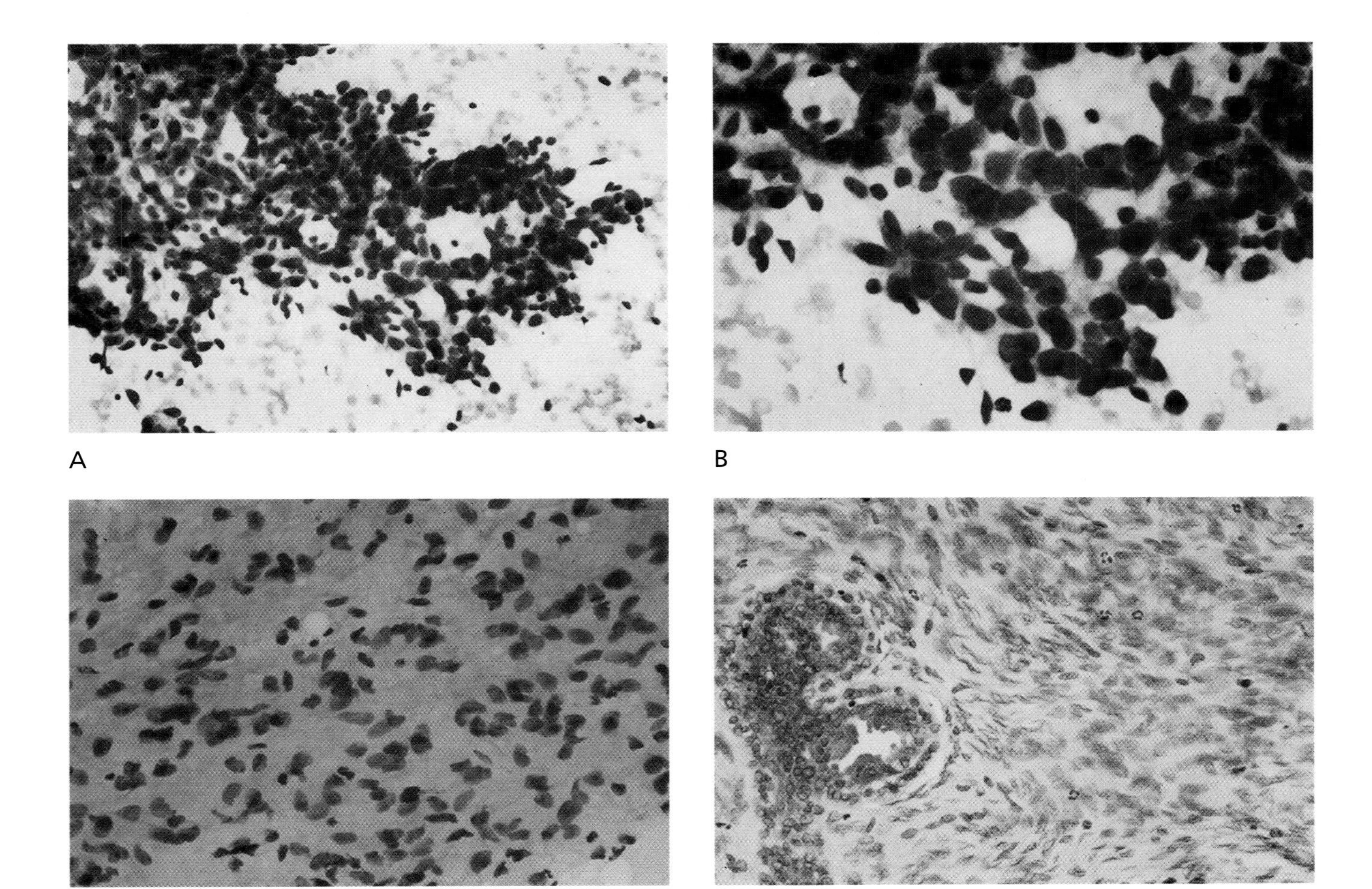

FIGURE 2–29. Phylloides tumor. (**A**) Low magnification demonstrates complex branching pattern of high cellularity with few stromal cells in background. (**B**) Higher magnification demonstrates marked cellular atypia with syncytial-like groupings and hyperchromatic nuclei with high N/C ratios. (**C**) Phylloides fragment characterized by hypercellular fragment of stroma with prominent myxoid matrix. (**D**) Markedly hypercellular stroma with atypical epithelial proliferation. (Tissue hematoxylin/eosin stain)

PHYLLOIDES TUMOR

The biphasic epithelial and stromal proliferation of phylloides type accounts for less than 0.5% of breast tumors. It can be distinguished from fibroadenomas and carcinomas on the FNA biopsy. In general, patients are in the fourth to fifth decade, compared to the younger patients with fibroadenomas.[83]

Cytologic findings may show epithelial ductal groups, islands of squamous epithelium, epithelial atypia, and variable stromal cellularity (Fig. 2–29). The cytologic diagnosis of phylloides tumor versus carcinoma or fibroadenoma rests predominately on the recognition of stromal hypercellularity and degree of stromal cell atypia.[213–216] Hypercellular stromal fragments of phylloides tumor stain metachromatically with modified Wright stains.[217] Phylloides stromal cells are elongated, have irregular nuclear membranes, and show nucleoli, compared to the thin bipolar stromal cells of fibroadenoma.[218] The term "phylloides fragments" has been given to the very cellular stromal fragments of this tumor.[213–216,219] Some investigators suggest that a diagnosis of malignant phylloides tumor can be entertained when smears show extreme hypercellularity with the presence of atypia.[220,221]

Dusenberry and Frable[222] and Silverman[2] found the diagnostic pitfall for phylloides tumor to be the pronounced epithelial proliferation that sometimes occurs, giving a false impression of carcinoma. For this reason, Silverman[2] recommended frozen-section confirmation before definitive treatment when phylloides tumor is suspected cytologically.

Cytomorphologic Features of Phylloides Tumor

- Hypercellular smears with phylloides fragments
- Biphasic epithelial/stromal cells
- Numerous elongated and irregular stromal cells with nucleoli
- Metachromatic stroma
- Epithelial hyperplasia may be marked

REFERENCES

1. The uniform approach to breast fine needle aspiration biopsy. A synopsis. Developed and approved at an NCI-sponsored conference, Bethesda, MD, Sept. 9–10, 1996. Acta Cytol 40:1120–1126, 1996.
2. Silverman JFS. Diagnostic accuracy, cost effectiveness and triage role of fine needle aspiration biopsy in the diagnosis of palpable breast lesions. Breast J 1:3–8, 1995.
3. Lanin DR, Silverman JFS, Pories WJ, et al. Cost effectiveness of fine needle aspiration biopsy of breast. Ann Surg 203:474–480, 1986.
4. Silverman JF, Lanin D, O'Brian K, et al. The triage role of fine needle aspiration biopsy of palpable masses. Acta Cytol 31:731–736, 1987.
5. Palombini L, Fulciniti F, Vetrani A, et al. Fine needle aspiration biopsy of breast masses. A critical analysis of 1956 cases in 8 years (1976–1984). Cancer 61:2273–2277, 1988.
6. Dominguez F, Riera JR, Tojo S, Junco P. Fine needle aspiration of breast masses: an analysis of 1398 patients in a community hospital. Acta Cytol 41:341–350, 1997.
7. Ellis H, Cox PJ. Brest problems in 1000 consecutive referrals to surgical outpatients. Postgrad Med J 60:653, 1984.
8. Feichter GE, Haberthür F, Gobat S, Dalqun P. Breast cytology: statistical analysis and cytohistologic correlations. Acta Cytol 41:327–332, 1997.
9. Gupta RK. Fine needle aspiration cytodiagnosis of recurrent carcinoma of the breast in operative scars. Diagn Cytopathol 16:14–16, 1997.
10. Sirkin W, Auger M, Donat E, et al. Cytospins—an alternative method from fine needle aspiration cytology of the breast: a study of 148 cases. Diagn Cytopathol 13:266–269, 1995.
11. Lee KR, Papillo JL, St. John T, Eyerer GJA. Evaluation of the ThinPrep processor for fine needle aspiration specimens. Acta Cytol 40:895–899, 1996.
12. Catanig S, Boccato P, Bono A, et al. Pneumothorax: a rare complication of fine needle aspiration biopsy of the breast. Acta Cytol 33:140, 1989.
13. Gately CA, Maddox PR, Mansel RE. Pneumothorax: a complication of fine needle aspiration biopsy of the breast. Br Med J 303:627–628, 1991.
14. Stevenson J, James AS, Johnston M, et al. Pneumothorax after fine needle aspiration biopsy of breast. Br Med J 303:924, 1991.
15. Stockdale FE. Questions and answers: mammography, needle biopsy and tumor spread. JAMA 272:895–896, 1994.
16. Harter LP, Curtis JS, Ponto G, et al. Malignant seeding of the needle track during stereotaxic core needle breast biopsy. Radiology 185:713–714, 1992.
17. Taxin A, Tartter PI, Zappetti D. Breast cancer diagnosis by fine needle aspiration and excisional biopsy: recurrence and survival. Acta Cytol 41:302–306, 1997.
18. Youngson BJ, Liberman L, Rosen PP. Displacement of carcinomatous epithelium surgical breast specimen following stereotaxic core biopsy. Am J Clin Pathol 103:598–602, 1995.
19. Lee KC, Chan JKC, Ho LC. Histologic changes in the breast after fine needle aspiration. Am J Surg Pathol 18:1039–1047, 1994.
20. Krasovec M, Golouh R, Avesperg M. Tissue damage after fine needle aspiration. Acta Cytol 36:456–457, 1992.
21. Youngson BJ, Cranor M, Rosen PP. Epithelial displacement in surgical specimens following needling procedures. Am J Surg Pathol 18:896–903, 1994.
22. Saarela AO, Kiviniemi HO, Rissanen TJ, et al. Nonpalpable breast lesions: pathologic correlation of ultrasonographically guided fine-needle aspiration biopsy. J Ultrasound Med 15:549–553, 1996.
23. Mitaick JS, Vazquez MF, Pressman PI, et al. Stereotactic fine needle aspiration biopsy for the evaluation of nonpalpable breast lesions: report of an experience based on 2988 cases. Ann Surg Oncol 3:185–191, 1996.
24. Layfield L, Mooney EE, Glasgow B, et al. What constitutes an adequate smear in fine needle aspiration cytology of the breast? CA Cytol 81(1):16–21, 1997.
25. Layfield LJ, Glasgow BI, Cramer TJ. Fine needle aspiration in the management of breast masses. Pathol Ann 24(pll):43–62, 1989.
26. Vural G, Hagmar B, Lilleng R. A one-year audit of fine needle aspiration cytology of breast lesions. Factors affecting adequacy and a review of delayed carcinoma diagnosis. Acta Cytol 39:1233–1236, 1995.
27. Raab SS, Bottles K, Cohen MB. Technology assessment in anatomic pathology: an illustration of test evaluation using fine needle aspiration biopsy. Arch Pathol Lab Med 118:1173–1180, 1994.
28. Stanley MW, Abele J, Kline T, et al. What constitutes adequate sampling of palpable breast lesions that appear benign by clinical and mammography criteria? Diagn Cytopathol 13:473–485, 1995.
29. Sneige N, Staerrkel GA, Conway ND, et al. A plea for uniform terminology and reporting of breast fine needle aspiration. M. D. Anderson Center proposal. Acta Cytol 38:971–972, 1994.
30. Sneige N. Should specimen adequacy be determined by the opinion of the aspirator or by the cells on the slides? CA Cytol 81(1):3–5, 1997.
31. Dodd LG. Cytologically low-grade malignancies; an important interpretive pitfall responsible for false-negative diagnoses in fine needle aspiration of breast. Diagn Cytopathol 15:250–259, 1996.
32. Pilott S, Rilke F, Delpiano C, et al. Problems in fine needle aspiration biopsy of clinically or mammographically uncertain breast tumors. Tumori 68:407–412, 1982.
33. Beidrzyck T, Dabsku M, Sikorowa L, et al. On cytologic vagaries in the diagnosis of breast tumors. Tumori 66:191–196, 1980.
34. Bell DA, Hajdu SI, Urban JA, et al. Role of aspiration cytology in the diagnosis and management of mammary lesions in office practice. Cancer 51:1182–1189, 1983.
35. Feldman PS, Covell JL. Breast and lung in fine needle aspiration cytology and its clinical application. Chicago: ASCP Press, 27–43, 1985.
36. Kline TS. Handbook of fine needle aspiration cytology. St Louis: CV Mosby, 1988.
37. Deschens L, Fabia J, Meisels A, et al. Fine needle aspiration in the management of palpable breast lesions. Can J Surg 21:417–419, 1978.
38. Cornillot M, Verhaeghe M, Cappelaere P, et al. Place de la cytologie par ponction dans le diagnostic des tumeurs du sien. Presse Med 79:1813–1819, 1971.
39. Kline TS, Joshi LP, Neal HS. Fine needle aspiration of the breast. Diagnoses and pitfalls. A review of 3545 cases. Cancer 44:1458–1464, 1979.
40. Pilott S, Rilke F, Delpiano C, et al. Problems in fine needle aspiration biopsy of clinically or mammographically uncertain breast tumors. Tumori 68:407–412, 1982.
41. Zajdela A, Ghossein NA, Pilleron JP, et al. The value of aspiration cytology in the diagnosis of breast cancer. Experience at the Foundation Curie. Cancer 35:499–506, 1975.
42. Wollenberg NJ, Caya JG, Clowry LJ. Fine needle aspiration cytology of the breast. A review of 321 cases with statistical evaluation. Acta Cytol 98:155–160, 1985.
43. Thomas PA, Vazquez MF, Waisman J. Comparison of fine needle aspiration and frozen sections of palpable mammary lesions. Mod Pathol 3:570–574, 1990.
44. Layfield LJ. Can fine needle aspiration replace open biopsy in the diagnosis of palpable breast lesions? Am J Clin Pathol 98:145–147, 1992.
45. Skoomal SM, Florell SR, Bydalek MK, et al. Malpractice protection: communication of diagnostic uncertainty. Diagn Cytopathol 14:385–389, 1996.
46. Masood S. Prognostic factors in breast cancer: Use of cytologic specimens. Diagn Cytopathol 13:388–395, 1995.

47. Robinson IA, McKee G, Kissin MW. Typing and grading breast carcinoma of fine needle aspiration: is this clinically useful information? Diagn Cytopathol 13:260–265, 1995.
48. Dabbs DJ. Role of nuclear grading of breast carcinomas in fine needle aspiration specimens. Acta Cytol 38:923–926, 1994.
49. Dabbs DJ, Silverman JF. Prognostic factors from the fine needle aspiration: breast carcinoma nuclear grade. Diagn Cytopathol 10:203–208, 1994.
50. Howell LP, Gandour-Edwards R, O'Sullivan D. Application of the Scarf-Bloom-Richardson tumor grading systems to fine needle aspirates of the breast. Am J Clin Pathol 101:262–265, 1994.
51. Page DL, Johnson JE, Dupont WD. Probabilistic approach to the reporting of fine needle aspiration cytology of the breast. CA Cytol 81:6–9, 1997.
52. Fisher ER, Redmond C, Fisher B. Histologic grading of breast cancer. Pathol Ann 15:239–251, 1980.
53. National Institutes of Health Consensus Development Conference Statement on the Treatment of Early Stage Breast Cancer. Oncology 5:120–124, 1991.
54. Fisher ER, Sass R, Fisher B. NSABP investigators. Pathologic findings from the National Surgical Adjuvant Project for Breast Cancers (Protocol 4). Cancer 50:712–723, 1984.
55. Fisher ER, Redmond C, Fisher B. Pathologic findings from the National Surgical Adjuvant Breast Project (Protocol 4). VI. Discriminants for five-year treatment failure. Cancer 46:908–918, 1980.
56. Fisher ER, Redmond C, Fisher B, et al. NSABP investigators. Pathologic findings from the National Surgical Adjuvant Breast and Bowel Projects. Prognostic discriminants for 8-year survival for node-negative invasive cancer patients. Cancer 65:2121–2128, 1990.
57. Doussal V, Tubiana M, Friedman S, et al. Prognostic value of histologic grade nuclear components of Scarf-Bloom-Richardson. Cancer 64:1914–1921, 1989.
58. Dabbs DJ. Ductal carcinoma of breast: nuclear grade as a predictor of S-phase function. Hum Pathol 24:652–656, 1993.
59. Dabbs DJ. Correlations of morphology, proliferation indices and oncogene activation in ductal breast carcinomas: nuclear grade, S-phase fraction, proliferating cell nuclear antigen, p53, EGFR and C-erb-B2. Mod Pathol 8:637–642, 1995.
60. Cajulis RS, Hessel G, Hyang S, et al. Simplified nuclear grading of fine needle aspiration of breast carcinoma: concordance with corresponding histologic nuclear grading and flow cytometric data. Diagn Cytopathol 11:124–130, 1994.
61. Moriquand J, Gozlan-Fior M, Villemain D, et al. Value of cytoprognostic classification in breast carcinoma. J Clin Pathol 39:489–496, 1986.
62. Sneige N. Nuclear grading in fine needle aspirates of the breast. Cytopathol Ann 161–172, 1992.
63. Tuczek H-V, Fritz P, Schwarzmann P, et al. Breast carcinoma. Correlations between visual diagnostic criteria for histologic gradings and features of image analysis. Anal Quant Cytol Histol 18:481–493, 1996.
64. Kamel OW, Hendrickson MR, Kempson RL. Breast biopsies: the content of the surgical pathology report. In: The mammographically directed biopsy. Philadelphia: Hanley and Belfos, 161–180, 1992.
65. Cajulis RS, Hessel RG, Frias-Hidvegi D, Yu GH. Cytologic grading of fine needle aspirates of breast carcinoma by private practice pathologists. Acta Cytol 41:313–320, 1997.
66. Cajulis RS, Hessel RG, Hwang S, et al. Simplified nuclear grading of fine needle aspirates of breast carcinoma: concordance with corresponding histologic nuclear grading and flow cytometric data. Diagn Cytopathol 11:124–130, 1994.
67. Hortobagy IGN, Ameş FC, Bozdar AU, et al. Management of stage III primary cancer with primary chemotherapy, surgery and radiation therapy. Cancer 62:2507–2516, 1988.
68. Sharkey FE, Addington SL, Fowler LJ, et al. Effects of preoperative chemotherapy on the morphology of resectable breast carcinoma. Mod Pathol 9:893–900, 1996.
69. Corkhill ME, Katz R. Immunocytochemical staining of C-erb-B2 oncogene in fine needle aspirations of breast carcinoma: a comparison with tissue sections and other breast prognostic factors. Diagn Cytopathol 11:250–254, 1994.
70. Dawson AE, Norton JA, Weinberg DJ. Comparative assessment of proliferation and DNA content in breast carcinoma by image analysis and flow cytometry. Am J Pathol 136: 1115–1124, 1990.
71. Sinha SK, Singh UR, Bhatia A. C-erb-B2 oncoprotein expression. Correlation with the Ki-67 labeling index and AgNOR counts in breast carcinoma on fine needle aspiration cytology. Acta Cytol 40:1217–1220, 1996.
72. Bozzetti C, Nizzol R, Naldi N, et al. Nuclear grading and flow cytometric DNA pattern in fine needle aspirates of primary breast cancer. Diagn Cytopathol 15:116–120, 1996.
73. Schmitt FC, Bento MJ, Amendoeira I. Estimation of estrogen receptor content in fine needle aspirations from breast cancer using the monoclonal antibody ID5 and microwave oven processes: correlation with paraffin-embedded and frozen-section determinations. Diagn Cytopathol 13:347–351, 1995.
74. Colecchia M, Frigo B, Zucchi A, et al. p53 protein expression in fine needle aspirations of breast cancer: an immunocytochemical assay for identifying high-grade ductal carcinoma. Diagn Cytopathol 13:128–132, 1995.
75. Dalquen P, Baschiera B, Chafford R, et al. MIB-1 (I-67) immunostaining in breast cancer cells in cytologic smears. Acta Cytol 41:229–237, 1997.
76. Kuenen-Boumeester V, Kwast THVD, Laarhaven HAJV, et al. Ki-67 staining in histological subtypes of breast carcinoma and fine needle aspiration smears. J Clin Pathol 44:208–210, 1991.
77. Martin AW, Davey DD. Comparison of immunoreactivity of *neu* oncoprotein in fine needle aspirations and paraffin-embedded materials. Diagn Cytopathol 12:142–145, 1995.
78. Masood S. Use of monoclonal antibody for assessment of estrogen receptor content in fine needle aspiration biopsy specimen from patients with breast cancer. Arch Pathol Lab Med 113:26–30, 1989.
79. Masood S, Dee S, Goldstein JD. Immunocytochemical analysis of progesterone receptors in breast cancer. Am J Clin Pathol 96:59–63, 1991.
80. Kuenen-Boumeester V, Blood DI, van der Kwast TH, et al. Ki-67 staining and histologic subtypes of breast carcinoma and fine needle aspiration smears. J Clin Pathol 44:208–210, 1991.
81. Felosi G, Bresaola E, Rodella S, Manfrin E, et al. Expression of proliferating cell nuclear antigen, Ki-67 antigen, estrogen receptor protein, and tumor suppressor p53 gene in cytologic samples of breast cancer: an immunohistochemical study with clinical, pathobiological, and histologic correlations. Diagn Cytopathol 11:131–140, 1994.
82. Alexiev BA. Localization of p53 and proliferating cell nuclear antigen in fine needle aspirates of benign and primary malignancy of the breast: an immunocytochemical study using supersensitive monoclonal antibodies and the biotin-streptavidin-amplified method. Diagn Cytopathol 15:277–281, 1994.
83. Azzopardi JG. Problems in breast pathology. In: Bennington JL (ed): Major problems in pathology, vol. II. Philadelphia: WB Saunders, 1979.
84. Haagensen CD. Diseases of the breast. Philadelphia: WB Saunders, 1986.
85. Silverman JF. FNA cytology of infectious and inflammatory diseases and other nonneoplastic disorders. In: Kline T (ed): Guides to clinical aspiration biopsy series. New York: Igaku-Shoin, 1991.
86. Jayaram G. Cytomorphology in tuberculous mastitis. A report of nine cases with FNAC. Acta Cytol 29:974–978, 1985.
87. Rosen PP, Oberman HA. Tumors of the mammary gland. In: Atlas of tumor pathology, 3rd series. Washington DC: Armed Forces Institute of Pathology, 1993.
88. Koss LG, Woyke J, Olszewski W. Aspiration biopsy: cytologic interpretation and histologic bases pp. 54–104. New York: Igaku-Shoin, 1984.

89. Dabbs DJ. Breast foam cells: macrophage immunophenotype. Hum Pathol 24:977–981, 1993.
90. Ciatto S, Cariaggi P, Bolgaresi P. The value of routine cytologic examination of breast fluids. Acta Cytol 31:301–304, 1987.
91. Taxeda EM, Suzuki M, Suto Y, et al. Aspiration cytology of breast cysts. Acta Cytol 26:37–43, 1982.
92. Silverman JF, Lanin DR, Medheim D, et al. Subareolar abscess of the breast: the role of fine-needle aspiration biopsy in the diagnosis and management. Contemp Surg 28:45–48, 1986.
93. Silverman F, Lanin DR. Unverferth M, et al. Fine needle aspiration cytology of subareolar abscess of the breast: spectrum of cytomorphologic findings and potential diagnostic pitfalls. Acta Cytol 30:413–419, 1986.
94. Kline TS, Joshi CP, Neal HS. Fine needle aspiration of the breast: diagnoses and pitfalls. A review of 3545 cases. Cancer 44:1458–1464, 1979.
95. Oertel YC, Galblum, CI. Fine needle aspiration of the breast. Diagnostic criteria. Pathol Ann (Part I) 18:375–407, 1983.
96. Hutter RVP. Goodbye to "fibrocystic disease." N Engl J Med 312:179, 1985.
97. Dabbs DJ. Mammary ductal foam cells: macrophage immunophenotype. Hum Pathol 24:977–981, 1993.
98. Wang NP, Wan BC, Skelly VE, et al. Monoclonal antibodies to novel myoepithelium associated proteins can distinguish between benign and malignant lesions of the breast. Mod Pathol 9:26A, 1996.
99. Russ JE, Winchester DP, Scanlon EF, et al. Cytologic findings of aspiration of tumors of the breast. Surg Gynecol Obstet 146:407–411, 1978.
100. Schondorf H [translated by Scheider V]. Aspiration cytology of the breast (p. 17). Philadelphia: WB Saunders, 1978.
101. Consensus Statement by the Cancer Committee of the College of American Pathologists: Is fibrocystic disease of the breast precancerous? Arch Pathol Lab Med 110:171, 1986.
102. Bodian CA. Prognostic significance by proliferative breast disease. Cancer 71:3896, 1993.
103. Rosen PP. Editorial. Cancer 71:3894, 1993.
104. Kreuzer G. Aspiration biopsy cytology in proliferating benign mammary dysplasia. Acta Cytol 22:128–132, 1978.
105. Sneige N, Staerkel GA. Fine needle aspiration cytology of ductal hyperplasia with and without atypia and ductal carcinoma *in situ*. Hum Pathol 25:485–492, 1994.
106. Masood S, Frykberg E, McZellan GL, et al. Cytologic differentiation between proliferative and nonproliferative breast disease in mammographically guided fine needle aspirates. Diagn Cytopathol 15:581–590, 1996.
107. Silverman JF, Masood S, Ductman BS. Can fine needle aspiration biopsy separate atypical hyperplasia, carcinoma *in situ* and invasive carcinoma of the breast? Cytomorphologic criteria and limitations in diagnosis. Diagn Cytopathol 9:713–728, 1993.
108. Thomas PA, Cangiarella J, Raab SS, et al. Fine needle aspiration biopsy of proliferative breast disease. Hum Pathol 6:130–136, 1995.
109. Thomas PA, Raab SS, Cohen MB. Is the fine needle aspiration biopsy diagnosis of proliferative breast disease feasible? Diagn Cytopathol 11:301–306, 1994.
110. Midulla C, Cenci M, Delorio P, et al. The value of fine needle aspiration cytology in the diagnosis of breast proliferative lesions. Anticancer Res 15:2619–2622, 1995.
111. Kline TS. Handbook of fine needle aspiration cytology. St. Louis: CV Mosby, 1988.
112. Bottles K, Chan JS, Holley EA, et al. Cytologic criteria for fibroadenoma: a stepwise logistic regression analysis. Am J Clin Pathol 89:707–713, 1988.
113. Al-Kaisi N. The spectrum of two "gray zones" in breast cytology. A review of 186 cases of a typical and suspicious cytology. Acta Cytol 38:898–908, 1994.
114. Kline TS. Masquerades of malignancy. A review of 4241 aspirates from the breast. Acta Cytol 25:263–266, 1981.
115. Stanley MW, Tani EM, Skoog L. Fine needle aspiration of fibroadenomas of the breast with atypia: A spectrum including cases that cytologically mimic carcinoma. Diagn Cytopathol 6:375–382, 1990.
116. Dejmek A, Lindholm K. Frequency of cytologic features in fine needle aspirates from histologically and cytologically diagnosed fibroadenomas. Acta Cytol 35:695–699, 1991.
117. Silverman JF, Dabbs DJ, Gilbert CF. Adenosis tumor of the breast: cytologic, histologic, immunocytochemical and ultrastructural observations. Acta Cytol 33:181–187, 1989.
118. Zajicek J, Caspersson T, Jakobsson P, et al. Cytologic diagnosis of mammary tumors from aspiration biopsy smears. Comparison of cytologic and histologic findings in 2111 lesions and diagnostic use of cytophotometry. Acta Cytol 14: 370–376, 1970.
119. Mulvany N, Lowhagen T, Skoog L. Fine needle aspiration cytology of tubular adenoma of the breast. A report of two cases. Acta Cytol 38:961–964, 1994.
120. Stormby N, Bondeson L. Adenoma of the nipple. Acta Cytol 28:729–732, 1984.
121. Jensen ML, Johansen P, Noer H, et al. Ductal adenoma of the breast: cytological features of six cases. Diagn Cytopathol 10:143–145, 1994.
122. Pinto RG, Mandreker S. Fine needle aspiration cytology of adenoma of the nipple. A case report. Acta Cytol 40:789–791 1996.
123. Bosch MMC, Boon ME. Fine needle aspiration cytology of an eccrine spiradenoma of the breast. Diagnosis made by a holistic approach. Diagn Cytopathol 8:366–368, 1992.
124. Sood N, Jayaram G. Cytology of papillary adenoma of the nipple: a case diagnosed on fine needle aspiration. Diagn Cytopathol 6:345–348, 1989.
125. Kumar N, Verma K. Clear cell hidradenoma simulating breast carcinoma: a diagnostic pitfall in fine needle aspiration of breast. Diagn Cytopathol 15:70–72, 1996.
126. Kanter MH, Sedeghi M. Pleomorphic adenoma of the breast: cytology of fine-needle aspiration and its differential diagnosis. Diagn Cytopathol 9:555–558, 1993.
127. Bottles K, Taylor RN. Diagnosis of breast masses in pregnant and lactating women by aspiration cytology. Obstet Gynecol 66:765–785, 1985.
128. Finley JL, Silverman JF, Lanin DR. Fine needle aspiration cytology of breast masses in pregnant and lactating women. Diagn Cytopathol 5:225–259, 1989.
129. Grenko RT, Lee KP, Lee KR. Fine needle aspiration cytology of lactating adenoma of the breast. A comparative light microscopic and morphometric study. Acta Cytol 34:21–26, 1998.
130. Gupta RK, McHutchison AGR, Dowle GS, et al. Fine needle aspiration cytodiagnosis of breast masses in pregnant and lactating women and its impact on management. Diagn Cytopathol 9:156–159, 1993.
131. Maygarden SJ, McCall JB, Frable WJ. Fine needle aspiration of breast lesions in women aged 30 and under. Acta Cytol 35:687–694, 1991.
132. Novotny DB, Maygarden SJ, Shermer RW, et al. Fine needle aspiration of benign and malignant masses associated with pregnancy. Acta Cytol 35:676–686, 1991.
133. Silverman JP. In: Bibbo M (ed): Comprehensive cytopathology (p. 749). Philadelphia: WB Saunders, 1997.
134. Garcia-Rostan y Perez GM, Ciriza LA, Demiguel Medina C, et al. Adenomyoepithelioma of the breast: a tumor frequently misdiagnosed in radiological and cytological evaluation. Breast 3:90–96, 1997.
135. Stanley MW, Tani EM, Skoog L. Fine needle aspiration of the breast with atypia: a spectrum including cases that cytologically mimic carcinoma. Diagn Cytopathol 6:375–382, 1990.
136. Tamai M, Nomura K, Hiyama H. Aspiration cytology of malignant intraductal myoepithelioma of the breast. A case report. Acta Cytol 38:435–440, 1994.

137. Loose JH, Pachefsky AS, Hollander J, et al. Adenomyoepithelioma of the breast. A spectrum of biologic behavior. Am J Surg Pathol 16:868–876, 1992.
138. Sirgi KE, Sneige N, Fanning TV, et al. Fine needle aspirates of granular cell lesions of the breast: report of three cases with emphasis on differential diagnosis and utility of immunostaining for CD68 (kp1). Diagn Cytopathol 15:403–408, 1996.
139. Silverman JF, Dabbs DJ, Norris HT, et al. Localized primary (AL) amyloid tumor of the breast: cytologic, histologic, immunocytochemical and ultrastructural observations. Am J Surg Pathol 10:539–545, 1986.
140. Lew W, Seymour A. Primary amyloid tumor of the breast: case report and literature review. Acta Cytol 29:7–11, 1985.
141. Das DK, Junaid TA, Mathews SP, et al. Fine needle aspiration cytology diagnosis of male breast lesions. A study of 185 cases. Acta Cytol 39:870–876, 1995.
142. Bhagyl P, Kline TS. The male breast and malignant neoplasms: diagnosis by aspiration biopsy cytology. Cancer 65:2338–2341, 1990.
143. Russin VL, Lachowicz C, Kline TS. Male breast lesions: gynecomastia and its distinction from carcinoma by aspiration biopsy cytology. Diagn Cytopathol 5:243–247, 1989.
144. Page DL, Anderson TJ. Diagnostic histopathology of the breast. New York: Churchill Livingstone, 1987.
145. Fisher ER, Gregorio RM, Fisher B, et al. The pathology of invasive breast cancer. A syllabus derived from the findings of the National Surgical Adjunct Breast Project (Protocol No. 4). Cancer 36:1–85, 1975.
146. Bloom HJG, Richardson WW. Histological grading and prognosis in breast cancer. A study of 1409 cases of which 359 have been followed for 15 years. Br J Cancer 11:359–377, 1957.
147. Sethi S, Cajulis RS, Gokaslan ST, et al. Diagnostic significance of signet ring cells in fine needle aspirates of the breast. Diagn Cytopathol 16:117–121, 1997.
148. Sneige N, Singletary SE. Fine needle aspiration of the breast: diagnostic problems and approaches to surgical management. Pathol Ann (Part I) 29:281–301, 1994.
149. Sneige N, Staerkel GA, Caraway NP, et al. A plea for uniform terminology and reporting of breast fine needle aspiration. M. D. Anderson Cancer proposal. Acta Cytol 38:971–972, 1994.
150. Wang HH, Ducatman BS, Eick D. Comparative features of ductal carcinoma *in situ* and infiltrating carcinoma of the breast on fine needle aspiration biopsy. Am J Clin Pathol 92:736–748, 1989.
151. Theocharous C, Greenberg ML. Cytologic features of ductal carcinoma *in situ*. Diagn Cytopathol 15:367–373, 1996.
152. Maygarden SJ, Brock MS, Novotny DB. Are epithelial cells in fat or connective tissue a reliable indicator of tumor invasion in fine needle aspiration of the breast? Diagn Cytopathol 16:137–142, 1997.
153. Fanning TV, Sneige N, Staerkel G. Mucinous breast lesions. Fine needle aspiration findings. Acta Cytol 34:754, 1990.
154. Gupta RK, McHutchinson AGR, Simpson JS, et al. Value of fine needle aspiration cytology of the breast, with emphasis on the cytodiagnoses of colloid carcinoma. Acta Cytol 34:703–709, 1991.
155. Stanley MW, Tani EM, Skoog L. Mucinous breast carcinoma and mixed mucinous infiltrating ductal carcinoma. A comparative cytologic study. Diagn Cytopathol 5:134–138, 1989.
156. Bhargav AV, Miller T, Cohen MB. Mucocele-like tumors of the breast: cytologic findings in two cases. Am J Clin Pathol 45:875–877, 1991.
157. McDivitt RW, Boyce W, Gersell D. Tubular carcinoma of the breast. Clinical and pathological observations concerning 135 cases. Am J Surg Pathol 6:401–410, 1982.
158. Bondeson L, Lindholm K. Aspiration cytology of tubular breast carcinoma. Acta Cytol 34:15–20, 1990.
159. Dawson AE, Logan-Young W, Mulford DK. Aspiration cytology of tubular carcinoma. Diagnostic features with mammographic correlation. Am J Clin Pathol 101:488–492, 1994.
160. Deitos AP, Giustina DD, Martin VD, et al. Aspiration biopsy cytology of tubular carcinoma of the breast. Diagn Cytopathol 11:146–150, 1994.
161. de le Torre M, Lindholm K, Liudgren A. Fine needle aspiration cytology of tubular breast carcinoma and radial scar. Acta Cytol 38:884–890, 1994.
162. Fischler DF, Sneige N, Ordonez NG, et al. Tubular carcinoma of the breast: cytologic features of FNA and application of monoclonal anti-a-smooth muscle actin in diagnosis. Diagn Cytopathol 10:120–125, 1994.
163. Gaffey MJ, Mills SE, Frierson HF, et al. Medullary carcinoma of the breast. Interobserver variability in histopathologic diagnosis. Mod Pathol 8:31–38, 1995.
164. Ellis IO, Galea M, Broughton N, et al. Pathological prognostic factors in breast cancer II. Histological type. Relationship with survival in a large study with long-term followup. Histopathology 20:479–489, 1992.
165. Dixon JM, Anderson TG, Page DL, et al. Infiltrating lobular carcinoma of the breast: an evaluation of the incidence and consequence of bilateral disease. Br J Surg 70:513–516, 1983.
166. Rosen PP. The pathological classification of human mammary carcinoma: past, present and future. Ann Clin Lab Sci 9:144–156, 1979.
167. Frost AR, Terahata S, Yeh IT, et al. The significance of signet ring cells in infiltrating lobular carcinoma of the breast. Arch Pathol Lab Med 119:64–68, 1995.
168. Leach C, Howell LP. Cytodiagnosis of classic lobular carcinoma and its variants. Acta Cytol 36:199–202, 1992.
169. Dabbs DJ, Grenko RT, Silverman JF. Fine needle aspiration cytology of pleomorphic lobular carcinoma of the breast. Duct carcinoma as diagnostic pitfall. Acta Cytol 38:923–926, 1994.
170. Kline TS, Kannan V, Kline IK. Appraisal and cytomorphologic analysis of common carcinomas of the breast. Diagn Cytopathol 1:188–193, 1995.
171. Greeley CF, Frost AR. Cytologic features of ductal and lobular carcinoma in fine needle aspirates of the breast. Acta Cytol 41:333–340, 1997.
172. Sahlaney KE, Page DL. Fine needle aspiration of mammary lobular carcinoma *in situ* and atypical lobular hyperplasia. Am J Clin Pathol 92:22–26, 1989.
173. Dodd LG, Layfield LJ. Fine needle aspiration of inflammatory carcinoma of the breast. Diagn Cytopathol 15:363–366, 1997.
174. Duggan MA, Young GK, Hwang WS. Fine needle aspiration of an apocrine breast carcinoma with multivacuolated lipid-rich giant cells. Diagn Cytopathol 4:62–66, 1988.
175. Mossler JA, Barton TK, Brinkhous AD, et al. Apocrine differentiation in human mammary carcinoma. Cancer 46: 2463–2471, 1980.
176. Nguyen G-K, Neifer R. Aspiration biopsy cytology of secretory carcinoma of the breast. Diagn Cytopathol 3:234–237, 1987.
177. Shingawa T, Tadokoro M, Kitamora H, et al. Secretory carcinoma of the breast. Correlation of aspiration cytology and histology. Acta Cytol 38:909–914, 1994.
178. Alexiev BA. Glycogen-rich clear cell carcinoma of the breast: report of a case with FNAC and immunocytochemical and ultrastructural studies. Diagn Cytopathol 12:62–66, 1995.
179. Aida Y, Takeuchi E, Shingawat, et al. Fine needle aspiration cytology of lipid-secretory carcinoma of the breast. A case report. Acta Cytol 37:547–551, 1993.
180. Cohen H, Szvalb S, Bickel A, et al. Myoblastomatoid carcinoma of the breast: an unusual variant of apocrine carcinoma: report of a case with fine needle aspiration cytology and immunohistochemical study. Diagn Cytopathol 16:145–148, 1997.
181. Yoshida K, Inoue M, Furuta S, et al. Apocrine carcinoma vs. apocrine metaplasia with atypia of the breast. Use of aspiration biopsy cytology. Acta Cytol 40:247–251, 1996.
182. Squires E, Betsill W. Intracystic carcinoma of the breast: a correlation of cytomorphology, gross pathology, microscopic pathology and clinical data. Acta Cytol 25:267–271, 1981.

183. Corkhill ME, Sneige N, Fanning T, et al. Fine needle cytology and flow cytometry of intracystic papillary carcinoma of the breast. Am J Clin Pathol 94:673–680, 1990.
184. Galed-Placed I, Garcia-Ureta E. Fine needle aspiration biopsy diagnosis of adenoid cystic carcinoma of the breast. A case report. Acta Cytol 36:364–366, 1992.
185. Culubret M, Roig I. Fine needle aspiration biopsy of adenoid cystic carcinoma of the breast. A case report. Diagn Cytopathol 15:431–434, 1996.
186. Hsiu J-G, Hawkins AG, D'Amato NA, et al. A case of pure primary squamous cell carcinoma of the breast diagnosed by fine needle aspiration biopsy. Acta Cytol 29:650–651, 1985.
187. Leiman G. Squamous carcinoma of the breast. Diagnosis by aspiration cytology. Acta Cytol 26:201–209, 1982.
188. Macia M, Ces JA, Becerra E, et al. Pure squamous carcinoma of the breast. Report of a case diagnosed by aspiration cytology. Acta Cytol 33:201–204, 1989.
189. Motoyama T, Watanabe H. Extremely well-differentiated squamous cell carcinoma of the breast. Report of a case with a comparative study of an epidermal cyst. Acta Cytol 40:729–733, 1996.
190. Kaufman MW, Marti JR, Gallager HS, et al. Carcinoma of the breast with pseudosarcomatous metaplasia. Cancer 53:1908–1917, 1984.
191. Huros AG, Lucas JC, Foote FW. Metaplastic breast carcinoma. Rare form of mammary cancer. NY State J Med 12:550–561, 1973.
192. Gupta RK, Wakefield SJ, Holloway LJ, et al. Immunocytochemical and ultrastructural study of the rare osteoclast-type carcinoma of the breast in a fine needle aspirate. Acta Cytol 32:79–82, 1988.
193. Gersell DJ, Katzenstein AL. Spindle cell carcinoma of the breast. A clinicopathologic and ultrastructural study. Hum Pathol 12:550–561, 1981.
194. Douglas-Jones AG, Barr WT. Breast carcinoma with tumor giant cells. Report of a case with fine needle aspiration cytology. Acta Cytol 33:109–114, 1989.
195. Boccato P, Briani G, d'Atri C, et al. Spindle cell and cartilaginous metaplasia in a breast carcinoma with osteoclastic-like stromal cells. Acta Cytol 32:75–78, 1988.
196. Wargotz ES, Norris HJ. Metaplastic carcinomas of the breast. Matrix-producing carcinoma. Hum Pathol 20:628–635, 1989.
197. Carson KF, Hirschowitz SL, Nieberg RK, et al. Pitfalls in the cytologic diagnosis of angiosarcoma of the breast by fine needle aspiration. A case report. Diagn Cytopathol 11:297–300, 1994.
198. Foust RL, Berry AD, Moinuddin SM. Fine needle aspiration cytology of liposarcoma of the breast. A case report. Acta Cytol 38:957–960, 1994.
199. Lansham MR, Mills AS, Demay RM, et al. Malignant fibrous histiocytoma of the breast. Cancer 54:558–563, 1984.
200. Stanley MW, Tani EM, Hurwitz CA, et al. Primary spindle cell sarcomas of the breast. Diagnosis by fine needle aspiration. Diagn Cytopathol 4:244–249, 1988.
201. Gupta RK. Aspiration cytodiagnosis of a rare carcinoma of breast with bizarre malignant giant cells. Diagn Cytopathol 15:66–69, 1996.
202. Silverman JF, Geisinger KR, Frable WJ. Fine needle aspiration cytology of mesenchymal tumors of the breast. Diagn Cytopathol 4:50–58, 1988.
203. Walaas L, Angervall L, Hagmar B, et al. A correlative cytologic and histologic study of malignant fibrous histiocytoma: an analysis of 40 cases examined by FNAC. Diagn Cytopathol 2:46–54, 1986.
204. Silverman JF, Feldman PS, Covell JL, et al. Fine needle aspiration cytology of neoplasms metastatic to the breast. Acta Cytol 31:291–300, 1987.
205. Schwarz JG, Clark FGI. Fine needle aspiration biopsy of mycosis fungoides presenting as an ulcerating breast mass. Arch Dermatol 124:409–413, 1988.
206. Matsuda M, Sone H, Ishigoro S, et al. Fine needle aspiration cytology of malignant schwannoma metastatic to the breast. Acta Cytol 33:372–376, 1989.
207. Kumar PV, Esfahari FN, Selimi A. Choriocarcinoma metastatic to the breast diagnosed by fine needle aspiration. Acta Cytol 35:239–242, 1991.
208. Balgia M, Holmquist ND, Espinoza CG. Medulloblastoma metastatic to breast diagnosed by fine needle aspiration biopsy. Diagn Cytopathol 10:33–36, 1994.
209. Hogge JP, Magnant CM, Lage JM, Zuurbier RA. Rhabdomyosarcoma metastatic to the breast. Breast 2:270–273, 1996.
210. Hajda SI, Urban JA. Cancers metastatic to the breast. Cancer 29:1691–1696, 1972.
211. Domanski HA. Metastases to the breast from extramammary neoplasms: a report of six cases with diagnosis by fine needle aspiration cytology. Acta Cytol 40:1293–1299, 1996.
212. Fabian C, Dabbs DJ. The immunohistochemical profile of breast carcinoma metastatic in the lung. Breast J 3:98–103, 1997.
213. Rao CR, Narasimhamurthy NK, Jaganathan K, et al. Cystosarcoma phylloides. Diagnosis by fine needle aspiration cytology. Acta Cytol 36:203–207, 1992.
214. Stanley MW, Tani EM, Rutquist LE, et al. Cystosarcoma phylloides of the breast: a cytologic and clinicopathologic study of 23 cases. Diagn Cytopathol 5:29–34, 1989.
215. Shimizu K, Majawa N, Yamada T, et al. Cytologic evaluation of phylloides tumors as compared to fibroadenomas of the breast. Acta Cytol 38:891–897, 1984.
216. Simi U, Moretti D, Iacconi P, et al. Fine needle aspiration cytology of phylloides tumor: differential diagnosis with fibroadenoma. Acta Cytol 32:63–66, 1988.
217. Linsk JA, Franzen S. Clinical aspiration cytology. Philadelphia: JB Lippincott, 105–137, 1983.
218. Sneige N, Singletary EE. Fine needle aspiration biopsy of the breast: diagnostic problems and approach to surgical management. Pathol Ann (Part I) 29:281–301, 1994.
219. Shabb NS. Phylloides tumor: fine needle aspiration cytology of eight cases. Acta Cytol 41:321–326, 1997.
220. Stawicki M, Hsiu J. Malignant cystosarcoma phylloides. Acta Cytol 23:61–64, 1979.
221. Koss LG, Woyke J, Olszewski W. Aspiration biopsy: cytologic interpretation and its histologic bases (pp. 53–104). New York: Igaku-Shoin, 1984.
222. Dusenberry D, Frable WJ. Fine needle aspiration cytology of phylloides tumor, potential diagnostic pitfalls. Acta Cytol 36:215–221, 1992.

CHAPTER 3

Thyroid and Parathyroid

Thyroid nodules are a relatively common occurrence, especially in geographic areas in the Northern Hemisphere, formerly considered goitrogenic zones. Despite the incidence of thyroid nodules, only a small portion are malignant.[1] However, before the advent of fine needle aspiration (FNA) biopsy, patients with thyroid nodules had a limited number of options for diagnosis and management. Surgery (usually hemithyroidectomy) was the procedure of choice for diagnosis and therapy. The development of FNA biopsy as a safe, cost-effective, and accurate diagnostic procedure currently allows conservative management of most patients with thyroid nodules (Table 3–1).[1–9] In addition, the use of suppression therapy and/or regular clinical follow-up of patients diagnosed with nonneoplastic disease has significantly reduced the need for surgery in these patients. As a result, the thyroid is one of the most common organs undergoing superficial FNA biopsy. FNA biopsy is also useful in the diagnosis of malignancy in patients unsuitable for surgery and in the documentation of recurrent or metastatic disease (Table 3–2).

CLINICAL ASPECTS

As with most FNA biopsies, whether performed by a surgeon, endocrinologist, or cytopathologist, appropriate specimen procurement and preparation are critical. Equally important for accurate and specific diagnosis is knowledge of all available clinical, radiologic, and laboratory data at the time of the procedure. Before performing thyroid FNA biopsy, it is important to review the medical history of the patient, particularly if there is a history of previous surgery or irradiation to the head and neck region, prior or current thyroid disease, and any other medical problems that could interfere with the procedure.[8] Patients should also be asked if they bruise easily or are taking any aspirin or anticoagulant medication. Detailed information should be obtained regarding the duration and rapidity of growth of the thyroid nodule and the radiologic findings obtained before the procedure. This information may suggest a cyst or cystic component to the mass, which is extremely useful information to have before FNA biopsy. Informed consent, either verbal or written, should be obtained, and potential complications should be explained. Bleeding and bruising are the most common problems associated with thyroid FNA biopsy. They can be minimized by awareness of potential problems (Table 3–3). Patients should also be instructed in the use of ice rather than heat to reduce any swelling. Pain medication, usually acetaminophen, is recommended for discomfort.

Although thyroid lesions represent one of the most common sites selected for superficial FNA biopsies, the incidence of serious complications is virtually nil.[10] When a fine-bore (23–27 g) needle is used, there

TABLE 3–1. Clinical Management of Thyroid Disease

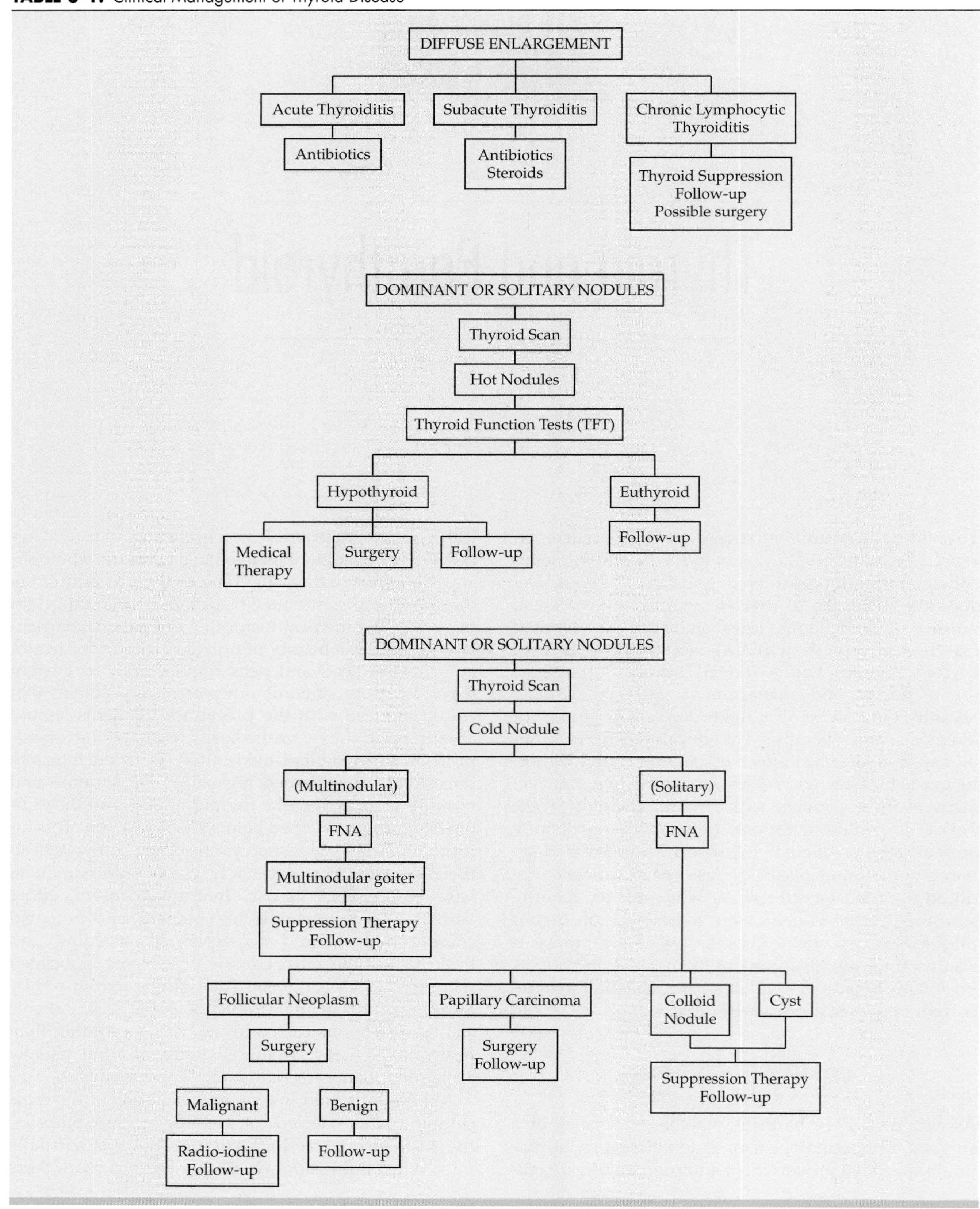

Modified from: Powers CN, Frable WF: Fine needle aspiration biopsy of the head and neck. Boston: Butterworth-Heinemann, 1996:52

TABLE 3–2. Indications of Thyroid FNA Biopsy

Localized Disease	Diffuse Disease
Single nodule Dominant nodule in multinodular goiter	Inflammatory (thyroiditis) Hyperplasia
	Rapid diagnosis of lymphoma or anaplastic carcinoma
Nodule in patient with familial medullary carcinoma or multiple endocrine neoplasia	

is virtually no danger of needle tract seeding. Another complication that is extremely rare but rather alarming to both patient and physician is acute enlargement of the thyroid gland.[11,12] The reason for this abrupt, transient swelling is unknown. It does not appear to result from hemorrhage, nor does it compromise the airway. Ice packs applied to the neck can help reduce the swelling, which generally resolves in a few hours. Occasionally, especially with small, mobile masses in the isthmus, the trachea may be entered. This is immediately detectable during the aspiration because of the loss of negative pressure. The patient may feel a tickling sensation or the urge to cough, but there are no other sequelae. Benign ciliated bronchial cells may be present in the smears. If the cilia are not apparent and the bronchial cells aggregate into clusters, the neophyte may overinterpret this as atypia or neoplasm.

Once informed consent is obtained, the aspirator should discuss with the patient the history and should review relevant clinical information. This should be followed by a thorough physical examination of the thyroid gland, its nodule(s), and the neck region. Even when the nodule is obvious, a complete examination before the FNA biopsy is warranted. This examination is best begun with the patient in a sitting position, with the aspirator behind the patient. Landmarks include the trachea and thyroid cartilage. Firm, gentle pressure usually allows the thyroid lobes to be palpated. Most benign nodules move when the patient swallows. Small sips of water swallowed at the physician's direction can help identify smaller, ill-defined nodules as being within the thyroid proper rather than adjacent lymph nodes. Although the sitting position may ultimately be chosen as the best for the biopsy procedure, the patient should also be examined in the supine position with a roll or pillow under the shoulder blades, thus extending the neck and accentuating the thyroid gland and any masses within it. This may be difficult to accomplish in older patients and in patients with arthritis, problems with the cervical spine, or respiratory compromise.

TABLE 3–3. Risks and Complications of Thyroid FNA

Minor	Major (rare)
Bruising Vasovagal reaction Local bleeding	Hematoma Acute enlargement, thyroid Acute swelling, prethyroidal soft tissues
Tracheal puncture	

TECHNIQUE

FNA biopsy of the thyroid gland is no different from other superficial aspiration biopsies and is described in detail elsewhere. Because the thyroid is a highly vascular organ, small-gauge needles (23–27 g) are preferred.[13] Immediately after the aspiration, firm pressure should be applied to prevent bleeding, not only at the puncture site but also within the gland. Bleeding causes the nodule to enlarge, much to the distress of both patient and clinician, and can terminate further sampling. If additional aspirations are attempted, the specimen may be diluted by excessive amounts of blood, rendering it virtually useless for cytomorphologic examination. Although the traditional FNA biopsy is performed with a syringe attached to the needle to provide negative pressure (vacuum), some advocate the use of needle-only techniques. These nonaspiration biopsies rely on capillary pressure to draw the specimen into the needle hub and may reduce the amount of trauma to the gland, ultimately decreasing the amount of bleeding and potential adverse sequelae. The number of aspirates necessary to ensure an adequate sample is variable and often dependent on whether a cytologist is present to evaluate the sample immediately.

ADEQUACY

There has always been debate regarding what constitutes an adequate sample. For the cytopathologist who must rely on the clinician to provide not only the sample but the history and physical findings, guidelines for adequacy that include cell number and type are useful. One such guideline suggests that six groups of follicular epithelium present on two slides is the minimum number required for adequacy.[14,15] However, when cytopathologists perform and interpret their own FNA biopsies, rigid standards may not be that useful. Cysts, a common occurrence, represent another challenge, because follicular cells may not be present. In these situations, the amount and consistency of the fluid should be documented. The presence of hemosiderin-laden macrophages suggests

TABLE 3–4. Diagnosis of Thyroid Disease Based on Cytomorphologic Features

- FINE-NEEDLE ASPIRATION
  - Low cellularity / Variable colloid
    - Scant colloid
      - Unsatisfactory
    - Abundant colloid
      - Colloid nodule
      - Nodular goiter
  - Moderate or high cellularity / Minimal colloid
    - Predominant cell type
      - Epithelioid histiocytes
        - Subacute thyroiditis
      - Lymphocytes
        - Hashimoto's thyroiditis
        - Lymphoma
        - Hyperplasia
      - Neutrophils
        - Acute thyroiditis
        - Anaplastic carcinoma
    - Follicular Cells
      - Papillary formations
        - Papillary carcinoma, Hyperplasia, Nodular goiter, Pregnancy
      - Oncocytic cells
        - Oncocytic cell neoplasm (Hürthle cell tumor), Nodular hyperplasia, Hashimoto's thyroiditis
      - Follicular cells
        - Follicular adenoma, Follicular carcinoma, Follicular variant of papillary carcinoma
      - Malignant cells
        - Medullary carcinoma (spindle, carcinoid-like cells)
        - Follicular carcinoma (follicular cells with round nuclei)
        - Anaplastic carcinoma (bizarre cells)
        - Metastases

Modified from: Powers CN, Frable WF: Fine needle aspiration biopsy of the head and neck. Boston: Butterworth-Heinemann, 1996:53.

prior hemorrhage into the nodule, a common occurrence and often the reason for sudden enlargement and presentation of a thyroid nodule. If the cyst is decompressed, the area should be re-examined for the presence of a residual mass; if found, it should be aspirated.[4,13,16] Algorithms based on cytomorphologic criteria are useful to categorize the findings, but the final diagnosis should be made in concert with the clinical data (Table 3–4).

NORMAL THYROID

Occasionally, FNA biopsy of a small or difficult-to-position nodule may cause normal thyroid to be sampled. Fragments of thyroid parenchyma are seen as variably sized, well-ordered follicles with central colloid. This colloid is usually dense and darkly stained by both Papanicolaou and Romanowsky stains (Fig. 3–1). Many of the fragments have depth of focus, which reveals the three-dimensional arrangement of the follicles. The follicular cells are arranged in a honeycomb pattern. Small clusters of follicular cells have a moderate amount of delicate cytoplasm and round, smooth nuclei that may vary in size within follicles or cellular sheets. This variability in nuclear size is a normal feature of endocrine glands. Background elements include dispersed stripped nuclei and colloid. Colloid can appear as thick, inspissated globules, viscous "waves," or thin sheets (Fig. 3–2). The latter formation, when present on air-dried preparations, often imparts a "cracked" appearance to the colloid that simulates a stained-glass window (Fig. 3–3). Colloid appears translucent green or pink on Papanicolaou stain and variable shades of blue on Romanowsky stains.

A

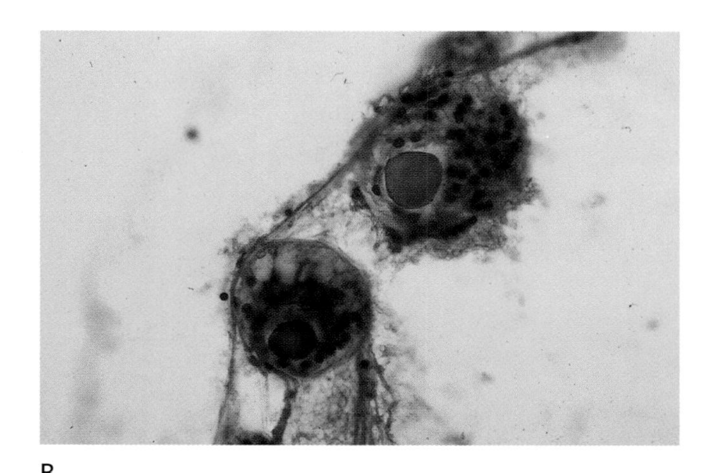

B

FIGURE 3–1. (**A**) Normal follicle with central colloid in a background of blood and foamy histiocytes. On Romanowsky-stained material, inspissated colloid appears dense blue. (**B**) Normal follicles appear three-dimensional when stained with Papanicolaou method. Central colloid appears orangeophilic. (**A,** Diff-Quik, ×400; **B,** Papanicolaou, ×400)

MULTINODULAR GOITER

Multinodular goiter (MNG) represents the most common nonneoplastic process evaluated by FNA biopsy. Patients have multiple nodules of varying size and consistency. Typically, patients with multiple nodules have one or more dominant nodules that are sampled. Aspirates are commonly composed of an admixture of colloid and bland-appearing follicular cells. The final cytomorphologic diagnosis of this type of benign nodule, whether termed goiter or nonneoplastic thyroid, usually prompts the clinician toward conservative management. In aspirates of MNG, colloid cysts and adenomatous nodules may demonstrate divergent clinical and cytomorphologic presentations. A nodule composed almost entirely of colloid is frequently encountered in patients with MNG. This highly viscous material may be difficult to express from the needle and smears with an oily consistency. Air-dried preparations may reveal cracked colloid or darker-staining "inspissated" colloid. The latter is seen especially if the nodule is longstanding. Alternatively, the contents of many of these colloid nodules have a watery consistency and aspirate rapidly with subsequent decompression of the nodule. Although the aspirated smears are similar, when exposed immediately to alcohol for Papanicolaou staining, much of the colloid is washed off the slide. For this reason, Romanowsky staining is preferred. The number of follicular cells obtained during aspiration may vary considerably from scant to numerous. Follicular cells may be arranged in folli-

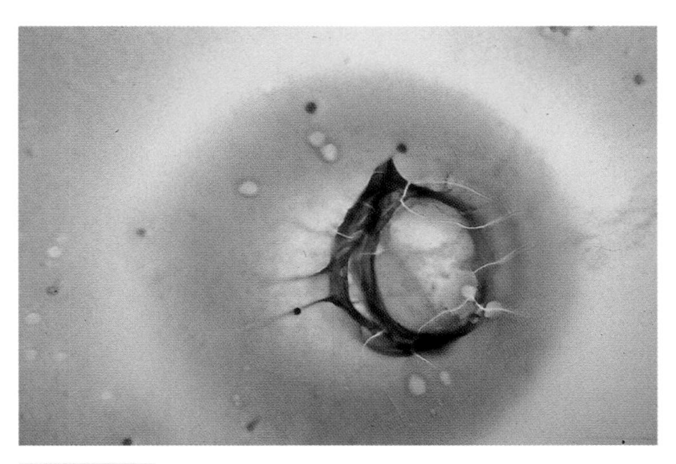

FIGURE 3–2. Colloid is sometimes difficult to appreciate on Papanicolaou-stained preparations, particularly in bloody specimens. Colloid may be visualized as viscous, cyanophilic waves or bubbles. (Papanicolaou, ×200)

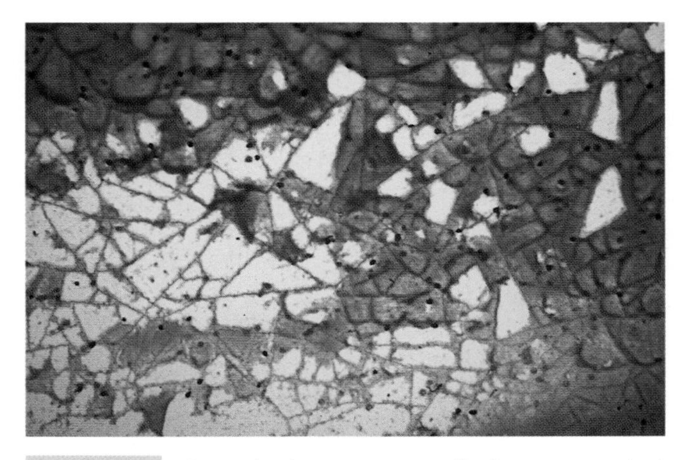

FIGURE 3–3. On air-dried preparations, colloid appears cracked and has been likened to a stained-glass window or a spider web. (Diff-Quik, ×150)

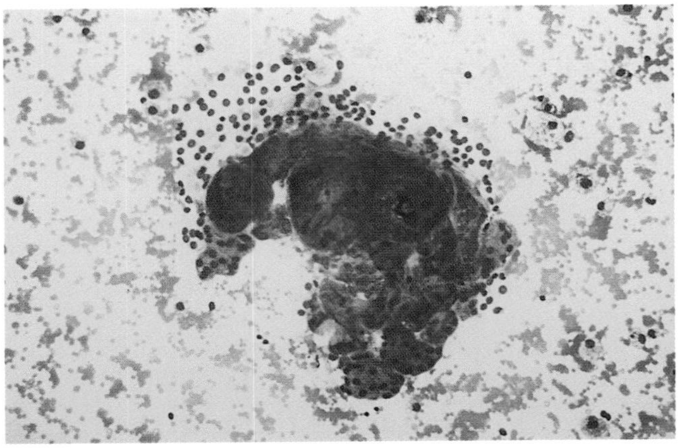

FIGURE 3–4. Nodular goiter may range in cellularity, with scattered follicles or groups of follicles arranged in a dense cluster, often with scattered histiocytes in the background. (Diff-Quik, ×200)

cles, honeycomb sheets, or isolated cells. The cell nuclei is bland, round, and regular, with indistinct nucleoli (Fig. 3–4).

Less frequent, but more worrisome, is the adenomatous nodule that presents as a highly cellular aspirate composed of clusters, follicles, and sheets of follicular cells, with variable amounts of colloid. In isolation, without clinical and radiologic data, it is difficult to predict whether this lesion represents a cellular nodule of MNG or a follicular neoplasm. Depending on the clinical situation, many of the nodules are excised. This is perhaps the most difficult cytomorphologic presentation to interpret accurately (Fig. 3–5). Compounding cytomorphologic findings may be the presence of Hürthle cells, pseudopapillary clusters, and stripped nuclei. The amount of colloid is often a deciding factor in diagnosis. Scant colloid in the presence of numerous follicular cells, particularly when patterns such as small follicles and/or overlapping nuclei are present, warrants further evaluation to confirm or exclude a follicular neoplasm.

A frequent finding in aspirates of nodules in MNG is the presence of cystic degeneration, often the result of prior hemorrhage. Aspirates from these nodules may contain an admixture of watery colloid and blood, hemosiderin-laden macrophages, and cholesterol crystals (Fig. 3–6). Romanowsky-stained slides reveal hemosiderin-laden macrophages characterized by variably sized histiocytes with engulfed dark-blue particles, often so numerous that the nucleus is obscured. Follicular cells, when present, are often degenerated with ragged cytoplasm. Regressive changes that include elongated or spindle-shaped follicular cells are typically present in longstanding nodules (Fig. 3–7). Cholesterol crystals, characterized by sharp, right-angle edges, are easily visualized when the substage condenser is "racked down." This effect highlights the refractivity of the crystals (Fig. 3–8). Most of these specimens are more liquid or semiliquid and are best evaluated in cytospin preparations. Although most cystic lesions are benign components of MNG, cystic degeneration may also be seen in papillary and to a lesser extent follicular carcinomas (Table 3–5). Large nodules (>3 cm) that have cystic and solid components are more likely to be neoplastic than their smaller counterparts.[17]

Cytomorphologic Features of Benign Nodular Goiter

- Colloid
- Follicular cells arranged in uniform follicles, honeycombed sheets, and single cells
- Bland round nuclei with smooth contours
- Fine, even chromatin
- Indistinct nucleoli
- Hemosiderin-laden macrophages

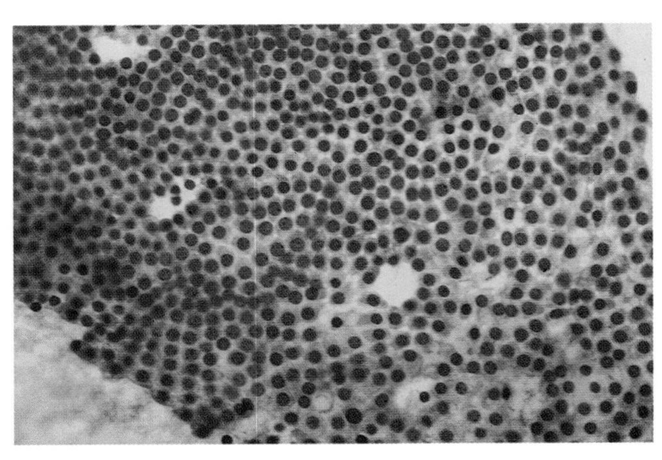

FIGURE 3–5. In cellular or adenomatous goiters, sheets of bland, evenly arranged follicular cells may be seen with scattered empty follicular lumens. (Papanicolaou, ×100)

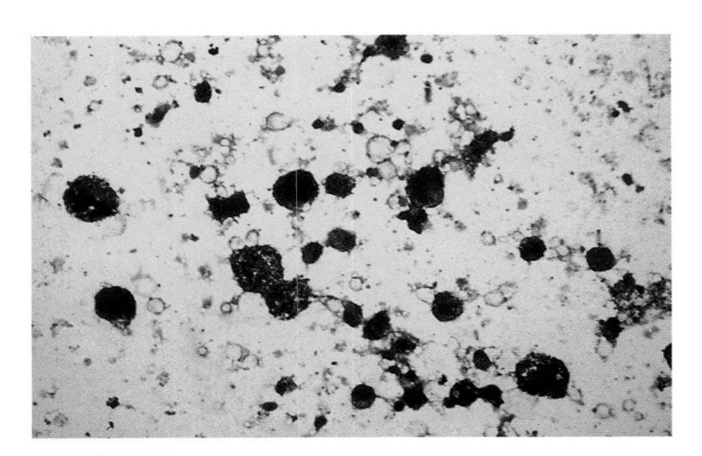

FIGURE 3–6. Cystic degeneration of nodules in goiter is common and may present as hemosiderin-laden macrophages and degenerating follicular cells in a background of blood and watery colloid. (Diff-Quik, ×250)

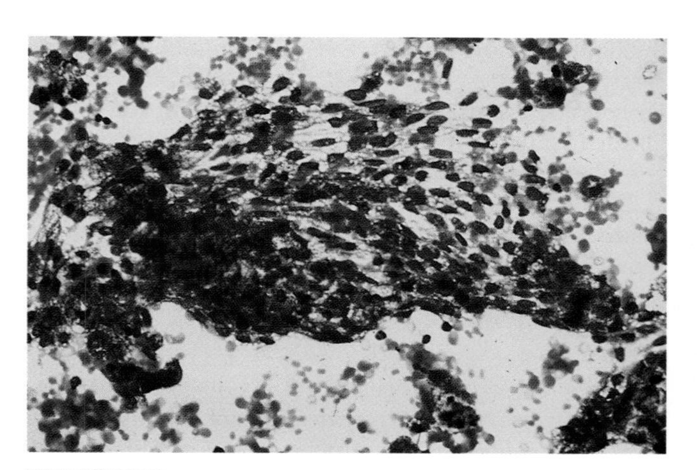

FIGURE 3–7. Regressive and degenerative changes can result in a spindle arrangement of follicular cells, resembling "reparative" sheets. (Diff-Quik, ×200)

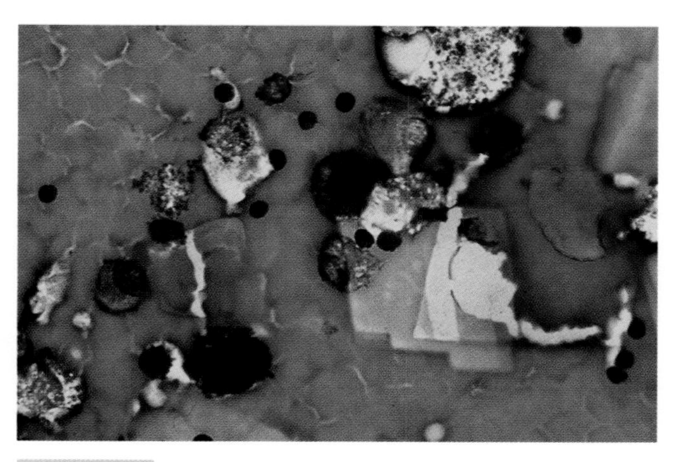

FIGURE 3–8. Prominent, longstanding cystic degeneration may result in the formation of cholesterol crystals. These sharp, right-angle crystals are best appreciated on Romanowsky-stained preparations, particularly if the substage condenser is "racked down" to provide contrast. (Diff-Quik, ×1000)

THYROIDITIS

Although MNG is the most commonly diagnosed nonneoplastic lesion, inflammatory conditions may also be encountered.[18,19] The most common of these processes is chronic lymphocytic inflammation, often termed lymphocytic (Hashimoto's) thyroiditis. Less common are acute or suppurative thyroiditis and subacute or granulomatous thyroiditis. Most patients with chronic lymphocytic thryoiditis have hypothyroidism and diffuse, often asymmetric thyroid enlargement. Clinical symptoms in conjunction with positive serologic tests of antithyroid (microsomal) antibodies may aid in making the correct diagnosis. However, sometimes the asymmetry is significant or nodules are present so that FNA biopsy is used. In addition, lymphocytic thyroiditis can be associated with neoplasia (especially lymphoma and, to a lesser extent, follicular and papillary carcinoma), requiring tissue evaluation. Aspirates of chronic thyroiditis reveal a mixed population of lymphocytes and follicular cells, with limited amounts of colloid (Fig. 3–9). Hürthle cells are also associated with chronic thyroiditis and to a lesser extent MNG (Fig. 3–10). The presence of numerous Hürthle cells may result in a false-positive diagnosis of Hürthle cell neoplasia, particularly if a Hürthle cell nodule is directly sampled in Hashimoto's. Even hyperplastic follicular epithelium may be misinterpreted as neoplastic in these situations.[13,20]

Depending on the degree of sampling and the relative proportions of lymphocytes within the area sampled, lymphoid cells may be scattered among small clusters of follicular cells or may dominate the smear. The presence of lymphocytes within the groups of Hürthle or follicular cells is especially helpful in making the correct diagnosis. When stripped follicular cell nuclei are numerous, they can be confused with lymphocytes. Air-dried preparations are quite useful because Romanowsky stains indicate the presence of lymphoglandular bodies, small round fragments of lymphocyte cytoplasm, and accentuate the thin, almost invisible rim of basophilic cytoplasm present in mature lymphocytes. The presence of lymphocytes suggests one of three possibilities. The first two are chronic inflammation associated with MNG or chronic lymphocytic thyroiditis; both show a heterogeneous population of lymphocytes, often with ger-

TABLE 3–5. Major Differential Diagnostic Features of Cystic Lesions

Cytologic Feature	Nodular Goiter	Thyroglossal Duct Cyst	Parathyroid Cyst	Papillary Carcinoma
Cyst fluid	Bloody Thin-watery	Proteinaceous Viscous	Water Clear	Bloody
Cellularity	Moderate or high	High	Acellular	Variable; often scant
Cholesterol crystals	Occasional	Frequent	None	Rare
Cell types	HLM Follicular (degenerating)	Macrophages Squamous/columnar	—	Follicular cells Papillae

HLM = hemosiderin-laden macrophages

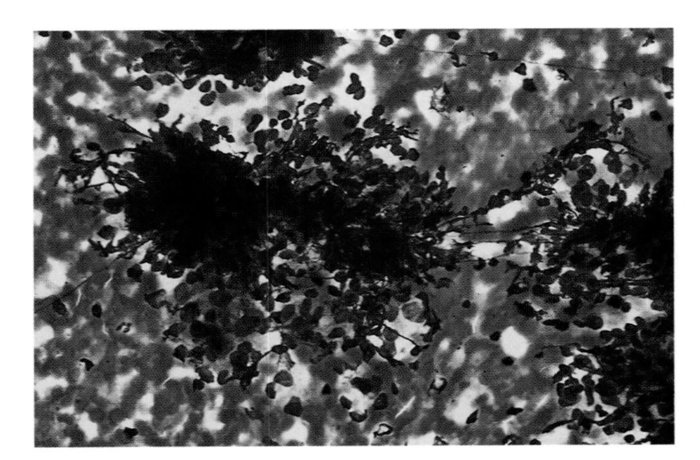

FIGURE 3–9. Aspirates from chronic lymphocytic thyroiditis (Hashimoto's) should have both a lymphoid and a follicular component. Often lymphocyte aggregates appear as darkly stained tangles with prominent DNA artifact and associated follicular cells. (Diff-Quik, ×250)

minal center fragments. The latter is more common with lymphocytic thyroiditis. The third possibility is malignant lymphoma, which usually presents as a monomorphic population of atypical lymphocytes. Flow cytometry or other means of immunophenotyping is often necessary to confirm or exclude lymphoma. Sampling is very important in the setting of Hashimoto's thyroiditis to avoid a false-negative diagnosis of lymphoma.

Cytomorphologic Features of Hashimoto's Thyroiditis

- Hürthle cells and reactive epithelial follicles
- Polymorphic population of lymphocytes
- Fibrosis
- Scant colloid

Subacute granulomatous (DeQuervain's) thyroiditis is rarely encountered but has a fairly distinctive cytomorphologic pattern that allows confident diagnosis. Unfortunately, patients tend to present with fever and a tender, often asymmetric gland, and the biopsy procedure may be painful. Without patient cooperation, inadequate sampling may occur. The key cytomorphologic feature for the diagnosis of subacute thyroiditis is the granuloma, recognized as a loose collection of epithelioid histiocytes and scattered multinucleated giant cells (Fig. 3–11). The presence of multinucleated giant cells alone is not diagnostic, because they are found in a variety of thyroid lesions. As with other inflammatory processes, colloid tends to be scant on smears; follicular cells are variably present. Sampling adequacy is often dependent on the stage of this self-limited process. In the late fibrotic stage, the biopsy will yield scant if any material for cytomorphologic evaluation. Patients with acute thyroiditis rarely present for FNA biopsy.

Cytomorphologic Features of Subacute Thyroiditis

- Scant follicular cells showing degeneration
- Mixed inflammation, background debris
- Granulomas and epithelioid histiocytes
- Mutlinucleated giant cells

As with their subacute counterparts, fever and swelling of the thyroid are prominent features of acute thyroiditis, and palpation and aspiration of the gland tend to be quite painful. Because the process is acute and self-limited because of its response to antibiotics, FNA biopsy is seldom performed. However, if the gland is aspirated, neutrophils and degenerated cellular material, including follicular cells, are seen

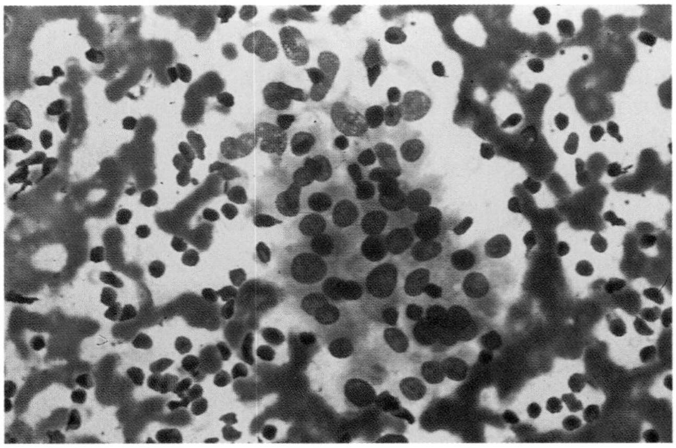

FIGURE 3–10. Follicular groups surrounded by small mature lymphocytes may show varying degrees of Hürthle cell change. Lymphocytes can also infiltrate the groups of follicular cells. (Diff-Quik, ×400)

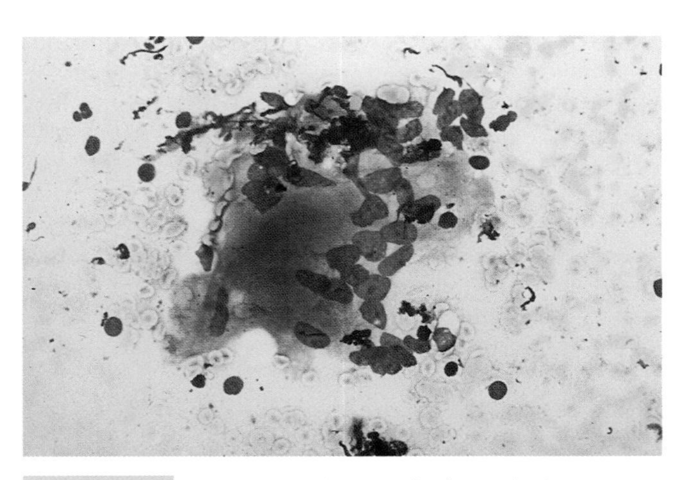

FIGURE 3–11. Multinucleated giant cells, foreign-body type are the prominent finding in subacute or granulomatous thyroiditis. (Diff-Quik, ×400)

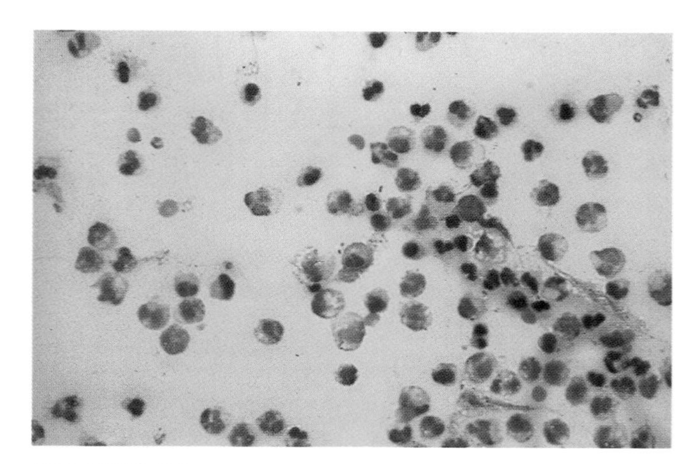

FIGURE 3–12. Neutrophils and degenerating follicular cells are suggestive of acute thyroiditis, and bacterial stains should be performed. (Diff-Quik, ×400)

(Fig. 3–12). Bacteria may be present but often require bacterial stains and material sent for culture for conclusive identification. Romanowsky stains may demonstrate bacteria in the smear background. Fungal infections due to *Aspergillus, Candida,* and *Cryptococcus* have also been reported as the etiology of acute thyroiditis.[21–23]

Cellular Features of Acute Thyroiditis

- Numerous neutrophils, fibrin, blood
- Bacteria (Romanowsky or Gram's stain)
- Rare follicular cells
- Histiocytes, fibroblasts (late stage)

HYPERTHYROIDISM

Another nonneoplastic condition that may be encountered is hyperthyroidism. Graves' disease can present as a diffuse lesion or a solitary mass (toxic nodule).[24] When radiologic evaluation precedes the biopsy, the increased uptake of radioactive iodine in patients with symptoms of hyperactivity (exophthalmos, increased appetite, tachycardia) is diagnostic. However, FNA biopsy is often used to evaluate the presence of a "hot nodule" to confirm rapidly the clinical suspicion of hyperthyroidism.

Once again, Romanowsky stains are extremely useful because of the characteristic appearance of follicular cells. The cytoplasm of these cells shows thick, magenta-tinged, frayed edges, often with peripheral vacuoles or blebs of detached cytoplasm.[25] These cytoplasmic changes have been variously referred to as fire flares, colloid suds, and marginal vacuoles and are not seen with any clarity on Papanicolaou-stained preparations (Fig. 3–13). However distinctive, marginal vacoules are not pathognomonic for hyperthyroidism and are seen, albeit to a lesser degree, in nontoxic goiters and occasionally in follicular neoplasms.[26] Because the amount of follicular cells and colloid is extremely variable, a diagnosis of hyperthyroidism should be based in large measure on the associated clinical information. Focal hyperplastic change can be seen in MNG; these areas, when sampled, yield identical cytomorphologic findings, although to a lesser degree. Occasionally, thyroid aspirates of patients with Graves' disease who have been treated with radioactive iodine are performed. The clinical history in these situations is imperative, because follicular cell atypia secondary to radiation can be impressive and can result in false-positive diagnoses.[27,28]

Cytomorphologic Features of Hyperthyroidism

- Moderate to high cellularity of follicular cells
- Fire flares or marginal vacuoles
- Hürthle cell metaplasia
- Scant colloid, abundant blood

FOLLICULAR NEOPLASMS

The spectrum of diagnoses associated with highly cellular aspirates composed of follicular cells is daunting and ranges from the adenomatous nodule of MNG to follicular neoplasms and even the follicular variant of papillary carcinoma. The inability of aspiration cytology to predict the biologic behavior of some follicular lesions does not detract from its utility in clinical management.[29–31] Most would agree that any neoplasm in the thyroid should be removed and thoroughly evaluated. Because the cytomorphologic criteria for a follic-

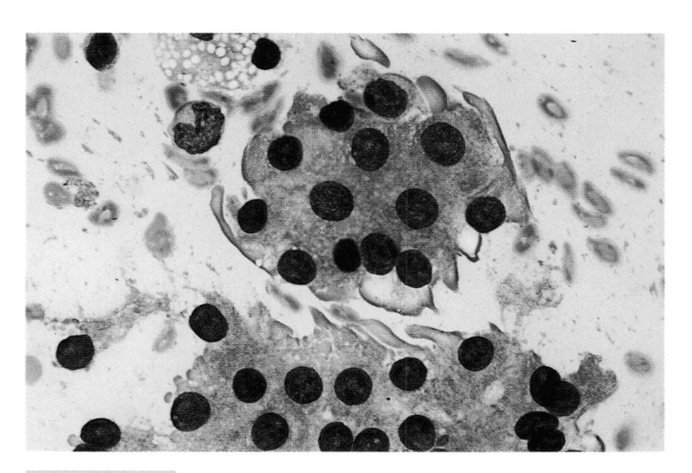

FIGURE 3–13. Scattered clusters of follicular cells with magenta-tinged vacuoles at the cytoplasmic periphery—the so-called fire flares or colloid suds. (Diff-Quik, ×200)

ular neoplasm can overlap with cellular MNG—high cellularity, cellular arrangement of the follicles, and a paucity of colloid can be seen in both lesions—there will always be a small percentage of nonneoplastic nodules that are excised.[32]

Hypercellular aspirates in which follicular neoplasia is suspected generally show micro- and macrofollicular arrangements admixed with single follicular cells and stripped nuclei. Many of the larger fragments have rounded or scalloped borders (Fig. 3–14). Most follicular neoplasms are benign. However, follicular carcinomas do represent 15–20% of thyroid malignancies and therefore should be excluded from diagnostic consideration. Unfortunately, the standard criteria for malignancy cannot always be relied on with follicular lesions; in particular, the use of nuclear morphology as a criterion is limited. Nuclear overlap and individual cells with increased nuclear-to-cytoplasmic ratios, hyperchromia, and small nucleoli may suggest malignancy, but most follicular carcinomas have deceptively bland cytology. Rarely, stromal fragments from follicular neoplasms resemble the fibrovascular cores seen in papillary carcinoma (Fig. 3–15). Excision and surgical pathology evaluation for capsular and vascular invasion is the definitive procedure for a correct diagnosis.

Cytomorphologic Features of Follicular Neoplasms

- High cellularity of follicular cells
- Monotonous, enlarged round nuclei
- Distinct nucleoli
- Micro-, macrofollicles, trabeculae
- Scalloped borders
- Minimal or absent colloid
- Nuclear overlapping
- Loss of honeycomb pattern

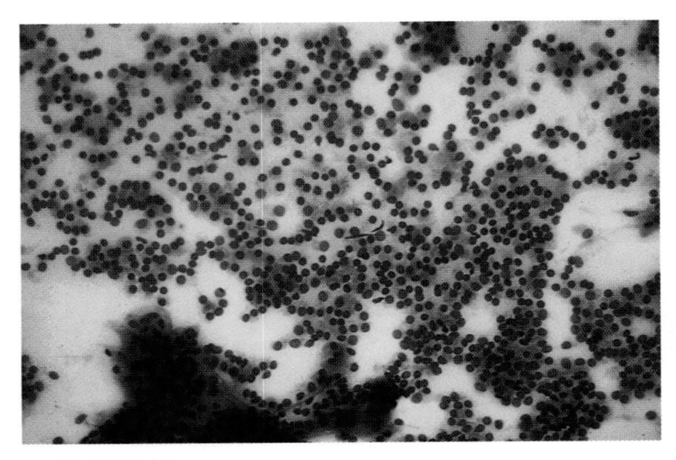

FIGURE 3–14. FNA of a follicular neoplasm is highly cellular, with a monotonous pattern of micro- or macrofollicular structures. Colloid is absent or sparse. (Diff-Quik, ×200)

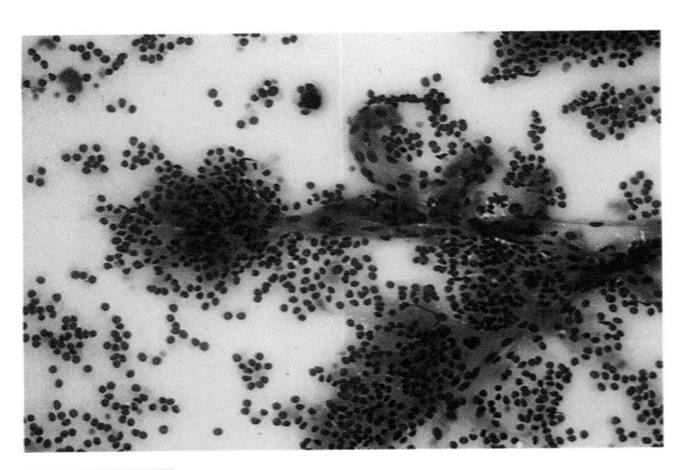

FIGURE 3–15. Stroma from follicular neoplasms may mimic fibrovascular cores of papillary carcinoma. (Papanicolaou, ×200)

A variant of the follicular neoplasm is the Hürthle cell tumor. Hürthle cells are characterized by their abundant granular cytoplasm, large round nuclei, and prominent central nucleoli (Fig. 3–16).[33,34] These cells may appear in a wide variety of thyroid conditions, including MNG, lymphocytic thyroiditis, and benign and malignant neoplasms (Fig. 3–17). Hürthle cell tumors are usually highly cellular and composed almost exclusively of Hürthle cells, with variable but usually minimal amounts of colloid. Hürthle cells occur singly, in sheets or clusters, and rarely with papillary formations. Their cytoplasm may stain a variety of colors such as blue, green, or orange on Papanicolaou stain and blue or purple on Romanowsky stains. Nuclear atypia in both benign and malignant Hürthle cell tumors is common. A recent morphometric study suggested that nucleolar morphology and size may be

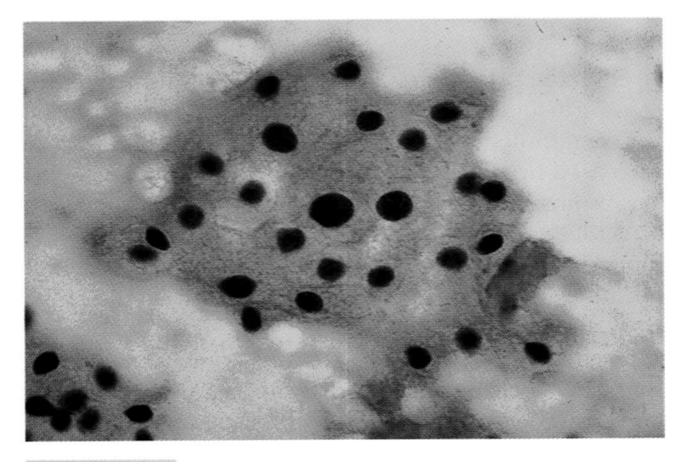

FIGURE 3–16. Follicular neoplasms composed predominantly of Hürthle cells show the characteristic oncocytic features: moderate amounts of granular cytoplasm with round, somewhat eccentric nuclei and large prominent nucleoli. Anisonucleosis is usually present. Papanicolaou-stained Hürthle cells can appear orange, green, or even blue. (Papanicolaou, ×400)

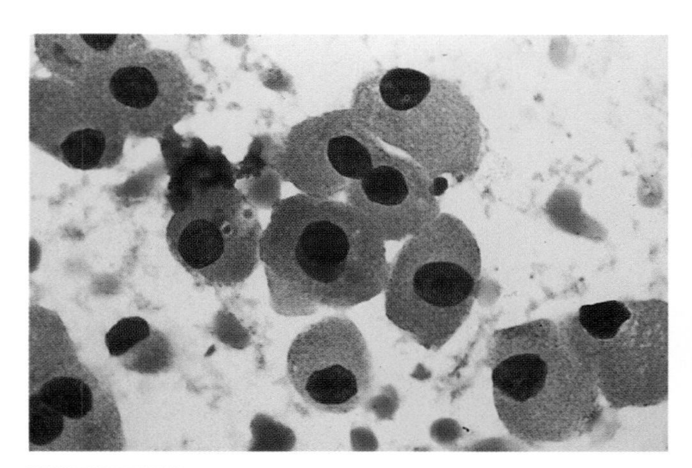

FIGURE 3–17. Hürthle cells stained with a Romanowsky stain appear purple and very large when compared to their Papanicolaou-stained counterparts. (Diff-Quik, ×1000)

FIGURE 3–18. Aspirates of papillary carcinoma are very cellular and often contain very obvious papillary fragments. On Romanowsky-stained material, the fibrovascular cores are pink and variably cellular. The neoplastic cells appear as cellular tufts or flat sheets adjacent to these central cores. (Diff-Quik, ×200)

useful in distinguishing these tumors.[35] As previously indicated, metaplastic nodules of Hürthle cells can occur in other thyroid conditions and if sampled mimic a Hürthle cell tumor. Multiple aspirations may be necessary to sample larger nodules, particularly in a multinodular thyroid. As with pure follicular lesions, excision and surgical pathology evaluation is recommended to classify this lesion as an adenoma or carcinoma.

Cytomorphologic Features of Hürthle Cell Tumor

- High cellularity of Hürthle cells, both individually and in groups
- Monomorphic, polygonal cells
- Cytoplasmic granularity
- Prominent, round nucleoli
- Minimal colloid

PAPILLARY CARCINOMA

The most commonly encountered thyroid malignancy is papillary carcinoma, which can be readily diagnosed by aspiration cytology in most cases. Aspirates are typically highly cellular, containing papillary arrays as well as sheets of neoplastic cells (Fig. 3–18).[36–38] Numerous cytomorphologic criteria have been associated with this tumor; the two most reliable and important criteria are nuclear grooves and intranuclear inclusions (Fig. 3–19). Electron microscopy has demonstrated that these two features are formed by irregularity and infolding of the nuclear membrane.[39] The Papanicolaou stain highlights the grooves, which should traverse the longitudinal axis of the nucleus.[40–42] Although intranuclear inclusions may also be seen using the Papanicolaou stain, Romanowsky stains are very useful because the pale cytoplasmic inclusion is highlighted against the deep-purple nucleus (Fig. 3–20). These nuclei typically are oval and often have a powdery chromatin with distinct nucleoli (better appreciated with the Papanicolaou stain). The presence of papillae with true vascular cores is also very helpful. Often the cells lining the papillae are sheared from the connective tissue core, resulting in large, sometimes rolled or overlapping sheets of cells and naked fibrovascular cores (Fig. 3–21). However, pseudopapillary formations may be associated with hyperplastic changes, particularly in women during pregnancy and occasionally in MNG. Although the overall architecture may mimic papillary carcinoma, the distinctive nuclear features are not readily present in the latter.

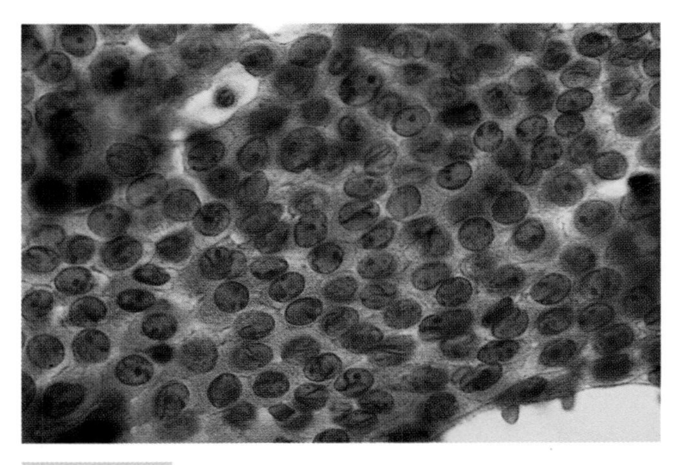

FIGURE 3–19. Nuclear grooves traverse the longitudinal axis of the nucleus. These delicate lines or folds are best appreciated on Papanicolaou-stained preparations. (Papanicolaou, ×1000)

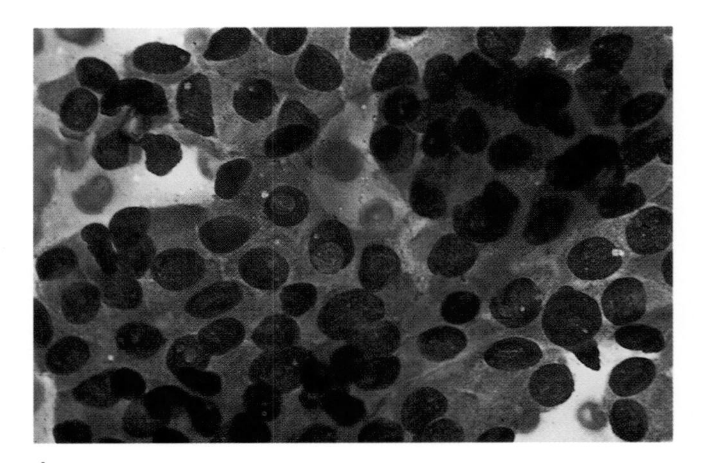

A

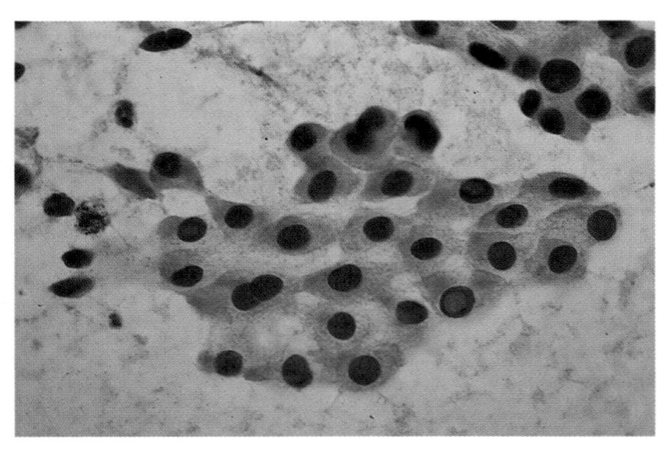

B

FIGURE 3–20. Nuclear inclusions are visualized as sharply demarcated holes within the nucleus. Because of the sharp contrast in the Romanowsky stains, these inclusions are easily appreciated at low power. (**A,** Diff-Quik ×600; **B,** Papanicolaou ×600)

Psammoma bodies, with their concentric lamellae, are very suggestive of a papillary neoplasm but are not often aspirated. This may be the result of fragmentation during aspiration or the blockage of the needle bore with these calcifications. With Romanowsky stains, small psammoma bodies or fragments may be seen as translucent calcifications (Fig. 3–22). Another distinctive feature, better appreciated on Romanowsky stains, is ropy or so-called bubble gum colloid. This dense, vivid metachromatic-staining colloid is thick and acellular and often appears as though it were pulled apart and placed on the smears (Fig. 3–23). Histiocytic multinucleated giant cells are frequently encountered in aspirates of papillary carcinoma; however, they may also be seen in inflammatory and nonneoplastic conditions, as well as in other malignancies. Their presence in papillary carcinoma seems to be a response to leakage of colloid into the interstitium.[43]

Cytomorphologic Features of Papillary Carcinoma

- High cellularity of follicular cells
- Monolayered sheets, papillary fragments
- Multinucleated giant cells
- Ropy colloid, psammoma bodies
- Distinct cell borders, enlarged nuclei
- Powdery chromatin, distinct nucleoli
- Intranuclear inclusions, nuclear grooves

Occasionally, papillary carcinomas undergo cystic degeneration. When hemorrhagic cyst fluid is aspirated and appears to contain small clusters of overlapping follicular cells, careful review for nuclear alterations is essential. Reaspiration of decompressed cysts is also valuable, particularly if there is a residual mass or if the cyst immediately reaccumulates.

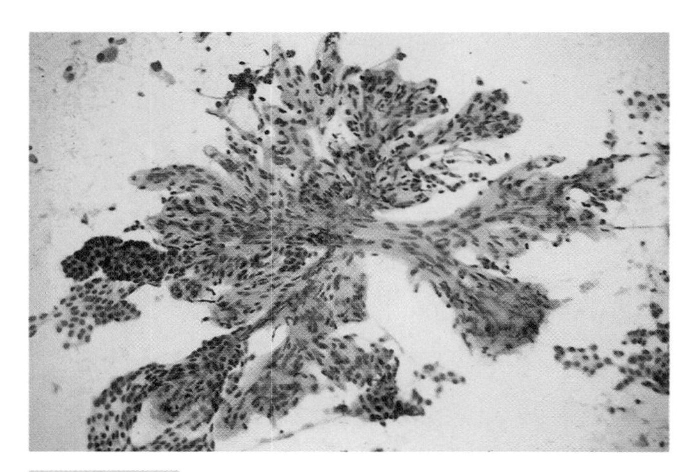

FIGURE 3–21. Neoplastic cells may be stripped from their papillae, leaving bare fibrovascular cores. (Papanicolaou, ×200)

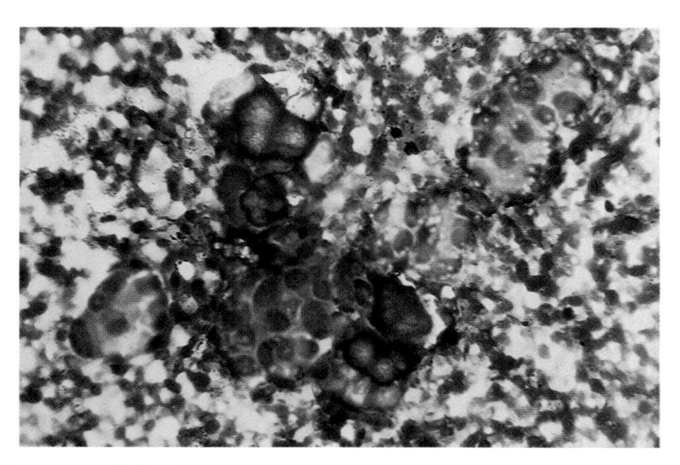

FIGURE 3–22. Psammoma bodies may appear as colorless, refractile calcifications intimately associated with the papillary fragments. (Diff-Quik, ×200)

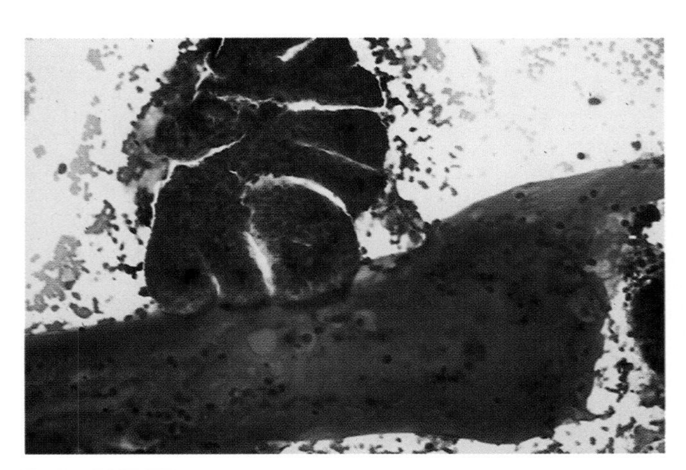

FIGURE 3–23. Thick or ropy colloid associated with papillary carcinoma stains magenta on Romanowsky stains. (Diff-Quik, ×200)

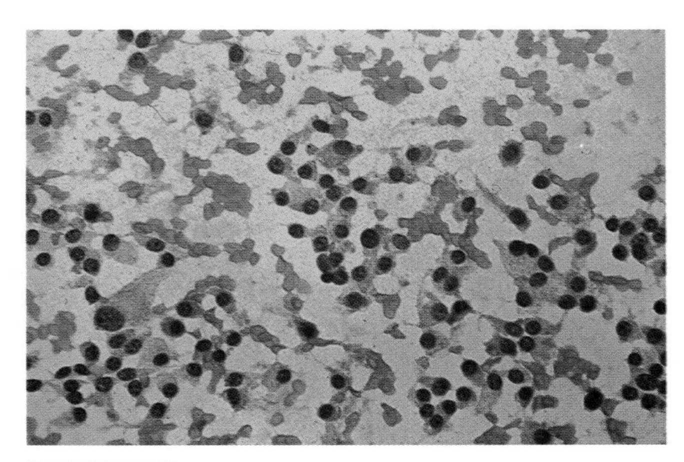

FIGURE 3–24. Discohesive, polygonal cells of "carcinoid"-like medullary carcinoma have a moderate amount of cytoplasm and granular chromatin pattern. At low magnifications this pattern resembles aspiration smears of neuroendocrine tumors. (Papanicolaou, ×200)

Variants of papillary carcinoma have been described. The follicular variant is often difficult to distinguish from a follicular neoplasm; in the absence of distinctive nuclear clues, only histologic evaluation will yield an accurate diagnosis.[44,45] Another variant is the Hürthle cell or oxyphil variant. In addition to their distinctive cytomorphology, the neoplastic cells frequently show nuclear grooves and pseudoinclusions as well as papillary structures.[46] As with the follicular variant, aspiration cytology should be able to detect the presence of a neoplasm and suggest its possible classification, but excision is necessary for definitive diagnosis. Conversely, distinguishing variants of papillary carcinoma from variants of other neoplasms, for example the papillary Hürthle cell tumor, may be virtually impossible using aspiration cytology alone.[47,48] Overlapping criteria, particularly inclusions, grooves, and papillary formations, may be problematic.[49,50] Papillary formations may also be seen in aspirates from foci of hyperplasia in nonneoplastic conditions such as multinodular goiter, or nodules arising during pregnancy (Table 3–6).[51,52]

MEDULLARY CARCINOMA

Medullary carcinoma may occur as a sporadic tumor or associated with multiple endocrine neoplasia (MEN). The development of thyroid nodules in relatives of patients with a history of medullary carcinoma or MEN is easily evaluated with FNA biopsy. The cytomorphology of medullary carcinoma is distinctive, particularly in Romanowsky-stained preparations. At low magnification, the highly cellular smears are composed of large discohesive cells (Fig. 3–24). These large, polygonal neoplastic C cells are frequently bi- and multinucleated, with prominent nucleoli, often with a plasmacytoid appearance.[53] Their cytoplasm is abundant and often manifests neurosecretory granules, present on Romanowsky stain as minute red granules in the perinuclear region or as a red blush at the cytoplasmic periphery (Fig. 3–25). The cytoplasmic borders are characteristically frayed or wispy. Amyloid is occasionally present and is eas-

TABLE 3–6. Differential Diagnoses of Papillary Lesions

Papillary carcinoma
Papillary hyperplasia
Nodular goiter
Pregnancy
Hürthle cell tumor, papillary variant

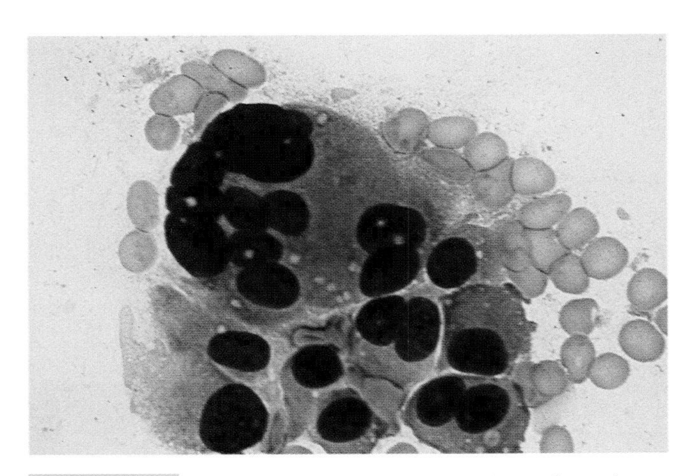

FIGURE 3–25. There can be marked polymorphism of neoplastic cells of medullary carcinoma. Bi- and multinucleation are common. Romanowsky-stained preparations may also show red neuroendocrine granules in the perinuclear region and/or a magenta blush at the cytoplasmic periphery. (Diff-Quik, ×1000)

ily detected on Romanowsky-stained smears as an amorphous metachromatic substance with a slight fibrillary appearance at higher magnifications. On Papanicolaou-stained material, amyloid is cyanophilic and as a result is often mistaken for colloid or even debris (Fig. 3–26). The spindle cell variant of medullary carcinoma may resemble regressive changes seen in degenerating nonneoplastic nodules; however, the latter tend to form cohesive, "reparative" sheets. In addition, spindle cell medullary carcinoma tends to have less cytoplasm and more hyperchromatic nuclei (Fig. 3–27).

Cytomorphologic Features of Medullary Carcinoma

- Moderate to high cellularity of discohesive, monomorphic C cells
- Polygonal, spindle, or "carcinoid"-like
- Bi- and multinucleated, eccentric nuclei
- Neuroectodermal chromatin pattern ("salt and pepper" chromatin)
- Neurosecretory granules (Romanowsky stain)
- Frayed cytoplasmic borders
- Intranuclear inclusions, nucleoli
- Amyloid often present

Occasionally, medullary carcinoma cells resemble Hürthle cells because of their cytoplasmic granularity, and as such can be confused with a Hürthle cell neoplasm. However, medullary carcinoma cells have frayed cytoplasmic borders, unlike the well-defined borders of Hürthle cells. Hürthle cells can also be spindle-shaped or "carcinoid"-like, resulting in confusion with the spindle-cell variant of medullary carcinoma. Rarely, medullary carcinoma may resemble nonthyroidal malignancies such as melanoma, myeloma, and even lymphoma. Positive immunostaining for calcitonin is very specific, particularly when used in conjunction with negative thyroglobulin staining, to detect the presence of medullary carcinoma.

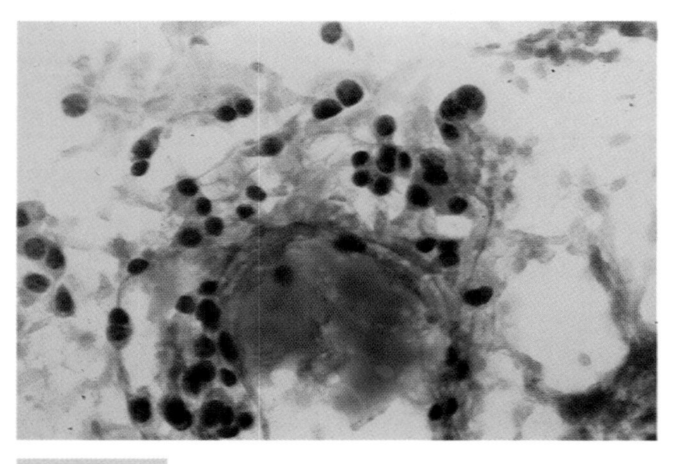

FIGURE 3–26. Amyloid appears as a dense amorphous, cyanophilic material that can resemble inspissated colloid. (Papanicolaou, ×250)

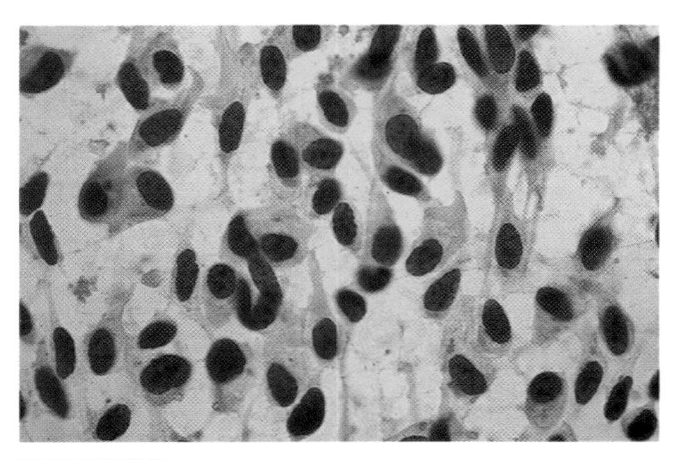

FIGURE 3–27. Spindle-cell medullary carcinoma shows relatively uniform oval nuclei with moderate amounts of cytoplasm. (Papanicolaou, ×400)

ANAPLASTIC CARCINOMA

Anaplastic carcinoma is one of the most aggressive thyroid malignancies. Typically occurring in older patients, this rapidly growing mass can be life-threatening when it compromises the airway. FNA biopsy provides a rapid and accurate diagnosis that in many cases obviates the need for an excisional biopsy and allows the patient to proceed immediately to radiation therapy. The major differential diagnosis, based on clinical presentation alone, is malignant lymphoma. Although the treatment is similar, it is rare that these two malignancies cannot be distinguished on aspiration cytology.

Anaplastic carcinoma tends to be highly cellular with dispersed large, pleomorphic, often bizarre cells scattered amid a necrotic and inflammatory background. Acute inflammatory cells are often seen transgressing the cytoplasm of these obviously malignant cells (Fig. 3–28). Diagnostic difficulties arise if areas of hemorrhage, necrosis, or fibrosis are sampled.[54] Tumor diathesis and regressive changes are uncommonly the sole component of the aspirate or the predominant finding that obscures rare diagnostic cells. In these situations, repeat sampling is encouraged; if aspirations continue to be unsatisfactory and clinical suspicion persists, surgical excision is recommended.

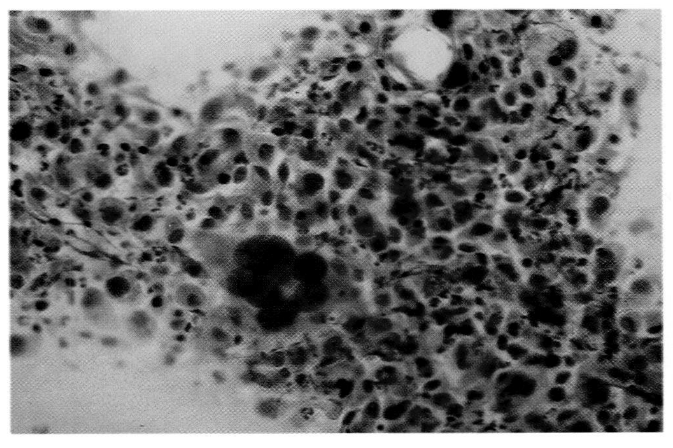

A

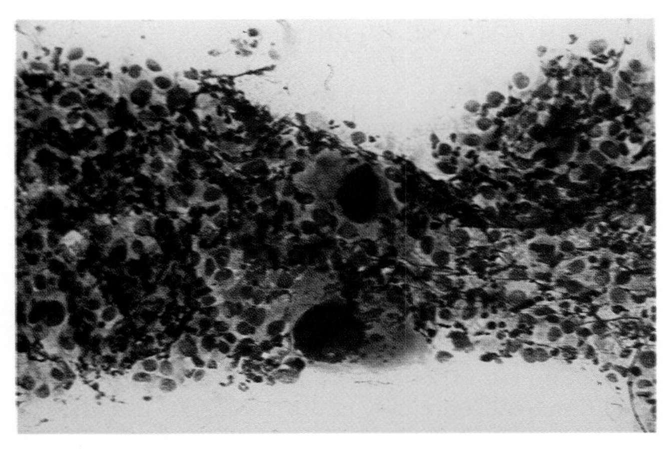

B

FIGURE 3–28. Anaplastic carcinoma. Multinucleated tumor giant cells and obviously malignant cells are scattered in a background of acute inflammation. Neutrophils may be seen within the cytoplasm of the tumor giant cells. (**A,** Papanicolaou, ×400; **B,** Diff-Quik, ×400)

Cytomorphologic Features of Anaplastic Carcinoma

- High cellularity
- Discohesive
- Pleomorphic, often multinucleated cells
- Giant cells and/or sarcomatoid cells
- Cytophagocytosis of neutrophils
- Acute inflammation, necrosis in background

MALIGNANT LYMPHOMA

Most malignant lymphomas present as emergent, rapidly enlarging masses in older patients with no other known predisposing factors. These lymphomas usually are large cell in type and have a B-cell phenotype. Aspirates are highly cellular, composed of a monomorphic population of neoplastic lymphocytes. Lymphoglandular bodies help identify these cells as lymphocytic in origin. An immunopanel composed of lymphoid markers as well as cytokeratins can be used. A second presentation is that of a lymphoma associated with chronic lymphocytic thyroiditis. Diagnosis of these lymphomas may not be as straightforward, because there may be an admixture of nonneoplastic lymphocytes. Flow cytometry or lymphocyte typing is usually essential to help confirm the diagnosis.

METASTATIC DISEASE

Metastases to the thyroid gland represent less than 10% of thyroid neoplasms sampled by FNA. Common primary tumors include those from the lung, gastrointestinal tract, and kidney, as well as melanoma (Figs. 3–29 and 3–30).[55] FNA biopsy is rarely responsible for the *de novo* diagnosis of a nonthyroidal malignancy:

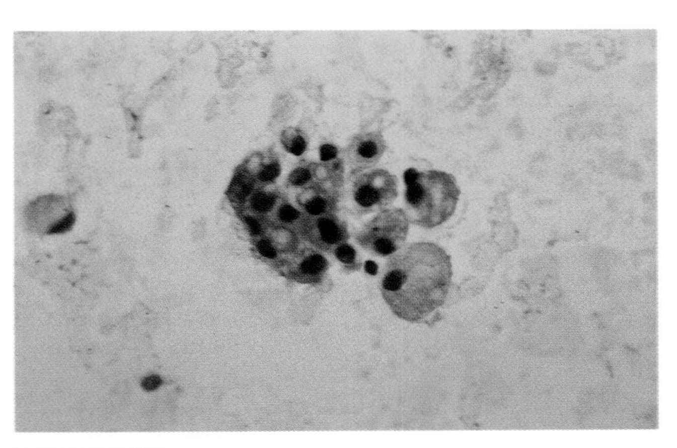

FIGURE 3–29. Renal cell carcinoma metastatic to the thyroid. Renal cell carcinoma is one of the most common metastases to the thyroid. The cells of clear cell renal cell carcinoma are bland, with vacuolated cytoplasms that resemble histiocytes. (Diff-Quik, ×400)

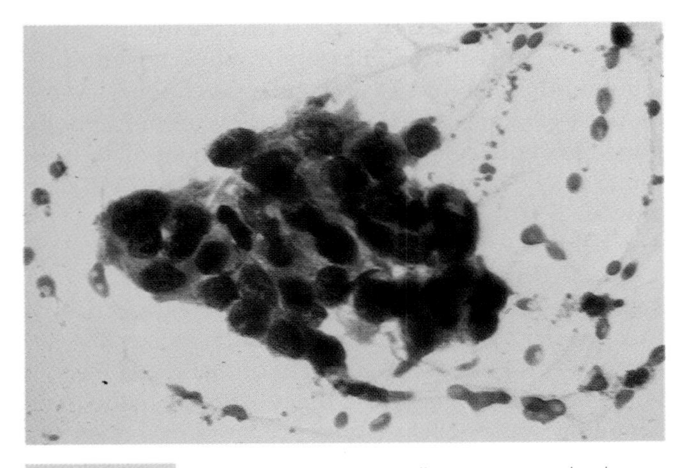

FIGURE 3–30. Metastatic squamous cell carcinoma to the thyroid. Occult head and neck cancer may involve the thyroid by direct extension or metastasis. Well- or moderately differentiated squamous cell carcinomas show evidence of keratin formation; poorly differentiated tumors have obvious malignant features. (Papanicolaou, ×400)

typically, patients present with a thyroid nodule and have a history or confirmed diagnosis of a primary neoplasm elsewhere.

UTILITY, COMPLICATIONS, AND ACCURACY

Most FNA diagnoses of thyroid nodules are benign, paralleling the experience of surgical pathology interpretations. Indeed, many endocrinologists use FNA in conjunction with radiologic and laboratory studies to confirm their clinical impression of a nonneoplastic process rather than to document the presence of malignancy in a clinically suspicious nodule. As a result, FNA of thyroid nodules is often used by clinicians as a triage technique to select patients who require surgery.[56] However, the most common indeterminate diagnosis is the "adenomatous nodule" that may represent a cellular component of nodular goiter or a follicular neoplasm. Many clinicians suggest surgical confirmation of these nodules, particularly if other clinical data are suggestive or equivocal.

Rarely, FNA biopsy results in specific histologic alterations of the area sampled.[57,58] These changes can include infarction of the nodule or extensive hemorrhage, both of which can interfere with histologic evaluation. Hürthle cell lesions appear particularly sensitive to infarction.[59–61] Indeed, oncocytic lesions in general seem prone to infarction and even spontaneous necrosis.[10,13,62–65] Another alteration that can be as disturbing during histologic examination is the reactive changes associated with FNA biopsy, including an exuberant proliferation of granulation tissue that leads ultimately to fibrosis.[66–68] During this process, follicles disrupted by the aspiration procedure may become entrapped in the "healing wound" in the area of the capsule. A history of FNA and the identification of its subsequent tract will usually avoid misinterpretation of these nodules as follicular carcinomas. Many of these changes can be avoided or minimized by good technique and immediate assessment of each aspirate pass. Using very small-gauge needles (23 g, 25 g, or even 27 g) and limiting the number of aspirates will also help.

The sensitivity and specificity of thyroid FNA can vary significantly depending on the categorization of "suspicious" and "indeterminate" diagnoses.[69,70] Accuracy is also heavily dependent on the experience of both the aspirator and the diagnostician. Inadequate sampling and poor specimen preparation remain the predominant reasons for false-negative diagnoses. FNA, when performed by inexperienced physicians, often results in paucicellular specimens or specimens diluted by blood.[71] In a study of nondiagnostic thyroid aspirates by MacDonald and Yazdi,[72] more than 98% of patients with this diagnosis had benign lesions confirmed by histopathology. Additional, often unavoidable, problems include occult tumors within or adjacent to nonneoplastic nodules. In a review of 13 published series, Bedrossian et al.[73] calculated the mean sensitivity and specificity to be 94% and 96%, respectively. Other recent studies have reached similar conclusions.[74–77]

The greatest problem facing the cytopathologist is the overlapping spectrum of cytomorphologic features of thyroid disease, and the lack of a single pathognomonic cytomorphologic criterion for any specific thyroid lesion. The degree of overlap of these criteria can even make it difficult to categorize lesions as nonneoplastic or neoplastic.[78–84] Therefore, correlation of clinical and radiologic data with aspiration cytology, including familiarity with the cytomorphologic pitfalls, is needed to obtain diagnostic accuracy.

PARATHYROID

FNA of the parathyroid glands is not routinely requested; however, parathyroid cysts and adenomas may be encountered as "thyroid" nodules. Although parathyroid is easily differentiated from thyroid on histologic sections, such is not always the case in cytomorphologic preparations. Touch imprints from parathyroidectomy specimens can help familiarize one with the subtle differences. As with thyroid and other endocrine organs, smears from parathyroid glands often contain nuclei arranged in loosely organized rosette-like structures, linear arrays of cells, and numerous stripped round nuclei. Colloid is conspicu-

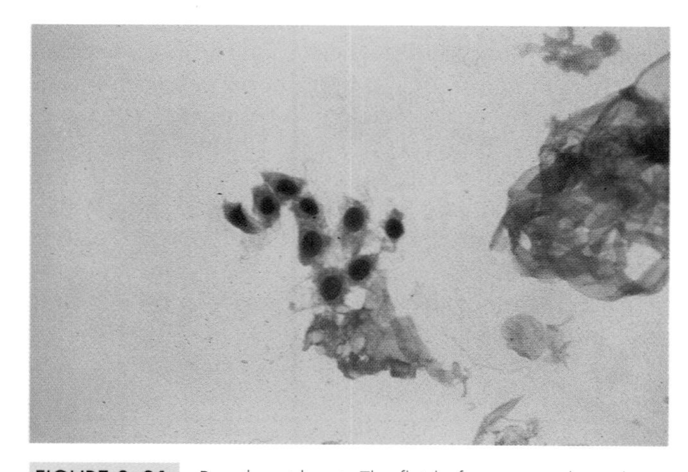

FIGURE 3–31. Parathyroid cyst. The fluid of most parathyroid cysts is water-clear and acellular. However, occasionally small, bland, degenerating cells from the cyst wall are identified. (Papanicolaou, ×400)

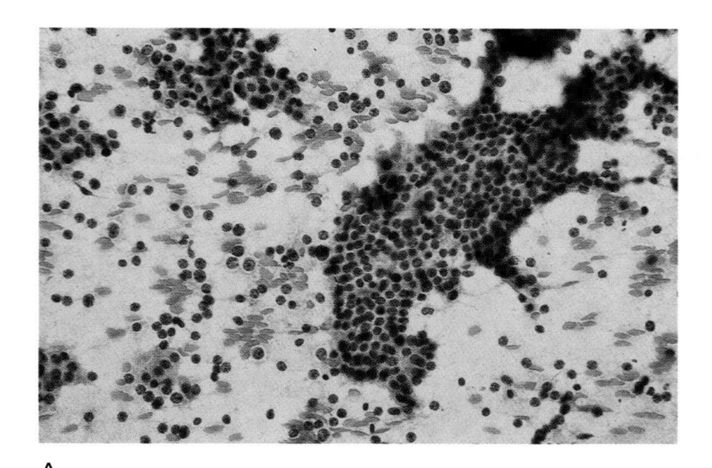
A

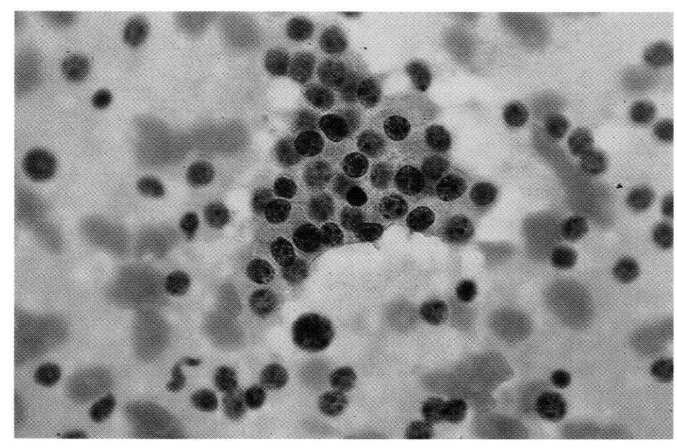
B

FIGURE 3–32. Parathyroid adenoma. Aspirations from parathyroid adenomas and parathyroid hyperplasia are indistinguishable. Smears are highly cellular with uniform small cells with finely granular cytoplasm that varies from scant to moderate; it is easily stripped from the nuclei during smearing. Nuclei show the typical endocrine stippled chromatin pattern with indistinct nucleoli. (Papanicolaou; **A,** ×200; **B,** ×400)

ously absent, but abundant lipid is noted in the background of the air-dried smears.

Parathyroid cysts are probably the most frequently aspirated parathyroid lesions.[85,86] When cyst fluid alone is aspirated, it is water-clear and acellular or contains small degenerating fragments of cells (Fig. 3–31). Although this appearance is virtually pathognomonic for parathyroid cyst, assay of parathormone levels will confirm this diagnosis.[87]

The distinction between parathyroid hyperplasia and adenoma is based on the principles used in surgical pathology. Radiologic studies to identify the presence and location of one or more masses are essential once aspirated material has been identified as parathyroid. A single nodule is consistent with an adenoma; multiple nodules reflect hyperplasia.[88,89] Aspirates are usually cellular, with dispersed small monotonous cells with scant cytoplasm and round regular nuclei. Anisonucleosis is common, and the chromatin is stippled, often with small distinct nuclei (Fig. 3–32).

REFERENCES

1. Silverman JF, West RL, Larkin EW, et al. The role of fine-needle aspiration biopsy in the rapid diagnosis and management of thyroid neoplasm. Cancer 57:1167, 1986.
2. Hamburger JI, Miller JM, Kini SR. Clinical-pathological evaluation of thyroid nodules—handbook and atlas. Private publication, 1979.
3. Frable WJ. Thin-needle aspiration biopsy. In: Bennington JL (ed): Major problems in pathology, vol 14. Philadelphia: WB Saunders, 1983.
4. Frable MA, Frable WJ. Thin-needle aspiration biopsy of the thyroid gland. Laryngoscope 90:1619, 1980.
5. Griffies WS, Donegan E, Abel ME. The role of fine needle aspiration in the management of the thyroid nodule. Laryngoscope 95:1103, 1985.
6. Frable WJ. The treatment of thyroid cancer: the role of fine needle aspiration cytology. Arch Otolaryngol Head Neck Surg 112:1200, 1986.
7. Asp AA, Georgitis W, Waldron EJ, et al. Fine needle aspiration of the thyroid: use in an average health care facility. Am J Med 83:489, 1987.
8. Caruso D, Mazzaferri EL. Fine needle aspiration biopsy in the management of thyroid nodules. Endocrinologist 1:194, 1991.
9. Bisi H, de Camargo RYA, Fihlo AL. Role of fine-needle aspiration cytology in the management of thyroid nodules: review of experience with 1,925 cases. Diagn Cytopathol 8:504, 1992.
10. Powers CN. Complications of fine needle aspiration biopsy. The reality behind the myths. In: Schmidt WA (ed): Review in pathology: cytopathology. Baltimore: ASCP Press, 1996:69.
11. Haas SN. Acute thyroid swelling after needle biopsy of the thyroid. N Engl J Med 307:1349, 1982.
12. Dal Fabbro S, Barbazza R, Fabris C, Perelli R. Acute thyroid swelling after fine needle aspiration biopsy. J Endocrinol Invest 10:105, 1987.
13. Powers CN, Frable WJ. Fine needle aspiration biopsy of the head and neck (pp. 49–73). Boston: Butterworth-Heinemann, 1996.
14. Hamburger JI, Husain M. Semiquantitative criteria for fine needle biopsy diagnosis: reduced false-negative diagnosis. Diagn Cytopathol 4:14, 1988.
15. Hamburger JI, Husain M, Nishiyama R, et al. Increasing the accuracy of fine-needle biopsy for thyroid nodules. Arch Pathol Lab Med 113:1035, 1989.
16. Kini SR. Needle aspiration biopsy of the thyroid: revisited. Diagn Cytopathol 9(3):249, 1993.
17. Meko JB, Norton JA. Large cystic/solid thyroid nodules: a potential false-negative fine-needle aspiration. Surgery 118:996, 1995.
18. Guarda LA, Baskin J. Inflammatory and lymphoid lesions of the thyroid gland: cytopathology of fine needle aspiration. Am J Clin Pathol 87:14, 1987.
19. Tani E, Skoog L. Fine needle aspiration cytology and immunocytochemistry in the diagnosis of lymphoid lesions of the thyroid gland. Acta Cytol 33:48, 1989.

20. Carson HJ, Castelli MJ, Gattuso P. Incidence of neoplasia on Hashimoto's thyroiditis: a fine-needle aspiration study. Diagn Cytopathol 14:38, 1996.
21. Szporn AH, Tepper S, Watson CW. Disseminated cryptococcosis presenting as thyroiditis. Acta Cytol 29:449, 1985.
22. Solary E, Rifle G, Chalopin JM. Disseminated aspergillosis revealed by thyroiditis in a renal allograft recipient. Transplant 44:839, 1987.
23. Vaidya KP, Lomvardias S. Cryptococcal thyroiditis: report of a case diagnosed by fine-needle aspiration cytology. Diagn Cytopathol 7:415, 1991.
24. Jayaram G, Singh B, Marwaha RK. Graves' disease: appearance in cytologic smears from fine needle aspirates of the thyroid gland. Acta Cytol 33:3640, 1989.
25. Nilsson G. Marginal vacuoles in fine needle aspiration biopsy smears of toxic goitres. Acta Pathol Microbiol Scand (A) 80:289, 1972.
26. Volavsek M, Us-Krasovec M, Auserspergn M, et al. Marginal vacuoles in fine-needle aspirates of follicular thyroid carcinoma. Diagn Cytopathol 15:93, 1996.
27. Centeno BA, Szyfelbein WM, Daniels GH, Vickery AL. Fine needle aspiration biopsy of the thyroid gland in patients with prior Graves' disease treated with radioactive iodine: morphologic findings and potential pitfalls. Acta Cytol 40:1189, 1996.
28. Granter SR, Cibas ES. Cytologic findings in thyroid nodules after ^{131}I treatment of hyperthyroidism. Am J Clin Pathol 107:20, 1997.
29. Frable WJ: Controversies in the pathology and cytology of well-differentiated thyroid carcinoma. Head Neck 2:170, 1990.
30. Suen KC: How does one separate follicular lesions of the thyroid by fine-needle aspiration biopsy? Diagn Cytopathol 4:78, 1988.
31. Ravinsky E, Safneck JR. Fine needle aspirations of follicular lesions of the thyroid gland: the intermediate type smear. Acta Cytol 34:813, 1990.
32. Layfield L. Editorial comments: fine-needle aspiration evaluation of the solitary thyroid nodule—the imprecision of diagnostic criteria. Diagn Cytopathol 9:355, 1993.
33. Kini SR, Miller JM, Hamburger JI. Cytopathology of Hürthle cell lesions of the thyroid gland by fine needle aspiration. Acta Cytol 25:647, 1981.
34. Gonzalez JL, Wang HH, Ducatman BS. Fine-needle aspiration of Hürthle cell lesions: a cytomorphologic approach to diagnosis. Am J Clin Pathol 100:231, 1993.
35. Pambuccian SE, Becker RL, Ali SZ, et al. Differential diagnosis of Hürthle cell neoplasms on fine needle aspirates: can we do any better with morphometry? Acta Cytol 41:197, 1997.
36. Kini SR, Miller JM, Hamburger JI, Smith MJ. Cytopathology of papillary carcinoma of the thyroid by fine-needle aspiration. Acta Cytol 24:511, 1980.
37. Miller JM, Hamburger JI, Kini SR. The needle biopsy diagnosis of papillary thyroid carcinoma. Cancer 48:989, 1981.
38. Akhtar M, Ali MA, Huq M, Bakry M. Fine-needle aspiration biopsy of papillary thyroid carcinoma: cytologic, histologic and ultrastructural correlations. Diagn Cytopathol 7:373, 1991.
39. Kaneko C, Shamoto M, Niimi H, et al. Studies on intranuclear inclusions and nuclear grooves in papillary thyroid cancer by light, scanning electron and transmission electron microscopy. Acta Cytol 40:417, 1996.
40. Gould, E, Watzak L, Chamizo W, Albores-Saavedra J. Nuclear grooves in cytologic preparations: a study of the utility of this feature in the diagnosis of papillary carcinoma. Acta Cytol 33:16, 1988.
41. Shurbaji MS, Gupta PK, Frost JK. Nuclear grooves: a useful criterion in the cytopathologic diagnosis of papillary thyroid carcinoma. Diagn Cytopathol. 4:91, 1988.
42. Rupp M, Ehya H. Nuclear grooves in the aspiration cytology of papillary carcinoma of the thyroid. Acta Cytol 33:21, 1988.
43. Guiter GE, DeLellis RA. Multinucleate giant cells in papillary thyroid carcinoma: a morphologic and immunohistochemical study. Am J Clin Pathol 106:765, 1996.
44. Martinez-Parra D, Campos Fernandez J, Hierro-Guilmain CC, et al. Follicular variant of papillary carcinoma of the thyroid: to what extent is fine-needle aspiration reliable? Diagn Cytopathol 15:12, 1996.
45. Gallagher J, Oertel YC, Oertel JE. Follicular variant of papillary carcinoma of the thyroid: fine-needle aspirates with histologic correlation. Diagn Cytopathol 16:207, 1997.
46. Dzieciol J, Musiatowicz B, Zimnoch L, et al. Papillary Hürthle cell tumor of the thyroid: report of a case with a cytomorphologic approach to diagnosis. Acta Cytol 40:311, 1996.
47. Doria MI, Attal H, Wang HH, et al. Fine needle aspiration cytology of the oxyphil variant of papillary carcinoma of the thyroid: a report of three cases. Acta Cytol 40:1007, 1996.
48. Kaur A, Jayaram G. Thyroid tumors: cytomorphology of Hürthle cell tumors, including an uncommon papillary variant. Diagn Cytopathol 9:135, 1993.
49. LiVolsi VA, Gupta PK. Thyroid fine-needle aspiration: intranuclear inclusions, nuclear grooves and psammoma bodies—paraganglioma-like adenoma of the thyroid. Diagn Cytopathol 8:82, 1992.
50. Fraser CR, Marley EF, Oertel YC. Papillary tissue fragments as a diagnostic pitfall in fine-needle aspirations of thyroid nodules. Diagn Cytopathol 16:454, 1997.
51. Betsill W. Thyroid fine needle aspiration in pregnant women. Diagn Cytopathol 1:53, 1985.
52. Fukuda K, Hachisuga T, Sugimori H, Tsuzuku M. Papillary carcinoma of the thyroid occurring during pregnancy: report of a case diagnosed by fine needle aspiration cytology. Acta Cytol 35:725, 1991.
53. Collins BT, Cramer HM, Tabatowski K, et al. Fine needle aspiration of medullary carcinoma of the thyroid. Cytomorphology, immunocytochemistry and electron microscopy. Acta Cytol 39:920, 1995.
54. Us-Krasovec M, Golouh R, Auersperg M, et al. Anaplastic thyroid carcinoma in fine needle aspirates. Acta Cytol 40:953, 1996.
55. Michelow PM, Leiman G. Metastases to the thyroid gland: diagnosis by aspiration cytology. Diagn Cytopathol 13:209, 1995.
56. Gharib H. Fine-needle aspiration biopsy of thyroid nodules: advantages, limitations and effect. Mayo Clin Proc 69:44, 1994.
57. LiVolsi VA, Merino MJ. Worrisome histologic alterations following fine needle aspiration of thyroid. Mod Pathol 3:59A, 1990.
58. Us-Krasovec M, Golouh R, Auesperg M, Pogacnik A: Tissue damage after fine needle aspiration biopsy. Acta Cytol 36:456, 1992.
59. Keyhani-Rofagha S, Kooner DS, Keyhani M, O'Toole RV. Necrosis of a Hürthle cell tumor of the thyroid following fine needle aspiration: case report and literature review. Acta Cytol 34:805, 1990.
60. Bauman A, Strawbridge HTG. Spontaneous disappearance of an atypical Hürthle cell adenoma. Am J Clin Pathol 80:399, 1983.
61. Ramp U, Pfitzer P, Gabbert HE. Fine needle aspiration (FNA)-induced necrosis of tumors of the thyroid. Cytopathol 6:248, 1995.
62. Jones JD, Pittman DL, Sanders LR. Necrosis of thyroid nodules after fine needle aspiration. Acta Cytol 29:29, 1985.
63. Kini SR, Miller JM. Infarction of thyroid neoplasms following aspiration biopsy. Acta Cytol 30:591, 1986.
64. Jayaram G, Aggarwal S. Infarction of thyroid nodule: a rare complication following fine needle aspiration. Acta Cytol 33:940, 1989.
65. Alejo M, Matias-Guiu X, de las Hera Duran P. Infarction of a papillary thyroid carcinoma after fine needle aspiration. Acta Cytol 35:478, 1991.
66. Axiotis CA, Merino MJ, Ain K, Norton JA. Papillary endothelial hyperplasia in the thyroid following fine-needle aspiration. Arch Pathol Lab Med 5:240, 1991.
67. Tsang K, Duggan MA. Vascular proliferation of the thyroid: a complication of fine-needle aspiration. Arch Pathol Lab Med 116:1040, 1992.

68. Ersoz C, Soylu L, Erkocak EU, et al. Histologic alterations in the thyroid gland after fine-needle aspiration. Diagn Cytopathol 16:230, 1997.
69. Akerman M, Tennvall J, Biorklund A, et al. Sensitivity and specificity of fine needle aspiration cytology in the diagnosis of tumors of the thyroid gland. Acta Cytol 29:850, 1985.
70. Suen KC, Quenville NF. Fine needle aspiration of the thyroid gland: a study of 304 cases. J Clin Pathol 36:1036, 1983.
71. Burch HB, Burman KD, Reed HL, et al. Fine needle aspiration of thyroid nodules. Determinants of insufficiency rate and malignancy yield at thyroidectomy. Acta Cytol 40:1176, 1996.
72. MacDonald L, Yazdi HM. Non-diagnostic fine needle aspiration biopsy of the thyroid gland: a diagnostic dilemma. Acta Cytol 40:423, 1996.
73. Bedrossian CWM, Martinez F, Silverberg AB. Fine needle aspiration. In: Gnepp DR (ed): Pathology of the head and neck. New York: Churchill Livingstone, 1988:25.
74. Mandreker SR, Nadkarni NS, Pinto RG, Menezes S. Role of fine needle aspiration cytology as the initial modality in the investigation of thyroid lesions. Acta Cytol 39:898, 1995.
75. Capri A, Ferrari E, DeGaudio C, et al. The value of needle aspiration biopsy in evaluating thyroid nodules. Thyroidology 6:5, 1994.
76. Vojvdich SM, Ballagh RH, Cramer H, Lampe HB. Accuracy of fine needle aspiration in the preoperative diagnosis of thyroid neoplasia. J Otolaryngol 23:360, 1994.
77. Sirpal YM. Efficacy of fine needle aspiration cytology in the management of thyroid disease. Indian J Pathol Microbiol 39:173, 1996.
78. Ananthakrishnan N, Rao KM, Narasimhan R, Veliath AJ. Problems and limitations of fine needle aspiration cytology of solitary thyroid nodule. Aust NZ J Surg 60:35, 1990.
79. Harach HR, Zusman SB, Day ES. Nodular goiter: a histo-cytological study with some emphasis on pitfalls of fine-needle aspiration cytology. Diagn Cytopathol 8:409, 1992.
80. Florella RM, Isley W, Miller LK, Kragel PJ. Multinodular goiter of the thyroid mimicking malignancy: diagnostic pitfalls in fine-needle aspiration biopsy. Diagn Cytopathol 9:351, 1993.
81. Caraway NP, Sneige N, Samaan NA. Diagnostic pitfalls in thyroid fine-needle aspiration: a review of 394 cases. Diagn Cytopathol 9(3):345, 1993.
82. Heinmann A, Gritsman A. Diagnostic problems and pitfalls in aspiration cytology of thyroid nodules. In: Schmidt WA (ed): Cytopathology annual. Baltimore: Williams and Wilkins, 1993.
83. Jayaram G. Fine needle aspiration cytologic study of the solitary thyroid nodule: profile of 308 cases with histologic correlation. Acta Cytol 29:967, 1985.
84. Hsu C, Boey J. Diagnostic pitfalls in fine needle aspiration of thyroid nodules. Acta Cytol 31:699, 1987.
85. Katz Ad, Dunkelman D. Needle aspiration of nonfunctional parathyroid cysts. Arch Surg 119:307, 1984.
86. Layfield LJ. Fine-needle aspiration cytology of cystic parathyroid lesions: a cytomorphologic overlap with cystic lesions of the thyroid. Acta Cytol 35:447, 1991.
87. Silverman JF, Khazanie PG, Norris HT, Fore WW. Parathyroid hormone (PTH) assay of parathyroid cysts examined by fine needle aspiration biopsy. Am J Clin Pathol 88:252, 1987.
88. Clark OH, Gooding GAW, Ljung BM. Locating a parathyroid adenoma by ultrasonography and aspiration biopsy cytology. West J Med 135:154, 1981.
89. Solbiati L, Montali G, Croce F, et al. Parathyroid tumors detected by fine-needle aspiration biopsy under ultrasound guidance. Radiology 148:793, 1983.

CHAPTER 4

Salivary Glands and Other Head and Neck Masses

A wide variety of tumefactive pathologic processes may be seated in the organs and tissues of the head and neck. Because exfoliative cytology of the oral cavity is little used at this time, cytologic examination of these entities is based on fine needle aspiration (FNA). This chapter will address clinical issues and lesions that are unique to the head and neck or that are most often considered in this area. We will emphasize the correlation of clinical and cytologic findings that is essential for success in FNA. Other authors address the cytology of lymph nodes, the thyroid, and the soft tissues; these will not be considered in detail here.

CLINICAL ISSUES

General Considerations

Approach to FNA. Techniques for FNA performance and specimen preparation have been described and illustrated in detail, and potential complications have been reviewed.[1] We will not discuss these subjects in depth, but several important issues should be emphasized in the context of aspirations directed to the head and neck (Table 4–1). Many of FNAs benefits and much of its safety can be directly attributed to the small caliber of the needles. These are usually 23 or 25 gauge (0.6- or 0.5-mm outside diameter, respectively). In our practice, we almost always use 25-gauge needles. Patient acceptance of these instruments is excellent, and repeat aspirations that might be required to secure adequate material for either diagnosis or special studies are rarely refused. Such requests for permission to perform additional aspirations are most common when one needs material for microbiologic culture or for flow cytometric analysis of lymphoid proliferations.

Wide sampling and abundant tissue recovery can be achieved when the needle is moved widely through a palpable abnormality. This type of sampling would be unsafe or intolerable if larger needles were used. Although small hematomas and local tenderness may follow the procedure, significant hemorrhage, nerve damage, and infection do not occur, even in sites as complex as the parotid gland and its intercourse with the seventh cranial nerve. This freedom from complications is also attributable to the use of very thin needles.[2–6] One example of sudden-onset facial nerve damage and pain following aspiration of a hemangioma was probably the result of an enlarging hematoma; seventh nerve function was completely recovered within 6 weeks.[7] Extensive clinical observa-

TABLE 4–1. Advantages of 23–25-Gauge Needles for FNA

- Complications are very rare and almost always minor.
- Patient acceptance is good.
- Patients readily accept repeat aspirations as needed.
- Generous tissue recovery with minimal hemorrhage.
- Large volumes of tissue can be sampled safely.

tions and numerous animal experiments demonstrate that metastases attributable to tumor cell dissemination by FNA do not occur. Further, local implantation and recurrence are almost always associated with larger-gauge instruments, such as those used for histologic core biopsies (reviewed in reference 3). Nonetheless, some continue to express concern about these issues.[8]

High-quality specimen preparation is essential for accurate diagnosis.[1] The details of smearing, cell block embedding, and allocation of material for ancillary testing (e.g., flow cytometry, immunocytochemistry, electron microscopy, microbiologic culture) are important. We do not routinely employ needle-rinse specimens.[9] We have found that this method is time-consuming and does little to improve the overall diagnostic yield.

Anatomic Considerations. Head and neck anatomy is very complex, and several factors must be kept in mind when examining patients. We have noted repeatedly that persons lacking experience in head and neck examinations occasionally mistake elements of the normal anatomy for pathologic masses. We have been asked to aspirate "lesions" that are shown to represent normal bones, muscles, or the carotid bulb. Clearly, skill and caution are required if one is to avoid inappropriate needle placement. Further, care must be taken to avoid misadventures occasioned by the vital structures that may lie near the needle's path. Some of these are unique to the head and neck (Table 4–2).

The complexity of head and neck anatomy also makes it difficult to determine the site of some lesions. Tumors may seem to be in the lower parotid, when they may actually represent the gland itself, a lymph node within the parotid, a cervical lymph node that is not within the parotid, a lateral neck cyst, a dermal-based mass, or a soft-tissue swelling. In some instances, FNA may direct clinical thinking by providing confirmation of the mass's location, even if a specific diagnosis is not obtained.[10]

Cervical ribs occur in up to 1.5% of patients[11] and may be mistaken for supraclavicular or low cervical lymphadenopathy.[12] We have seen one example in which clinically suspected Hodgkin's disease was not confirmed by FNA of a supraclavicular mass; excised tissues were normal, and the antecedent FNA had shown only unremarkable soft tissues. Bones of the cervical spine can be mistaken for lymph nodes or cysts of the neck.[12] In contrast to cervical ribs, which are too deep to touch with the aspirating needle, bony elements of the spine are usually palpated through the needle. This can resolve questions about the nature of the palpable "mass." Radiographic evaluation adds confidence to this evaluation; thus, careful consultation among the pathologist, the clinician, and the radiologist may prevent unnecessary surgery.

The pulmonary cupola lies above the clavicle, and this raises the danger of pneumothorax when targets located low in the neck or in the supraclavicular fossa are aspirated.[3] Similar problems attend aspiration of the breast, the chest wall, or the axilla in thin patients. This complication of FNA is very uncommon and has been discussed in detail elsewhere.[1] Fortunately, lymph nodes that are palpable in the supraclavicular fossa lie superficial to the lung; careful attention to the needle's placement and depth will suffice to avoid pleural puncture. It appears that most FNA-induced pneumothoraces cause pain but do not require chest tube placement, as long as a very thin needle is responsible for the injury. However, the risk of severe complications certainly exists in patients with compromised pulmonary function. This is most often a concern when thin lung cancer patients with advanced smoking-induced pulmonary injury undergo aspiration.

In patients with head and neck cancer, one frequently performs aspirations after the expected anatomy has been altered considerably by surgery and radiation therapy. In this setting, physical examination of the neck may be very difficult. After radical neck dissections, one often aspirates lesions lying very close to the carotid artery. Such masses seem disturbingly prominent, with transmitted pulses. Other postoperative alterations in which one may be asked to evaluate the possibility of recurrent malignancy include lumps that are shown to be fibrous tissue, damaged skeletal muscle, suture granulomas, or even postoperative neuromas. The cytologic manifestations of such masses are considered subsequently.

TABLE 4–2. Normal Anatomic Structures Sometimes Mistaken for Pathologic Masses

- Normal skeletal structures
- Cervical ribs or prominent lateral vertebral elements
- Prominent muscular or connective tissue structures
- Carotid bulb
- Laryngeal cartilages

Special Considerations

Cyst Aspiration. Many head and neck masses are cystic or partially cystic (Table 4–3). Commonly aspirated examples include congenital cysts (midline or lateral), dermal-based cysts of cutaneous origin, several types of salivary gland neoplasms, and cystic lymph node metastases. The latter are most commonly the result of either squamous cell carcinoma or papillary thyroid carcinoma. Specific cytodiagnostic problems related to these entities are discussed subsequently.

Optimal evaluation of cysts by FNA requires good techniques for aspiration and specimen preparation (Table 4–4).[1,13–17] Unlike the breast, in which many cysts can be completely drained, leaving no palpable residuum, the head and neck frequently host masses that are only partially cystic. Although some fluid can be easily recovered, it can be nondiagnostic, even in the case of cystic neoplasms. However, more solid areas of the lesion will remain after partial drainage, and these should be the target of subsequent aspirations. During aspiration of cyst fluid, the needle should be moved about within the lesion very little. If vigorous motions similar to those used to sample solid masses are used, a cyst can refill with blood. When this happens, clinical and cytologic assessment, as well as repeat aspiration, becomes problematic. This problem of feeling a cyst literally refill under one's fingertips is most common in the thyroid, where the surrounding tissue is extremely vascular. However, it can occur in other sites as well.

All fluids obtained by aspiration of masses in the head and neck must be studied carefully because they may contain very few diagnostically significant cells. (This is in contrast to breast aspiration, after which many surgeons discard nonbloody cyst fluids, eschewing cytologic examination of this material.) Cytologic examination is facilitated when centrifugation or filtration is used to concentrate the fluid's cellular elements onto a manageable glass slide area. Thus, adequate evaluation of cyst fluids need not entail laborious screening of numerous slides prepared as direct smears from the unconcentrated specimen.

TABLE 4–3. Cystic or Partially Cystic Head and Neck Masses

- Congenital cysts, midline
- Congenital cysts, lateral
- Dermal-based cutaneous cysts
- Nonneoplastic salivary gland duct obstructive lesions
- Salivary gland neoplasms of several types
- Metastatic squamous cell carcinoma
- Metastatic papillary thyroid carcinoma

TABLE 4–4. Evaluation of Cystic Head and Neck Lesions

- Remove as much fluid as possible.
- Any repositioning of the needle during aspiration must be very gentle to avoid hemorrhage and refilling of the cyst.
- After aspiration, check for any residual mass.
- Any residual mass should be aspirated.
- All cyst fluid should be centrifuged.
- Cell pellets can be used for smear preparation.
- Cell pellets can also be used to make paraffin-embedded cell blocks.

Microbiologic Evaluation of Aspirated Material. Aspiration of grossly purulent or malodorous material should lead to immediate reaspiration for microbiologic culture.[18] Other signs of possible infection must also be considered, such as redness, tenderness, a history of immunosuppression, or other factors that may raise the possibility of infection rather than neoplasm.[19] In other instances, the need for cultures may be apparent only after initial cytologic examination of the aspirated material (Table 4–5).

Specimen requirements and methods for thorough evaluation of potentially infectious aspirations have been described in detail.[20] We often consult our colleagues in microbiology before performing a repeat aspiration for culture. In this way, we receive the best possible advice regarding any special handling requirements that may be needed for productive culture of head and neck samples. Specific guidelines attend the search for mycobacteria, actinomyces, fungi, or anaerobic bacteria. Potentially infectious aspirates must also be examined carefully at the microscope; even purulent material should be searched for organisms. Some, including actinomyces, may be missed because few diagnostic organisms are present.[21] In other instances, such as leishmaniasis, the organisms may be very small and easily overlooked.[22,23]

Transoral FNA. Some of the lesions considered in this chapter (such as Warthin's tumor) are confined almost exclusively to the major salivary glands, where they

TABLE 4–5. Microbiologic Evaluation of Aspirates

- Masses that are red, tender, or warm may require culture.
- Immunosuppressed patients frequently have tumefactive infections that require culture and consideration of a wide range of organisms.
- Be aware of special handling requirements for certain organisms.
- Consult the microbiologist as needed.
- Always check for organisms when smears are examined.
- Inflammation may be associated with a necrotic neoplasm.

can be approached by transcutaneous FNA in the usual manner. However, a wide variety of lesions can be located in the oral cavity, tongue, or parapharyngeal space, where they are best approached by transoral FNA.[24,25] Salivary gland tumors, metastatic malignancies of various types, and soft-tissue tumors can all be diagnosed by this method. Initially, many such lesions present little in the way of visual or palpable abnormality. However, one can learn a great deal by review of relevant CT scans. This can provide information about the precise location of the mass, the best direction of approach, the depth that the needle can safely penetrate, and the proximity of vital anatomic structures. This type of radiologic review commonly provides considerable confidence in the safety of aspiration, when a large mass is seen to have pushed important normal structures aside and away from the advancing needle. Less commonly, we decline the request to perform the procedure when this type of reassurance is not forthcoming. Aspiration with some type of radiologic guidance is usually an acceptable alternative in these circumstances.

Patients tolerate transoral FNA with 25-gauge needles very well. Even physicians who do not use local anesthesia for percutaneous FNA typically spray topical anesthetic agents onto the target mucosa before transoral sampling. (Details of local anesthesia in FNA have been reviewed.[1])

In a variation on this procedure, bimanual palpation is used to facilitate aspiration of submandibular masses. These lymph nodes commonly recede behind the mandible, where they are difficult to palpate and virtually impossible to approach with a needle. A gloved hand can be placed in the patient's mouth and used to push the mass downward, where it becomes a stable target for aspiration.[1] Patients often find this maneuver uncomfortable but not intolerable. Improvement is described by some if the gloved hand is moistened with tap water and washed free of talc before the procedure is initiated.

FNA-Associated Tissue Damage. Problems with histologic interpretation of head and neck tissues excised after FNA are uncommon. Rare instances of nearly total infarction shortly after FNA leave only a thin peripheral rim of viable tissue. We have aspirated several thousand tumors and have observed this phenomenon on only five occasions. The following masses were involved: a benign hyperplastic lymph node (Fig. 4–1), a presumably cervical lymph node-based large cell malignant lymphoma, a thyroid adenoma that showed complete encapsulation in histologic sections, a thymoma, and a breast fibroadenoma. Most agree that post-FNA infarction is rarely a cause of difficulties in the histologic interpretation of excised masses.[26] Other reports describe similar occurrences in Warthin's tumors and pleomorphic adenomas,[27–29] as well as in thyroid nodules[30] and lymph nodes.[31]

There is no way to predict which masses may be at risk for this rare complication of FNA. Thrombosis with circulatory compromise has been suggested as a cause for such occurrences. Further, squamous metaplasia in pleomorphic adenomas may be due in part to ischemia.[28] However, few post-FNA infarcts have been associated with identifiable causes for ischemia, such as ill-fitting dental appliances, antecedent myocardial infarction, or cocaine use. Thus, a pathogenetic sequence similar to that postulated for necrotizing sialometaplasia may be operative.[32] Infarction of salivary gland tumors and necrotizing sialometaplasia also share the sudden onset of pain.[32] Spontaneous infarction in the absence of antecedent FNA has been described in pleomorphic adenomas and monomorphic adenomas, as well as in lymph nodes replaced by malignant lymphoma.[32,33]

A history of previous aspiration may lead to recognition of small areas of hemorrhage or granulation tissue when histologic sections are examined. Such foci can be associated with cytologic atypia and reparative increases in mitotic activity that may rarely cause diagnostic difficulties. A thin peripheral rim of viable tissue usually remains, but in cases of total infarction, only the FNA results will be available and must form the basis for diagnosis. Rarely, the mass is virtually replaced by a florid, fasciitis-like proliferation of reparative connective tissue that may be accomplished by varying amounts of identifiable but necrotic tumor tissue. We have seen one such case in which, without being aware of an FNA diagnosis of large cell malignant lymphoma rendered 2 weeks earlier, the surgical pathology service interpreted such a reparative spindle cell proliferation as a sarcoma. This problem was quickly resolved by consultation among the clinicians involved, and a repeat surgical biopsy confirmed the cytologically sampled lymphoma.

Clinical Application of FNA. Most masses can be approached by FNA, and in many instances initial evaluation by cytology quickly directs clinical thinking along the most efficient routes. This is true even when precise tumor subclassification is not possible based solely on the aspirate sample. The impact of aspiration cytology is greatest when high-quality collection and preparation of specimens result in rapid reporting of results.[34] Several investigators indicate that the results obtained when cytopathologists both perform and interpret aspirations are better than when these functions are split between a clinician and a pathologist who has not seen the patient.[1,3] Examples of the clinical impact by FNA include its early use in AIDS patients, in whom many masses are not malignant, and

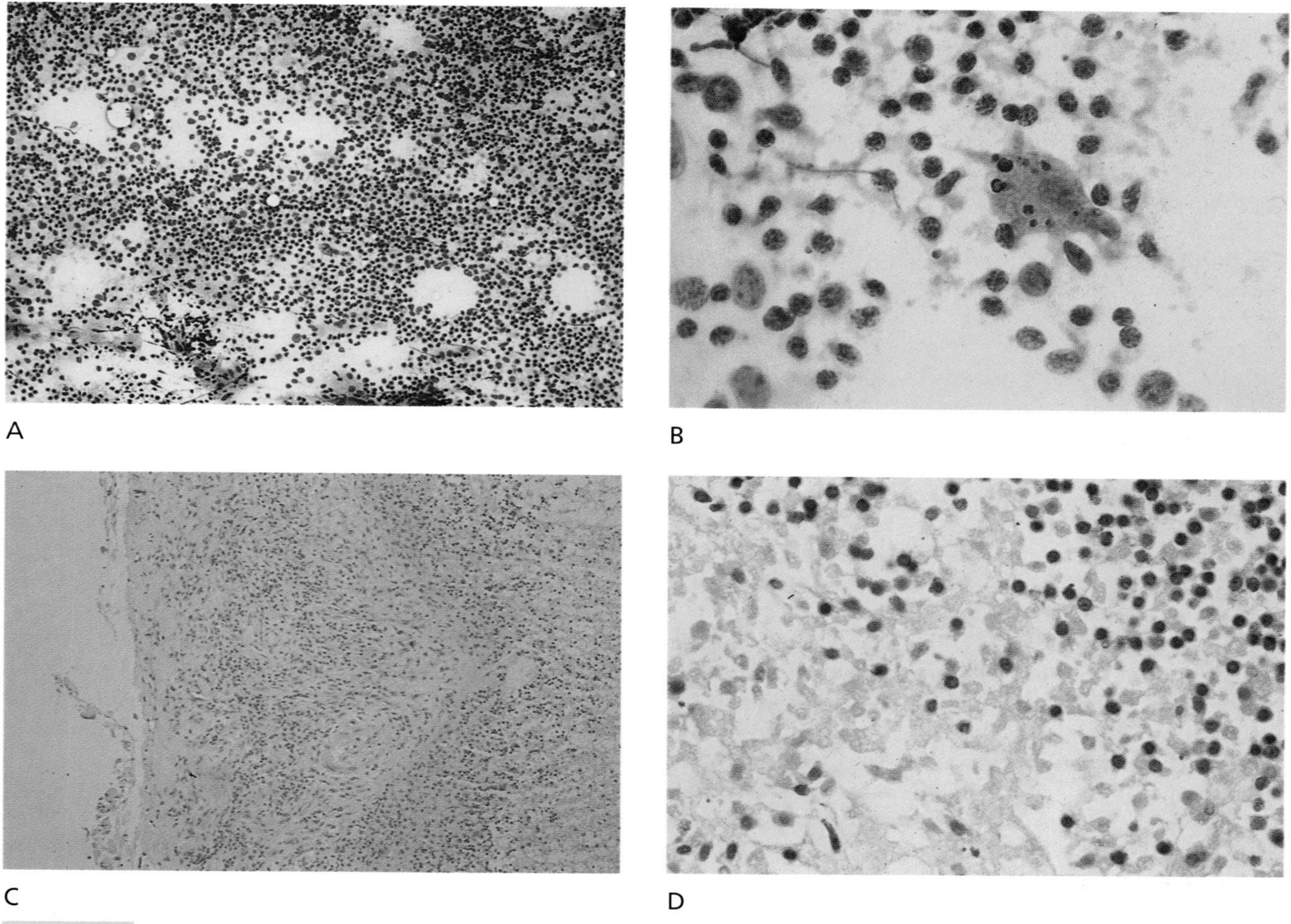

FIGURE 4–1. Post-FNA infarction in a cervical lymph node. (**A**) The preoperative aspiration yielded abundant lymphoid tissue, as shown by this low-magnification image. (Diff-Quik, ×40) (**B**) At higher magnification, the smears showed abundant polymorphous benign-appearing lymphoid tissue; a tingible-body macrophage is also illustrated. (Papanicolaou, ×600) (**C**) Histologic sections of the excised lymph node show diffuse coagulative necrosis, through which the architecture is still apparent. (Hematoxylin and eosin, ×100) (**D**) At higher magnification, the nodal parenchyma shows coagulative necrosis and a few nonviable-appearing lymphocytes. (Hematoxylin and eosin, ×600)

in whom malignant lymphoma can often be readily diagnosed in cytologic samples.[35,36] Also, salivary gland area masses in children are most often benign lymph nodes that require only clinical follow-up after diagnosis by aspiration.[37–39]

Non-Neoplastic Conditions and Differential Diagnostic Considerations

Inflammatory Masses. In many instances, inflammatory or infectious masses in the head and neck are related either to lymph nodes (discussed in other chapters) or to the salivary glands (discussed later). Infectious diseases that are studied in FNA material from the head and neck are cytologically similar to the same conditions in other sites, and include mycobacterial infections (tuberculous or atypical), Brucellosis, Toxoplasmosis, syphilis, fungal diseases, and leprosy (tuberculoid or lepromatous).

Leishmaniasis rarely presents as isolated nodal disease, but this condition is well suited to diagnosis by aspiration.[22, 23] Smears show a mixture of acute and chronic inflammatory cells, with giant cells and granulomas. Organisms should be sought in the smear background, and within macrophage cytoplasm. Cytologic examination of centrifuged fluid from cervical Echinococcal hydatid cysts shows the characteristic hooklets.[40] Actinomycotic masses can be clinically mistaken for malignant neoplasms. Aspiration shows acute inflammation with a few admixed chronic inflammatory cells.[21, 41] The diagnostic organisms can be easily overlooked in smears or cell block preparations, as they are overwhelmed by the outpouring of inflammatory cells. The filaments of Actinomyces are positive with both Gram stains and silver stains, but they are not acid fast. This characteristic staining profile which can be applied easily to cell block sections, is useful in difficult cases.

Congenital Cysts and Their Distinction from Cystic Metastases. In various series, the proportion of head and neck aspirates that are at least partially cystic ranges from 1 to 10%.[16 24,42–44] Most cysts in the neck can be initially categorized as either midline or lateral.[16] Those in the midline are often related to the thyroid and are discussed more fully in other chapters. Cytologically they differ from lateral cysts of the neck in that they are more likely to show columnar epithelial cells than squamous cells.[14] A midline neck cyst in a young patient has been considered to have a very low probability of malignancy, even when aspiration yields an acellular fluid (Table 4–6).[16]

Lateral neck cysts frequently have their origin in misadventures of branchial arch embryology and may pose significant clinical problems, as noted below. At the time of FNA, the cystic nature of such lesions is usually apparent; the previously cited guidelines for aspiration of cystic or partially cystic masses should be followed.[1,13–17] It is usually necessary to concentrate the sparsely cellular fluid by centrifugation, but occasional examples have sufficient cellularity to allow examination as direct smears. The expected cytologic findings include precipitated proteinaceous debris and mature-appearing squamous cells. Lymphocytes are common, and some inflamed examples may also feature neutrophils and fibrin strands. Alternatively, some cyst aspirates are virtually acellular (Table 4–7).[14,42]

Diagnostic problems relate to the fact that metastases of squamous cell carcinoma frequently involve lymph nodes that appear clinically as lateral neck masses and are often cystic. Cytologic evidence of malignancy should be sought[14,42,44]; some metastatic carcinomas show unequivocal malignancy that should not be found in congenital cysts. The nuclear smudging and hyperchromasia that indicate degeneration in the fluid's cells must not be mistaken for malignant nuclear changes. However, careful evaluation of the squamous elements in a cyst fluid may not simplify this difficulty, because many examples of cystic metastatic squamous cell carcinoma are very well differentiated and shed only mature-appearing cells.[14] Some cystic lesions yield only debris. In these cases, it is difficult to distinguish between the necrosis that usually represents evidence of carcinoma and the protein precipitate expected in benign cysts. Occasional aspirates show only lymphocytes; this nonspecific finding leaves open the possibility of several cystic lesions.[14]

Cysts that do not recur after aspiration are often considered benign.[24] Further, patients younger than 30 to 40 years of age are more likely to have a congenital cyst than a metastatic carcinoma. However, a large degree of overlap in the age distribution of these two conditions makes age an unreliable diagnostic criterion.[14,42] Thus, metastatic squamous cell carcinoma and lateral congenital cysts are sometimes impossible to distinguish clinically or cytologically. It has been suggested that all lateral neck cysts showing squamous cells of any type (or, we would add, lacking an epithelial component) on aspiration cytology should be excised.[14]

Nodal metastases of papillary thyroid carcinoma often present as lateral neck masses and are often cystic. We have seen several cases in which the lateral neck mass was identified as metastatic papillary carcinoma before the intrathyroidal primary lesion was known to exist. Further, this situation can be encountered in young adults and even in children, so that clinical factors may be of no diagnostic help whatsoever.

The aspirate background features various combinations of protein precipitate, colloid-like material, and hemosiderin-laden macrophages. Cell groups indicative of this diagnosis show the typical cytologic features of papillary carcinoma, as described in other chapters. As expected, these cells are of rather low nuclear grade. The diagnostic difficulties in these cases stem from the fact that even several milliliters of the fluid may yield only a few diagnostic groups. This

TABLE 4–6. Midline Congenital Neck Cysts

- Many are related to the thyroid or the thymus.
- More often show columnar cells than do lateral neck cysts
- Low probability of malignancy, especially in younger patients

TABLE 4–7. Lateral Congenital Neck Cysts

- Many are related to branchial arch embryology.
- Squamous cells frequently encountered
- Some cells may show advanced degenerative atypia.
- Can be very difficult to distinguish from cystic metastases of squamous cell carcinoma
- Metastases of papillary thyroid carcinoma may be detected before the primary tumor is found.
- Age ranges for carcinoma and benign cysts overlap.
- Excision is sometimes required for confident diagnosis.

emphasizes the need for centrifugation, with preparation of smears and cell blocks from sediment representing the entire volume of submitted fluid. Papillary carcinoma cells may show extensive degenerative cytoplasmic vacuolization; this can lead to a false-negative diagnosis when these cells are thought to represent macrophages. Intranuclear inclusions, sharp cell borders, dense cytoplasm, and convincing cell-to-cell cohesion all point toward papillary carcinoma as the correct diagnosis.

Aspiration of Therapy-Related Soft-Tissue Masses

Surgery and radiation therapy to the head and neck result in a number of alterations that may be addressed by FNA. One does so in the light of previously discussed derangements in the expected anatomy of this complex region. The most common indication for this type of study is either to confirm or refute a clinical impression of residual or recurrent carcinoma. However, other masses may be encountered. Among these, diffuse soft-tissue thickenings, suture granulomas, granulation tissue, and postoperative neuromas are common. Inflammation of various types and damaged skeletal muscle can also be aspirated. In our experience, infections approached by FNA are not common in this clinical setting (Table 4–8).

Diffuse Thickenings. Ill-defined soft-tissue thickenings can result from radiation therapy and are very difficult to evaluate clinically. These can be quite firm and may suggest recurrent carcinoma. These lesions can be very sclerotic, requiring several needle passes, with thorough sampling of a cone-shaped tissue volume during each pass (reviewed in reference 1). The yield of malignancy is very low; one usually recovers small fragments of benign soft tissue. These aspirates can be surprisingly bloody. The tissue seems to have a spongelike consistency. Several minutes of firm pressure can be required to stop the slow but steady seeping of blood from the puncture site. When this type of tissue is aspirated, the aspirator should be sure that either the patient or the assistant begins application of sufficient pressure before turning his or her attention to preparation of smears (Table 4–9).

TABLE 4–8. Therapy-Associated Head and Neck Masses That May Clinically or Cytologically Suggest Recurrent Carcinoma

- Soft-tissue thickening, often diffuse
- Suture granulomas
- Granulation tissue
- Other types of inflammation
- Postoperative neuromas
- Damaged skeletal muscle

TABLE 4–9. Postirradiation Soft-Tissue Thickening

- Firm fibrous masses that are hard to access clinically.
- Aspiration yields small bits of soft tissue.
- Aspirates may be quite bloody.
- The tissue may continue to hemorrhage after the aspiration.
- Some sclerotic tumor recurrences probably lead to faslely negative FNA evaluations in this setting.

Those who do frozen sections of margins when recurrent squamous cell carcinomas are excised are very aware of the fact that these tumors can recur as tiny, almost capillary-like nests and cords set in copious desmoplastic stroma. We suspect that this type of recurrence is underdiagnosed by FNA of radiated tissue, but no study has adequately addressed this issue. The cytologic problems occasioned by radiation sialoadenitis are similar and will be considered subsequently.

Suture Granulomas. Suture granulomas are easily palpated in the thinned tissues that often remain after extensive neck surgery. In this setting, they appear as small hard subcutaneous nodules that may be highly suspicious for recurrent carcinoma. Smears from such lesions are readily interpreted and show a foreign-body reaction, often with numerous multinucleated giant cells (Fig. 4–2). Other types of inflammatory cells may be present, and contribute to the very busy appearance of these aspirates, but the cytologic features of malignancy are not encountered. Bits of foreign material or even recognizable suture fragments are often present. These can be highlighted by examination with polarized light.

Granulation Tissue. This type of specimen is encountered in aspirates from a variety of body sites and presents much the same picture wherever it is seen. Variations in cytologic content can be expected and generally reflect the lesion's maturity, which in turn is a function of its age. More mature examples are less cellular, showing a shift away from inflammatory cells and prominent vessels. However, younger lesions are often extremely cellular. In addition to neutrophils and histiocytes, one may see lymphocytes, plasma cells, and multinucleated giant cells. Blood vessels are a conspicuous feature of granulation tissue (Fig. 4–3). These slender gracile structures branch often and sometimes create a retiform network within an aspirated tissue particle. They are lined by large, elongated endothelial cells with vesicular nuclei and small nucleoli. The luminal borders of a vessel form two parallel lines that course through a tissue frag-

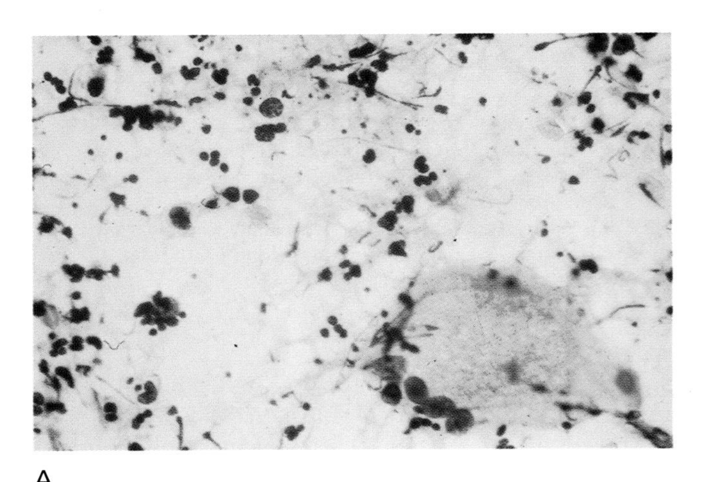

A

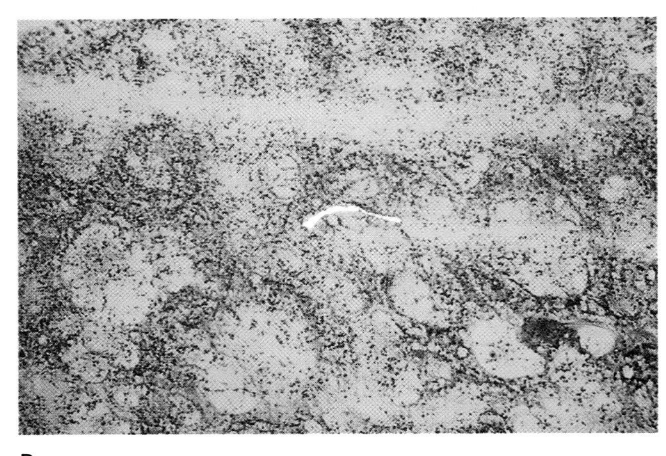

B

FIGURE 4–2. (**A**) Smear material from a suture granuloma. In addition to a single foreign-body giant cell, there is protein precipitate and nuclear debris. (Papanicolaou, ×600) (**B**) The same smear at very low magnification, as viewed through partially crossed polarizing filters. A short broad linear fragment of suture material glows brightly. (Papanicolaou, ×40)

ment like railroad tracks. Numerous acute and chronic inflammatory cells spread out from these central cores to be dispersed over the smear background.

Cytomorphologic Features of Granulation Tissue

- Cytology can vary with the lesion's age
- Mixed acute and chronic inflammation
- Histiocytes and neutrophils prominent
- Nuclear dust and other debris
- Tissue particles organized around large, branching blood vessels
- Mesenchymal-repair types of atypia can be encountered
- Microorganisms or tumor cells can be present

Although highly cellular and architecturally complex, this picture is quite reproducible and can be recognized dependably. Having interpreted the aspirated material as granulation tissue, one should search for rare tumor cells or recognizable microorganisms. It may be useful to submit such specimens for culture using the previously discussed guidelines. Reactive spindle cells, endothelial cells, and even histiocytes may raise the possibility of malignancy, unless interpreted carefully in the context of the entire smear pattern.

Damaged Skeletal Muscle. One sometimes aspirates masses that apparently consist of damaged skeletal muscle with its associated connective tissues. The residual muscle may be altered considerably, so that it

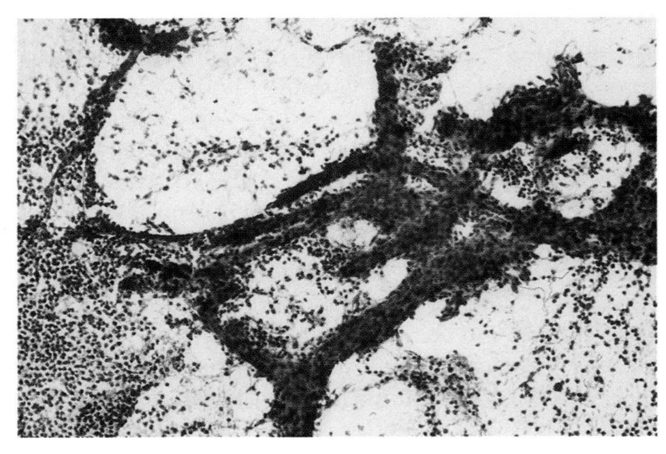

A

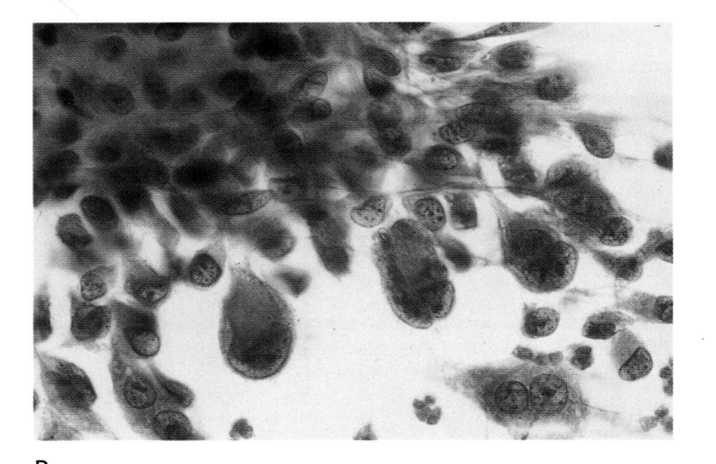

B

FIGURE 4–3. (**A**) Aspiration of granulation tissue often yields large tissue fragments, as shown here. These seem to be organized around branching central blood vessels from which fan out numerous inflammatory and connective tissue cells. (Papanicolaou, ×40) (**B**) Prominent nucleoli in reparative mesenchymal cells, macrophages, or reactive epithelial elements should not be mistaken for malignancy in this inflammatory background. (Papanicolaou, ×600) resembles the degenerated tissue that surgical specimens often show in areas of ischemia or adjacent to an invasive malignancy. The cells are large, and their abundant cytoplasm will have lost its characteristic cross-striations in the degenerative process. The nuclei are internalized rather than apposed closely to the cell membrane, and are frequently multiple. The chromatin may show degenerative smudging, and nucleoli are often very prominent (Fig. 4–4). Recognition of this characteristic cytology as a picture familiar from surgical pathology will prevent an erroneous diagnosis of malignancy. This picture can be seen in many body sites; in our practice, it is most common in the neck.

Cytomorphologic Features of Damaged Skeletal Muscle Fibers

- Internalized nuclei
- Bizarre cell shapes may be seen
- Multiple nuclei per cell
- Nucleoli may be very prominent
- Abundant cytoplasm imparts a low overall nuclear to cytoplasmic (N/C) ratio
- Cross-striations are lost

Postoperative Neuroma. Postoperative neuroma is another lesion that we most often encounter in the neck after extensive surgery. It presents clinically as a small, round, firm subcutaneous nodule that is frequently not tender to palpation. (There is often a considerable sensory deficit in the operative area.) Thus, it resembles the previously described suture granulomas. Attempts at aspiration immediately lead to intense pain that often radiates to the arm. The patient finds this experience intolerable and always declines any suggestion of further aspirations. Smears show either no material or a tiny clear droplet containing rare bland spindle cells. Most experienced head and neck surgeons consider this provocative test diagnostic; there is little need to pursue the matter further.

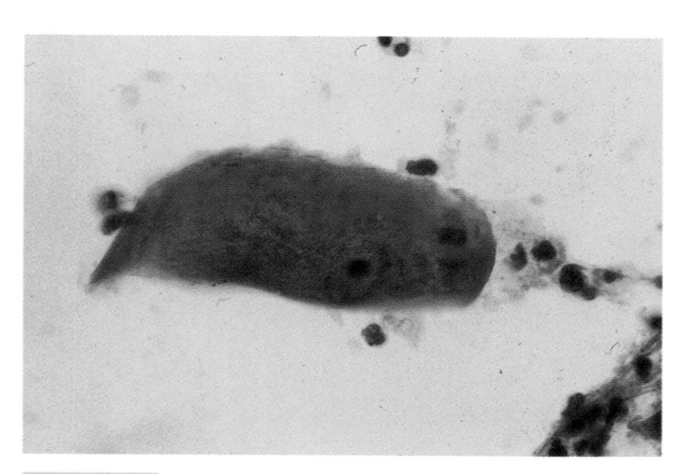

FIGURE 4–4. Damaged skeletal muscle cells undergo a number of alterations that may be alarming. The expected cytoplasmic tinctorial qualities and cross-striations are lost; the peripherally located nuclei are internalized. Reactive nuclear changes and prominent nucleoli are common. These cells may be associated with other mesenchymal reparative changes to give a picture that can suggest malignancy. (Papanicolaou, ×600)

Cytomorphologic Features of Postoperative Neuromas

- Small subcutaneous nodule
- May clinically suggest recurrent carcinoma
- Aspiration is extremely painful
- Usually no material is recovered
- Patients always refuse repeat FNA
- FNA is a provocative test considered by most to be diagnostic of neuroma

Neoplastic Conditions

Most commonly encountered primary neoplastic masses of the head and neck originate in the salivary glands, the lymph nodes, the thyroid, the mucosal surfaces, or the soft tissues. As noted previously, several of these are discussed in other chapters, and the mucosal lesions are not commonly studied by cytologic techniques. Salivary gland masses will be described subsequently.

This section addresses a host of primary masses occasionally described in cytologic samples, as well as metastatic malignancies. The latter are usually aspirated from lymph nodes and yield cells similar to those described in other chapters for the cytology of the primary lesions.

Lesions of the Skin and Subcutaneous Tissues. The most common application of FNA to lesions of the integument is diagnosis of metastatic deposits.[45] These may be widely distributed, but in our practice those most likely to be referred for FNA are dermal recurrences of breast carcinoma[1] and lesions involving the scalp.[46] When primary skin neoplasms are aspirated, most involve the head and neck and represent squamous cell carcinoma or basal cell carcinoma.[47] Less commonly, recurrences or metastases of malignant melanoma, cutaneous angiosarcoma, or sebaceous carcinoma undergo FNA. Primary dermal or adnexal tumors are occasionally encountered.[39,44,45,48–52] The FNA features of squamous cell carcinoma and malignant melanoma can be extrapolated from our knowledge of their cytology in other sites. The cytology of basal cell carcinoma will be discussed subsequently in our consideration of salivary gland lesions characterized by uniform small "blue" cells (Table 4–10).

TABLE 4–10. Head and Neck Neoplasms in the Skin and Subcutaneous Tissues

- Metastases or local recurrences: malignant melanoma and squamous cell carcinoma; rarely angiosarcoma, sebaceous carcinoma or others
- Recurrent skin neoplasms: basal cell or squamous cell carcinoma
- Skin adnexal tumors
- Keratinous cysts, with or without rupture

Recurrent breast carcinoma and most other metastases, including neuroblastoma in the scalp,[39] cause few diagnostic difficulties when the history and physical findings are evaluated in concert with the smear pattern. Primary or metastatic sebaceous carcinoma shows large cells with abundant vacuolated cytoplasm and prominent nucleoli.[13,45] When nodules of recurrent cutaneous angiosarcoma are aspirated, the cells appear clearly malignant, but they may be round and epithelioid, thus belying their origin in a mesenchymal neoplasm (Fig. 4–5).[45,53]

Dermal or adnexal-based primary skin tumors of the head and neck, including keratinous cysts,[44,45] pilomatrixoma,[45,49–52] and dermal eccrine cylindroma,[48] have been described in FNA cytology. Often, these masses are aspirated either to confirm or to refute the clinical impression that the mass in question represents one of the primary or recurrent malignancies discussed above. Finding that a cutaneous mass thought to be recurrent breast carcinoma or some other malignancy is actually a keratinous cyst represents a wonderful contribution to the patient. FNA can render this service inexpensively, with minimal pain and only a few minutes of clinic time.

Diagnosis of dermal lesions is not always straightforward. Keratinous cysts occur in many sites and clinically may raise the possibility of recurrent malignancy. When epithelial inclusions are situated in surgical scars, they are especially suspicious. Aspiration yields a small semisolid droplet of white material. This caseous debris exhibits the unpleasant odor so well known to all who have handled these tumors in surgical pathology. Microscopy reveals large numbers of mature, frequently anucleate squamous cells. Masses that fit this typical description are diagnosed with great accuracy.

After rupture, however, the squamous cells may show atypia, and they may be accompanied by exuberant inflammation. This response includes reactive spindle cells and granulation tissue similar to those discussed previously. Further, the diagnostic squamous cells may be difficult to identify because they are widely distributed through a large volume of inflammatory exudate. This cytologic picture is identical to that encountered with subareolar breast abscess. Cysts that lie in a radiation field may show striking atypia that can suggest squamous cell carcinoma to the pathologist who has been deprived of the appropriate clinical information.

Ideally, aspirates of pilomatrixoma are polymorphous and show material representing each of its several components (Fig. 4–6). These include small hyperchromatic basal cells, anucleate shadow ("ghost") cells, and multinucleated giant cells. In some examples, bits of calcified material are appreciated. The shadow cells may lie singly and clearly show their squamous features. Alternatively, they can be largely compacted into hypereosinophilic clumps[49,51] in which the squamous nature of individual cells can be

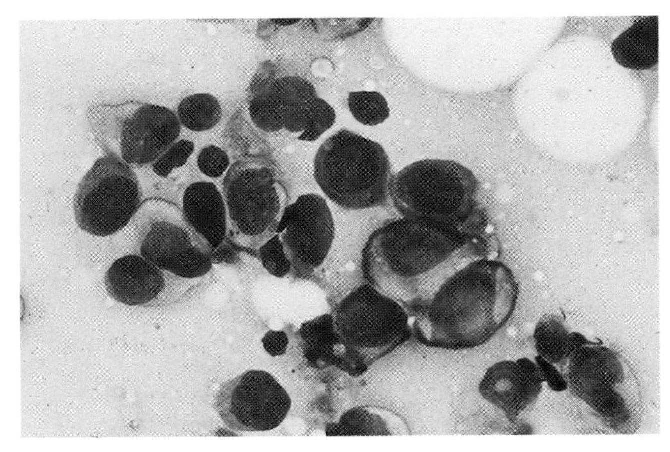

A

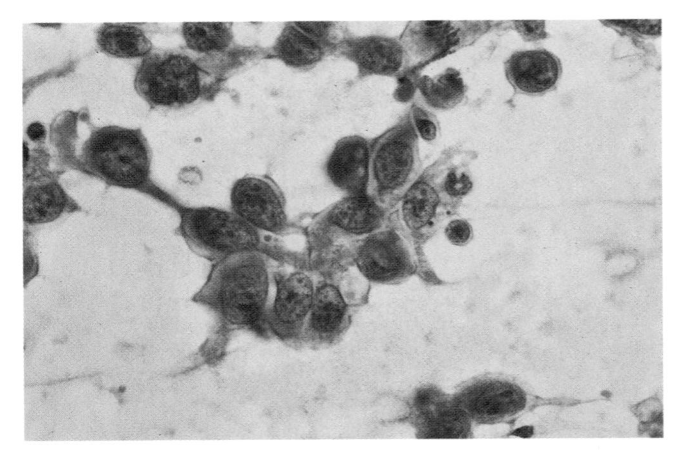

B

FIGURE 4–5. (**A**) This recurrent angiosarcoma shows rounded cells that are more suggestive of a carcinoma or even a hematopoietic process than a mesenchymal malignancy. (Diff-Quik, ×600) (**B**) Some slight spindling can be detected in the fixed smear. (Papanicolaou, ×600)

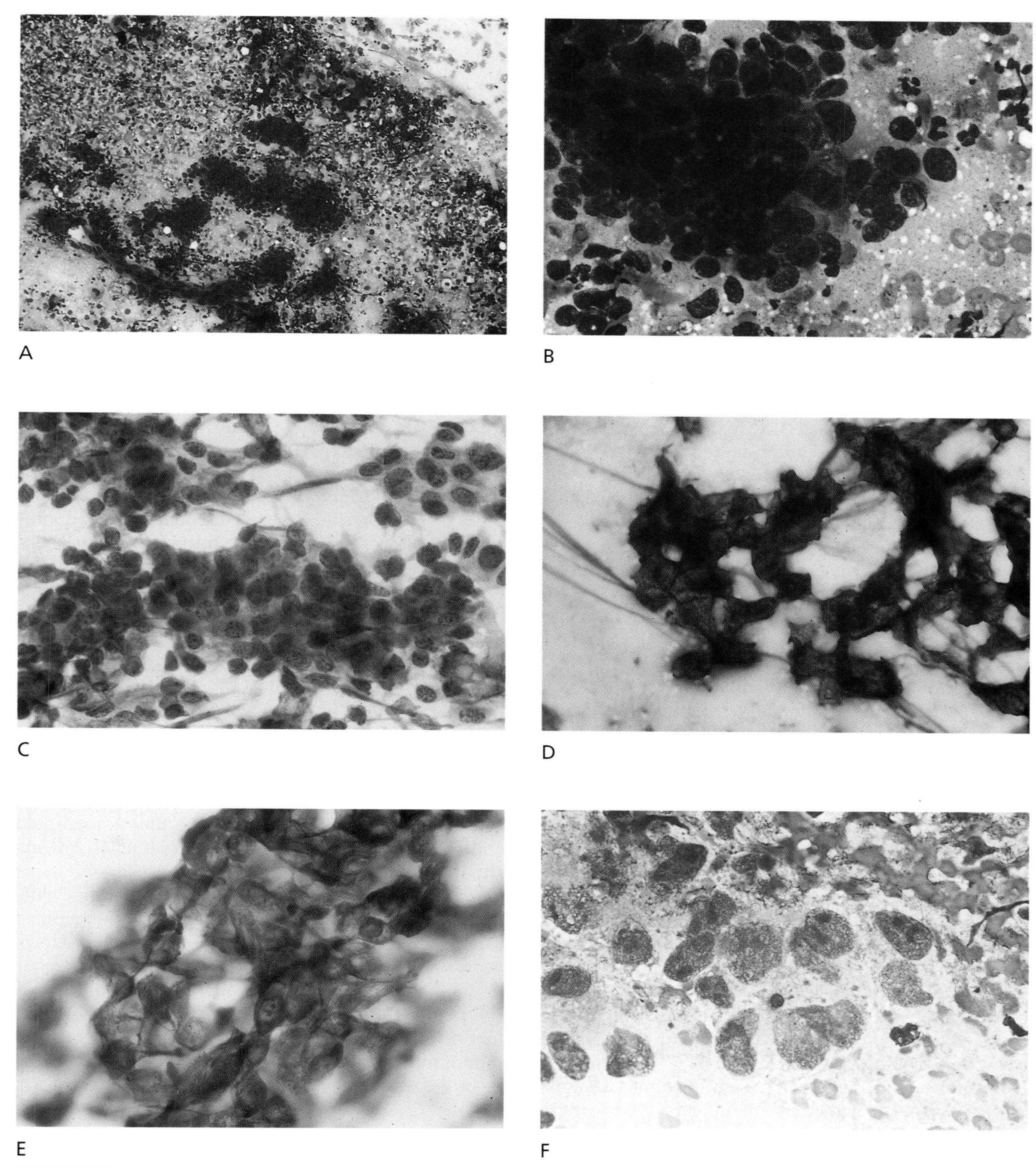

FIGURE 4–6. (**A**) Aspirates of pilomatrixoma can be highly cellular. When the small cell component is prominent, one immediately begins to consider various types of malignancy. (Diff-Quik, ×40) (**B**) The small cell component consists of crowded uniform cells in three-dimensional aggregates. Debris in the background may suggest necrosis. (Diff-Quik, ×600) (**C**) There is little to distinguish these pilomatrixoma basal cells from representatives of numerous other lesions. The risk of a false-positive diagnosis is very great unless other features of this benign neoplasm are identified. (Papanicolaou, ×600) (**D**) In air-dried slides, the ghost cells typical of pilomatrixoma stain very darkly and may be difficult to recognize. (Diff-Quik, ×100) (**E**) The squamoid nature of the ghost cell is more easily detected in fixed smears. These cells often occur in thick masses, as illustrated here. (Papanicolaou, ×400) (**F**) Occasional examples of pilomatrixoma contain larger atypical cells. (Diff-Quik, ×600)

difficult to recognize. Mitotic figures may be identified among the small basaloid cells.

Shadow cells are essential for positive identification of this tumor in aspirate samples. Often, however, they are difficult to identify or absent altogether. This is the major reason that other diagnoses are considered in up to 75% of pilomatrixomas described in the FNA literature (reviewed in reference 49). Those in difficulty with a pilomatrixoma are often drawn toward differential diagnostic considerations suggested by a prominent small blue cell component. Basal cell carcinoma, metastatic small cell anaplastic carcinoma, various salivary gland tumors, and metastatic nasopharyngeal carcinoma will be likely misdiagnoses. Nasopharyngeal carcinoma has been emphasized as a pitfall for those working with Chinese pilomatrixoma patients, in whom there is a relatively high incidence of this neoplasm. Because this malignancy often presents as a high cervical lymph node metastasis, it can clinically simulate a pilomatrixoma.[49]

Pilomatrixomas are very firm to palpation, and this can further elevate clinical suspicion, so that when a troublesome aspirate is encountered, the likelihood of an incorrect diagnosis of carcinoma may be very great indeed. Other problems occur when this tumor is located in sites other than the head and neck. Finally, some pilomatrixomas can show striking atypia, so that the expected small blue cell pattern is replaced by a proliferation of atypical cells that suggest large cell types of carcinoma. If unfortunate circumstances combine several of these difficulties, the correct diagnosis of pilomatrixoma may be extremely difficult. We encountered one such neoplasm with striking large cell-type cytologic atypia and a paucity of shadow cells on the forearm of a young man recently treated for testicular endodermal sinus tumor. We incorrectly interpreted the aspirate as probable evidence of metastatic malignancy, albeit before serum tumor marker studies were available. The cytologic atypia was also noted in the excised tumor. In this case, one could speculate that prominent enlargement and atypia of the usually small blue cells might have been induced by chemotherapy. If so, aspiration of pilomatrixomas in cancer patients with suspected tumor recurrences might represent a special problem.

Dermal eccrine cylindroma is an uncommon mimic of adenoid cystic carcinoma of salivary gland origin and will be discussed subsequently in that context.

Parathyroid Masses. Parathyroid lesions rarely become targets for FNA. When large enough to be palpable, they may clinically mimic masses in the thyroid or cervical lymph nodes. The aspiration cytology of both adenomas[54–56] and carcinoma[57] has been described. Aspirates from adenomas are highly cellular and often yield clear cyst fluid. Smears show single cells and numerous naked nuclei, often with multiple nucleoli. As in other benign endocrine neoplasms, the nuclei may show striking variation in size that is of no prognostic significance. Architecturally, the cell arrangements can mimic follicular thyroid neoplasms or, less commonly, show flat sheets similar to those in papillary thyroid carcinoma. The nuclear features of papillary thyroid carcinoma have been absent in cases described to date.

If the origin of an aspirate is unclear, parathormone evaluations in tumor cells (by immunocytochemistry) or in cyst fluids can be helpful. In comparison with adenomas, the cytology of parathyroid carcinoma has not been adequately described. One might expect a trabecular cell arrangement, spindle cells, or readily demonstrable mitotic activity. Radiographic or surgical evidence of attachment to surrounding structures would also suggest carcinoma. These lesions are extremely rare and usually present with clinical findings related to severe hyperparathyroidism.

Odontogenic Tumors. These masses are rarely described in FNA samples, but aspirations of odontogenic keratocysts,[44] Pindborg tumor (calcifying epithelial odontogenic tumor),[58] and ameloblastoma[44,59] have been reported. Clinical and radiographic findings are very important in reaching a correct diagnosis. Smears from keratocysts resemble the previously described cytology of dermal keratinous cysts. Pindborg tumors may recur locally but are less aggressive than ameloblastomas; a single cytologic case report has been published.[58] Cytologic findings included squamoid cells, vacuolated cells, psammoma-like calcifications, and a background of amorphous material. This picture can suggest mucoepidermoid carcinoma, but the clinical findings, the radiographic features, and negative stains for epithelial mucins should discourage this interpretation.

Aspirations of ameloblastoma are frequently cystic, and the fluid must be centrifuged if the sometimes rare squamous elements are to be appreciated. However, once identified, the variably acantholytic squamous elements may be mistaken for squamous cell carcinoma. Cohesive clusters of small palisading basaloid cells may also be present.[44,59] One must consider the tumor's location and radiology if the correct diagnosis is to be suggested from cytologic findings.

Chordoma. The cytology of this uncommon neoplasm is most often described for groups of cases, or for the more common sacral masses.[60,61] FNA findings include varying combinations of short spindle cells, small round cells, and large vacuolated ("physaliferous") cells. The latter component, although characteristic of this entity, may be absent, especially from metastatic deposits.[62] Cytoplasmic vacuoles and background myxoid material are better appreciated in air-dried, Romanowsky-stained smears than in fixed Papanicolaou-

or hematoxylin and eosin-stained preparations. Electron microscopy and immunocytochemistry (positive for cytokeratins, vimentin, epithelial membrane antigen, and S-100 protein) can be useful ancillary techniques in distinguishing chordoma from its mimics. These include chondrosarcoma, metastatic malignancy, and myxopapillary ependymoma.[60,61,63–69]

Four interesting clinical presentations of chordoma suggest that we must broaden our differential diagnostic thinking if we are not to mistake this neoplasm for its imitators. Some clival tumors may present as masses suitable for the previously described technique of transoral FNA.[61] In this setting, one may strongly consider diagnoses of minor salivary gland neoplasms, glial masses, or even meningioma. "Chordoma cutis" may approach the skin by direct extension, regional recurrence, or metastatic spread.[62,65,66] Further, primary sacral tumors may eventuate in cutaneous metastases in the head and neck. "Extranotochordal" examples of chordoma may present as lesions confined to the soft tissues of the neck, where they mimic various types of myxoid sarcoma.[60,67] Finally, one clivus-based chordoma recurred in the lateral neck, where it was confused with a deep pleomorphic adenoma of salivary gland origin.[64]

Meningioma and Craniopharyngioma. Intraoperative specimens evaluated as crush preparations underlie our understanding of meningioma cytology.[70] However, when these masses occur in sites accessible to FNA, material may be aspirated for cytologic evaluation. Meningiomas are uncommonly encountered as parapharyngeal[71] or orbital[72] masses, so one may not suspect the diagnosis at the time of aspiration.

Surgical pathologists recognize several variants of meningioma, most of which have not been described adequately in cytologic terms. However, the typical case can be diagnosed with confidence. One sees a cellular preparation with numerous uniform spindle cells that may be arranged in whorls. Round intranuclear invaginations of cytoplasm are common, and psammoma bodies are identified in some cases.[71–73] An example of pulmonary metastases from meningioma studied by FNA showed similar cytologic features.[74] Many of the metastatic sarcomas with which such material could be confused show more nuclear atypia than is usually appreciated in meningioma. On rare occasions, however, nuclear atypia and mitotic activity that can raise the possibility of carcinoma, melanoma, sarcoma, or a high-grade glial neoplasm are noted. Immunocytochemistry can be helpful in this situation.

Nonoperative aspirates from craniopharyngiomas are described rarely.[75] These specimens yield "dirty" cyst fluid with background debris, macrophages, calcific material, and squamous cells. The latter may be degenerated and ghostlike, or well preserved. Thus, although such material cytologically resembles ameloblastoma, pilomatrixoma, or other partially cystic lesions with a squamous component, correlation of the smear pattern with radiographic and clinical findings can lead to the correct diagnosis.

Paraganglioma. These neoplasms are widely distributed and are discussed more completely in chapters devoted to the cytology of other body sites. However, they can occur in several head and neck locations. Most paragangliomas suitable for evaluation by FNA represent carotid body tumors. At the time of physical examination, these masses can be moved horizontally but not vertically.[76] Aspirates are almost always bloody. Smears show cellular and nuclear pleomorphism, and naked nuclei may be numerous (Fig. 4–7). Round intranuclear cytoplasmic invaginations are commonly noted. The cytoplasm is abundant with ill-defined borders and may show fine red granularity on air-dried, Romanowsky-stained preparations.[76–78] These cells often grow in a follicular pattern, and this can be mistaken for a follicular thyroid neoplasm or for metastatic adenocarcinoma of various types. Alternatively, tumor cell spindling may suggest neurilemmoma or other mesenchymal neoplasms. Occasionally, clear cell change may raise the possibility of metastic renal cell carcinoma. Malignant paragangliomas can show necrosis or mitotic figures,[78] but nuclear variability alone is without prognostic significance.

Cytomorphologic Features of Paraganglioma

- Aspirates are always bloody
- Cellular and nuclear pleomorphism
- Naked nuclei may be seen
- Follicle-like grouping may suggest adenocarcinoma
- Spindle cells, myxoid material, and clear cells can be seen
- Large atypical nuclei can occur in benign paragangliomas

When a carotid body tumor is suspected clinically, some physicians prefer to avoid FNA and to seek support for the diagnosis from angiography, fearing hemorrhagic, thrombotic, or embolic sequelae of the procedure. In one reported case, carotid artery thrombosis progressed to cerebral infarct followed by death after FNA; a 22-gauge needle was used for that procedure.[76] Cytopathologists who offer FNA services to head and neck physicians occasionally evaluate carotid body tumors in which the diagnosis has not been suspected before the procedure. In our experience, this has been without incident; again, we emphasize the safety inherent in using very thin needles (25 gauge; 0.5 mm outside diameter).

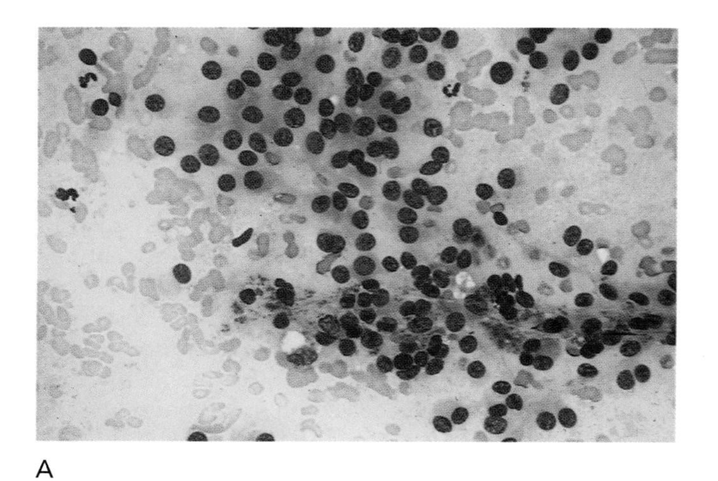

A

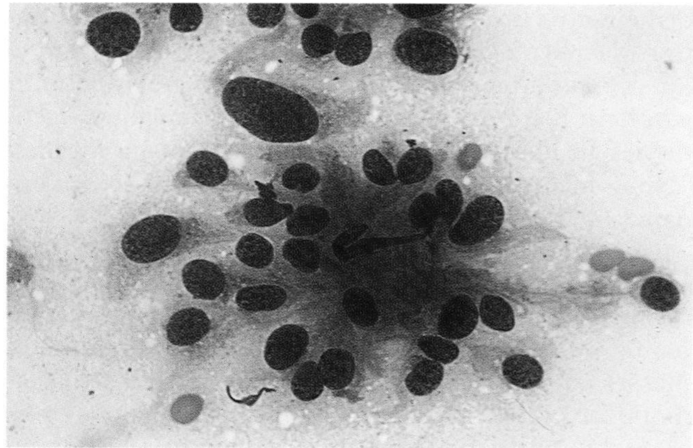

B

C

D

FIGURE 4–7. (**A**) The smear from a carotid body tumor shows a loosely cohesive flat sheet of rather uniform cells with several poorly formed follicle-like structures. Red blood cells and a few naked nuclei are present in the background. (Diff-Quik, ×100) (**B**) This field shows more cellular pleomorphism with a striking variation in nuclear size. Wispy cytoplasm, indistinct cell borders, a tendency for follicle formation, and spindling of individual cells complete the picture of paraganglioma cytology. (Diff-Quik, ×600) (**C**) This fixed smear from the same tumor shows chromatin clumping, small nucleoli, and wispy cytoplasmic extensions. (Papanicolaou, ×600) (**D**) Aspirates from paragangliomas occasionally show extremely atypical cells. As in surgical pathology, these do not permit prognostic assessment of a tumor. (Papanicolaou, ×600)

Benign Soft-Tissue Masses. Most head and neck masses that clinically appear to be located in the soft tissues represent metastases, either from squamous cell carcinomas of the head and neck or from more distant sites. Occasionally, a soft-tissue mass yields inflammatory cells without identifiable etiologic agents; a few show organisms such as actinomyces.[24]

When neoplasms are considered, one must realize that the great variety of soft-tissue lesions recognized in surgical pathology is not clearly delineated in the FNA literature. It is likely that many aspirates will continue to be described in somewhat generic terms (e.g., myxoid, high-grade, low-grade), resulting in a refined or narrowed clinical differential diagnosis rather than in confident recognition of some specific histopathologic entity. In many instances, the primary value of FNA will be in ruling out recurrence of a previously well-characterized malignancy or a specific lesion such as infection or malignant lymphoma for which effective treatment may be available.

Lipomas sometimes occur in the head and neck and show the expected clinical and cytologic features.[4] However, these lesions are much less common in these sites than are malignancies of various types and salivary gland tumors. Also, if the aspiration target is missed, the smears will show fat or skeletal muscle. For these reasons, one must always question the significance of pure adipose tissue on FNA smears from the head and neck. The diagnosis of lipoma should be entertained *only* if the clinical findings have been evaluated by an experienced physician and are entirely consistent with that interpretation. Further, the aspiration must have been performed by a practiced clinician who can provide assurance that the palpable mass has been accurately punctured. If these conditions are not completely satisfied, one is in danger of missing a significant lesion by mistaking an unsatisfactory aspirate for a specific benign entity.

Neurilemmoma may present as a parapharyngeal mass of impressive proportions. FNA material can usually be interpreted as a benign mesenchymal neoplasm.[31] If Verocay bodies can be readily identified, a specific diagnosis is possible (Fig. 4–8).

Melanotic neuroectodermal tumor of infancy (progonoma, or retinal anlage tumor) is most common in the head and neck; many are located in the maxilla or the mandible. Most of these tumors are benign, but a few recur or metastasize. The alveolar arrangement that characterizes this tumor's histopathology was not apparent in one case studied by FNA.[79] However, a combination of neuroblast-like elements and melanin-containing cells suggests this diagnosis. Rosai warns that malignant progonoma should not be confused with immature teratomas showing melanin-containing neural cells.[80] This issue has not been addressed in cytologic terms.

The adult type of rhabdomyoma is commonly seen in the head and neck of older patients, where it may be multifocal and can recur after excision.[81–83] The large cells that characterize this tumor feature a low N/C ratio because of abundant finely granular eosinophilic cytoplasm. The nuclei are eccentric and occasional cells show cross-striations. Thus, an important differential diagnostic consideration is aspiration of normal skeletal muscle. Clusters of needle-like cytoplasmic inclusions correspond to Z-band material when studied by electron microscopy, so that ultrastructural evaluation is useful for reaching a secure diagnosis. This is especially true in cases where cross-striations are not visible in routine histologic or cytologic preparations.

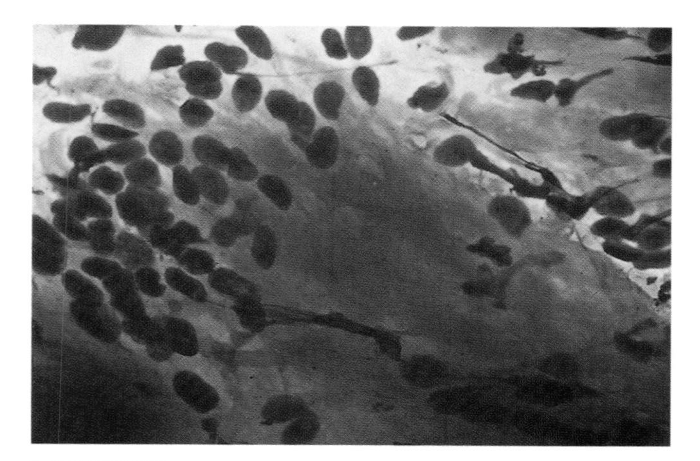

FIGURE 4–8. This neurilemmoma presented as a larger parapharyngeal mass in a 13-year-old boy. It was aspirated in the clinic, using a transoral approach. Smears show cellular areas alternating with collagenous foci in the pattern of Verocay bodies. (Diff-Quik, ×600)

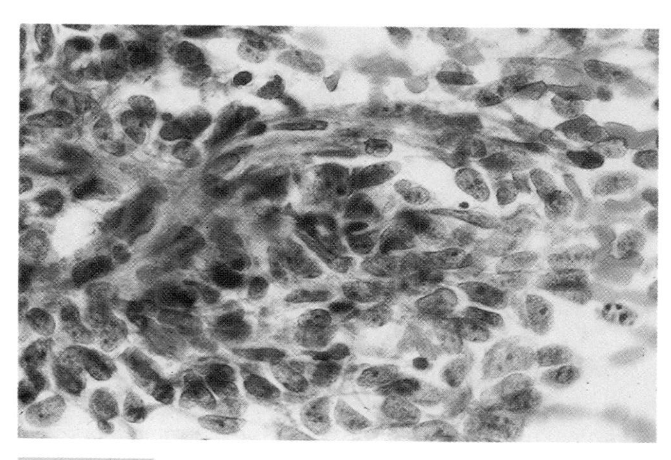

FIGURE 4–9. Metastatic sarcomas are occasionally aspirated from sites in the head and neck. This example shows metastatic synovial sarcoma from a previously diagnosed primary lesion in the chest wall. It is characterized by a highly cellular proliferation of uniform, malignant-appearing spindle cells. The differential diagnosis would include a synovial sarcoma primary in the head and neck, other metastatic sarcomas, spindle-cell ("sarcomatoid") carcinoma, and malignant melanoma. (Papanicolaou, ×600)

Spindle-cell pseudotumors related to mycobacterial infections are discussed subsequently, when head and neck FNA in the context of AIDS is addressed.

Sarcomas. Malignant soft-tissue tumors are not commonly represented in FNA samples from the head and neck. Rare examples of synovial sarcoma have been described; they show both spindle cells and three-dimensional glandlike clusters, indicating a biphasic neoplasm. These two components are immunocytochemically distinctive, with positive staining for vimentin and cytokeratin, respectively.[84] We have also aspirated examples of synovial sarcoma appearing as metastatic deposits in patients whose primary tumors were located elsewhere (Fig. 4–9).

As a head and neck neoplasm, rhabdomyosarcoma occurs primarily in children, so it rarely enters our differential diagnostic thinking in most head and neck oncology clinic situations. With the exception of nephroblastoma, other small blue cell tumors that may show rhabdomyoblastic differentiation are rare. Recent series have expanded our understanding of rhabdomyosarcoma's cytology, immunocytochemistry, and electron microscopy[85–87]; the latter methods are very useful in reaching the correct diagnosis and are applicable to cytologic samples. Reliable subclassification of the tumors as embryonal, alveolar, or undifferentiated is difficult when only cytologic material is available.[86]

The cells of rhabdomyosarcoma range from small with very little cytoplasm to larger with abundant cytoplasm that can be either dense or variably vacuolated. The larger cells can be round or elongated; in the latter instance, they resemble embryologic myotubes. Nucleoli are often inconspicuous in the larger cells. The classical tadpole or ribbon-shaped cells are seen in a minority of aspirates. Scattered large multinucleated cells are more typical of rhabdomyosarcoma than other small blue cell tumors of childhood.[86]

Other interesting features of rhabdomyosarcoma cytology may inform our differential diagnostic considerations. Nuclear molding has been considered more indicative of neuroblastoma but can also be seen in rhabdomyosarcoma. Small cells with prominent vacuoles may suggest either small noncleaved cell malignant lymphoma or Ewing's sarcoma. Further, small fragments of cytoplasmic debris that resemble the lymphoglandular bodies of malignant lymphoma smears may be present. However, in rhabdomyosarcoma, the vacuoles stain for neither the neutral lipids nor the glycogen that would typify the other tumors. The "tigroid" background of dispersed cytoplasm from damaged cells has been described in some rhabdomyosarcoma smears.[86,87] This is an expected finding in smears from glycogen-rich neoplasms and is most often associated with seminoma; it can also be seen in Ewing's sarcoma.

Kaposi's sarcoma will be discussed when AIDS-related lesions in head and neck cytology are considered.

Metastatic Malignancies. One of the most successful FNA applications is rapid, inexpensive, atraumatic identification of metastatic malignancies, especially those originating in one of the common carcinomas.[5] Even lesions of low clinical suspicion often represent appropriate targets, frequently with surprising results.

Problems related to distinguishing between cystic metastases of squamous cell carcinoma or papillary thyroid carcinoma and congenital cysts of the neck were discussed previously. These lesions often contain few tumor cells that may be degenerated or very well differentiated, leading to false-negative interpretations.[14,16,88–90] Further, it may be difficult to distinguish cytologically between precipitated debris in the benign cysts and the true necrosis seen in some of the malignancies.[91] This problem is complicated by the fact that some of these metastases may be associated with clinically occult primary tumors, especially in the tonsil.[89]

Inflammation and the reactive atypia with which it may be associated, as well as the overlapping age ranges between cancer patients and those with congenital cysts, have been discussed. Further, cystic lymph node metastases may yield only fluid and lymphocytes, or may show these components accompanied by a few large epithelial cells. Such findings may suggest a diagnosis of primary salivary gland neoplasms, such as Warthin's tumor or low-grade mucoepidermoid carcinoma, among others.[92] This problem is discussed more fully when salivary gland tumors are considered. Thus, correct diagnosis of cystic lesions in the lateral neck may ultimately require surgical excision.

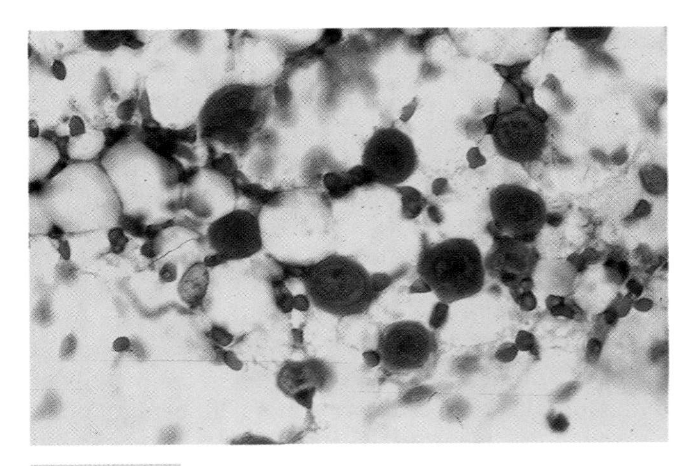

FIGURE 4–10. This example of acantholytic squamous cell carcinoma shows single cells. Their squamous nature is indicated by dense cytoplasm with concentric laminations and sharp cell borders. (Papanicolaou, ×600)

The glandlike aspects of acantholytic squamous cell carcinoma can be shared by the primary lesion and both its recurrence and metastases.[93] Differential diagnostic considerations include adenocarcinoma and other acantholytic squamous processes. Cytologically, aspiration of glandlike three-dimensional cell groups may lead to the former interpretation. Correct interpretation requires a search for pyknotic nuclei and the dense laminated cytoplasm typical of keratin formation (Fig. 4–10).

Numerous types of malignancy can give rise to metastases in the head and neck. Most of these are located in lymph nodes, as discussed more fully elsewhere in this text. Special cases, in which a metastatic deposit can mimic a primary salivary gland neoplasm, are considered subsequently. In other instances, sinonasal carcinoma can extend locally to present as a nodule on the face. Once a diagnosis of carcinoma has been reached, one must often rely on radiographic or historical information to help clarify the tumor's origin.

ASPIRATION OF HEAD AND NECK LESIONS ASSOCIATED WITH AIDS

As noted previously, FNA is a very useful way to address mass lesions in patients with AIDS; many of these do not represent malignancy. Further, cytology can usually identify the minority of patients in whom surgery is indicated.[35,36,94] One of the most common indications for biopsy in these patients is to evaluate the possibility of malignant lymphoma. As discussed in other chapters, these malignancies can be approached by cytologic means, especially if morphology is supplemented by cell marker studies based on either immunocytochemistry or flow cytometry.

AIDS-related salivary gland disorders have been reviewed.[95] Many of the benign conditions present as xerostomia. Nonneoplastic enlargement of the parotid is much more common in children (up to 58%) than in adults (approximately 1%).[94,95] FNA diagnoses of fungal infections (*Histoplasma, Candida,* and *Aspergillus*)[96] as well as cytomegalovirus parotiditis,[97,98] have been reported. The latter may be tumefactive or may present as pain without enlargement. Fixed smears are preferable to air-dried material for identification of viral inclusions. This was highlighted by one case in which nuclear enlargement without well-defined viral cytopathic features was erroneously interpreted as evidence of malignancy. This lead to surgical excision that otherwise might have been avoided (Table 4–11).[98]

In some AIDS-associated mycobacterial infections, macrophages assume a spindled shape, and their proliferation mimics a mesenchymal neoplasm.[99,100] The list of differential diagnostic possibilities occasioned by this picture is long, but the most important consideration is Kaposi's sarcoma, as we will discuss below when aspiration of AIDS-related neoplasms is considered. Further, cytoplasmic congregations of mycobacteria can mark positively with immunostains designed to identify cytoskeletal components, including actin, keratin, and tubulin. This can heighten the potential for diagnostic confusion. Stains for glial fibrillary acid protein and neurofilaments, however, are negative.

"Mycobacterial spindle cell pseudotumors" have been described in the literature. They can occur in the neck, where they are usually located within lymph nodes. These masses may or may not show the collections of lymphocytes and typical polyhedral macrophages that would suggest the correct diagnosis. Special stains for acid-fast bacilli show numerous organisms, so this diagnosis must be considered and the stain performed.

TABLE 4–11. AIDS-Related Head and Neck Masses

- Nonneoplastic salivary gland enlargements are more common in children than in adults.
- Fungal and cytomegalovirus-associated masses
- Kaposi's sarcoma
- Mycobacterial spindle-cell pseudotumor
- Benign lymphoepithelial cysts
- Malignant lymphoma of various types

Benign lymphoepithelial cysts of the parotid gland were known before the AIDS epidemic but were uncommon. It has been suggested that these lesions originate in salivary gland inclusions within cervical lymph nodes or as branchial cleft developmental abnormalities.[101,102] The cyst is surrounded by hyperplastic lymphoid tissue, and its lining can consist of squamous, mucinous, columnar, ciliated, or sebaceous cells in various patterns. We must distinguish benign lymphoepithelial lesions that are seen mostly in autoimmune disease (discussed below) from these lymphoepithelial cysts that are now relatively common in AIDS patients.[35] In the former, the gland parenchyma is replaced by a lymphohistiocytic infiltrate with residual epimyoepithelial islands and variable degrees of cystic alteration. The latter lesion is predominantly cystic with surrounding lymphoid tissue, and it is frequently situated within more normal-appearing salivary gland tissue than usually accompanies the diffuse pathology of benign lymphoepithelial lesions in Sjögren's syndrome.[102]

The benign lymphoepithelial cysts of AIDS are most common in the parotid, but they may also be found in submandibular salivary glands.[102] They may be unilateral or bilateral and are often multicystic. The apparent extent of involvement increases when physical examination is supplemented by computed tomography scans.[36,102,103] These cysts are often associated with extensive cervical lymphadenopathy, and it has been suggested that progressive intraglandular lymphoid hyperplasia leads from entrapment to obstruction to cystic dilatation of the gland's ducts.[94,103] They may recur whether simply drained by a needle or excised.[35] For this reason, repeat aspiration rather than excision has been recommended for palliation of recurrences.[36,94]

Cytologically, benign lymphoepithelial cyst aspirates show various epithelial cells of the previously listed types, variably vacuolated macrophages, and polymorphous lymphocytes. Lymphohistiocytic aggregates indicating germinal center formation, plasma cells, crystals, and parakeratotic cells have also been described.[35,36,102–104] In some cases, epithelial cells are not identified, so that one may diagnose a benign lymph node. Alternatively, this picture can lead to consideration of several primary salivary gland masses, including sialoadenitis, Warthin's tumor, or the classic type of benign lymphoepithelial lesion. If squamous cells are prominent, one may consider pleomorphic adenoma with metaplasia, mucoepidermoid carcinoma, squamous cell carcinoma, or a congenital cyst. Clinical and radiographic factors will contribute to the correct diagnosis.

When one considers salivary gland masses in AIDS patients, cysts and lymphoid hyperplasia are more common than malignancy.[35,36,94] Further, the probability of finding malignancy is much lower when a parotid mass is cystic than when it is solid. Most salivary gland malignancies associated with AIDS represent either Kaposi's sarcoma or malignant lymphoma.[95] The latter is discussed in other chapters devoted to lymph node cytology.

Kaposi's sarcoma has been described in FNA material from lymph nodes, mucosal sites, and the soft tissues.[53,95] Many of the patients to whose lesions FNA is applied have previous diagnoses of this tumor in other sites. Aspirations tend to be bloody and to show cohesive tissue fragments with overlapping spindle cells. A few free-lying cells and naked nuclei may also be present. In intact cells, the nuclei are surrounded by ill-defined cytoplasm, and the chromatin is finely divided. Eosinophilic globules described in histologic sections are not usually appreciated in smear material.

The differential diagnosis of Kaposi's sarcoma includes many low-grade spindle cell lesions. In mucosal or cutaneous sites, the characteristic clinical appearance should be present before the diagnosis is rendered without equivocation. Further, conservative interpretation of these often subtle and cytologically bland lesions is in order if the diagnosis of AIDS is not clearly established. As previously noted, acid-fast bacilli stains will be useful in excluding the cytologically similar mycobacterial spindle-cell pseudotumor.[99] One should not mistake crushed inflammatory cells or spindled enothelial cells for evidence of Kaposi's sarcoma.[53] The spindle cells of granulation tissue and other reactive processes show prominent nucleoli that are not a feature of Kaposi's sarcoma. The cytology of granulation tissue was considered earlier; its cellular mixed inflammatory background and well-formed capillaries are not features of Kaposi's sarcoma. High-grade sarcomas should not enter the differential diagnosis with the low-grade–appearing cells typical of Kaposi's sarcoma.

FNA CYTOLOGY OF SALIVARY GLAND MASSES

General Considerations

Approaches to Salivary Gland FNA. Historical aspects of FNAs application to salivary gland masses have been reviewed.[3,13,24,34,37,43,105] FNA enjoys its maximum impact on patient care when the results are available rapidly. If the patient is still in the clinic, follow-up or treatment options can be discussed and arranged immediately.[34]

We have noted previously that complications of the procedure are limited to tenderness and small local hematomas.[2,3,5–7,34,35] We have considered the parotid gland to be an area in which the theoretical problem of FNA-induced tumor recurrences can be studied.[1] This is because many aspirations show a pleomorphic adenoma, which is a neoplasm prone to recur if its capsule is violated by the surgeon. Penetration by the very thin aspirating needle, on the other hand, is not associated with this complication. Further, damage to the facial nerve does not result from FNA, either directly or through an enlarging hematoma.

When FNA shows metastatic malignancy, an inflammatory mass, or evidence of malignant lymphoma, the patient's workup can often be either truncated considerably or altered in other significant ways; in many instances, surgery is not necessary.[43,106] If a benign neoplasm such as a Warthin's tumor can be identified, surgery can be avoided or delayed as needed in older patients or in those with serious medical illnesses.

When a malignancy is unequivocally recognized in cytologic samples from a salivary gland, the surgeon's discussions with the patient regarding any proposed procedure and its possible complications (including damage to the facial nerve) may be facilitated considerably.[3] In some series, early application of FNA has led to decreases of up to 30% in the rate of surgical exploration for salivary gland masses.[18] Other investigators suggest that this method alters the clinical approach in up to 35% of masses.[6] It is often reasonable to evaluate the mass in a stepwise fashion, beginning with FNA, the least invasive procedure.[25,43] Imaging studies provide useful information regarding the size and precise location of the mass. However, unless evidence of local invasion or metastases is obtained, these studies will not clearly distinguish between benign and malignant masses.[107]

Considered in its entirety, the literature outlines three general approaches for using FNA to address salivary gland masses. (To some extent, these may be applied to head and neck FNA in general.) The optimistic information just presented suggests to some that FNA should be used early and often, perhaps playing a role in most masses in this area. However, citing significant false-negative and false-positive rates for the diagnosis of salivary gland neoplasms, others argue that this technique should be reserved almost exclusively for triage of new masses arising in patients with an established history of head and neck malignancy.[108] (Accuracy figures for salivary gland aspiration will be discussed subsequently. FNA will then be compared with frozen sections of excised masses for rapid diagnosis and treatment planning.)

Other workers are even more restrictive in their application of FNA to the salivary glands.[109] These physicians not only cite limited diagnostic accuracy, but also mention the types of post-FNA infarction and repair that they think might ultimately limit the histologic interpretation of tissues excised after preoperative FNA. Thus, while agreeing that FNA has a place in the evaluation of lymph node enlargements and thyroid disease, confirmation of suspected infectious or inflammatory masses, and demonstration of recurrent malignancy, these investigators would argue that FNA does not merit inclusion in schemes for salivary gland mass evaluation. We doubt that any unanimity of approach can be achieved in the near future; we expect the current variability among institutions to continue. Our experience indicates that when efficient, high-quality FNA services are available, many head and neck surgeons embrace the method. Many eventually come to use it very early in the evaluation of most masses.

Accuracy of Salivary Gland Mass Diagnosis by FNA. Several studies illustrate the improvements in diagnostic accuracy for FNA of salivary gland masses that come with increased experience. Training and experience are required in both collecting and interpreting samples.[1,10,13,37,110]

Reliable accuracy measures can be difficult to obtain, because some series are small. Further, some authors exclude unsatisfactory or nondiagnostic cases from summary calculations; others do not. Several papers tabulate relatively numerous aspirations showing only normal tissues. Without more detailed information than is frequently published, many reviewers would consider these unsatisfactory, or at least "nondiagnostic." Definitions of false-negative results also vary. For example, some argue that a Warthin's tumor interpreted cytologically as chronic inflammation would represent a false-negative diagnosis; others would not. A sparsely cellular aspirate showing mostly mucus suggests low-grade mucoepidermoid carcinoma and should direct further clinical investigations toward the appropriate interventions. However, the FNA report for such an aspirate usually does not contain an unequivocal diagnosis of malignancy, so that this aspiration would be considered as a false-negative reading for the purpose of publishing statistical accuracy assessments.[110] However, is this clinically useful report a false-negative result? Would the patient benefit from having the FNA considered unsatisfactory?

Negative interpretations that are truly false, however, remain a problem.[109] Most involve malignancies that yield sparsely cellular smears or show only cells of low cytologic grade, or from which rare diagnostic elements are overwhelmed by copious benign tissue. Low-grade mucoepidermoid carcinoma, small cell malignant lymphoma, acinic cell carcinoma, and Hodgkin's disease may result in samples showing these problems.[13,32,106,107,110] Others result from sampling error,[111] particularly when small metastatic deposits are evaluated.[24]

Thus, one should cautiously interpret published accuracy figures for salivary gland FNA with these caveats in mind. Using detection of malignancy as the diagnostic goal, sensitivity rates range from 64–100%; those for specificity are usually higher (94–100%).[8,110] Given the low-grade mucoepidermoid scenario cited above, one can see that rates for unsatisfactory aspirations are also subject to interpretation. Such figures in part reflect the procedural and interpretive skills as well as the experience and orientation of the persons involved. Overall nondiagnostic rates for salivary gland or general head and neck FNA are approximately 10% in most series,[2,5,34,108,111] but some report up to 21%.[111]

It is useful to compare FNA with the only other means for rapid tissue diagnosis, the intraoperative frozen section applied to surgically excised tissues.[112] Aspiration actually allows more time for reflection, consultation, and special studies than does frozen-section analyses that are performed during an operative procedure.[37] The rates for correct recognition of malignancy by these two procedures are similar in many studies and can be as low as 60%.[3,5,13,37,106,110,111,113–116] Some series show FNA to be slightly more accurate; others suggest that frozen sections lead to fewer diagnostic errors. Both methods can lead to false-positive[116] and false-negative[112] diagnoses; both have been blamed for unnecessarily radical surgery resulting from false-positive interpretations.[112]

Higher-grade malignancies are more readily interpreted both in FNA and on frozen sections, and accurately figures for both tests depend in part on the case mix in the material available for review.[113] Uncommon malignancies, low-grade (mucoepidermoid) carcinomas, cystic (or partially cystic) masses, and complex lesions with multiple components (some mixed tumors and most carcinomas ex pleomorphic adenoma) may be very difficult to interpret using either medium of rapid diagnosis. Inflammatory and fibrotic masses are also likely to be difficult.

One important consequence of these observations is that diagnostic problems at the time of FNA may not be addressed effectively by intraoperative frozen sections. Even though recommending cryostat "confirmation" for difficult cytologic diagnosis is a common practice,[117] resolution of diagnostic problems may not be possible until well after completion of the surgery, when paraffin-embedded permanent sections become available. The parallels between FNA and surgical pathology are so extensive and so compelling that a case difficult at FNA is likely to be just as difficult at the time of frozen section. Both methods have limitations, and these frequently converge on the same residuum of very difficult diagnoses.

Some have found that using both FNA and frozen sections improves the overall accuracy of salivary gland diagnosis before definitive resection.[111,112] Heller et al.[113] suggested that both methods should be applied to "virtually all patients in whom surgery for salivary gland tumors is performed." Because the complication rate of FNA is negligible, repeated aspiration of clinically worrisome masses can be safely used as necessary.[17] Further, even if FNA fails to provide useful information, it will have cost very little and caused virtually no delay in the application of other diagnostic approaches.

Many physicians now find FNA to be a useful test for the preoperative evaluation of salivary gland masses. However, this brief overview of the abundant published information indicates that this test has areas of considerable interpretational difficulty, as well as occasional false-positive and false-negative results. Reasons for these problems include the extraordinary diversity of salivary glands neoplasms, and the fact that many are rare. The low-grade cytology of some malignancies and the frequent superimposition of inflammation or cystic change on many masses exacerbate this already difficult situation. The former is in contrast to the fact that the most common cause of false-positive diagnoses is "atypia" in benign mixed tumors.[105,118] Thus, we argue that salivary gland FNA is useful if performed and interpreted with skill and experience, and if its limitations are kept in mind.

INTRODUCTION TO SALIVARY GLAND CYTOLOGY

As noted previously, the salivary glands host an extraordinary array of mass lesions. Many of these can be complicated by cystic alterations, inflammation, clear cell change, or metaplasias of various types. For these reasons, the cytologic range of any given type of mass can be very broad. Interpretation is made even more difficult by the fact that many tumors are so uncommon that we never acquire adequate experience (Table 4–12).

Our approach to the cytology of these diverse masses is based on the differential diagnosis of certain key findings that are usually apparent early in the evaluation of a case. Some of these are common to a number of lesions, so that a list of differential diagnostic possibilities flows from our recital of each key finding. The next step is identification of additional features that indicate a more specific diagnosis, or that permit refinement of our list of possible diagnoses. If neither of these is possible, we must recognize that one of the limits inherent in salivary gland cytology has been reached.

TABLE 4–12. Factors Complicating Salivary Gland Tumor Diagnosis by FNA

- Great tumor diversity
- Many tumors are rare.
- Some malignancies are cytologically of low grade.
- A variety of secondary alterations are shared by several tumors.
- Can occur in many sites, often outside the major glands
- Can arise in aberrantly situated salivary gland tissue

This section is organized around these key findings that often form the first impression of an aspirate. Lesions that are cytologically similar or even indistinguishable (e.g., mixed tumor, benign metastasizing mixed tumor, and most examples of carcinoma ex pleomorphic adenoma) are discussed together. We have found this scheme more useful in organizing our differential diagnostic thinking than the traditional catalogue of benign lesions followed by a list of malignancies.

In the discussion that follows, we have tried to recapitulate and illustrate the thinking that goes into our cytologic interpretation of these complex masses. Our discussions include occasional lesions that are not salivary gland tumors, but that must be considered in certain cytologic situations. Pleomorphic adenoma and Warthin's tumor are sufficiently common and (usually) distinctive to be listed as primary considerations around which our thinking can frequently be organized. The other categories result from our initial impressions of an aspirate, and may eventuate in a variety of final diagnoses. Some tumor types appear in more than one category, reflecting the complexity of certain differential diagnoses. We discuss such lesions in the context of their most frequent or vexing mimics.

Clinical Expectations

Most salivary gland masses are located in the parotid. A smaller number can be found in the submaxillary gland, and occasional lesions occur in the sublingual gland or in submucosal minor salivary glands.[10,18,24,34,110] Further, approximately 20% of persons harbor accessory parotid tissue along Stensen's duct. This large gland may also have an anteriorly situated facial lobe. Heterotopic salivary gland tissue can occur in several sites around the head and neck.[1] Thus, a wide range of salivary gland masses, including malignancies, may be found in unusual locations. One must consider these tumors frequently, despite clinical suggestions that the lesion may be based in other tissues. Up to 50% of aberrantly located salivary gland neoplasms are malignant, with mucoepidermoid carcinoma being the most common type.

Referral patterns and patients' ages influence the types of salivary gland masses one is likely to encounter.[43] Pathologists who frequently study material from AIDS patients are more likely to see masses from younger persons, with a correspondingly low rate of malignancy. It has been said that 80% of parotid tumors are benign, and that 80% of these represent pleomorphic adenoma.[119] In more detailed studies, it was noted that 71–96% of parotid tail masses unaccompanied by clinical signs of malignancy (pain, facial nerve damage, trismus, or fixation) represented mixed tumors (sensitivity and specificity of 88% and 91%, with specificities of 50% and 84%, respectively).[120] A rare instance of pleomorphic adenoma infarction associated with pain was discussed previously.

Cytologic Findings Shared by Several Lesions

Several cytologic findings overlap a number of diagnoses. Lymphoid stroma,[92] sebaceous differentiation,[121–125] cystic change,[6,10,13,16,18,37,101,114,121,126] squamous metaplasia,[18,27,28,44,94,109,110,121,127–130] clear cell change,[18,57,78,131–135] and oncocytic features[136] are all cytologic findings distributed across several types of masses. Care must clearly be exercised in using these as important diagnostic criteria. Taken alone, each outlines only a differential diagnosis. This overlap contributes significantly to the difficulty of salivary gland cytology, and serves to emphasize the need for excellent specimen collection as well as the diagnostic limitations of the method. Although each of these may be cytologically impressive or may even be the predominant finding, each is useful only in concert with other features. As evidence of this approach, we note that none of these findings figure in our first-level, rapid-triage approach to salivary gland FNA (Tables 4–13 to 4–18).

Aspiration of Normal Salivary Gland Tissue or Adipose Tissue

We have previously commented on the fact that in some series, up to 20% of salivary gland aspirates yield only normal tissue.[10,34,43] Cytologically, this shows acinar cells in characteristic round, basket-like arrangements. These oval to wedge-shaped cells have small, uniform, eccentrically placed nuclei surrounded by abundant granular or vacuolated cytoplasm. Ductal tissue fragments show the honeycomb cell arrangement typical of benign glandular tissues from many body sites. Often, large branching duct fragments are present. These two types of tissue may be closely associated or may be found separately. Variable numbers of adipocytes accompany these epithelial elements (Fig. 4–11). Acinar cells are large and

TABLE 4–13. Cytologic Findings Shared by Several Salivary Gland Masses

- Lymphoid stroma
- Cystic change
- Clear cell alterations
- Oncocytic cells
- Sebaceous differentiation

TABLE 4–14. Salivary Gland Masses That May Have Lymphoid Stroma

- Sialoadenitis
- Cystic lesions due to duct obstruction
- Warthin's tumor
- Sebaceous lymphadenoma
- Lymphoepithelial cyst
- Lymphoepithelial lesion
- Oncocytic lesions
- Acinic cell carcinoma
- Mucoepidermoid carcinoma
- Primary lymphoepithelioma-like carcinoma
- Metastatic lymphoepithelioma
- Other metastatic malignancies

TABLE 4–15. Salivary Gland Masses That May Have Cystic Change

- Congenital cysts
- HIV-associated cysts
- Stones or other obstructive lesions
- Pleomorphic adenoma
- Monomorphic adenoma
- Warthin's tumor
- Mucoepidermoid carcinoma
- Acinic cell carcinoma
- Metastatic carcinoma

TABLE 4–16. Salivary Gland Masses That May Have Clear Cell Alterations

- Mucoepidermoid carcinoma
- Acinic cell carcinoma
- Oncocytic lesions
- Epithelial-myoepithelial carcinoma
- Adenocarcinoma
- Clear cell carcinoma
- Some metastatic carcinomas

TABLE 4–17. Salivary Gland Masses That May Have Oncocytic Cells

- Warthin's tumor
- Pleomorphic adenoma
- Mucoepidermoid carcinoma (rare)
- Acinic cell carcinoma
- Oncocytic lesions
- Oncocytoid adenocarcinoma

TABLE 4–18. Salivary Gland Masses That May Have Sebaceous Differentiation

- Normal salivary tissue (all sites)
- Sebaceous adenoma
- Sebaceous lymphadenoma
- Monomorphic adenoma
- Pleomorphic adenoma
- Warthin's tumor
- Mucoepidermoid carcinoma

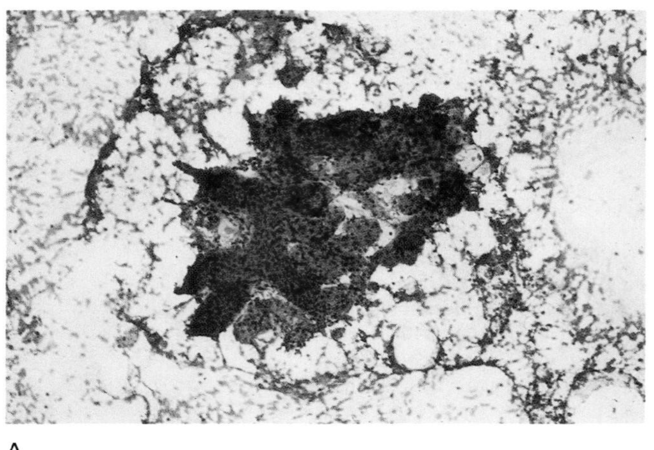

A

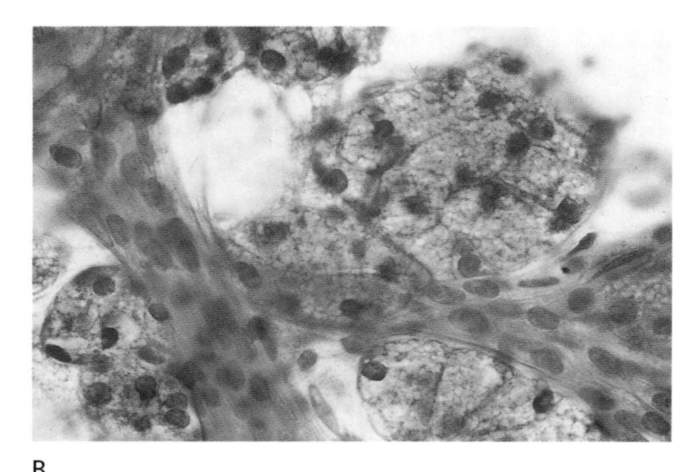

B

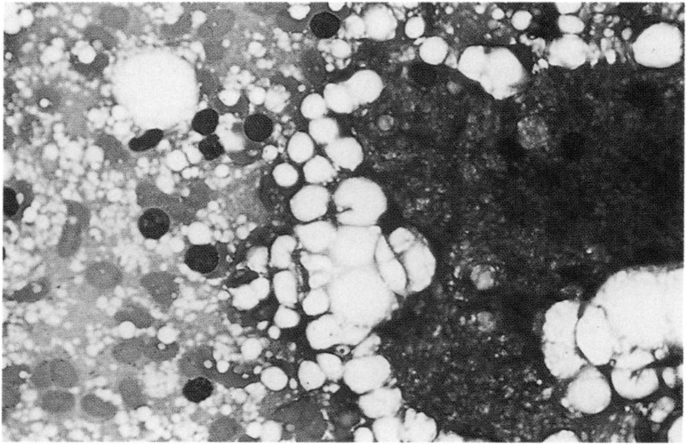

C

FIGURE 4–11. (**A**) This low-magnification image shows a large cohesive fragment of normal parotid tissue. An arborizing network of ducts is surrounded by acinar tissue. (Papanicolaou, ×40) (**B**) The ducts in this normal parotid tissue are composed of small uniform cells. The acinar cells are larger, with abundant granular cytoplasm and innocuous-appearing nuclei. (Papanicolaou, ×400) (**C**) In air-dried smears, the acinar cells' cytoplasm stains darkly. During the smearing process, some cells are damaged. Their cytoplasm forms a granular background, while their nuclei are stripped. The latter should not be mistaken for lymphocytes. (Diff-Quik, ×400)

fragile, so smears usually show some cell damage. As a result, naked acinar cell nuclei and granulovacuolar cytoplasmic debris litter the smear background. One should not interpret these naked nuclei as lymphocytes. Hematopoietic cells should be required to show a thin intact rim of basophilic cytoplasm for positive identification.[137] Both lymphocytes and the cytoplasmic residua of damaged acinar cells are more easily assessed in air-dried than in fixed material.

Sialosis. This nonneoplastic salivary gland enlargement shows no evidence of inflammation and does not cause pain. The increase in size is slowly progressive, and the lesion has frequently been present for months or years before diagnosis. Physical examination shows a doughy swelling without distinct borders.[137] Most cases are bilateral, and the parotids are more commonly involved than the submandibular glands.[18,138–140] Sialosis may be associated with nutritional deficiencies, hence the term "nutritional mumps."[138] Other patients have diabetes mellitus, alcoholism, cirrhosis, or endocrine deficiencies. Still other examples have been associated with drugs such as phenylbutazone, iodine-containing compounds, some antibiotics, and adrenergic agents.[18,139–141]

The affected glands show normal histologic architecture, but the acinar cells are said to be hypertrophic, with abundant cytoplasm.[139,140] In one study, aspiration of four bilateral parotid examples of sialosis showed no lymphocytes and were more cellular than most aspirates of normal tissue.[141] (Our experience suggests that the latter observation would be very difficult to use as a diagnostic criterion in an individual case.) The mean acinar cell diameter in sialosis (75 μm) was significantly greater than that of normal parotid tissue (56 μm). In contrast to the bilateral nature of sialosis, symptomatic stones are usually unilateral.[142]

Salivary Gland Lipomatosis and Lipomas. Aspiration of normal adipose tissue from salivary gland enlargements may suggest lipoma.[10,18,34,105,138] Clinically, these masses are soft and boggy, resembling Warthin's tumors. The cytologic impression of a lipoma can be confirmed when computed tomography or magnetic resonance imaging images show a lesion with the density of fat that is located within the gland.

Bilateral fatty infiltration (lipomatosis) is similar to sialosis at physical examination and can share some of the same endocrine, metabolic, or drug-related associations. Acinar enlargement can also be noted.[138] Radiographic studies show diffuse infiltration of the glands with tissue of fatty density, rather than a mass. Sialosis and lipomatosis may have identical FNA findings, and it has been suggested that lipomatosis represents an end stage of sialosis.[141]

Aspirates Showing Only Normal Salivary Gland Tissue. Sialosis, lipoma, and lipomatosis are uncommon conditions rarely encountered in most FNA practices. Without strong clinical and radiographic support of the types previously described, recovery of such material from a mass lesion is usually interpreted as nondiagnostic or unsatisfactory. This stance is often reasonable,[18,138] and it is motivated by a desire to minimize false-negative diagnoses that result from sampling error. However, given excellent aspiration technique and careful clinical correlation, selected examples of sialosis can be diagnosed or at least suggested by FNA. Further, small masses that would previously have been followed without sampling often make very challenging FNA targets. Some of these will not be successfully aspirated, and this may explain the relatively frequent identification of normal tissue in some series.[10,34,43,137]

When only normal or hypertrophic acinar tissue is recovered by FNA, one may consider a diagnosis of acinic cell carcinoma. If the aspirate is reasonably cellular, the admixture of ductal elements with acinar tissue will eliminate this from consideration.[143] Whether or not a small lesion has been missed altogether is more difficult to ascertain. In our series reviewing normal tissue aspirated from 18 unilateral and two bilateral parotid or submandibular gland enlargements, the only missed neoplasm that subsequently came to light was a single pleomorphic adenoma 0.5 cm in diameter[137]; this lesion was correctly diagnosed by a second FNA. An example of false-negative findings in a metastatic squamous cell carcinoma has also been reported.[13]

These data suggest that if physical examination, FNA, and follow-up are expertly performed and carefully coordinated, the chance of missing a significant neoplasm is probably low. However, in situations that are less than ideal, aspiration of normal tissue from a salivary gland mass should still be considered evidence that the lesion has probably not been targeted accurately. FNA is easy to repeat and enjoys excellent patient acceptance. In many instances, a second aspiration may be all that is needed to resolve problems related to sampling (Table 4–19).[1]

Interestingly, the findings presented above suggest that the clinical definition of sialosis might be expanded to include masses that are unilateral, submandibular, or not associated with other clinical problems. Acinar diameter measurements deserve further evaluation in this regard.

TABLE 4–19. Differential Diagnosis of Aspirates Showing Normal Acinar, Ductal, or Adipose Tissue

- The aspirator missed the target lesion
- Sialosis
- Lipoma
- Lipomatosis
- Except for the first, each of these is rare.

Inflammatory Salivary Gland Masses

One of the immediate benefits of FNA is that it can quickly triage mass lesions into either inflammatory or neoplastic categories.[13] In the salivary glands, relatively few masses referred for FNA are inflammatory.[10,34,111] However, many of these swellings do not require surgery.[42] The most common clinical causes of sialoadenitis are mumps and sialolithiasis.[144] Mumps does not require cytologic diagnosis, and stone disease is rarely addressed by these means (see below). Although inflammatory in their primary manifestations, duct obstructive lesions (including sialolithiasis) will be discussed later with the cystic lesions, as this is the category in which most of the relevant differential diagnostic considerations are to be found.

Acute Inflammation. Diagnosis of infectious salivary gland lesions by FNA is uncommon.[142,144,145] However, these may be encountered by AIDS patients or in a lesion secondary to duct obstruction, such as that caused by stones. Most purulent aspirates are regarded as nonspecific, but material for cultures should be set aside in accordance with the guidelines previously discussed. Aspirates of a purulent mass show a mixture of inflammatory cells, fibrin, and debris.[18,105] Necrotic neoplasms can give rise to acute inflammation, so a careful search for malignant cells is also required. However, in the setting of marked acute inflammation, cells shed from the normal salivary glands can show striking cytologic atypia. Cautious and conservative interpretation is required to avoid false-positive diagnoses of malignancy. In most cases, a diagnosis of salivary gland carcinoma with acute inflammation is in error. This principle applies to FNA of other body sites as well, especially the breast.

Chronic Inflammation and Its Differential Diagnosis. A wide variety of salivary gland lesions can be accompanied by chronic inflammatory cells. When other, usually epithelial, diagnostic elements are not seen or are scanty, many of these aspirates are interpreted cytologically as nonspecific chronic inflammation. Some aspirates with chronic inflammation represent hyperplastic intraparotid lymph nodes that have come to clinical attention because of enlargement.[13] Lymphoid proliferations that persist after FNA are clinically suggestive of a neoplasm.[110] The neoplasm most commonly mistaken for inflammation is Warthin's tumor, from which oncocytes may be difficult to recover because of sampling error in a partially cystic lesion.[43]

In addition to mixed acute and chronic inflammatory cells and foam cells, aspirates from chronic inflammatory lesions can show squamous metaplasia or stringy extracellular mucus, as well as epithelial atypia of the type described in acute inflammation.[6,13,110,146] Such findings may lead to an erroneous diagnosis of mucoepidermoid carcinoma that can be very difficult to avoid, given the occasional occurrence of considerable chronic inflammation in aspirates from this malignancy.

Fibrotic masses may give sparsely cellular aspirates that are nondiagnostic. Alternatively, fragments of metachromatic, fibrillary, collagenous stroma may be reminiscent of the matrix of pleomorphic adenomas.[146–148] (This and other errors in matrix identification are discussed subsequently.) The previously described distinction between acinar cell nuclei and lymphocytes is useful in avoiding an erroneous diagnosis of acinic cell carcinoma in cases of chronic sialoadenitis.[116]

Postirradiation sialoadenitis is characterized by chronic inflammation, fibrosis, and acinar atropy; the duct system is usually preserved to some extent. Clinically, this process can produce very firm glands that are difficult to distinguish from carcinoma. Aspirates are often scanty and may show variable degrees of cytologic atypia.[13,18] Cohesion within cell groups usually suggests a benign condition, but clusters of duct cells can mimic squamous cell carcinoma or low-grade mucoepidermoid carcinoma. Other features of these neoplasms should be sought in difficult cases. Examples with marked radiation-associated cytologic atypia have been mistaken for undifferentiated carcinoma.[4]

The benign lymphoepithelial cyst in AIDS patients and the benign lymphoepithelial lesion associated with Sjögren's syndrome have previously been contrasted,[13,149] but the two may coexist. Aspirates show polymorphous lymphocytes, some plasma cells, macrophages, and germinal center fragments (so-called lymphohistiocytic aggregates). Lacking epimyoepithelial cell clusters or at least groups of spindled myoepithelial cells, this cytologic picture is nonspecific,[10,43] and the diagnosis must be supported by clinicopathologic correlation.[150] The most common alternative interpretation is Warthin's tumor.[17,112] The lymphoid tissue of this neoplasm is more monomorphous and less likely to show germinal center fragments than expected in the benign lymphoepithelial lesion.[17] Numerous microliths have been described in aspirates from this lesion.[150] The finding of chronic inflammation and stonelike fragments requires distinction from sialolithiasis by clinical criteria (see below). The malignant lymphoepithelial lesion is discussed subsequently, when it is contrasted with other malignancies that show mostly small cells.

Hematopoietic neoplasms are discussed in other chapters, and only specific observations related to the salivary glands are summarized here. We would like to identify these lesions in FNA samples, so that the necessary studies can be instituted; in many cases, surgery can be avoided.[111] Reactive lymphocytes may be confused with cells representing a low-grade malignant lymphoma. Usually, however, they are more polymorphous than their neoplastic counterparts (see Fig. 4–11).[105,116] Nevertheless, aspiration of well-differentiated lymphoid neoplasms frequently leads to false-negative diagnoses, as they are thought to represent lesions such as Warthin's tumor, congenital cyst, or benign inflammation.[10,13,14,107,110] Mixtures of reactive and neoplastic lymphoid elements may further complicate diagnosis, with polymorphism that belies the nature of the underlying process.[18] In salivary gland lymphomas of the mucosa-associated lymphoid tissue (MALT) type, residual entrapped epithelial cells can lead to an erroneous diagnosis of benign lymphoepithelial lesion.[151] High-grade hematopoietic neoplasms are more readily recognized as malignant[107] but may be mistaken for small cell carcinoma.[21,152,153] Difficulties are minimized if very high technical standards attend specimen preparation. Consideration of the patient's previous history and the application of marker studies can be used to resolve most difficult cases.

Hodgkin's disease is encountered rarely in salivary gland aspirates. The possibility of false-negative diagnoses has been mentioned; these occur when Reed-Sternberg cells are rare or absent. Such cases are usually thought to represent nonspecific chronic inflammation.[13] Other examples have been confused with pleomorphic adenoma, metastatic carcinoma, malignant melanoma, and undifferentiated carcinoma.[13,34,105] Acute leukemia in salivary gland aspirates is rarely described.[15]

Granulomatous Inflammation. Granulomatous sialoadenitis can be either unilateral or bilateral, but it is an uncommon FNA diagnosis in salivary gland sites.[13] Sarcoidosis may involve the salivary glands or intraglandular lymph nodes.[154] Rupture of salivary gland ducts or cysts leads to extravasation of secretory material that can cause a giant cell reaction and changes of mesenchymal repair.[110] Further, this reac-

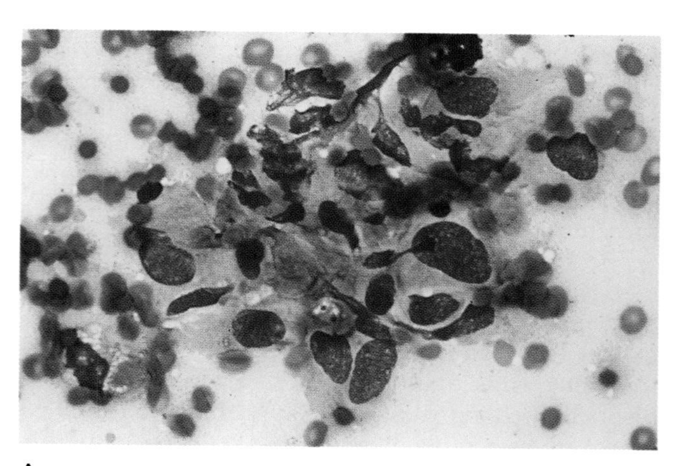

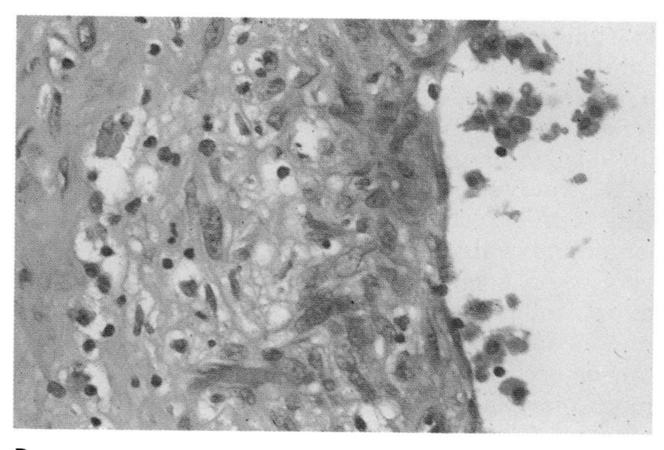

FIGURE 4–12. (**A**) This group of atypical spindle cells was present in an otherwise sparsely cellular aspirate from a low-grade mucoepidermoid carcinoma. (Diff-Quik, ×600) (**B**) Histologic sections of the excised tumor subsequently showed rupture of a cystic space and this exuberant mesenchymal repair reaction to the cyst contents. (Hematoxylin and eosin, ×400)

tive pattern can be superimposed on a variety of benign and malignant processes. In low-grade mucoepidermoid carcinomas, for example, we have seen atypia in reactive stromal cells that far outweighs that of the carcinoma cells, leading to a rather confusing FNA picture (Fig. 4–12).

Sialoadenitis with Crystal Formation. This special type of sialoadenitis is uncommon.[155,156] Crystals of various types can be seen in normal salivary gland tissue as well as in cysts, nonspecific chronic sialoadenitis, pleomorphic adenoma (see below), and Warthin's tumor. The crystals in this condition consist of amylase. They are 20–300 μm long and up to 100 μm wide (Fig. 4–13). They are associated with various combinations of duct cells, acinar cells, and granulation tissue. Bacteria are not identified, and this process may recur after the administration of antibiotics. Clinically, this is a lesion of adults, mostly over 70 years of age. Sialoadenitis with crystal formation can involve either the parotid or the submandibular gland and can cause physical findings that suggest carcinoma. Patients have usually been symptomatic for weeks or months.

Cytology and Differential Diagnosis of Pleomorphic Adenoma

Pleomorphic Adenoma (Benign Mixed Tumor). One approaches salivary gland FNA knowing that a substantial majority of aspirates represent pleomorphic adenomas.[119–121] Physical examination shows a well-circumscribed mass that is usually very firm. Many have a lobulated surface. This presents a different sensation to the fingertips than the softer surface of a

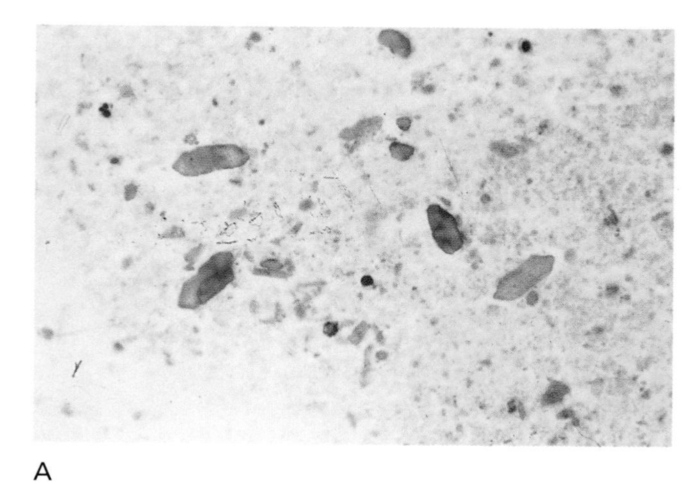

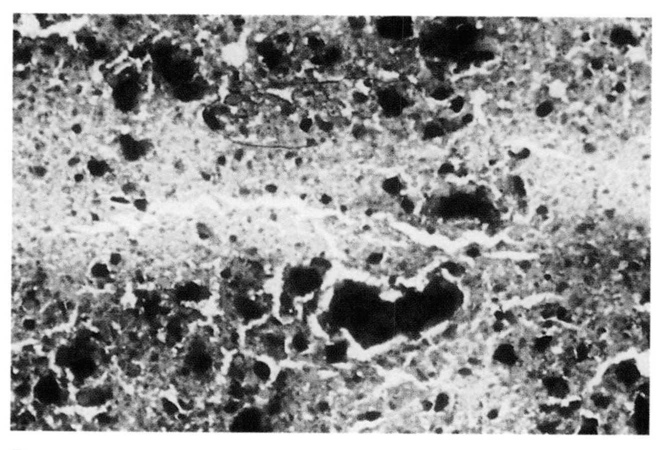

FIGURE 4–13. (**A**) When seen in fixed smears, sialoadenitis with crystal formation is very distinctive. One may encounter duct cells or acinar cells, but much of the material consists of debris and numerous crystals, as illustrated here. (Papanicolaou, ×600) (**B**) Air drying accentuates the background material but renders the crystals very difficult to visualize. (Diff-Quik, ×600)

Warthin's tumor.[1] The cytologic diagnosis is usually straightforward[18]; 80% of cases or more are correctly interpreted at the time of FNA.[10,13,107]

Smears show three components and differ largely in the relative proportions of these basic elements (Fig. 4–14). Duct cells are small and cuboidal and can form flat sheets, glandular associations, trabeculae, or larger branching tissue fragments. They usually show very bland nuclear morphology, but moderate pleomorphism in a minority of cells is common.[106] Mixed tumors are given to a wide variety of alterations that can be seen singly or in combinations. Thus, squamous metaplasia, oncocytic change, cystic alteration, mucin production, or sebaceous differentiation may complicate the cytologic picture.[122,136,157] Myoepithelial cells are usually present and may be either spindled or plasmacytoid. The latter often show marked dissociation, and the single plasmacytoid cells are sometimes mistaken for hematopoietic cells. Myoepithelial cells can also be trapped in the chondroid matrix of the mixed tumor.

The epithelial and myoepithelial elements of pleomorphic adenomas are sufficiently complex and diverse to raise many diagnostic considerations. In general, however, the chondromyxoid matrix is a more specific and helpful feature. As discussed elsewhere in this chapter, this material can be mimicked by other types of collagenous stroma. This is most significant when the desmoplastic stroma of a carcinoma is mistaken for mixed tumor matrix. When associated with a malignancy of low nuclear grade, this can result in severe diagnostic difficulties.

When fixed and prepared by the method of Papanicolaou, the matrix of pleomorphic adenomas is pale and gray to cyanophilic and closely resembles the chondroid material as it appears in histologic sections. When air-dried and stained by a Romanowsky method, this matrix is metachromatic and shows a distinctly fibrillary substructure. The latter is often appreciated best at the edge of tissue particles, where it presents as frayed edges.

The Papanicolaou stain shows myoepithelial cells sequestered within the matrix to good advantage; they appear stellate or spindled and have small dense nuclei. They can be widely dispersed through a large fragment of the matrix material, and this imparts a distinctly chondro-osseous appearance. Air-dried preparations render these cells as pale smudgy blue outlines largely obscured by the densely staining matrix in which they lie. Our inability to see the nuclear details of these cells on air-dried smears causes no diagnostic difficulties. The appearance of these tissue particles with their ghostlike myoepithelial inhabitants is characteristic and diagnostic (Fig. 4–15).

Cytomorphologic Features of Pleomorphic Adenomas

- Clusters of ductal cells
- Spindled or plasmacytoid myoepithelial cells
- Extracellular matrix material
- Some myoepithelial cells entrapped within matrix
- Squamous metaplasia
- Oncocytic change
- Mucin production
- Sebaceous differentiation

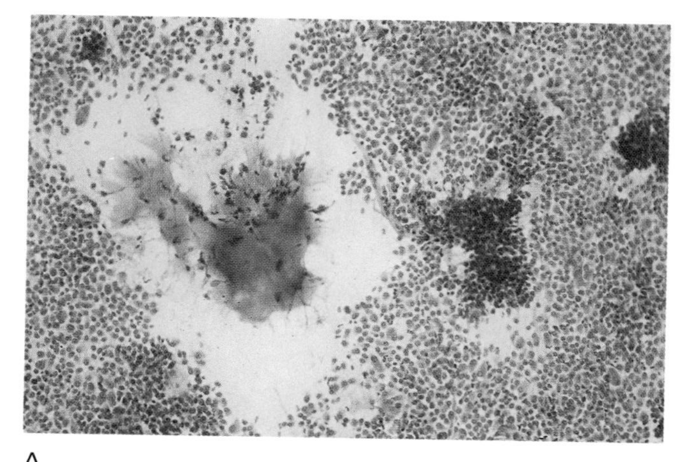

A

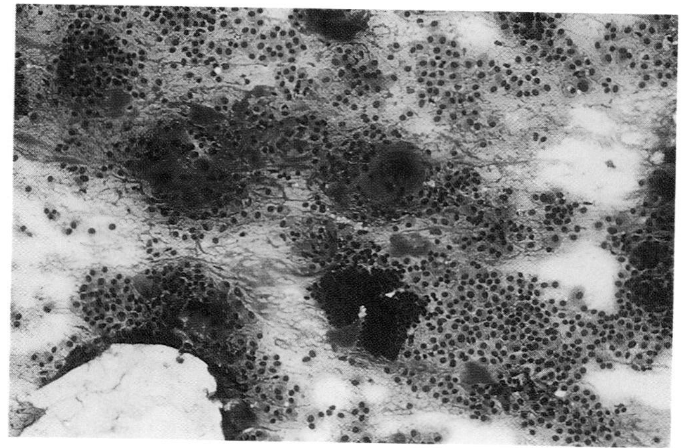

B

FIGURE 4–14. In both the fixed (**A**) and air-dried (**B**) smears, this mixed tumor shows all three types of tissue found in this neoplasm. Darkly staining cohesive clusters of ductal cells are three-dimensional, whereas the myoepithelial cells lie singly. Fragments of stroma are pale on the fixed preparation and brightly metachromatic on the dried smear. (**A,** Papanicolaou, ×100; **B,** Diff-Quik, ×100)

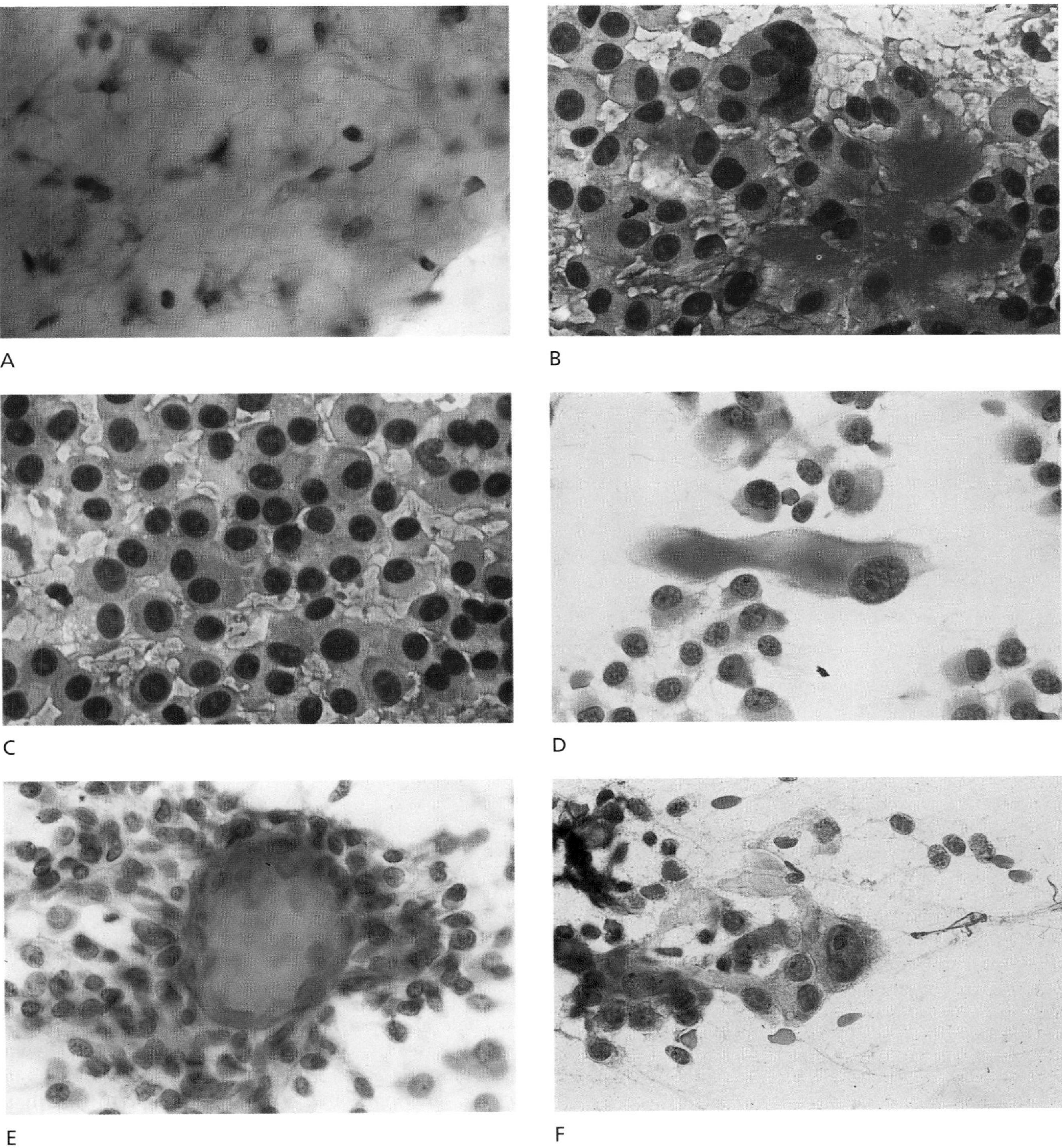

FIGURE 4–15. (**A**) This fragment of matrix material from a pleomorphic adenoma has a distinctly chondro-osseous appearance that closely resembles its histologic counterpart. Scattered through the matrix are numerous stellate myoepithelial cells. (Papanicolaou, ×600) (**B**) In air-dried smears, the same material is metachromatic and has a fibrillary quality. The latter is most apparent at the edge of large tissue fragments. Myoepithelial cells that are trapped in the matrix are pale and show little detail. (Diff-Quik, ×600) (**C**) Myoepithelial cells can be plasmacytoid (so-called hyaline cells) and have been mistaken for a hematopoietic neoplasm. (Diff-Quik, ×600) (**D**) When fixed, these plasmacytoid myoepithelial cells lack the chromatin structures usually associated with lymphoid malignancies. Occasional large forms should not be considered evidence of malignancy. (Papanicolaou, ×600) (**E**) Morular squamous metaplasia can complicate the cytology of mixed tumors. In this example, the involved duct is surrounded by numerous myoepithelial cells. (Papanicolaou, ×400) (**F**) Oncocytic metaplasia can occur in mixed tumor. The large nuclei and prominent nucleoli of these cells may be mistaken for carcinoma. However, cells with oncocytic cytoplasmic features are more likely to be metaplastic than neoplastic. (Diff-Quik, ×600)

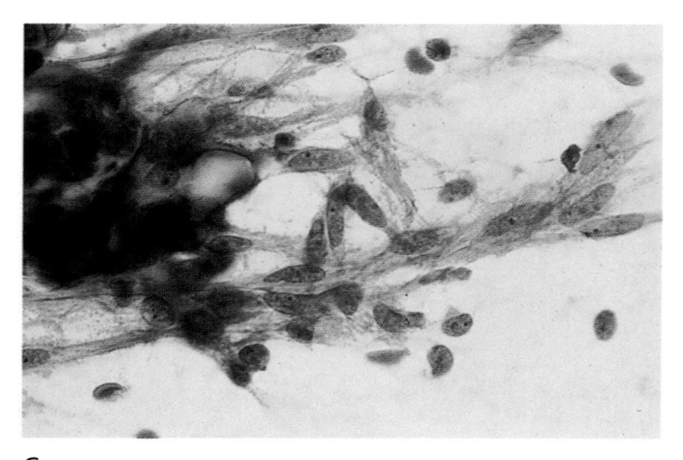

G

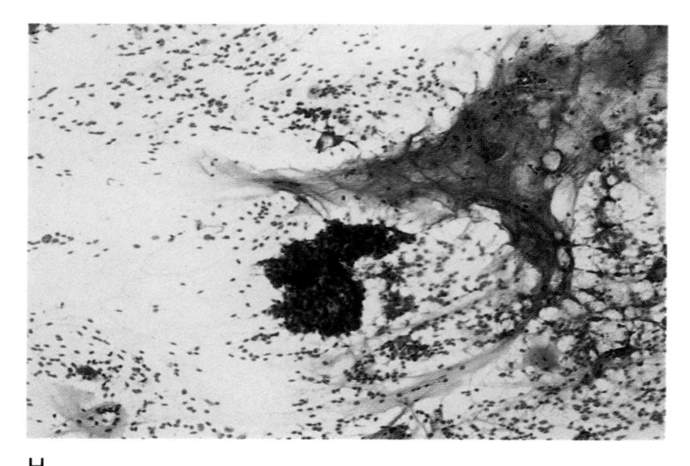

H

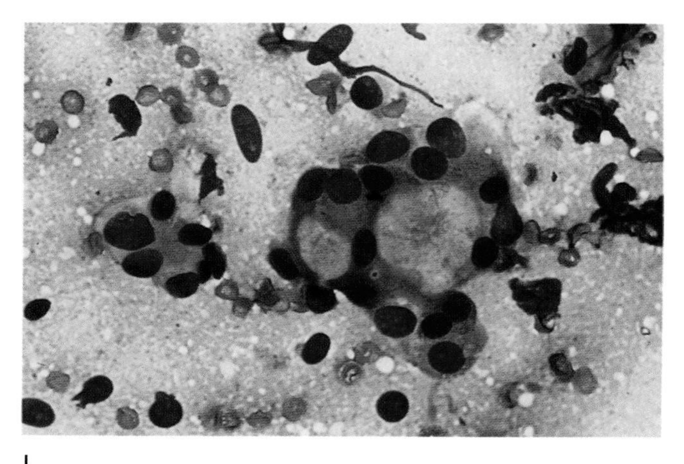

I

FIGURE 4–15. *Continued* (**G**) Myoepithelial cells can also be spindled. (Papanicolaou, ×600) (**H**) This mixed tumor aspirate shows cystic change, with epithelial cell clusters scattered through mucinous-appearing fluid. The resemblance to low-grade mucoepidermoid carcinoma can be quite misleading. (Papanicolaou, ×100) (**I**) Cellular degeneration with vacuole formation furthers the similarity of this cystic mixed tumor to mucoepidermoid carcinoma. (Diff-Quik, ×600)

Matrix of Mixed Tumors

- Fixed, Papanicolaou-stained smears pale gray or cyanophilic may resemble chondroid material spindled or stellate myoepithelial cells well seen small amounts of matrix difficult to see
- Air-dried, Romanowsky-stained slides metachromatic staining dense and fibrillary entrapped myoepithelial cells partially obscured small amounts of matrix easily seen

When a mixed tumor aspirate shows relatively little matrix material, the risk of making an incorrect diagnosis increases considerably. Small amounts of this material can be more easily detected by reviewing air-dried, Romanowsky-stained slides. In a preparation of this type, its presence is trumpeted by bright metachromasia rather than muted by the translucent quality it shows in fixed material.[1]

Other findings occasionally embellish the cytology of mixed tumors. Intranuclear cytoplasmic inclusions can be seen.[158] These distinctive structures cause us to think about papillary thyroid carcinoma. However, we have also noted previously their occurrence in paragangliomas and in meningiomas, so that their utility as a diagnostic criterion is not great, even in an aspirate from the neck. Tyrosine crystals can be found in sections of pleomorphic adenomas, reflecting the normal gland's ability to concentrate this amino acid. However, these needle-like or rosette-forming crystals are rarely found in FNA samples. Tyrosine crystals are not, as previously suggested, diagnostic of pleomorphic adenoma; they have been identified in cysts, carcinomas ex pleomorphic adenoma, and one example of polymorphous low-grade adenocarcinoma.[126,159,160]

We have previously discussed FNA-associated infarction and squamous metaplasia in pleomorphic adenomas. Further, we have noted the potential parallels between these events and the pathogenesis of necrotizing sialometaplasia.[27–29,32,103,109] Warthin's tumors are more frequently affected by post-FNA infarction than pleomorphic adenomas.[28] The sudden onset of dental-like pain associated with enlargement that is not responsive to antibiotics may herald this type of infarction.[27,32] These signs can be mistaken clinically for a post-FNA infection. However, this is a vanishingly rare event and is usually not the cause of unexpected patient difficulties.

Spontaneous infarction of pleomorphic adenomas in the absence of antecedent FNA is an even rarer event. Layfield et al.[28] reported an example in which pain was associated with tongue paresthesia and cystic change on computed tomography scan. The aspirate showed necrosis with atypia. All of these findings pointed toward malignancy. Before excision, FNA diagnoses of squamous cell carcinoma and mucoepidermoid carcinoma were considered.

Pleomorphic Adenoma—Differential Diagnostic Considerations. Problems in cytologic diagnosis of mixed tumors can originate in any of this tumor's three components (Table 4–20). Distinction from low-grade mucoepidermoid carcinoma is a common issue. The matrix can be mistaken for mucus, especially if squamous metaplasia is present. Cystic change with foam cells and a predominance of epithelium over matrix heighten this concern.[17,28,37,107,110,112,116,146,157,161–163] However, careful attention to the extracellular material can help resolve this diagnostic difficulty; mucus lacks the fibrillary quality of true matrix material and usually stains much less densely.[17]

The nonspecific collagenous stroma of many lesions can resemble mixed tumor matrix. It too is metachromatic and can show both dense staining and a fibrillary substructure. Less commonly, the necrosis of metastatic carcinoma has been mistaken for this matrix material.[13] In this way, lesions as diverse as sialoadenitis, sialolithiasis with fibrosis, metastases from cutaneous basal cell or squamous cell carcinomas, and a single example of neurofibrosarcoma have all been interpreted as mixed tumor at the time of FNA.[116,147,148,164] As noted above, a chordoma that invades the parotid may also be mistaken for mixed tumor.[64]

TABLE 4–20. Problems in the Diagnosis of Pleomorphic Adenoma

- Matrix mistaken for mucus, resulting in a diagnosis of mucoepidermoid carcinoma
- Squamous metaplasia, foam cells, and cystic change can also suggest mucoepidermoid carcinoma.
- Collagenous stroma of many lesions can be mistaken for matrix.
- Atypical epithelial cells may suggest malignancy.
- Atypical myoepithelial cells may suggest malignancy.
- Cylindromatous foci may mimic adenoid cystic carcinoma.
- More cellular examples look like various small blue cell tumors.
- Pleomorphic adenoma is very difficult to distinguish from polymorphous low-grade adenocarcinoma.
- Spindled myoepithelial cells may suggest a sarcoma.
- Plasmacytoid myoepithelial cells can mimic lymphoma.

One of the most common problems leading to false-positive diagnoses in salivary gland FNA is interpretation of atypical epithelial or myoepithelial cells aspirated from pleomorphic adenomas.[37,116,118,162] Cells suggestive of various malignancies, including carcinoma ex pleomorphic adenoma, feature enlarged nuclei with irregular contours, macronucleoli, and clumped chromatin. Such cells may originate in either the ductal or the myoepithelial areas.[165]

The first step in avoiding diagnostic errors is looking at the most characteristic feature of mixed tumor cytology, which is the matrix. When this component of the smear suggests the diagnosis of pleomorphic adenoma, this finding should override concern over focal atypia, even if it is severe. In the vast majority of instances, mixed tumor is the correct interpretation. Mixed tumors with striking but meaningless atypia are relatively common. On the other hand, carcinoma ex pleomorphic adenoma is rarely recognized in FNA samples.[5,10,37,113,117,163] This issue is discussed more fully below.

Whether studied by frozen sections or in FNA samples, pleomorphic adenomas can be mistaken for adenoid cystic carcinoma. This usually occurs when the matrix is scanty or absent so that the mixed tumor is hypercellular. Further, duct lumina can mimic cylinder formation, or frankly cylindromatous foci may be present.[6,17,18,40,110,115,117,148,166–169] This cytology and the histology of these cellular mixed tumors approach the appearance of basal cell adenomas in that they show little of the diagnostic chondromyxoid matrix. The term "minimally pleomorphic adenoma" has been used to describe this histologic circumstance. Distinction of these cases from adenoid cystic carcinoma, particularly its solid (anaplastic) variant, is very difficult at the time of FNA. Because these cases are cytologically similar to basal cell adenomas, we prefer to discuss them more fully with the other small cell epithelial neoplasms.[13,18]

Polymorphous low-grade adenocarcinoma can resemble mixed tumor quite closely.[170] A tumor's location can provide an important clue to the diagnosis, because the carcinomas usually occur in the minor salivary glands, whereas mixed tumors are often located in the major glands. When polymorphous low-grade adenocarcinoma does arise in the parotid, it is commonly found within carcinoma ex pleomorphic adenoma. Histologically, the encapsulation of mixed tumors contrasts with the infiltrative nature of the carcinomas and with their proclivity for perineural growth. These key histologic signs of malignancy cannot be assessed in cytologic samples. The epithelial-myoepithelial carcinoma is another rare low-grade malignancy that may show mixed tumor-like foci. Adequate sampling based on several FNA passes can resolve some diagnostic problems.[135]

The differential diagnosis of myoepithelial cells

with a spindle shape, including myoepithelioma and malignant myoepithelioma, is discussed subsequently with other spindle cell proliferations. Plasmacytoid myoepithelial cells may be mistaken for plasmacytoma or plasmacytoid malignant lymphoma. Individual myoepithelial cells may be large and multinucleated. The suspicion of malignancy can be very great unless other evidence that the aspirate represents a mixed tumor is evaluated carefully. We have even seen this type of scattered large myoepithelial cell with multiple nuclei and a background of small uniform myoepithelial cells mistaken for Hodgkin's disease.

Cytologic diagnosis of mixed tumors is usually straightforward, but the previous descriptions emphasize the complexity and variability of this neoplasm. They also illustrate that fact that the most secure cytologic diagnoses are those based on multiple criteria. Cells within the chondromyxoid matrix of mixed tumors are frequently positive with immunostains for glial fibrillary acid protein. This may form the basis for resolution of some diagnostic difficulties related to myoepithelial cells.[148,169] Further, the epithelia of mixed tumors can often be decorated by immunoreagents that recognize the breast antigen GCDFP-15 (gross cystic disease fluid protein). This might be exploited to exclude diagnoses of polymorphous low-grade adenocarcinoma and adenoid cystic carcinoma, both of which are negative for this marker.[171]

Benign Metastasizing Pleomorphic Adenoma. Histologically typical examples of pleomorphic adenoma result rarely in distant metastases. Pulmonary and osseous deposits have been approached by FNA.[172,173] Myxoid matrix and bland-appearing epithelial cells may be confused with pulmonary hamartoma. Bone lesions may be destructive and radiographically mistaken for malignant fibrous histiocytoma or chondrosarcoma. As noted by Pitman et al.,[173] historical, radiographic, and cytologic information is required for correct interpretation of these cases.

Malignant Mixed Tumor (True Carcinosarcoma) and Carcinoma ex Pleomorphic Adenoma. Mixed tumors that are large or that show prognostically meaningless cytologic atypia are much more common than carcinoma ex pleomorphic adenoma. The malignant areas in carcinoma ex pleomorphic adenoma can be very focal. Thus, false-negative results are common, and smears show only the mixed tumor portion of the mass with no evidence of malignancy.[5,10,37,113,117,163] A much less common sampling problem arises when only the carcinoma is aspirated, with no cytologic evidence of the associated mixed tumor.[174]

The malignancy encountered in carcinoma ex pleomorphic adenoma is variable and may be poorly differentiated, with little resemblance to the common carcinomas of this organ. Two rare examples of low-grade mucoepidermoid carcinoma ex pleomorphic adenoma have been described in FNA samples.[157,163] Because the neoplastic cells resemble the cystic change or mucus production that can be seen in otherwise unremarkable mixed tumors, this diagnosis is problematic at the time of FNA.

True carcinosarcoma (malignant mixed tumor) occurs much less often than the carcinoma ex pleomorphic adenoma, which is itself a rare condition. Most are recognized as malignant when aspirated,[18,24,175] but some are not.[10] Concurrent cytologic demonstration of both sarcomatous and carcinomatous components is unusual. One case studied by FNA showed both poorly differentiated squamous cell carcinoma and chondrosarcoma.[155]

Warthin's Tumor

Cytology. Warthin's tumors are softer than mixed tumors at physical examination; they often feel like soft fluid-filled cysts.[1,138] Aspirates can show three components in varying proportions: oncocytic epithelium and lymphocytes are associated with cyst fluid. Any of these may be absent. Some examples show only nonspecific cyst fluid or lack the diagnostic oncocytes. Intact papillary tissue fragments with epithelium overlying lymphoid tissue are sometimes seen and recapitulate this tumor's histology.[1,13]

Oncocytes typically form flat cohesive sheets (Fig. 4–16). The nuclei of these cells are subject to the same enlargement, variability, and nucleolar prominence that attend oncocytic metaplasia in other tissues, such as in chronic lymphocytic thyroiditis.[18,106] The abundant cytoplasm is usually sufficient to impart a low N/C ratio. In air-dried smears, the cytoplasm is dense; on fixed preparations, it often appears finely granular. Oncocytes are often cyanophilic in Papanicolaou-stained FNA samples, thus failing to recapitulate the eosinophilia typical of histologic preparations. For this reason, eosinophilia should not be a requirement for designation of cells as oncocytic. In addition to oncocytes, Warthin's tumors may show sebaceous, squamous, mucinous, or pleomorphic adenoma-like foci.[34,122]

The lymphoid tissue may consist mostly of free-lying mature cells that feature thin rims of basophilic cytoplasm surrounding densely hyperchromatic nuclei. As previously discussed, the lymphoid cells in Warthin's tumors tend to be more monotonous and to show fewer germinal center fragments than usually seen in lymph node aspirates. In some instances, however, they are polymorphous and resemble reactive lymphoid tissue. Lymphoid tangles represent another manifestation of lymphoid tissue and appear as smeared fragments of darkly staining chromatinic material. Lymphoglandular bodies are small detached

A

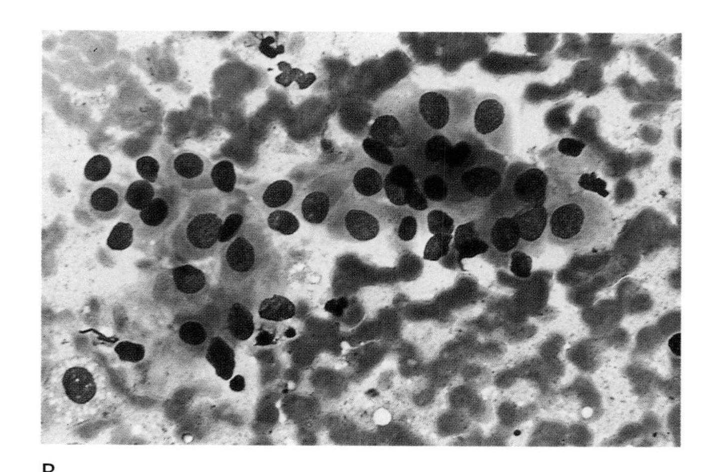

B

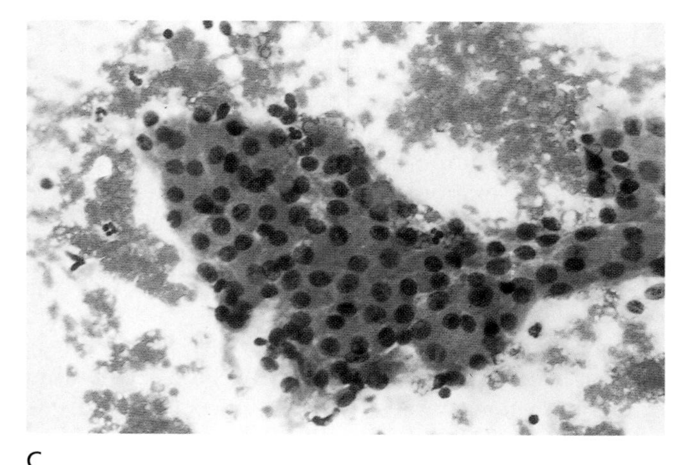

C

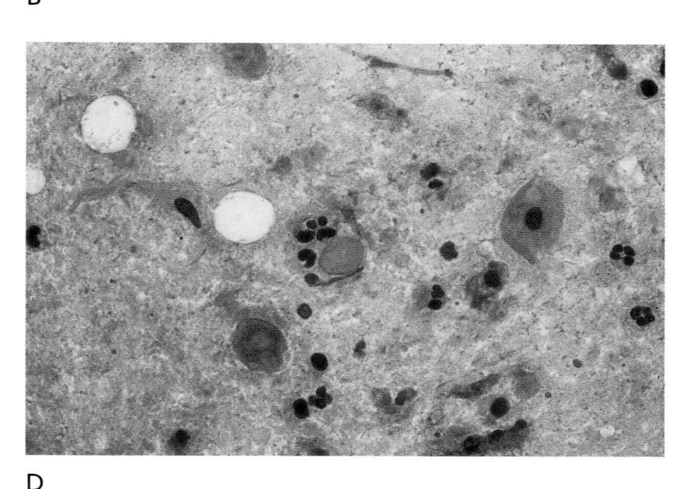

D

FIGURE 4–16. (**A**) Aspirates from Warthin's tumor often consist of cyst fluid. This must be concentrated. It gives the smears a background of proteinaceous material, scattered inflammatory cells, and occasional foam cells. (Papanicolaou, ×100) (**B**) Oncocytes usually form flat sheets and feature nuclei of varying size, abundant cytoplasm, and occasional nucleoli. (Diff-Quik, ×400) (**C**) Oncocytes from a fixed smear of a Warthin's tumor are shown here. Their cytologic details are as illustrated in more detail subsequently when oncocytoma is considered. In Warthin's tumor aspirates, it is important to scan the entire smear at intermediate magnification to identify what can be very few clusters of these diagnostic cells. (Papanicolaou, ×400) (**D**) Squamous metaplasia is common in Warthin's tumors. The keratinized cells shown here are evidence of this process. Significant degenerative atypia can be noted in such cells and may lead to a mistaken diagnosis of a cystic squamous cell carcinoma metastasis. (Papanicolaou, ×400) (**E**) This type of sparsely cellular fluid with foam cells and ill-defined metachromatic material, possibly representing mucus, has caused Warthin's tumors to be misdiagnosed as mucoepidermoid carcinoma. (Diff-Quik, ×400)

F

cytoplasmic droplets. These probably represent cells that have been damaged in the smearing process and are identical to the small fragments in lymph node aspirates.

Cytomorphologic Features of Warthin's Tumor

- Cyst fluid with foam cells, macrophages, and debris
- Oncocytes usually occur in flat sheets
- Oncocytes show abundant cytoplasm and may have prominent nucleoli
- Free lymphocytes that may appear monomorphous
- Lymphoid tangles
- Germinal center fragments
- Lymphoglandular bodies, if the lymphoid tissue is abundant

Warthin's tumor aspirations yield from a few drops to a milliliter or more of brown turbid cyst fluid. These tumors are only partially cystic and cannot be drained completely at the time of FNA. The fluid can be sparsely cellular, so that concentration by centrifugation is essential if occasional oncocyte clusters are to be identified. Further, several smears from a centrifuge pellet may be required to find even a few diagnostic cell clusters. In addition to lymphoid tissue and oncocytes, the fluid itself is represented by debris and precipitated protein that give the slide a "dirty" background. We have previously discussed complication of this picture by spontaneous or post-FNA infarction.[28,29]

Differential Diagnostic Considerations. When all the elements discussed above are present, diagnosis of Warthin's tumors is straightforward, and 60–88% are correctly diagnosed at the time of initial FNA (Table 4–21). Cases that yield fluid lacking oncocytes and lymphoid tissue must be distinguished from several other cystic salivary gland lesions, discussed more fully below. Unfortunately, epithelial cells required for diagnosis of most other cystic masses are also absent. This means that one is left with a nondiagnostic cyst fluid specimen, similar to those sometimes obtained by thyroid aspiration. A second aspiration may be useful, because most Warthin's tumors are only partially cystic and the initial sampling will have left a palpable mass. However, such fluids should not be interpreted as unequivocally benign without other compelling evidence that leads to a specific diagnosis. Further, the association of neutrophils with a cyst fluid does not automatically connote an infection or other primarily inflammatory process.[101,105,110]

If the clusters and sheets of oncocytes are extremely prominent, this may suggest an oncocytic neoplasm,[112] especially because these lesions may show abundant lymphoid stroma.[176] Conversely, a cytologic diagnosis of Warthin's tumor may be suggested after study of a cyst fluid with debris and lymphocytes. Ultimately, however, it is impossible to make this diagnosis without positive identification of oncocytic epithelium.[13]

The fluid aspirated from Warthin's tumors may also show foam cells, as well as stringy precipitated material that can resemble extracellular mucus. Thus, in the setting of an oncocyte-deficient aspirate, one may strongly consider a diagnosis of low-grade mucoepidermoid carcinoma.[17,113,146,177] When lymphocytes are the most prominent element of the aspirate, this may suggest chronic inflammation, follicular hyperplasia, or a low-grade malignant lymphoma.[34,43,110,112] Conversely, a salivary gland aspirate showing small cell malignant lymphoma may be thought to represent the much more common Warthin's tumor.[13] Cell marker studies can resolve this diagnostic problem.

The papillary fronds in Warthin's tumor frequently undergo squamous metaplasia with marked individual cell keratosis. In some cases this is associated with considerable cytologic atypia or degenerative nuclear hyperchomasia that can mimic squamous cell carcinoma.[18,44,110,127–130,146,177] The problem of cystic metastases of squamous cell carcinoma has been discussed previously in relation to lateral congenital cysts of the neck. The most important maneuvers in preventing a diagnosis of carcinoma are to repeat any questionable or sparsely cellular aspirates, and to concentrate any fluid that is recovered. A diligent search will usually show oncocytes, and this should lead to the correct diagnosis.

Lymphocytes and cystic change may both be features of acinic cell carcinoma. Further, the cells of this carcinoma often closely resemble oncocytes and enhance its similarity to Warthin's tumor in aspirate smears.[18,122,143,178] In our experience, however, most aspirates of acinic cell carcinoma show many more epithelial cells than are usually recovered from Warthin's tumors. Further, the zymogen granules of acinic cells are a useful differential diagnostic feature: they are more coarse than the fine cytoplasmic granularity of oncocytes. As illustrated subsequently, we have found cell block sections to be very useful in demonstrating these granules.

Distinction between Warthin's tumor and acinic cell carcinoma is an uncommon but occasionally difficult problem, as is the distinction between other oncocytic lesions and acinic cell carcinoma. Oncocytic neoplasms, including rare malignant examples, and oncocytoid adenocarcinoma are discussed below, where they are placed in the context of other large cell epithelial neoplasms of low nuclear grade.

TABLE 4–21. Differential Diagnosis of Warthin's Tumor

- Nonspecific cyst fluid
- Oncocytic neoplasms
- Low-grade mucoepidermoid carcinoma
- Chronic inflammation
- Lymph node hyperplasia
- Well-differentiated malignant lymphoma
- Squamous cell carcinoma
- Acinic cell carcinoma

Cystic Salivary Gland Masses

Cystic change in excised surgical specimens is often a secondary embellishment of diagnoses based primarily on other criteria. At the time of FNA, however, the presence of cyst fluid is frequently the first apparent morphologic feature of a patient's mass not ascertained solely by physical examination. Thus, it frequently initiates and begins to organize the pathologist's differential diagnosis. The simple fact that a mass is cystic or partially cystic requires that a large array of diagnostic possibilities be considered.[10,13,15–18,34,107,113,161,162,167] Problems are compounded by the fact that any diagnostic cellular elements may be diluted in a considerable specimen volume. This emphasizes the need for careful aspiration and for repeating the aspiration as needed.

Several lesions that are occasionally cystic are not often given over to extensive cystic alteration (squamous cell carcinoma, other metastatic carcinomas, pleomorphic adenoma, monomorphic adenoma, acinic cell carcinoma) and can usually be recognized by the expected criteria. These are discussed in other sections, where they present more frequent or more urgent dilemmas. Warthin's tumors are often cystic and were considered in the previous section.

In this section, we address the remaining lesions that are commonly cystic. These include obstructive masses, of which sialolithiasis will be used as the prototype, and low-grade mucoepidermoid carcinoma. (High-grade mucoepidermoid carcinoma is in most schema much less cystic by definition, and will be considered subsequently.)

Duct Obstruction as a Cause of Cystic Salivary Gland Enlargement. Longstanding duct obstruction has several causes. Those usually operative in patients referred for FNA include sialothiasis, compression by an adjacent mass, and radiation fibrosis (discussed previously). At physical examination, these glands are usually well circumscribed and very firm. These findings can mimic recurrent or metastatic malignancy and seem to represent excellent targets for FNA.

The surgical pathology of this condition is well known. Obstruction leads to atrophy of the gland parenchyma, extensive fibrosis, chronic periductal inflammation, and dilatation of residual ducts. Further, the epithelium that lines entrapped ducts may be flattened or can show various combinations of squamous, mucinous, or ciliated metaplasia. The lumina of these structures are frequently filled with mucoid or granular secretory material.[128] If duct rupture occurs, a foreign-body reaction and cytologic atypia associated with mesenchymal repair can be part of an exuberant inflammatory mass. This progression of pathologic changes resembles that seen in the lymphoepithelial cysts, most commonly in AIDS patients. (The concept that the latter masses result from salivary gland duct obstruction by hyperplastic lymphoid tissue has been discussed previously.[94])

The combination of cystlike spaces with complex lining epithelium, a background of fibrous tissue, inflammation, and abundant extracellular mucus is strongly suggestive of low-grade mucoepidermoid carcinoma on both cytology and frozen section.[114,121] The distinction between these two conditions on cytologic grounds is frequently impossible (Fig. 4–17).

Sialolithiasis. Sialolithiasis is a specific type of tumefactive duct obstruction (Table 4–22). Varying degrees of the pathology described above can be seen in longstanding cases. In general, sialolithiasis is second only to mumps as a cause of sialoadenitis. However, because very few stones occur in children, it is the most

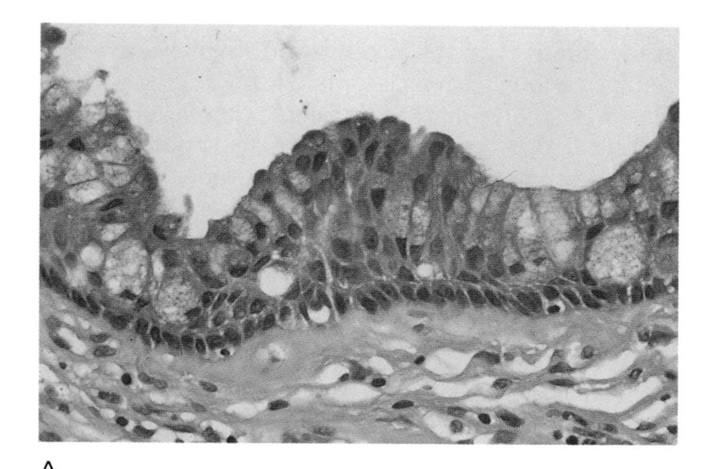
A

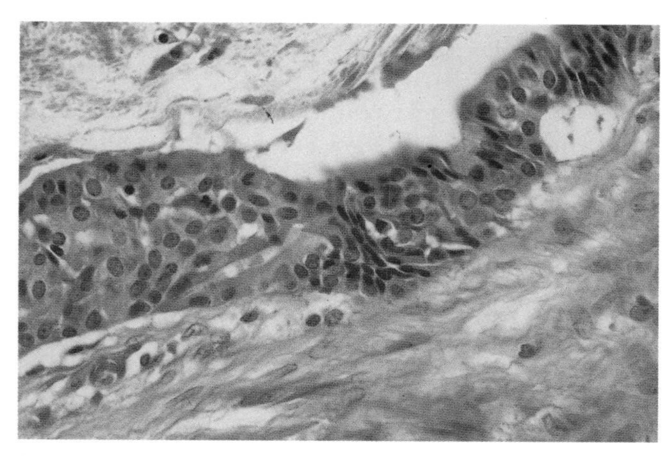
B

FIGURE 4–17. (**A**) This histologic section from a dilated salivary gland duct was taken from a firm enlarged gland that showed diffuse radiation damage. The surrounding fibrous tissue contains no acini and shows a few chronic inflammatory cells. The duct lining is thickened by a complex epithelium with squamous, mucinous, and immature metaplastic cells. The latter resemble the intermediate cells of low-grade mucoepidermoid carcinoma. The central mucinous contents and foam cells are not well seen in this illustration. (Hematoxylin and eosin, ×400) (**B**) This cystic space represents a portion of a low-grade mucoepidermoid carcinoma. Its lining epithelium and mucinous contents are very similar to those in nonneoplastic duct obstruction, hence the difficulty in distinguishing these two processes at the time of FNA. (Hematoxylin and eosin, ×400)

TABLE 4–22. Sialolithiasis

- More common in women than in men
- Most common in the submandibular gland
- Most are unilateral.
- Some incidentally discovered on dental radiographs
- Some not radiopaque
- Presentations: pain, a mass, or a secondary infection
- Cytology: cystic aspirate with mucus and a few cells
- FNA differential diagnosis: low-grade mucoepidermoid carcinoma

common cause in older persons. Stones are more common in women than in men; the parotids are involved more often than the submandibular glands. They are more likely to be unilateral than bilateral. Some are discovered incidentally on dental radiographs; a significant minority (20%) are not radiographically visible. The presence of stones does not imply an associated abnormality of calcium metabolism. The classic symptoms of stones include pain and swelling at mealtime. Some patients may be symptomatic for many years. Alternatively, a stone may present as a hard mass that suggests malignancy. Others come to medical attention only after a secondary infection supervenes. In the latter instance, FNA for diagnosis and possibly for cultures may be requested.[121,139,144,179]

Only occasional applications of FNA to evaluate sialolithiasis have been reported.[121,146,150,180] The diagnosis is straightforward if stone fragments are identified in aspirate material. Also, if ciliated metaplastic cells are present, this can suggest a benign condition, but in the absence of stone fragments a specific diagnosis may not be possible. We have seen examples of sialolithiasis interpreted as congenital cysts based on the finding of ciliated cells (Fig. 4–18).

The greatest danger with aspirates from sialolithiasis occurs when both stones and ciliated metaplasia are absent. In these cases, extracellular mucus, through which are scattered foam cells and occasional epithelial cell groups, strongly suggests low-grade mucoepidermoid carcinoma. The danger is greatest when the cell clusters show complex and variable mixtures of squamous and vacuolated cells. It is paradoxical that aspiration sialolithiasis often yields more epithelial cells with greater nuclear atypia than sampling of the typical low-grade mucoepidermoid carcinoma.

Further, only one patient in a series of five sialolithiasis cases studied by FNA had the typical history.[121] This is an important clinical consideration and suggests that patients referred for FNA are not a typical sample of stone-bearing persons. Rather, they are most likely to be those in whom a diagnosis of malignancy is under clinical consideration. Because inflammation can be a prominent component of low-grade mucoepidermoid cytology, this finding is not a helpful differential diagnostic consideration.

Psammomatous calcifications can be found in histologic sections of the normal or inflamed submandibular gland, as well as in its neoplasms.[180] These have not yet been described in cytologic material, but this information should sharpen our thinking about sialolithiasis and low-grade mucoepidermoid carcinoma.

Low-Grade Mucoepidermoid Carcinoma. In the experience of many investigators, this is a very difficult diagnosis in cytologic samples. One reason for this is its similarity to the duct obstructive conditions discussed above. Cystic change, inflammation, and various epithelial metaplasias all contribute difficulty in distinguishing low-grade mucoepidermoid carcinoma from some examples of lymphoepithelial cyst, pleomorphic adenoma, Warthin's tumor, chronic sialoadenitis, and sialolithiasis.[6,10,17,18,101,107,110,113,121,127,146,148,162,167,181]

Histologic grading schema for mucoepidermoid carcinoma must form the basis for our consideration of its cytologic manifestations. Several have been advanced, and a complete review of this subject is beyond the scope of this chapter. However, Evans[182] suggests that grading schemes with more than two levels do not improve our ability to predict the clinical outcome of a given case. High-grade lesions more frequently result in local recurrence, metastasis, positive surgical margins, or tumor-related death. In this study, however, tumor grade was not useful in predicting which patients with lesions smaller than 2.5 cm would do well.

In Evans' classification, mucoepidermoid carcinomas that are largely cystic would be considered low grade; those that are mostly solid (>90%) and show similar cytology are designated high grade. Neoplasms showing marked cytologic atypia, numerous mitotic figures, or extensive necrosis should be considered representative of categories other than high-grade mucoepidermoid carcinoma.

This type of classification is readily adopted for use with FNA. The older cytology literature, however, is frequently unclear about grading, and various histologies are combined under the term "mucoepidermoid carcinoma." These publications are thus very difficult to interpret when one wishes to address problems with diagnosis and classification of this neoplasm. It is our impression that the term "high-grade mucoepidermoid carcinoma" has been rather loosely applied, and that application of the Evans' criteria could form the basis for a reasonable approach to this diagnosis.

Based on this reasoning, we begin with the concept that low-grade mucoepidermoid carcinoma is predominantly cystic. This is our initial impression at the

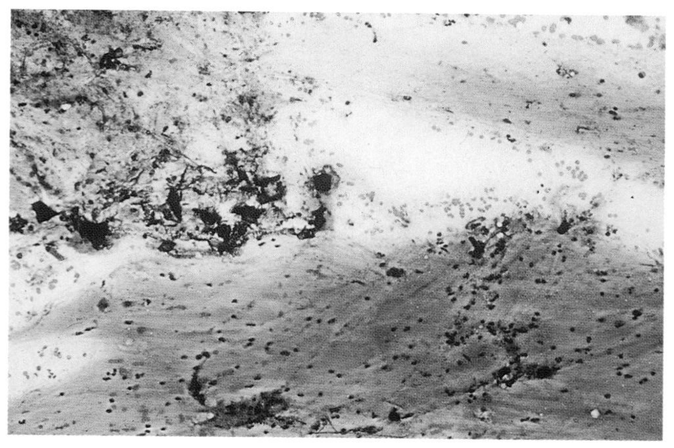

A

B

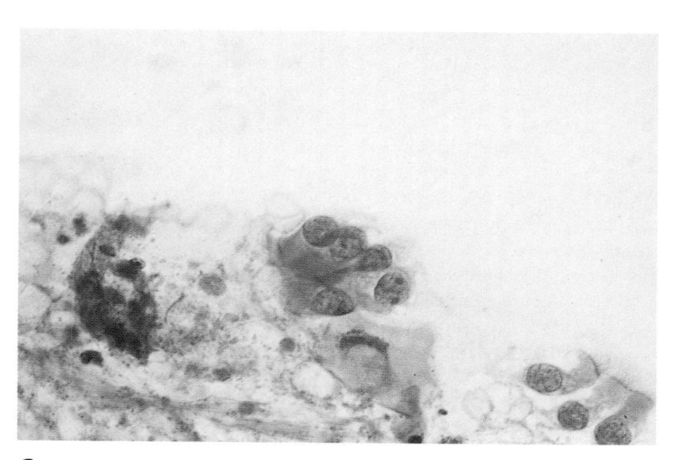

C

D

FIGURE 4–18. (**A**) At low magnification, this aspirate of sialolithiasis shows abundant mucus, through which are scattered several cell clusters. At this level, and at the bedside where mucinous fluid will have been recovered, this process closely mimics low-grade mucoepidermoid carcinoma. (Diff-Quik, ×40) (**B**) Some aspirates of sialolithiasis show stone fragments such as these. (Papanicolaou, ×400) (**C**) In some cases the fluid of sialolithiasis contains ciliated cells. Although nonspecific, such elements indicate a benign metaplasia. (Papanicolaou, ×400) (**D**) The epithelium of sialolithiasis shows combinations of immature squamous cells and frothy mucinous cells, thus closely resembling the cyst linings in low-grade mucoepidermoid carcinoma. (Diff-Quik, ×600) (**E**) In fixed material, the cells of sialolithiasis may show more atypia than usually seen in low-grade mucoepidermoid carcinoma. These cells show crowding, nuclear hyperchromasia, and some degree of chromatin clearing. (Papanicolaou, ×600)

time of FNA, as aspiration generally yields clear to turbid mucoid fluid. The smears are sparsely cellular and show a few foam cells and epithelial cell clusters in a background of abundant extracellular mucus. The surgical pathology of this neoplasm's epithelium leads us to search for combinations of mature squamous cells, the less mature intermediate cells, and mucinous cells.[146,183] The latter are most often single foam cells that are cytologically indistinguishable from the macrophages that inhabit almost any cyst fluid. Vacuolated cells can also be situated within the epithelial cell clusters. In this setting, one can be more confident that they truly represent the epithelium. Intermediate squamous cells are a very important component of this tumor as it is seen in histologic sections. Cytologically, these cells have a relatively high N/C ratio; in our experience, they can best be conceptualized as looking like squamous metaplastic cells familiar from many types of cytology (Fig. 4–19).

Cytologic diagnosis of low-grade mucoepidermoid carcinoma is further confused by the presence of inflammation; it shares this finding with several of its mimics.[92,163] Oncocytic variants are rare. However, they are important, because most other oncocytic masses are benign.[136,184] Clear cell change and foci of sebaceous differentiation may be present.[131,132,181] Rare examples of low-grade mucoepidermoid carcinoma ex pleomorphic adenoma have been described.[157,163]

The mucoid cyst fluid may be completely acellular or sparsely cellular. These aspirates often lead to false-

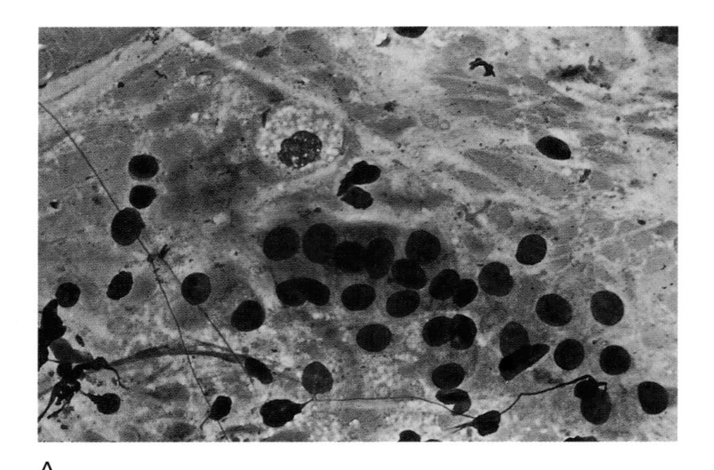

A

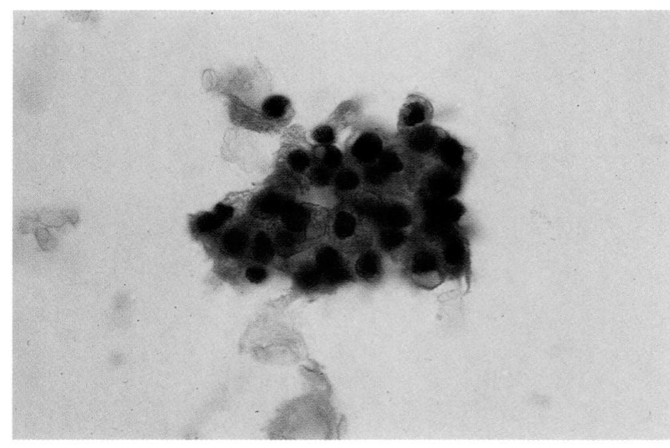

B

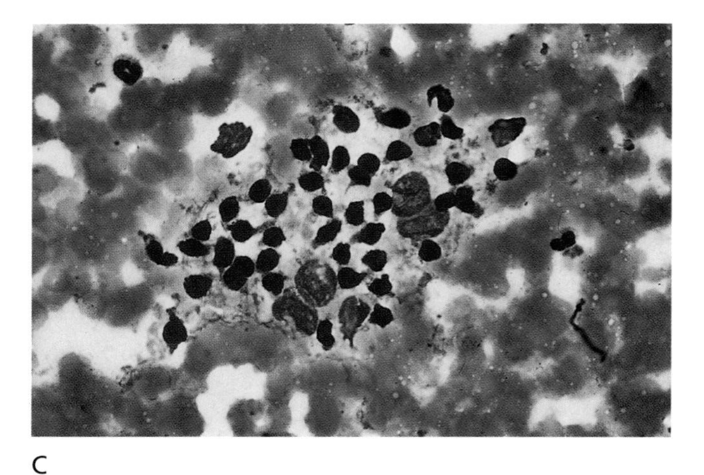

C

FIGURE 4–19. (**A**) This mucinous material was aspirated from a low-grade mucoepidermoid carcinoma. Cell groups such as the one shown here are usually scattered widely through abundant cyst fluid. The free-lying foam cells are typical but nonspecific. The remaining cells have a somewhat squamoid appearance and probably correspond to the intermediate cells in histologic preparations. (Diff-Quik, ×600) (**B**) The background mucus is much less apparent in fixed material. These epithelial cells are much more bland than those depicted previously from a case of sialolithiasis. (Papanicolaou, ×600) (**C**) Polymorphous, benign-appearing lymphoid tissue can be seen in many salivary gland neoplasms. This example was aspirated from a low-grade mucoepidermoid carcinoma. (Papanicolaou, ×600)

negative interpretations.[16,101,127,162,163,167,183] When cells are present, the diagnosis is still difficult because they show none of the traditional nuclear features of malignancy. We suggest that any mucoid salivary gland aspirate should be considered at least suspicious for carcinoma regardless of the microscopic findings. This includes those that are acellular or that contain only isolated foam cells. One should certainly not conclude that the mass is benign based on the absence of malignant-appearing cells.

Small Cell Epithelial Neoplasms

Most small cell epithelial neoplasms of the salivary glands are uncommon. However, as a group they represent a significant portion of the most difficult differential diagnostic dilemmas in salivary gland FNA, hence our conceptualization of these lesions as a group despite their great diversity in clinical presentation and outcome. An interesting and incompletely explored facet of this subject is that several of the primary salivary gland lesions in this category seem to share an origin from the intercalated duct (Table 4–23).[185]

Cutaneous neoplasms, including pilomatrixoma and dermal eccrine cylindroma, should be considered with these lesions, particularly if adenoid cystic carcinoma is cytologically suspected. They will be discussed briefly in that context. Clinical and radiographic findings usually suffice to exclude

TABLE 4–23. Epithelial Neoplasms With Small Cells

- Monomorphic adenoma
- Pleomorphic adenoma with little chondromyxoid matrix
- Carcinoma ex monomorphic adenoma
- Adenoid cystic carcinoma
- Dermal eccrine cylindroma
- Pilomatrixoma
- Basal cell adenocarcinoma
- Primary small cell carcinoma
- Primary lymphoepithelioma-like carcinoma
- Metastatic carcinoma
    - Small cell carcinoma
    - Cutaneous basal cell carcinoma
    - Nasopharyngeal carcinoma
    - Merkel cell carcinoma
- Malignant lymphoma

ameloblastoma from consideration with this cytologic category.[59]

Monomorphic Adenoma, Carcinoma ex Monomorphic Adenoma and Distinction of These Lesions from Adenoid Cystic Carcinoma. The term "monomorphic adenoma" was previously applied to a disparate group of salivary gland neoplasms. Their most important shared features were embodied by the concept that they were benign neoplasms of a glandular organ and were not "pleomorphic." Thus, the major import of this term seems to be in distinguishing these lesions from mixed tumors. More recently, however, oncocytomas, Warthin's tumors, and neoplasms with sebaceous differentiation have been classified separately. Currently, the term "monomorphic adenoma" is usually applied to basal cell adenomas.[186]

Basal cell adenomas are composed of uniform small cells.[187,188] Architectural subclassification as tubular, trabecular, solid, or canalicular is based on the predominant pattern, as seen in histologic sections. (The dermal analogue or membranous type is a special entity that more closely resembles dermal eccrine cylindroma or adenoid cystic carcinoma than other salivary gland tumors.) We noted previously that some mixed tumors show only small foci of chondromyxoid matrix; these tumors closely resemble basal cell adenomas.[18,34,117]

We have attempted to reconcile the cytologic diagnosis of basal cell adenoma with the ultimate histopathologic interpretation as mixed tumor by use of the term "minimally pleomorphic adenoma" to describe the histologic entity. As discussed subsequently, however, the most clinically relevant differential diagnosis is not between adenomas that may or may not be pleomorphic. Rather, it involves distinctions between the monomorphic (or minimally pleomorphic) adenomas and certain carcinomas with which they share a cytologic presentation as monotonous small cells.

FNA of a basal cell adenoma shows numerous small blue cells that are associated with variable amounts of collagenous stroma.[13,107,110,188] The latter component is cyanophilic in fixed preparations and metachromatic with Romanowsky stains. Tumors with a more solid growth pattern may show little or no stroma on aspirate smears.[13] Single cells and naked nuclei may be numerous. Basosquamous whorls and sebaceous differentiation are seen rarely.[122] Spontaneous infarction of basal cell adenomas has been described.[106]

The collagenous stroma seen in aspirates of basal cell adenoma represents the interstitial connective tissue between nests and cords of epithelial cells. Thus, is it equipped with scattered fibroblast-like spindle cells and scattered capillaries. The interface between this stroma and the epithelial tumor cells is irregular. Wisps of frayed collagenous material interdigitate among the epithelial cells in a tissue particle (Fig. 4–20).

Cytomorphologic Features of Basal Cell Adenoma

- May be pure or may be the predominant pattern in a pleomorphic adenoma with very little matrix material (a "minimally pleomorphic" adenoma)
- Small uniform cells
- Naked nuclei may be numerous
- Basosquamous-like whorls are occasionally present
- Arborizing stroma may show vessels or spindle cells
- Metachromatic stroma and small blue cells can closely mimic adenoid cystic carcinoma

These stromal components and the features of the cell–stroma interface are important when we consider the differential diagnosis of basal cell adenoma with the cribriform type of adenoid cystic carcinoma. Aspiration of this carcinoma also yields uniform small blue cells associated with stroma. In both neoplasms, the stroma is brightly metachromatic in air-dried preparations. This is probably the reason that basal cell adenomas often lead to false-positive diagnoses of adenoid cystic carcinoma in FNA material.[6,18,34,110,112,113,117]

The extracellular material in adenoid cystic carcinoma consists of whorling reduplicated basal lamina material elaborated by the tumor cells. Because it is not truly a connective tissue stroma, it is always acellular and avascular. Further, it has a very sharp, linear interface with the surrounding small blue epithelial cells. These features of the cell–stroma interface are useful in distinguishing between frequently quite similar cytologic presentations of basal cell adenoma and cribriform adenoid cystic carcinoma.

Unfortunately, careful evaluation of these features does not solve all the problems occasioned by cytologic similarities between these adenomas and adenoid cystic carcinoma. The utility of these criteria breaks down when we are confronted with the solid (anaplastic) variant of adenoid cystic carcinoma. This aggressive tumor consists of small blue cells, but it lacks the spheres of extracellular material typical of its cribriform counterpart. Further, aspiration may yield small bits of stroma that can mimic the stroma of basal cell adenomas or even pleomorphic adenomas. Some authors suggest that adenoid cystic carcinoma can be recognized if sufficient attention is given to subtle nuclear features. We have not found this to be the case, as recently reviewed.[188] This differential diagnosis will be discussed in more detail when anaplastic adenoid cystic carcinoma is considered.

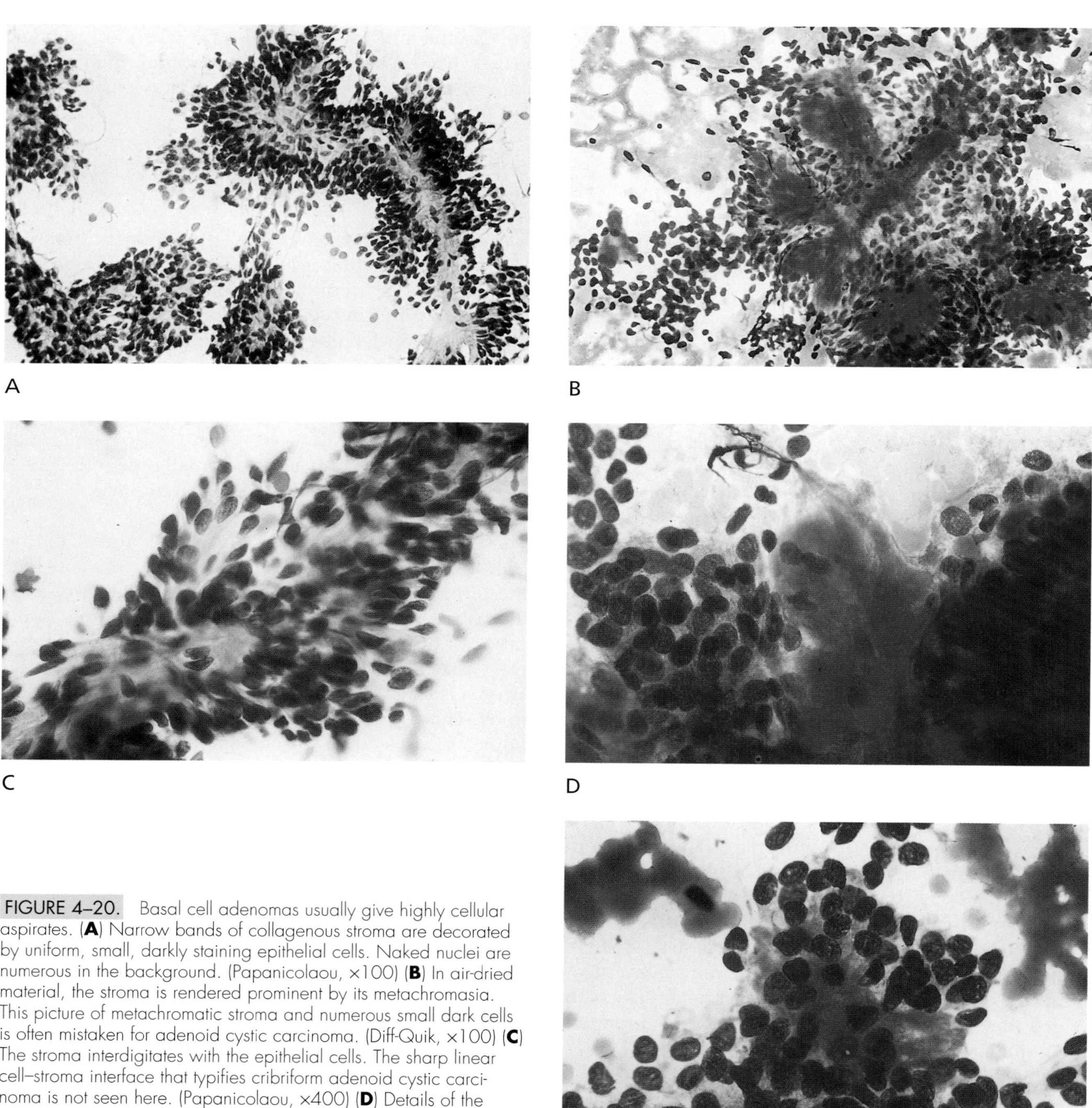

FIGURE 4–20. Basal cell adenomas usually give highly cellular aspirates. (**A**) Narrow bands of collagenous stroma are decorated by uniform, small, darkly staining epithelial cells. Naked nuclei are numerous in the background. (Papanicolaou, ×100) (**B**) In air-dried material, the stroma is rendered prominent by its metachromasia. This picture of metachromatic stroma and numerous small dark cells is often mistaken for adenoid cystic carcinoma. (Diff-Quik, ×100) (**C**) The stroma interdigitates with the epithelial cells. The sharp linear cell–stroma interface that typifies cribriform adenoid cystic carcinoma is not seen here. (Papanicolaou, ×400) (**D**) Details of the cell–stroma interface are clearly defined in air-dried smears. (Diff-Quik, ×400) (**E**) Some examples of basal cell adenoma have very little stroma. Several small cell neoplasms must be considered in the differential diagnosis of such cases. (Papanicolaou, ×400)

The membranous form of basal cell adenoma is also known as the dermal analogue type of monomorphic adenoma.[147,187,189] Its histology closely resembles that of dermal cylindroma, and it can be associated with either benign or malignant cutaneous adnexal tumors in a synchronous or metachronous fashion. Cytologically, this tumor can mimic adenoid cystic carcinoma because it grows as large rounded cell nests that are surrounded by bands of basement membrane material. The tissue particles recovered by FNA consist of small blue cells surrounded by sheaths of hyaline material. These are virtually impossible to distinguish from adenoid cystic carcinoma.

Carcinoma ex monomorphic adenoma arises in the parotid, where it usually appears approximately 10 years after diagnosis of the parent adenoma.[187] This neoplasm behaves in a locally aggressive fashion. It has been described histologically as a caricature of the adenoma that shows infiltrative growth and variable combinations of necrosis, mitotic activity, and nuclear pleomorphism. Five of eight cases in one series arose in a background of dermal analogue tumor. Examples of carcinoma ex monomorphic adenoma have not yet been described in the FNA literature.

The most common small blue cell tumor encountered in salivary gland cytology is adenoid cystic carcinoma. The celebrated spheres and cylinders of metachromatic extracellular matrix material surrounded by uniform small blue cells reflect the prominent cribriform pattern of this tumor's histology (Fig. 4–21). Naked tumor cell nuclei litter the smear background. The extracellular spheres are sharply demar-

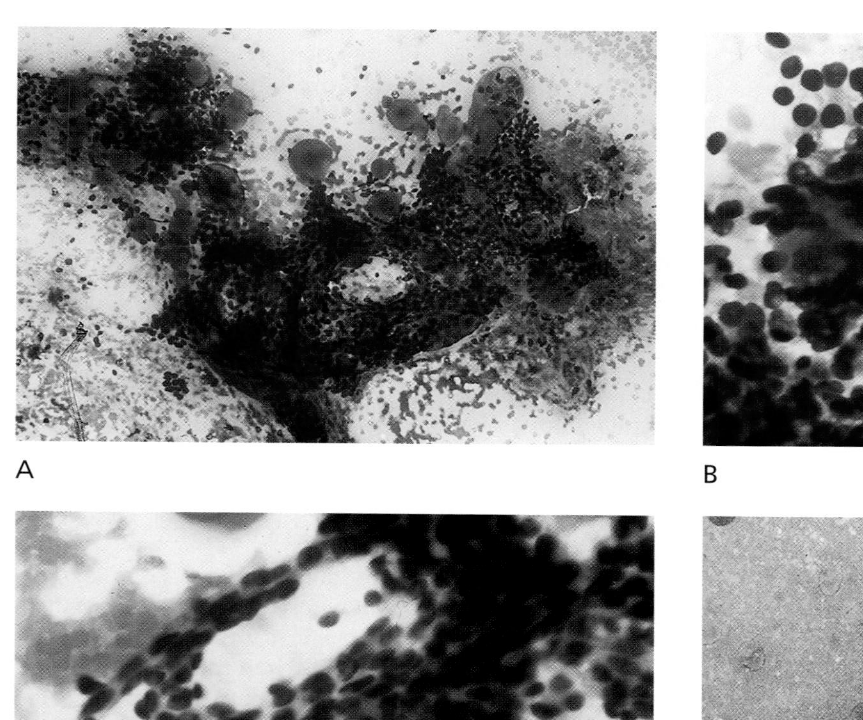

A

B

C

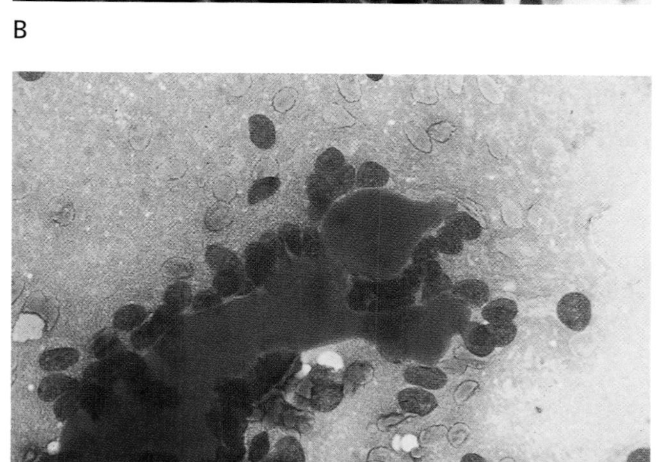

D

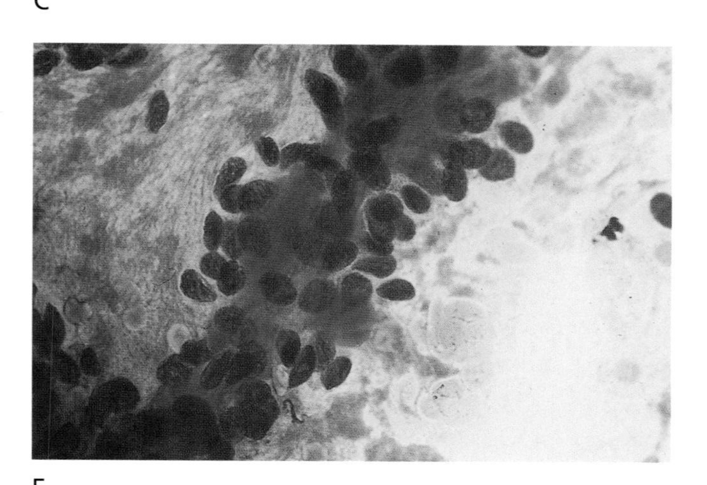

E

FIGURE 4–21. (**A**) At low magnification, smears from the cribriform type of adenoid cystic carcinoma are highly cellular. This pattern of small blue cells and metachromatic stroma is mimicked by basal cell adenomas. (Diff-Quik, ×40) (**B**) The celebrated appearance of blue cells and red balls typical of cribriform adenoid cystic carcinoma. (Diff-Quik, ×400) (**C**) In fixed smears, the extracellular matrix is translucent. It is easily missed if present in small quantities. (Papanicolaou, ×400) (**D**) In adenoid cystic carcinoma, the interface between the cells and the stroma is extremely sharp. In some areas, there is even a linear clearing between the two. (Diff-Quik, ×600) (**E**) The irregular cell–stroma interface typical of basal cell adenomas. (Diff-Quik, ×600) cated; none of their constituent material interdigitates with the surrounding tumor cells.[188] However, other "noncylinder" connective tissue may interdigitate with the epithelial cells in a manner that mimics the adenomas.

Cytomorphologic Features of Adenoid Cystic Carcinoma

- Small uniform cells with very little cytoplasm
- Most cells lack nucleoli
- Naked nuclei may be numerous
- Metachromatic stromal spheres and cylinders
- Spheres are transparent or pale with the Papanicolaou stain
- Spheres contain neither cells nor vessels
- Sharp ("pencil-drawn") cell–stroma interface

Some authors consider this cylindromatous appearance to be pathognomonic for adenoid cystic carcinoma. Further, some earlier writings on the subject suggested that this diagnosis should be straightforward. However, we now know that when studied in FNA samples, several other neoplasms can mimic adenoid cystic carcinoma. Pleomorphic adenomas, basal cell adenomas, dermal analogue tumor, basal cell adenocarcinoma, epithelial-myoepithelial carcinoma, and polymorphous low-grade adenocarcinoma can all show small uniform cells and may even feature areas with frankly cylindromatous architecture.[6,13,17,18,134,146,148,160,166,169,170,190,191] This problem is not unique to FNA cytology; the same difficulties can arise at the time of frozen section.[10,107] The small uniform cells of adenoid cystic carcinoma can result in false-negative diagnoses. This is especially true if the sample is sparsely cellular or if the matrix component is not well represented.[112,115,148] This pattern of uniform small cells has also been mistaken for small cell anaplastic ("oat cell") carcinoma.[152,153]

Dermal eccrine cylindroma is cytologically identical to adenoid cystic carcinoma of salivary gland origin.[48] Either neoplasm can occur as a primary lesion in the external auditory canal. This is also the most common location for primary adenoid cystic carcinoma of cutaneous origin. FNA thus appears to have a limited role in this setting, and clinical distinction is also difficult. Pilomatrixoma is another benign neoplasm of cutaneous adnexal origin that presents cytologically with a small blue cell pattern. This tumor was previously discussed.

We have previously noted the cytologic similarity between basal cell adenomas and the uncommon solid type of adenoid cystic carcinoma. This anaplastic variant is less often described in FNA reports than the cribriform type.[2,13,18,27,43,188,194] Smears are usually very cellular and show small uniform blue cells that may lie singly or in groups. Cylinders and spheres of extracellular matrix are absent or extremely rare. Fibrillary collagenous fragments of metachromatic desmoplastic stroma may be encountered. This material has been mistaken for the matrix component of pleomorphic adenoma, as well as for the extracellular connective tissue of basal cell adenoma.

This highlights the fact that details of the cell–stroma interface that can be useful in distinguishing cribriform adenoid cystic carcinoma from basal cell adenoma are not helpful when the solid variant is encountered (Fig. 4–22). Some authors have suggested that these tumors can be distinguished by careful attention to subtle nuclear features that indicate malig-

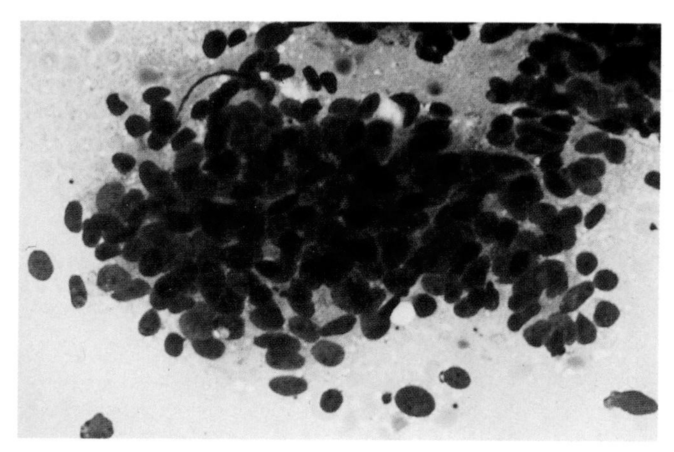

A

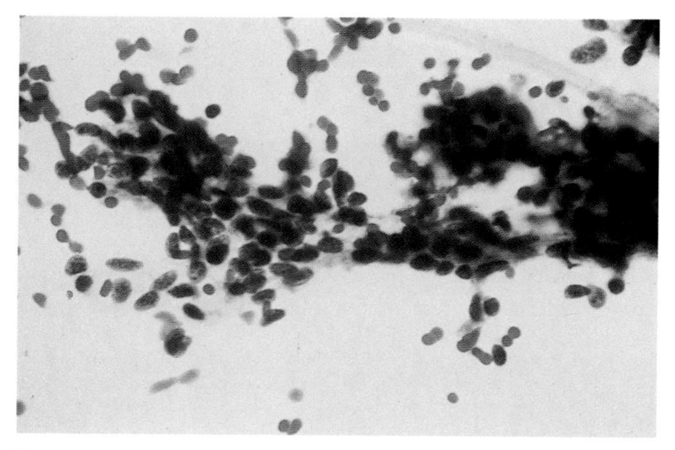

B

FIGURE 4–22. (**A**) Aspirates from the solid (anaplastic) variant of adenoid cystic carcinoma show the same uniform small blue cells seen in the cribriform type. However, the metachromatic spheres and cylinders are either absent or extremely rare. (Diff-Quik, ×400) (**B**) This solid adenoid cystic carcinoma can be indistinguishable from basal cell adenomas unless tumor necrosis is unequivocally identified. (Papanicolaou, ×400) nancy. Other investigators have not found this to be the case and leave us with the important but disconcerting concept that these two prognostically very different entities are virtually identical in cytologic material. (Reference 188 compiles and contrasts the identification of various cytologic features in both tumors.) If one can unequivocally identify necrosis, adenoid cystic carcinoma should be the preferred diagnosis. In our experience, only about half of anaplastic adenoid cystic carcinoma aspirates show necrosis in smear material.

Reviewing these considerations and their own extensive experience, Löwhagen et al.[195] are reluctant to render an unequivocal diagnosis of adenoid cystic carcinoma on cytologic evidence alone. They write, "in our institution, we refuse to take the full diagnostic responsibility for a radical surgical procedure in which sacrifice of the facial nerve may be necessary in cases where there may be classic cytologic findings of adenoid cystic carcinoma but the patient is symptom free."[195] We would add that if this stance is justified for the cribriform pattern, it is also appropriate for the solid variant.

Uncommon Primary Salivary Gland Carcinomas with a Small Cell Cytologic Pattern. Basal cell adenocarcinoma usually occurs in the parotid but can also be seen in the submandibular and minor salivary glands. Histologically, it closely resembles metastatic basal cell carcinoma of cutaneous origin (see below). This carcinoma shares its immunocytochemical profile with basal cell adenoma but shows an infiltrative growth associated with local recurrence. Metastases can occur in up to 10% of cases.

Aspirates of basal cell adenocarcinoma have the same pattern noted earlier in basal cell adenoma and solid adenoid cystic carcinoma.[190,196,197] Some three-dimensional cell clusters may be papillary or filiform, with palisading of peripheral nuclei. Individually, the cells show a high N/C ratio, fine chromatin, and prominent nucleoli. Mitotic figures may be identified. The smear background can be myxoid or can show necrosis; the latter finding suggests a possible diagnosis of solid adenoid cystic carcinoma. A relation of basal cell adenocarcinoma with carcinoma ex monomorphic adenoma has been postulated.[187]

Small cell anaplastic carcinomas can arise as primary salivary gland neoplasms. Most occur in the parotid, but examples have been described in the submandibular glands and in the minor salivary glands. Ultrastructural studies may disclose neuroendocrine, ductal, or squamous features. However, virtually all cases show immunocytochemical evidence of neuroendocrine differentiation.[152,153,198–200] Cytologically, primary small cell carcinoma recapitulates the pattern that is familiar from our study of similar neoplasms in the lung and other sites. Paranuclear blue cytoplasmic inclusions similar to those to be discussed in Merkel cell carcinoma have been noted.[201]

Metastases from one of the other organs that more commonly host small cell carcinomas are more common than primary small cell anaplastic carcinoma in a salivary gland site. After this type of FNA diagnosis, search for a primary tumor in the lung or the head and neck (Merkel cell carcinoma) should be considered.[152,202] Merkel cell carcinoma is considered in more detail subsequently.

A single case of well-differentiated neuroendocrine carcinoma has been reported in FNA of a parotid mass.[174] Free-lying plasmacytoid tumor cells resulted in a smear pattern that was strongly reminiscent of medullary thyroid carcinoma. These cells gave a positive reaction when stained with anticytokeratin immunoreagents.

Primary lymphoepithelioma-like carcinoma of the parotid or submandibular gland is also known as malignant lymphoepithelial lesion, or as the Eskimo tumor. Poorly differentiated malignant cells with abundant lymphocyte-rich stroma cause this tumor to resemble nasopharyngeal carcinoma closely.[149,203,204] FNA of metastatic nasopharyngeal carcinoma with an occult primary is more common than cytologic evaluation of a primary salivary gland tumor of this type. The former often presents as an intraparotid or high cervical lymph node and may appear clinically to represent a primary parotid lesion.

Primary lymphoepithelioma-like carcinoma and metastatic lymphoepithelioma are morphologically identical, and distinction should be based on clinical and radiographic findings.[174] The cells are pleomorphic and show a high N/C ratio, finely granular chromatin, and a single prominent nucleolus. Mitotic figures are often prominent. The associated lymphoid tissue varies in quantity but should be polymorphous, in keeping with its benign nature. Both mucoepidermoid carcinoma and acinic cell carcinoma show large cells and may have prominent lymphoid infiltrates. However, both of these tumors lack the high-grade cytology of lymphoepithelioma.[129]

Metastatic Carcinomas That Involve the Salivary Glands and Show a Small Cell Pattern. Metastatic basal cell carcinoma of cutaneous origin has an FNA smear pattern of small blue cells (Fig. 4–23). Variable amounts of necrosis and occasional squamous pearls can be superimposed on this picture. Individual cells show scanty cytoplasm, fine chromatin, and a single small nucleolus. Peripheral palisading of nuclei has been noted in some cases as a feature in favor of basal cell carcinoma, but it is not uniformly present.[164,205] These findings are very similar to those noted in aspirations of the primary cutaneous lesions.[205]

In a salivary gland site, it may be clinically impossible to distinguish between metastases to the gland it-

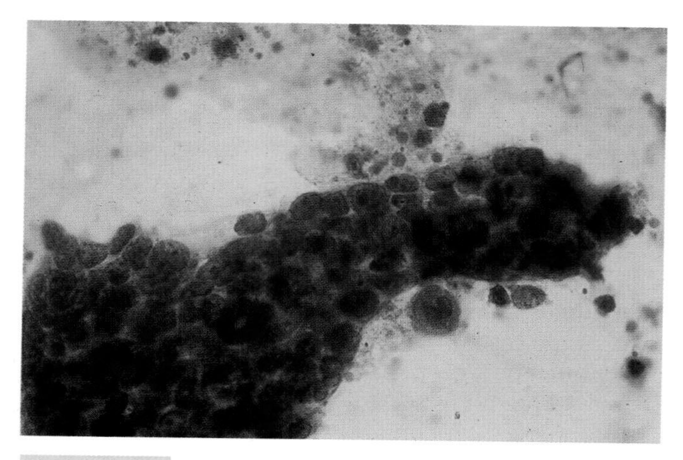

FIGURE 4–23. Without a good clinical history, there is nothing specific about this small blue cell neoplasm to indicate its origin in a cutaneous basal cell carcinoma. The necrotic background helps prevent a false-negative diagnosis. (Papanicolaou, ×400)

self or to a lymph node. If a previous history of basal cell carcinoma is available at the time of FNA, the material should be reviewed. However, metastases have been noted up to 30 years after excision of a primary skin lesion. Further, cutaneous basal cell carcinomas are very common, but metastases are rare. Thus, patients, clinicians, and pathologists may not relate a new salivary gland area mass to the original diagnosis.[164] One case was initially mistaken for pleomorphic adenoma. The history of basal cell carcinoma was not available, and desmoplastic tumor stroma was thought to represent mixed tumor matrix.

FNA of Merkel cell carcinoma has been described.[206,207] Metastatic deposits show the small cell anaplastic carcinoma cytology familiar from other sites. Perinuclear blue inclusions ("buttons") are a characteristic finding on air-dried smears but are not seen in fixed preparations.[23,152,199,206,208] Ultrastructurally, they consist of intermediate filaments and are immunocytochemically positive for cytokeratin and neurofilaments. In smears they can be seen as crescentic intracytoplasmic bodies or as free-lying round structures in the background. These inclusions are apparently rare in non-small cell carcinoma and malignant lymphoma.[201] Unfortunately, they are less specific than previously suggested: similar structures have been described in primary small cell carcinoma of the salivary glands,[199] as well as in small cell lung carcinoma and rhabdomyosarcoma.[23] Isolated psammoma bodies should not be mistaken for this type of inclusion.[209]

Malignant Lymphoma. Problems related to lymphoid proliferations in salivary gland sites have been discussed previously. Other chapters describe the cytology of lymphoid lesions in greater detail.

Epithelial Neoplasms Characterized by Large, Low-Grade Cells

Table 4–24 summarizes these neoplasms.

Acinic Cell Carcinoma. Aspirates from acinic cell carcinoma are usually quite cellular, with numerous large cells that lie singly and in groups (Fig. 4–24). Well-differentiated acinic cell carcinoma differs in part from normal salivary gland tissue by the absence of duct cells and adipocytes. However, any admixture of neoplastic cells with tissue from the surrounding normal gland can considerably complicate interpretation.[143] A few rounded acinar arrangements may be present. Larger tissue particles are often organized around a wide, thin-walled, venule-sized central vessel that may branch and ramify through microbiopsy fragments of considerable size. These fragments have irregular borders, as tumor cells fall away to lie singly in the background.

Cytomorphologic Features of Acinic Cell Carcinoma

- Ductal cells and adipose tissue absent
- Most aspirates are highly cellular
- Large tumor cells easily damaged by smearing
- Numerous naked nuclei result from cell damage
- Granular background material represents the cytoplasm of cells damaged during the smearing process
- Large tissue particles show intact cells
- Prominent blood vessels often traverse the larger particles
- Cells show abundant cytoplasm and variably prominent nucleoli

Many of these large fragile tumor cells can be damaged in the smearing process, so that numerous naked nuclei litter the smear background. These are occasionally mistaken for lymphocytes representing sialoadenitis, Warthin's tumor, or metastatic acinic cell carcinoma aspirated from a lymph note.[2,13,116,143] The background debris of Warthin's tumor is not pre-

TABLE 4–24. Epithelial Neoplasms With Large Cells of Low Nuclear Grade

- Acinic cell carcinoma
- Oncocytic lesions
- Polymorphous low-grade carcinoma
- Epithelial-myoepithelial carcinoma
- Predominantly sebaceous neoplasms
- Squamous cell carcinoma
- Clear cell carcinoma
- Metastases

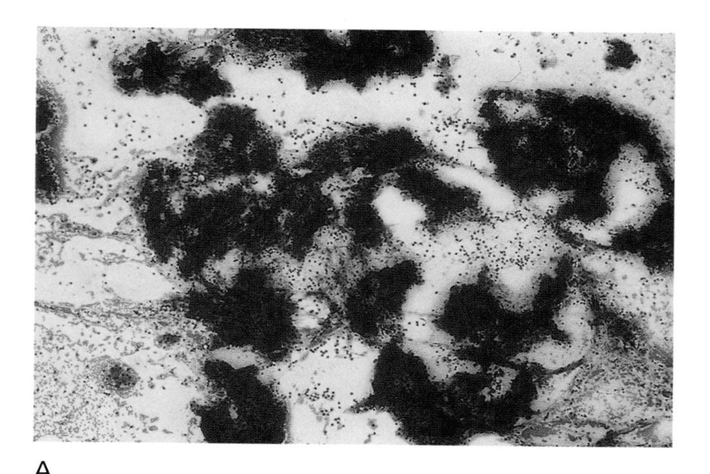

A

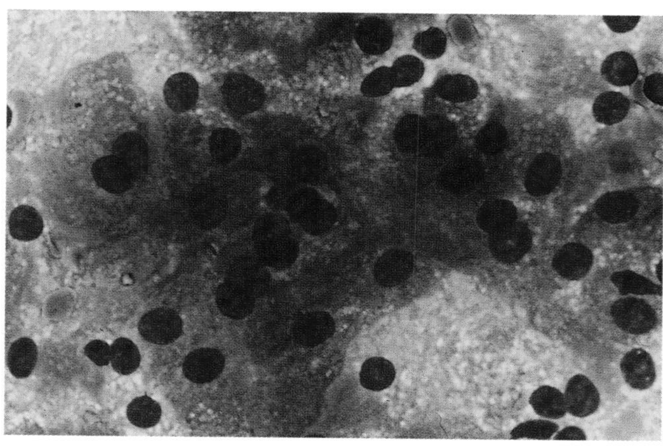

B

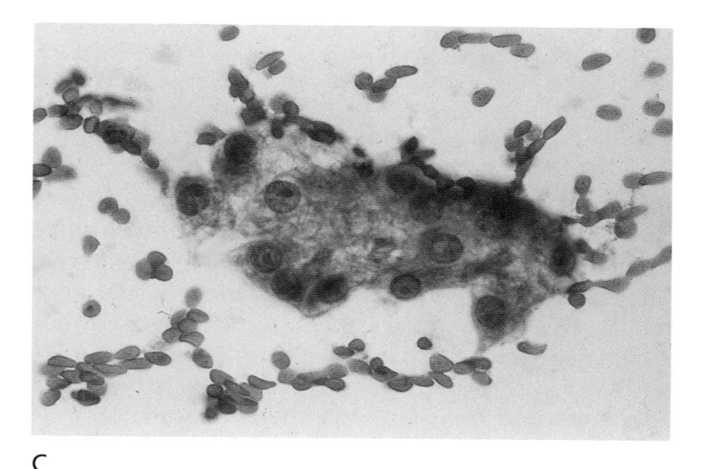

C

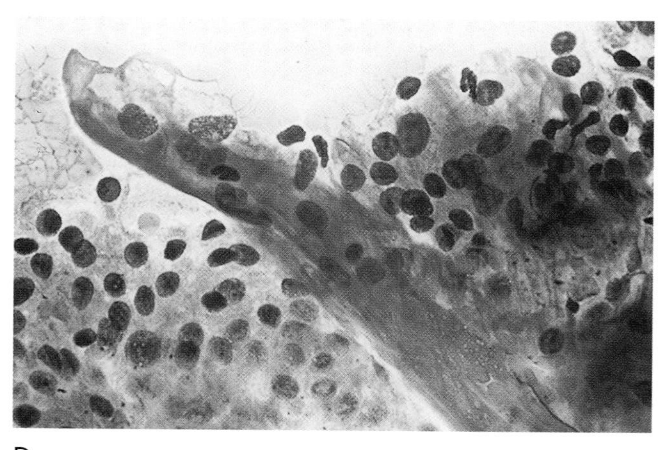

D

FIGURE 4–24. (**A**) This acinic cell carcinoma aspirate is highly cellular. (Diff-Quik, ×100) (**B**) The cells are large, with abundant finely granular cytoplasm, and resemble oncocytes. (Diff-Quik, ×600) (**C**) In a fixed smear, the cytoplasm appears more clear than granular, and the nuclei are round. (Papanicolaou, ×600) (**D**) A common feature of this tumor is the presence of large blood vessels around which tumor cells cluster. (Diff-Quik, ×400)

sent, unless one is dealing with a cyst variant of acinic cell carcinoma (see below). The distinction between naked epithelial cell nuclei and lymphocytes was discussed earlier. These difficulties are compounded when true lymphocytic infiltrates accompany an acinic cell carcinoma.[92] In other cases, the cells' granular cytoplasm has been thought to indicate the vacuoles of mucoepidermoid carcinoma.[110]

The intact cells of this carcinoma show abundant finely granular cytoplasm and can be difficult to distinguish from oncocytes. Nucleoli range from inconspicuous to prominent, but striking cytologic features of malignancy are not usually noted. For this reason, some cases are falsely interpreted as negative.[10,13,107,110,143] Pulmonary metastases of acinic cell carcinoma are occasionally diagnosed by FNA.[210]

Several variants in the pattern of acinic cell carcinoma have been described in histopathology and in FNA material. Clear cell change may be very extensive and suggest metastatic renal cell carcinoma. Usually, however, cells more typical of this neoplasm are associated with the clear cell component.[131,132] The papillary cystic variant is difficult to diagnose in aspirate samples.[116,178] FNA yields cyst fluid containing flat sheets of large granular cells that suggest a Warthin's tumor; vacuolated cells resemble low-grade mucoepidermoid carcinoma. Some aspirates show psammoma bodies, but other features that might indicate papillary thyroid carcinoma are not present.[40,209]

Oncocytic Neoplasms. Oncocytes occur in mature organs and are large cells with abundant granular cytoplasm containing numerous mitochondria. Some cells without this ultrastructural finding can still appear oncocytoid at the light microscopic level.[176,211,212] Cells of this type can be seen in normal salivary glands and in a variety of neoplasms.[213] Their cytologic features were described during our consideration of Warthin's tumor (Fig. 4–25).

Most oncocytomas are benign, regardless of the

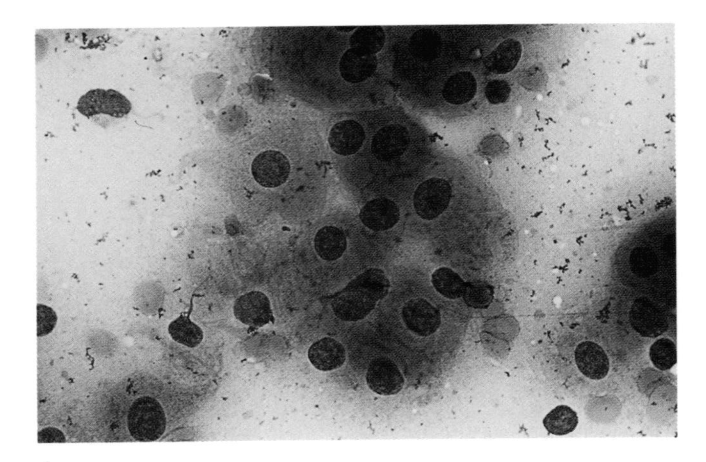

A

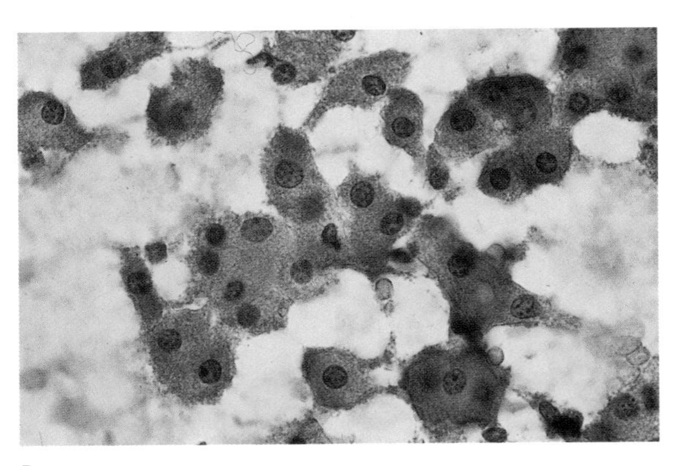

B

FIGURE 4–25. (**A**) This oncocytoma shows a monotonous population of large cells with abundant dense cytoplasm. Cell borders are difficult to discern, and there are occasional nucleoli. (Diff-Quik, ×600) (**B**) In contrast to their appearance in histologic sections, oncocytes are not always eosinophilic in cytologic samples. (Papanicolaou, ×400)

pleomorphism that the smears or sections occasionally show.[106,212,214] Differential diagnostic problems relate to the similarity of oncocytes to acinic cells and to other neoplasms that show oncocytic foci or features. Oncocytic neoplasms can be mistaken for Warthin's tumor, especially if lymphocytes are present.[92,112,176,177] This is not a problem that is likely to alter patient care adversely. The dirty cyst fluid of Warthin's tumor is usually not seen in oncocytoma.[136,211]

Distinction between oncocytoma and acinic cell carcinoma may require electron microscopy and must often await surgical excision; FNA gives only a differential diagnosis in difficult cases. Clear cell change in one case was associated with more typical oncocytic features in other areas.[132] Squamous metaplasia, mucinous change, and necrosis of cell nests have been described in surgical specimens.[177]

Oncocytomas often show nuclear pleomorphism but still behave in a benign fashion. Malignant oncocytoma is rare, but has been described.[136,212,215] Histologic or clinical evidence of aggressive behavior is required to override our tolerance for cytologic atypia in oncocytic neoplasms. Thus, it is not possible to diagnose malignant oncocytoma from FNA samples. Oncocytoid adenocarcinoma is probably more common than true malignant oncocytomas.[174,213] Such tumors show lesser degrees of mitochondrial hyperplasia.

Polymorphous Low-Grade Adenocarcinoma. Polymorphous low-grade adenocarcinoma is also known as terminal duct carcinoma or as lobular carcinoma. It is most common in minor salivary gland sites, but rare cases are seen in the submandibular or parotid glands.[170,191,193,216] In the latter instance, this tumor is most likely to represent a component of carcinoma ex pleomorphic adenoma.[217]

Histologically, uniform cells with bland nuclear features grow in architecturally diverse combinations of tubular, solid, cribriform, and papillary patterns. These can be easily mistaken for adenoid cystic carcinoma. Follow-up shows local recurrences, but regional node metastases are rare. Tumors that show extensive areas of papillary growth tend to be more aggressive and are usually classified separately.[170,217]

Cytology of the few reported cases confirms that this tumor is very difficult to distinguish from adenoid cystic carcinoma.[160,170] Uniform cells show fine chromatin, tiny nucleoli, and moderate amounts of cytoplasm. These are arranged in sheets, glands, and solid rounded groups. Tyrosine crystals may be identified.

Epithelial-Myoepithelial Carcinoma. The histology of this neoplasm is reflected in aspirate smears that show a combination of large myoepithelial cells and tubules formed by small, dark, uniform duct cells.[134,135] Numerous naked nuclei are present. Cell clusters are associated with acellular hyaline material. In some cases, small cells and hyaline predominate, and one may consider a diagnosis of adenoid cystic carcinoma. Clear myoepithelial cells may suggest clear cell carcinoma, acinic cell carcinoma with clear cell change, and sebaceous carcinoma.[80,81] None of these has a component of small duct cells. Further, epithelial-myoepithelial carcinoma-like areas can occur in pleomorphic adenoma. Confident diagnosis should be based on thorough sampling.[135]

Other Tumor Types. Normal salivary glands and a variety of neoplasms can show sebaceous differentiation.[122] The latter are recovered rarely in FNA samples.[124,125]

TABLE 4–25 Epithelial Neoplasms With Large Cells of High Nuclear Grade

- High-grade mucoepidermoid carcinoma
- High-grade carcinoma, not otherwise specified
- Squamous cell carcinoma
  - Primary
  - Metastatic
  - Direct extension from adjacent sites
- Ductal carcinoma
- Metastatic carcinoma
- Metastatic malignant melanoma

Epithelial Neoplasms Characterized by Large, High-Grade Cells

Table 4–25 summarizes these neoplasms.

The salivary glands can be involved by direct extension of carcinomas from the head and neck mucosal surfaces or the skin.[133] Metastatic lesions may involve an intraglandular lymph node or a major salivary gland. The parotid is more often involved than the submandibular gland.[174] Malignant melanoma and squamous cell carcinoma are the most common tumor types.[10,19,43,166] FNA of malignant melanoma is discussed in other chapters. Additional metastases that are occasionally reported include small cell carcinoma, seminoma, sebaceous carcinoma, retinoblastoma, and astrocytoma. Adenocarcinomas from the lung, breast, colon, stomach, and pancreas may also be found in head and neck sites.[13,24,132,133]

Distinction between primary and metastic carcinoma is of great clinical importance, and FNA can often address this issue very well.[111] Those who perform the aspiration and interpret the smears should always be provided with any available history of previous malignancy. However, in our series of nine metastatic high-grade carcinomas, only six patients had malignancies that were recognized in their primary sites before the salivary gland FNA.[174] One must always consider the possibility of a metastatic deposit, especially if the sample does not seem typical of a well-characterized type of salivary gland tumor.

High-grade mucoepidermoid carcinoma was discussed previously. The highly cellular aspirate of large carcinoma cells must be distinguished from metastatic carcinoma and from other primary high-grade carcinomas.[6,13,18] Extensive clear cell change can suggest other diagnoses.

Salivary gland duct carcinoma is operatively defined by its resemblance to breast carcinoma. The cells of this tumor can be arranged in solid, cribriform, or papillary patterns (Fig. 4–26).[83,218–220] The histology of this tumor indicates that we expect occasional psammoma bodies, keratinized cells, or comedo-like necrosis. Occasional cases yield low-grade malignant cells and collagenous stroma in a pattern mistaken for a pleomorphic adenoma (Fig. 4–27). Based on cytologic findings alone, this can be a formidable diagnostic dilemma. Clinical findings can be helpful in this regard; ductal carcinoma usually occurs in elderly men, whereas mixed tumors are more common in women in the middle years.

Spindle Cell Lesions

Primary spindle cell lesions of the salivary glands are not commonly studied by FNA (Tables 4–26 and 4–27). Most examples are reported in large series, where they receive limited treatment. Those that have been described include hemangioma,[17,18,43] lymphangioma,[221] and hemangiopericytoma.[18] A single case of

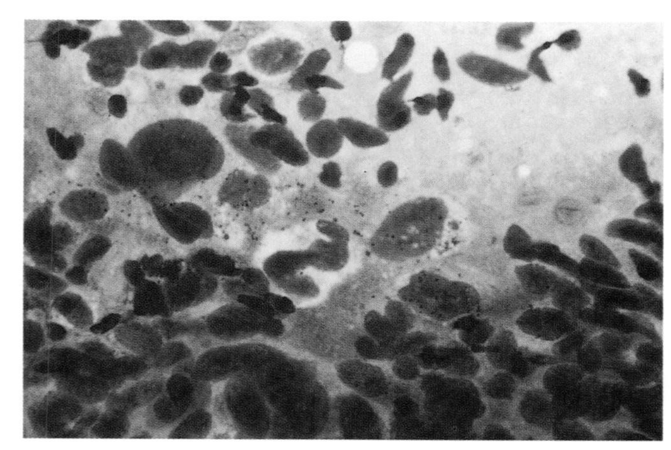

A

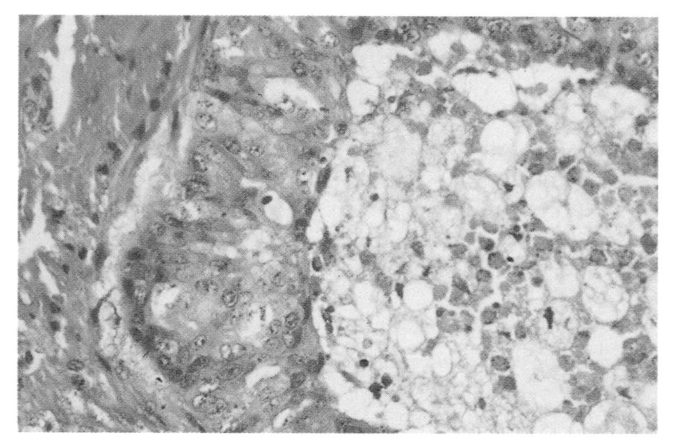

B

FIGURE 4–26. (**A**) This aspirate from a parotid mass in an elderly man shows a non-small cell carcinoma with no distinguishing features. (Diff-Quik, ×600) (**B**) Histologically, the resemblance of this neoplasm to breast carcinoma is striking and results in a diagnosis of ductal carcinoma of the parotid. (Hematoxylin and eosin, ×100)

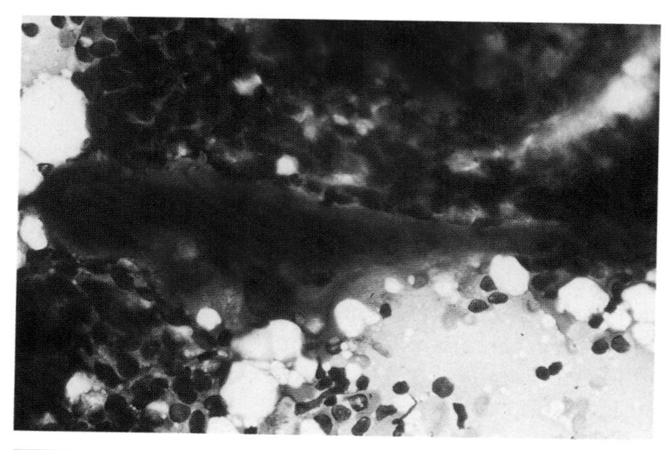

FIGURE 4–27. This example of salivary gland duct carcinoma shows low-grade nuclei and fragments of desmoplastic stroma. This aspirate was mistaken for pleomorphic adenoma at the time of FNA. (Diff-Quik, ×400)

nodular fasciitis was followed nonsurgically after FNA and resolved.[222] FNA of interstitial amyloidosis presenting as a submandibular gland has been described.[223]

The cells of myoepithelioma can be spindled or, less commonly, hyaline (plasmacytoid). However, plasmacytoid cells in pleomorphic adenomas lack ultrastructural and immunocytochemical features of myoepithelial differentiation.[165,168,224]

Most myoepitheliomas are located in the parotid, but submandibular and rare extraglandular cases occur. Cytologically, these neoplasms usually show monotonous spindle cells that resemble those of a leiomyoma and are identical to the cell type occasionally aspirated from mixed tumors (see Fig. 4–15).[107,168] When this picture is encountered in smears, one must consider other low-grade spindle cell neoplasms because the cytologic picture is nonspecific. Electron microscopic or immunocytochemical confirmation can be sought for difficult cases. The plasmacytoid type of myoepithelioma can show striking nuclear enlargement and atypia of single cells that we have noted in similar cells derived from pleomorphic adenomas. Confusion of these cells with hematopoietic processes was discussed earlier.

Malignant myoepithelioma can arise *de novo* or within a pre-existing pleomorphic adenoma. Only a few examples have been described in the FNA literature. These have shown a poorly cohesive proliferation of mitotically active spindle cells with high-grade nuclear features (Fig. 4–28).[17,107,224,225] Definitive diagnosis requires ultrastructural or immunocytochemical demonstration of myoepithelial differentiation. Hemangiopericytoma has been cytologically mistaken for malignant myoepithelioma.

Aspiration of a single salivary gland anlage tumor (congenital pleomorphic adenoma) was reported by Bondeson et al.[226] Smears showed a combination of small uniform spindle cells and ductal elements. This neoplasm occurs exclusively in the neonatal period.[227] The differential diagnosis must include teratoma; most of the small blue cell tumors of childhood do not share this cytologic picture.

Spindle cell malignancies can also reach the salivary glands by metastasis or direct extension. Malignant melanoma may present in this manner (Fig. 4–29).

TABLE 4–26. Spindle Cell Lesions of Low Nuclear Grade

- Reactive, not otherwise specified
- Nodular fasciitis
- Hemangioma
- Myoepithelioma
- Kaposi's sarcoma

TABLE 4–27. Spindle Cell Lesions of High Nuclear Grade

- Primary sarcoma
- Sarcoma invasion from adjacent tissues
- Malignant myoepithelioma
- Squamous cell carcinoma
- Metastatic carcinoma
- Metastatic malignant melanoma

CONCLUSION

Preoperative diagnosis of salivary gland masses by FNA can be clinically useful; common inflammatory and neoplastic conditions are readily diagnosed.

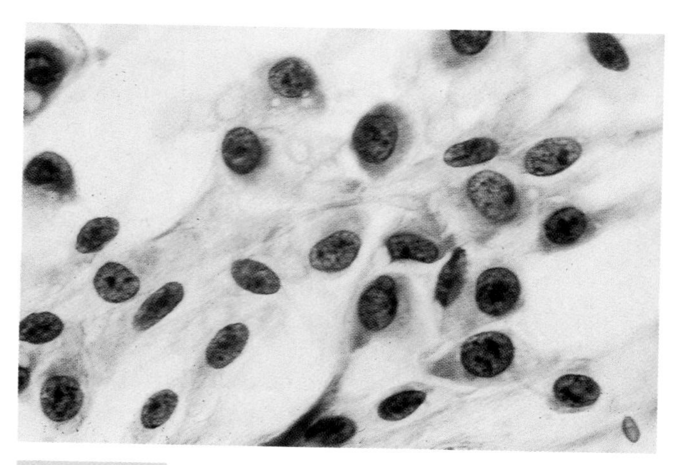

FIGURE 4–28. This malignant myoepithelioma shows loosely cohesive malignant spindle cells. (Papanicolaou, ×600)

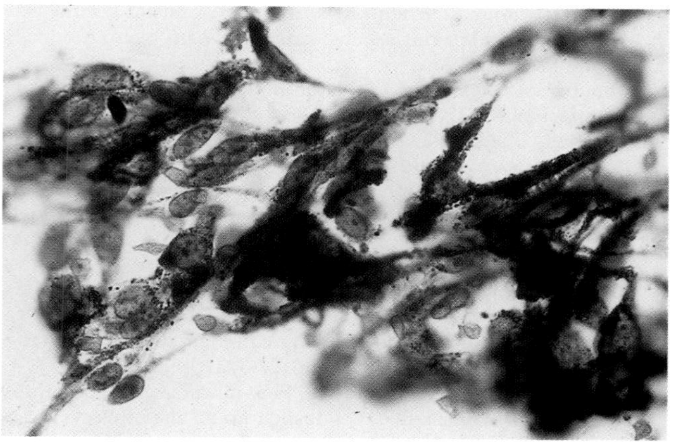

A

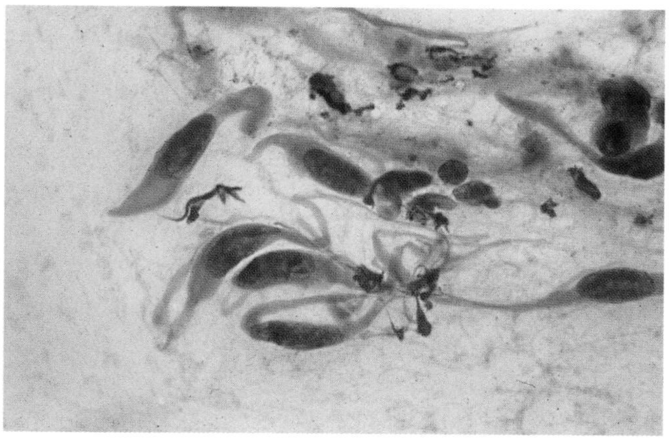

B

FIGURE 4–29. (**A**) Metastatic malignant melanoma can show a spindle-cell pattern. In this case, heavy pigment makes the diagnosis simple. (Diff-Quik, ×600) (**B**) Other metastatic melanomas may present a diagnostic problem if pigment is not identified. (Papanicolaou, ×600)

Thus, patients can be rapidly and efficiently directed toward the most effective treatment and follow-up. However, our discussions have highlighted the significant difficulties and limitations that exist in this field.

Aspirates that are highly cellular, lack diagnostic matrix materials, and show no cytologic features of malignancy are very troubling. Both false-positive and false-negative diagnoses result from this cytologic picture. The very useful term "salivary gland neoplasms" to be further investigated by other means has been suggested by Layfield et al.[18] We agree with this practical directive. Such cases are analogous to diagnosing follicular neoplasms when a thyroid FNA is thought to represent either a follicular adenoma or a follicular carcinoma. When we offer this type of interpretation, we recognize the limitations inherent in FNA. However, we do so without decreasing its value in areas where more definitive diagnoses are possible. Further, this type of diagnosis represents only a small number of cases. Each one will have been preceded by a large number of cases in which FNA was diagnostic and made a meaningful contribution to patient care. The care of most patients with a head and neck mass should be improved by FNA, notwithstanding the occasional deferred diagnosis.

REFERENCES

1. Stanley MW, Lowhagen T. Fine needle aspiration of palpable masses. Stoneham, MA: Butterworth-Heinnemann, 1993.
2. Flynn MB, Wolfson SE, Thomas S, Kuhns JG. Fine-needle aspiration biopsy in clinical management of head and neck tumors. J Surg Oncol 44:214–217, 1990.
3. Zakowski MF. Fine-needle aspiration cytology of tumors: diagnostic accuracy and potential pitfalls. Cancer Invest 12(5):505–515, 1994.
4. Schelkun PM, Grundy WG. Fine-needle aspiration biopsy of head and neck lesions. J Oral Maxillofac Surg 49:262–267, 1991.
5. Schwarz R, Chan NH, MacFarlane JK. Fine needle aspiration cytology in the evaluation of head and neck masses. Am J Surg 159:482–485, 1990.
6. Heller KS, Dubner S, Chess Q, Attie JN. Value of fine needle aspiration biopsy of salivary gland masses in clinical decision-making. Am J Surg 164:667–670, 1992.
7. Kameswaran M, Abu-Eshy S, Hamdi J. Facial palsy following fine needle aspiration biopsy of parotid hemangioma: a case report and review of the literature. Ear Nose Throat J 70:801–803, 1991.
8. Carter MH Jr. Letters. J Miss State Med Assoc 35(9):265, 1994.
9. Henry-Stanley MJ, Stanley MW. Laboratory processing of needle-rinse material rarely identifies malignancy not recognized in smear material. Diagn Cytopathol 8:538–540, 1992.
10. O'Dwyer P, Farrar WB, James AG, Finkelmeier W, McCabe DP. Needle aspiration biopsy of major salivary gland tumors. Cancer 57:554–557, 1986.
11. Felson B. Chest roentgenology. Philadelphia: WB Saunders, 1973.
12. Stanley MW, Knoedler JP. Skeletal structures that clinically simulate lymph nodes: encounters during fine needle aspiration. Diagn Cytopathol 9:86–88, 1993.
13. Qizilbash AH, Sianos J, Young JEM, Archibald SD. Fine needle aspiration biopsy cytology of major salivary glands. Acta Cytol 29(4):503–512, 1985.
14. Engzell U, Zajicek J. Aspiration biopsy of tumors of the neck. I. Aspiration biopsy and cytologic findings in 100 cases of congenital cysts. Acta Cytol 14(2):51–57, 1970.
15. Bhatia A. Fine needle aspiration cytology in the diagnosis of mass lesions of the salivary gland. Indian J Cancer 30(1):26–30, 1993.
16. Dejmek A, Lindholm K: Fine needle aspiration biopsy of cystic lesions of the head and neck, excluding the thyroid. Acta Cytol 34(3): 443–448, 1990.
17. MacLeod CB, Frable WJ. Fine-needle aspiration biopsy of the salivary gland: problem cases. Diagn Cytopathol 9(2):216–225, 1993.
18. Layfield LJ, Glasgow BJ. Diagnosis of salivary gland tumors by fine-needle aspiration cytology: a review of clinical utility and pitfalls. Diagn Cytopathol 7(3):267–272, 1991.
19. Layfield LJ, Glasgow BJ, DuPuis MH. Fine-needle aspiration of lymphadenopathy of suspected infectious etiology. Arch Pathol Lab Med 109:810, 1985.

20. Silverman JF. Guides to clinical aspiration biopsy: infectious and inflammatory diseases and other nonneoplastic disorders. New York: Igaku-Shoin Medical Publishers, 1991.
21. Das DK, Gulati A, Bhatt NC, Mandal AK, Khan VA, Bhambhani S. Fine needle aspiration cytology of oral and pharyngeal lesions: a study of 45 cases. Acta Cytol 37(3):333–342, 1993.
22. Martin JJR, Salaverri CO, Gonzalez-Cámpora R. Intraparotid *Leishmania donovani* lymphadenitis: diagnosis by fine needle aspiration. Acta Cytol 37:843–845, 1993.
23. Tallada N, Raventós A, Martinez S, Compañó C, Almirante B. Leishmania lymphadenitis diagnosed by fine-needle aspiration biopsy. Diagn Cytopathol 9(6):673–676, 1993.
24. Klijaniennko J, Vielh P. Fine-needle samping of salivary gland lesions II: Cytology and histology correlation of 71 cases of Warthin's tumor (adenolymphoma). Diagn Cytopathol 16:221–225, 1997.
25. Mondal A, Gupta S. The role of peroral fine needle aspiration cytology (FNAC) in the diagnosis of parapharyngeal lesions: a study of 51 cases. Indian J Pathol-Microbiol 36(3):253–259, 1993.
26. Cramer H, Layfield L, Lampe H. Letters to the editor. Ann Otol Rhinol Laryngol 102:483–484, 1993.
27. Gottschalk-Sabag S, Glick T. Necrosis of parotid pleomorphic adenoma following fine needle aspiration: a case report. Acta Cytol 39:252–254, 1995.
28. Layfield LJ, Reznicek M, Lowe M, Bottles K. Spontaneous infarction of a parotid gland pleomorphic adenoma: report of a case with cytologic and radiographic overlap with a primary salivary gland malignancy. Acta Cytol 36(3):381–386, 1992.
29. Kern SB. Necrosis of a Warthin's tumor following fine needle aspiration. Acta Cytol 31:207–208, 1987.
30. Jones JD, Pittman DL, Sanders LR. Necrosis of thyroid nodules after fine needle aspiration. Acta Cytol 29(1):29–32, 1985.
31. Davies JD, Webb AJ. Segmental lymph-node infarction after fine-needle aspiration. J Clin Pathol 35:855–857, 1982.
32. Allen CM, Damm D, Neville B, Rodu B, Page D, Weathers DR. Necrosis in benign salivary gland neoplasms: not necessarily a sign of malignant transformation. Oral Surg Oral Med Oral Pathol 78:455–461, 1994.
33. Batsakis JG. Letters to the editor. Ann Otol Rhinol Laryngol 102:484–485, 1993.
34. Roland NJ, Caslíln AW, Smith PA, Turnbull LS, Panarese A, Jones AS. Fine needle aspiration cytology of salivary gland lesions reported immediately in a head and neck clinic. J Laryngol Otol 107:1025–1028, 1993.
35. Casiano RR, Cooper JD, Gould E, Ruiz P, Uttamchandani R. Value of needle biopsy in directing management of parotid lesions in HIV-positive patients. Head Neck 13:411–414, 1991.
36. Huang RD, Pearlman S, Friedman WH, Loree T. Benign cystic vs. solid lesions of the parotid gland in HIV patients. Head Neck 13:522–527, 1991.
37. Cohen MB, Ljung B-ME, Boles R. Salivary gland tumors: fine-needle aspiration vs. frozen-section diagnosis. Arch Otolaryngol Head Neck Surg 112:867–869, 1986.
38. Howell LP, Russell LA, Howard PH, Teplitz RL. Fine needle aspiration biopsy of superficial masses in children. West J Med 155:33–38, 1991.
39. Wakely PE, Kardos TF, Frable WJ. Application of fine needle aspiration biopsy to pediatrics. Hum Pathol 19(12):1383–1386, 1988.
40. Amice J, Sparfel A, Pétillon F, Amice V, Jézéquel J, Rivière MR. Hydatid cyst of the neck: diagnosis by fine needle aspiration. Acta Cytol 36(3):454–455, 1992.
41. Das DK, Bhatt NC, Khan VA, Luthra UK. Cervicofacial actinomycosis: diagnosis by fine needle aspiration cytology. Acta Cytol 33(2):278–280, 1989.
42. Lampe HB, Cramer HM. Advances in the use of fine-needle aspiration cytology in the diagnosis of palpable lesions of the head and neck. J Otolaryngol 20:108–115, 1991.
43. Candel A, Gattuso P, Reddy V, Matz G, Castelli M. Is fine needle aspiration biopsy of salivary gland masses really necessary? ENT J 72(7):485–489, 1993.
44. Ramzy I, Rone R, Schantz HD. Squamous cells in needle aspirates of subcutaneous lesions: a diagnostic problem. Am J Clin Pathol 85:319–324, 1986.
45. Layfield LJ, Glasgow BJ. Aspiration biopsy cytology of primary cutaneous tumors. Acta Cytol 37(5):679–688, 1993.
46. Bardales RH, Stanley MW. Subcutaneous masses of the scalp and forehead: diagnosis by fine needle aspiration. Diagn Cytopathol 12:131–134, 1995.
47. Daskalopoulou D, Maounis N, Kokalis G, Liodandonaki P, Belezini E, Markidou S. The role of fine needle aspiration cytology in the diagnosis of primary skin tumors. Arch Anat Cytol Pathol 41(2):75–81, 1993.
48. Bondeson L, Lindholm K, Thorstenson S. Benign dermal eccrine cylindroma: a pitfall in the cytologic diagnosis of adenoid cystic carcinoma. Acta Cytol 27(3):326–328, 1983.
49. Wong MP, Yuen ST, Collins RJ. Fine-needle aspiration biopsy of pilomatrixoma: still a diagnostic trap for the unwary. Diagn Cytopathol 10(4):365–369, 1994.
50. Pisharodi LR. Editorial comments: fine-needle aspiration biopsy of pilomatrixoma. Diagn Cytopathol 10(4):369–370, 1994.
51. Domanski HA, Domanski AM. Cytology of pilomatrixoma (calcifying epithelioma of Malherbe) in fine needle aspirates. Acta Cytol 41:771–777, 1997.
52. Ma KF, Tsui MS, Chan SK. Fine needle aspiration diagnosis of pilomatrixoma: a monomorphic population of basaloid cells with squamous differentiation not to be mistaken for carcinoma. Acta Cytol 35:570–574, 1991.
53. Hales M, Bottles K, Miller T, Donegan E, Ljung B-M. Diagnosis of Kaposi's sarcoma by fine-needle aspiration biopsy. Am J Clin Pathol 88:20–25, 1987.
54. Mincione GP, Borrelli D, Cicchi P, Ipponi PL, Fiorini A. Fine needle aspiration cytology of parathyroid adenoma: a review of seven cases. Acta Cytol 301:65–69, 1986.
55. Bondeson L, Bondeson A-G, Nissborg A, Thompson NW. Cytopathological variables in parathyroid lesions: a study based on 1,600 cases of hyperparathyroidism. Diagn Cytopathol 16:476–482; 1997.
56. Friedman M, Shimaoka K, Lopez CA, Shedd DP. Parathyroid adenoma diagnosed as papillary carcinoma of thyroid on needle aspiration smears. Acta Cytol 27(3):337–340, 1983.
57. Guazzi A, Gabrielli M, Guadagni G. Cytologic features of a functioning parathyroid carcinoma: a case report. Acta Cytol 26(5):709–713, 1982.
58. Fulciniti F, Vetrani A, Zeppa P, Califano L, Palombini L. Calcifying epithelial odontogenic tumor (Pindborg's tumor) on fine-needle aspiration biopsy smears: a case report: Diagn Cytopathol 12:71–75, 1995.
59. Radhika S, Nijhawan R, Das D, Dey P. Ameloblastoma of the mandible: diagnosis by fine-needle aspiration cytology. Diagn Cytopathol 9:310–313, 1993.
60. Walaas L, Kindblom L-G. Fine-needle aspiration biopsy in the preoperative diagnosis of chordoma: a study of 17 cases with application of electron microscopic, histochemical, and immunocytochemical examination. Hum Pathol 22(1):22–28, 1991.
61. Hazarika D, Kumar RV, Muniyappa GD, et al. Diagnosis of clival chordoma by fine needle aspiration of an oropharyngeal mass: a case report. Acta Cytol 39(3):507–510, 1995.
62. Elliott EC, McKinney S, Banks H, Fulks RM. Aspiration cytology of metastatic chordoma: a case report. Acta Cytol 27(6):658–662, 1983.
63. Gherardi G, Marveggio C, Cola C, Redaelli G. Decisive role of immunocytochemistry in aspiration cytology of chordoma of the clivus: a case report with review of the literature. J Laryngol Otol 108:426–430, 1994.
64. Martin H, Janda J, Werbs M, Dorste P. Unusual chordoma of the neck region simulating salivary gland pleomorphic adenoma. Head Neck Oncol 38(12):462–464, 1990.
65. Peramezza C, Cellini A, Berardi P, Benvenuti S, Offidani A. Chordoma with multiple skin metastases. Dermatology 186:266–268, 1993.

66. Daniel WP, Louback JB, Gagne EJ, Scheithauer BW. Chordoma cutis: a report of nineteen patients with cutaneous involvement of chordoma. J Am Acad Dermatol 29:63–66, 1993.
67. Thakar A, Tandon DA, Bahadur S, Vijayaraghavan M. Extranotochordal chordoma presenting as multiple neck masses: report of a case. J Laryngol Otol 107:942–945, 1993.
68. Kontozoglou T, Quizilbash AH, Sianos J, et al. Chordoma: cytologic and immunocytochemical study of four cases. Diagn Cytopathol 2:55–61, 1986.
69. Finley JI, Silverman JF, Dabbs JD, et al. Chordoma. Diagnosis by fine needle aspiration biopsy with histology, immunocytochemical and ultrastructural confirmation. Diagn Cytopathol 2:330–337, 1986.
70. Nguyen G-K, Johnson ES, Mielke BW. Cytology of meningiomas and neurilemomas in crush preparations: a useful adjunct to frozen section diagnosis. Acta Cytol 32(3):362–366.
71. Rorat E, Yang W, De La Torre R. Fine needle aspiration cytology of parapharyngeal meningioma. Acta Cytol 35(5):497–500, 1991.
72. Cristallini EG, Bolis GB, Ottaviano P. Fine needle aspiration biopsy of orbital meningioma: report of a case. Acta Cytol 34(2):236–238, 1990.
73. Iazmontazer N, Bedayat G. Cytodiagnosis of meningioma with atypical cytologic features. Acta Cytol 35(5):501–504, 1991.
74. Tao L-C. Pulmonary metastases from intracranial meningioma diagnosed by aspiration biopsy cytology. Acta Cytol 35(5):524–528, 1991.
75. Mincione GP, Mincione F, Mennonna P. Cytological features of a craniopharyngioma. Pathologica 83:191–196, 1991.
76. Engzell U, Franzén S, Zajicek J. Aspiration biopsy of tumors of the neck. II. Cytologic findings in 13 cases of carotid body tumor. Acta Cytol 15(1):25–29, 1971.
77. Kapila K, Tewari MC, Verma K. Paragangliomas—a diagnostic dilemma on fine needle aspirates. Indian J Cancer 30(4):152–157, 1993.
78. Hood IC, Qizilbash AH, Young JEM, Archibald SD. Fine needle aspiration biopsy cytology of paragangliomas: Cytologic, light microscopic and ultrastructural studies of three cases. Acta Cytol 27(6):651–657, 1983.
79. Rao CR, Visweshwaraiah LD, Veerapaiah KS, Satpute SD, Hazarika D, Bhargava MK. Melanotic neuroectodermal tumor of infancy initially diagnosed by fine needle aspiration biopsy. Acta Cytol 345:681–684, 1990.
80. Rosai J. Ackerman's Surgical Pathology, 8th edition (pp. 2093–2094). St. Louis, C. V. Mosby Co., 1996.
81. Walker WP, Laszewski MJ. Recurrent multifocal adult rhabdomyoma diagnosed by fine-needle aspiration cytology: Report of a case and review of the literature. Diagn Cytopathol 6:354–358, 1990.
82. Bertholf MF, Frierson HF, Feldman PS. Fine-needle aspiration cytology of an adult rhabdomyoma of the head and neck. Diagn Cytopathol 4:152–155, 1988.
83. Bondeson L, Andreasson L. Aspiration cytology of adult rhabdomyoma. Acta Cytol 30(6): 679–682, 1986.
84. Gherardi G, Marveggio C. Immunocytochemistry in head and neck aspirates: Diagnostic application on direct smears in 16 problematic cases. Acta Cytol 36(5):687–696, 1992.
85. Akhtar M, Asshraf M, Bakry M, Hug M, Sackey K. Fine-needle aspiration biopsy diagnosis of rhabdomyosarcoma: cytologic, histologic, and ultrastructural correlations. Diagn Cytopathol 8:465–474, 1992.
86. de Almeida M, Stastny JF, Wakely PE, Frable WJ. Fine-needle aspiration biopsy of childhood rhabdomyosarcoma: reevaluation of the cytologic criteria for diagnosis. Diagn Cytopathol 11:231–236, 1994.
87. Silverman JF, Joshi VV. FNA biopsy of small round cell tumors of childhood: cytomorphologic features and the role of ancillary studies. Diagn Cytopathol 10:245–255, 1994.
88. Thompson HY, Fulmer RP, Schnadig VJ. Metastatic squamous cell carcinoma of the tonsil presenting as multiple cystic neck masses: report of a case with fine needle aspiration findings. Acta Cytol 38:605–607, 1994.
89. Granström G, Edström S. The relationship between cervical cysts and tonsillar carcinoma in adults. J Oral Maxillofac Surg 47:16–20, 1989.
90. Burgess KL, Hartwick RWJ, Bedard YC. Metastatic squamous carcinoma presenting as a neck cyst: differential diagnosis from inflamed branchial cleft cyst in fine needle aspirates. Acta Cytol 37(4):494–498, 1993.
91. Warson F, Blommaert D, De Roy G. Inflamed branchial cyst: a potential pitfall in aspiration cytology. Acta Cytol 30:201–202, 1986.
92. Auclair PL. Tumor-associated lymphoid proliferation in the parotid gland: a potential diagnostic pitfall. Oral Surg Oral Med Oral Pathol 77:19–26, 1994.
93. Dodd LG. Fine-needle aspiration cytology of adenoid (acantholytic) squamous-cell carcinoma. Diagn Cytopathol 12:168–172, 1995.
94. Sperling NM, Lin P-T, Lucente FE. Cystic parotid masses in HIV infection. Head Neck 12:337–341, 1990.
95. Schiødt M. HIV-associated salivary gland disease: a review. Oral Surg Oral Med Oral Pathol 73:164–167, 1992.
96. Raab SS, Thomas PA, Cohen MB. Fine-needle aspiration biopsy of salivary gland mycoses. Diagn Cytopathol 11(3):286–290, 1994.
97. Redleaf MI, Bauer CA, Robinson RA. Fine-needle detection of cytomegalovirus parotitis in a patient with acquired immunodeficiency syndrome. Arch Otolaryngol Head Neck Surg 120:414–416, 1994.
98. Wax TD, Layfield LJ, Zaleski S, et al. Cytomegalovirus sialadenitis in patients with the acquired immunodeficiency syndrome: a potential diagnostic pitfall with fine-needle aspiration cytology. Diagn Cytopathol 10:169–174, 1994.
99. Corkill M, Stephens J, Bitter M. Fine needle aspiration cytology of mycobacterial spindle cell pseudotumor: a case report. Acta Cytol 39:125–128, 1995.
100. Umlas J, Federman M, Crawford C, O'Hara CJ, Fitzgibbon JS, Modeste A. Spindle cells pseudotumor due to *Mycobacterium avium-intracellulare* in patients with acquired immunodeficiency syndrome (AIDS): positive staining of mycobacteria for cytoskeleton filaments. Am J Surg Pathol 15(12):1181–1187, 1991.
101. Weidner N, Geisinger KR, Sterling RT, Miller TR, Yen TSB. Benign lymphoepithelial cysts of the parotid gland: a histologic, cytologic, and ultrastructural study. Am J Clin Pathol 85:395–401, 1986.
102. Elliott JN, Oertel YC. Lymphoepithelial cysts of the salivary glands: histologic and cytologic features. Am J Clin Pathol 93:39–43, 1990.
103. Tao L-C, Gullane PJ. HIV infection-associated lymphoepithelial lesions of the parotid gland: aspiration biopsy cytology, histology, and pathogenesis. Diagn Cytopathol 7:159–162, 1991.
104. Finfer MD, Gallo L, Perchick A, Schinella RA, Burstein DE. Fine needle aspiration biopsy of cystic benign lymphoepithelial lesion of the parotid gland in patients at risk for the acquired immunodeficiency syndrome. Acta Cytol 34(6): 821–826, 1990.
105. Kline TS, Merriam JM, Shapshaw SM. Aspiration biopsy cytology of the salivary gland. Am J Clin Pathol 76:263–269, 1981.
106. Patt BS, Schaefer SD, Vuitch F: Role of fine-needle aspiration in the evaluation of neck masses. Med Clin North Am 77(3):611–623, 1993.
107. Zurrida S, Alasio L, Tradati N, Bartoli C, Chiesa F, Pilotti S. Fine-needle aspiration of parotid masses. Cancer 72:2306–2311, 1993.
108. Guyot J-P, Obradovic D, Krayenbuhl M, Zbaeren P, Lehmann W. Fine-needle aspiration in the diagnosis of head and neck growths: is it necessary? Otolaryngol Head Neck Surg 103(5):697–701, 1990.
109. Batsakis JG, Sneige N, El-Naggar AK. Fine-needle aspiration of salivary glands: its utility and tissue effects. Ann Otol Rhinol Laryngol 101:185–188, 1992.

110. Layfield LJ, Tan P, Glasgow BJ. Fine-needle aspiration of salivary gland lesions: comparison with frozen sections and histologic findings. Arch Pathol Lab Med 111:346–353, 1987.
111. Cross DL, Gansler TS, Morris RC. Fine needle aspiration and frozen section of salivary gland lesions. South Med J 83(3):283–286, 1990.
112. Megerian CA, Maniglia AJ. Parotidectomy: a ten-year experience with fine needle aspiration and frozen-section biopsy correlation. ENTJ 73(6):377–380, 1994.
113. Heller KS, Attie JN, Dubner S. Accuracy of frozen section in the evaluation of salivary tumors. Am J Surg 166:424–427, 1993.
114. Wheelis RF, Yarington CT. Tumors of the salivary glands. Arch Otolaryngol 110:76–77, 1984.
115. Miller RH, Calcaterra TC, Paglia DE. Accuracy of frozen-section diagnosis of parotid lesions. Ann Otol 88:573–576, 1979.
116. Weinberger MS, Rosenberg WW, Meurer WT, Robbins KT. Fine-needle aspiration of parotid gland lesions. Head Neck 14:483–487, 1992.
117. Hajdu SI, Melamed MR. Limitations of aspiration cytology in the diagnosis of primary neoplasms. Acta Cytol 28(3):337–345, 1984.
118. Eneroth CM, Zajicek J. Aspiration biopsy of salivary gland tumors. III: Morphologic studies on smears and histologic sections from 368 mixed tumors. Acta Cytol 10:440–454, 1966.
119. Shaheen OH. Benign salivary gland tumors. In: Scott Brown's Otolaryngology, 5th edition, Vol. 5. London: Butterworth Publications, 1987.
120. Phillips DE, Jones AS. Reliability of clinical examination in the diagnosis of parotid tumours. J R Coll Surg Edinb 39:100–102, 1994.
121. Stanley MW, Bardales RH, Beneke J, Korourian S, Stern SJ. Sialolithiasis: differential diagnostic problems in fine needle aspiration cytology. Diagn Cytopathol 106:229–233, 1996.
122. Batsakis JG, El-Naggar AK. Sebaceous lesions of salivary glands and oral cavity. Ann Otol Rhinol Laryngol 99:416–418, 1990.
123. Hayes MMM, Cameron RD, Jones EA. Sebaceous variant of mucoepidermoid carcinoma of the salivary gland: a case report with cytohistologic correlation. Acta Cytol 37(2):237–241, 1993.
124. Derias NW, Chong WH, Pambakian H. Sebaceous adenoma of parotid gland—a rare tumour diagnosed by the fine needle aspiration cytology. Cytopathology 5(6):392–395, 1994.
125. Pai RR, Bharathi S, Naik R, Raghuveer CV. Unilocular cystic sebaceous lymphadenoma of the parotid gland. Indian J Pathol Microbiol 37(3):327–330, 1994.
126. Carson HJ, Raslan WF, Castelli MJ, Gattuso P. Tyrosine crystals in benign parotid gland cysts: report of two cases diagnosed by fine-needle aspiration biopsy with ultrastructural and histochemical evaluation. Am J Clin Pathol 102:699–702, 1994.
127. Mavec P, Eneroth CM, Franzen S, Moberger G, Zajicek J. Aspiration biopsy of salivary gland tumors. Acta Otolaryngol 58:471–484, 1964.
128. Laucirica R, Leopold SK, Kalin GB. False-positive diagnosis in fine-needle aspiration of an atypical Warthin's tumor: histochemical differential stains for cytodiagnosis. Diagn Cytopathol 5:412–415, 1989.
129. Chen KTK. Letter to the editor: aspiration cytology of metaplastic Warthin's tumor mimicking squamous-cell carcinoma. Diagn Cytopathol 7:330–331, 1991.
130. Olsen KD, Goellner JR. False-positive cytologic findings in Warthin's tumor: a report of two cases. ENTJ 71(9):417–421, 1992.
131. Leiman G. Editorial comments: clear cell tumors of the salivary glands. Diagn Cytopathol 9(6):711–712, 1993.
132. Layfield LJ, Glasgow BJ. Aspiration cytology of clear-cell lesions of the parotid gland: morphologic features and differential diagnosis. Diagn Cytopathol 9(6):705–712, 1993.
133. Gattuso P, Castelli MJ, Shah PA, Kron T. Fine needle aspiration cytologic diagnosis of metastatic Merkel cell carcinoma in the parotid gland. Acta Cytol 32(4):576–578, 1988.
134. Carrillo R, Poblet E, Rocamora A, Rodriguez-Peralto JL. Epithelial-myoepithelial carcinoma of the salivary gland: fine needle aspiration cytologic findings. Acta Cytol 34(2):243–247, 1990.
135. Arora VK, Misra K, Bhatia A. Cytomorphologic features of the rare epithelial-myoepithelial carcinoma of the salivary gland. Acta Cytol 34(2):239–242, 1990.
136. Abdul-Karim FW, Weaver MG. Needle aspiration cytology of an oncocytic carcinoma of the parotid gland. Diagn Cytopathol 7(4):420–422, 1991.
137. Henry-Stanley MJ, Beneke J, Bardales RH, Stanley MW. Fine needle aspiration of normal tissue from enlarged salivary glands: sialosis, or missed target? Diagn Cytopathol 13:300–303, 1995.
138. Layfield LJ, Glasgow BJ, Goldstein N, Lufkin R. Lipomatous lesions of the parotid gland: potential pitfalls in fine needle aspiration biopsy diagnosis. Acta Cytol 35(5):553–556, 1991.
139. Speight PM, Tinkler S. The salivary glands: sialosis. In: McGee J, Isaacson PG, Wright NA (eds): Oxford textbook of pathology, Vol. 2a, Pathology of Systems. Oxford: Oxford University Press, 1992, p. 1080.
140. Waldron CA. Face, lips, teeth, oral soft tissue, jaws, salivary gland and neck: enlargements related to malnutrition, hormonal disturbances, and alcoholic cirrhosis (sialosis). In: Kissane JM (ed): Anderson's pathology, Vol. 2, 9th ed. St. Louis: CV Mosby, 1990, p. 1128.
141. Ascoli V, Albedi FM, De Blasiis R, Nardi F. Sialadenosis of the parotid gland: report of four cases diagnosed by fine-needle aspiration cytology. Diagn Cytopathol 9(2):151–155, 1993.
142. Speight PM, Tinkler S. The salivary glands: salivary calculi. In: McGee J, Isaacson PG, Wright NA (eds): Oxford textbook of pathology, Vol. 2a, Pathology of Systems. Oxford: Oxford University Press, 1992, p. 1070.
143. Nagel H, Laskawi R, Büter JJ, Schröder M, Chilla R, Droese M. Cytologic diagnosis of acinic-cell carcinoma of salivary glands. Diagn Cytopathol 16:402–412, 1997.
144. Blatt IM. Studies in sialolithiasis. III. Pathogenesis, diagnosis and treatment. South Med J 57:723–729, 1964.
145. Oyafuso MS, Ikeda MK, Longatto-Filho A, Shirata NK. Fine needle aspiration cytology in the diagnosis of non-neoplastic diseases of head and neck masses. Pathologica 83(1085):311–316, 1991.
146. Cohen MB, Fisher PE, Holly EA, Ljung B-M, Löwhagen T, Bottles K. Fine needle aspiration biopsy diagnosis of mucoepidermoid carcinoma: statistical analysis. Acta Cytol 34(1):43–49, 1990.
147. Sparrow SA, Frost FA. Salivary monomorphic adenomas of dermal analogue type: report of two cases. Diagn Pathol 9:300–303, 1993.
148. Ostrzega N, Cheng L, Layfield L. Glial fibrillary acid protein immunoreactivity in fine-needle aspiration of salivary gland lesions: a useful adjunct for the differential diagnosis of salivary gland neoplasms. Diagn Cytopathol 5(1):145–149, 1989.
149. Saw D, Lau WH, Ho JHC, Chan JKC, Ng CS. Malignant lymphoepithelial lesion of the salivary gland. Hum Pathol 17:914–923, 1986.
150. Günhan O, Celasun B, Dogan N, Önder T, Pabuscu Y, Finci R. Fine needle aspiration cytologic findings in a benign lymphoepithelial lesion with microcalcifications: a case report. Acta Cytol 36(5):744–747, 1992.
151. Takahashi H, Cheng J, Fujita S, et al. Primary malignant lymphoma of the salivary gland: a tumor of mucosa-associated lymphoid tissue. J Oral Pathol Med 21(7):318–325, 1992.
152. Cameron WR, Johansson L, Tennvall J. Small cell carcinoma of the parotid: fine needle aspiration and immunochemical findings in a case. Acta Cytol 34(6):837–841, 1990.
153. Rollins CE, Yost BA, Costa MJ, Vogt PJ. Squamous differentiation in small-cell carcinoma of the parotid gland. Arch Pathol Lab Med 119:183–185, 1995.

154. Pérez-Guillermo M, Pérez JS, Espinosa Parra FJ. Asteroid bodies and calcium oxalate crystals: two infrequent findings in fine-needle aspirates of parotid sarcoidosis. Diagn Cytopathol 8:248–252, 1992.
155. Johnson FB, Oertel YC, Ammann K. Sialadenitis with crystalloid formation: a report of six cases diagnosed by fine-needle aspiration. Diagn Cytopathol 121:76–80, 1995.
156. Jayaram G, Khurana N, Basu S. Crystalloids in a cystic lesion of parotid salivary gland: diagnosis by fine-needle aspiration. Diagn Cytopathol 9:70–71, 1993.
157. Stanley MW, Lowhagen T. Mucin production by pleomorphic adenomas of the parotid gland: a cytologic spectrum. Diagn Cytopathol 6(1):49–52, 1990.
158. Murty DA, Sodhani P. Intranuclear inclusions in pleomorphic adenoma of salivary gland: a case report. Diagn Cytopathol 9:194–196, 1993.
159. Bottles K, Ferrell LD, Miller TR. Tyrosine crystals in fine needle aspirates of a pleomorphic adenoma of the parotid gland. Acta Cytol 28(4):490–492, 1984.
160. Cleveland DB, Cosgrove MM, Martin SE. Tyrosine-rich crystalloids in a fine needle aspirate of a polymorphous low-grade adenocarcinoma of a minor salivary gland: a case report. Acta Cytol 38(2):247–251, 1994.
161. Cohen MB. Editorial comments: FNAB of salivary gland. Diagn Cytopathol 9(2):224–225, 1993.
162. Viguer JM, Vicandi B, Jiménez-Heffernan JA, López-Ferrer P, Limeres MA. Fine needle aspiration of pleomorphic adenoma: an analysis of 212 cases. Acta Cytol 41:786–794, 1997.
163. Jacobs JC. Low-grade mucoepidermoid carcinoma ex pleomorphic adenoma: a diagnostic problem in fine needle aspiration biopsy. Acta Cytol 38(1):93–97, 1994.
164. Stanley MW, Horwitz CA, Bardales RH, Stern SJ, Korourian S. Basal cell carcinoma metastatic to the salivary glands: differential diagnosis in fine-needle aspiration cytology. Diagn Cytopathol 16:247–252, 1997.
165. Sciubba JJ, Brannon RB. Myoepithelioma of salivary glands: report of 23 cases. Cancer 49:562–572, 1982.
166. Eneroth CM, Zajicek J. Aspiration biopsy of salivary gland tumors: IV. Morphologic studies on smears and histologic sections from 45 cases of adenoid cystic carcinoma. Acta Cytol 13:59–63, 1969.
167. Lindberg LG, Åkerman M. Aspiration cytology of salivary gland tumors: diagnostic experience from 6 years of routine laboratory work. Laryngoscope 86:584–594, 1976.
168. Dodd LG, Caraway NP, Luna MA, Byers RM. Myoepithelioma of the parotid: report of a case initially examined by fine needle aspiration biopsy. Acta Cytol 38:417–421, 1994.
169. Domagala W, Halczy-Kowalik L, Weber K, Osborn M. Coexpression of glial fibrillary acid protein, keratin and vimentin: a unique feature useful in the diagnosis of pleomorphic adenoma of the salivary gland in fine needle aspiration biopsy smears. Acta Cytol 32(3):403–408, 1988.
170. Ritland F, Lubensky I, LiVolsi VA. Polymorphous low-grade adenocarcinoma of the parotid salivary gland. Arch Pathol Lab Med 117:1261–1263, 1993.
171. Swanson PE, Pettinato G, Lillemoe TJ, Wick MR. Gross cystic disease fluid protein-15 in salivary gland tumors. Arch Pathol Lab Med 115:158–163, 1991.
172. Landolt U, Zöbeli L, Pedio G. Letter to the editors: pleomorphic adenoma of the salivary glands metastatic to the lungs: diagnosis by fine needle aspiration cytology. Acta Cytol 34:101–112, 1990.
173. Pitman MB, Thor AD, Goodman ML, Rosenberg AE. Benign metastasizing pleomorphic adenoma of salivary gland: diagnosis of bone lesions by fine-needle aspiration biopsy. Diagn Cytopathol 8(4):384–387, 1992.
174. Stanley MW, Bardales RH, Farmer CE, et al. Primary and metastatic high-grade carcinomas of the salivary glands: a cytologic-histologic correlation study of twenty cases. Diagn Cytopathol 13:37–43, 1995.
175. Granger JK, Houn H-Y. Malignant mixed tumor (carcinosarcoma) of parotid gland diagnosed by fine-needle aspiration biopsy. Diagn Cytopathol 7:427–432, 1991.
176. Gray SR, Cornog JL, Seo IS. Oncocytic neoplasms of salivary glands: A report of fifteen cases including two malignant oncocytomas. Cancer 38:1306–1317, 1976.
177. Taxy JB. Necrotizing squamous/mucinous metaplasia in oncocytic salivary gland tumors: a potential diagnostic problem. Am J Clin Pathol 97:40–45, 1992.
178. Sauer T, Jebsen PW, Olsholt R. Cytologic features of papillary-cystic variant of acinic-cell adenocarcinoma: a case report. Diagn Cytopathol 10:30–32, 1994.
179. Levy DM, ReMine WH, Devine KD. Salivary gland calculi: pain, swelling associated with eating. JAMA 181(13):1115–1119, 1962.
180. Frierson HF, Fechner RE. Chronic sialadenitis with psammoma bodies mimicking neoplasia in a fine-needle aspiration specimen from the submandibular gland. Am J Clin Pathol 95:884–888, 1991.
181. Hayes MMM, Cameron RD, Jones EA. Sebaceous variant of mucoepidermoid carcinoma of the salivary gland: a case report with cytohistologic correlation. Acta Cytol 37(2):237–241, 1993.
182. Evans HL. Mucoepidermoid carcinoma of salivary glands: a study of 69 cases with special attention to histologic grading. Am J Clin Pathol 81:696–701, 1984.
183. Kumar N, Kapila K, Verma K. Fine needle aspiration cytology of mucoepidermoid carcinoma: a diagnostic problem. Acta Cytol 35(3):357–359, 1991.
184. Hamed G, Shmookler BM, Ellis GL, Punja U, Feldman D. Oncocytic mucoepidermoid carcinoma of the parotid gland. Arch Pathol Lab Med 118:313–314, 1994.
185. Batsakis JG, El-Naggar AK, Luna MA. "Adenocarcinoma, not otherwise specified": A diminishing group of salivary carcinomas. Ann Otol Rhinol Laryngol 101:102–104, 1992.
186. Rosai J. Ackerman's surgical pathology. St. Louis: CV Mosby Co., 1989, pp. 652–687.
187. Luna MA, Batsakis JG, Tortoledo ME, del Junco GW. Carcinomas ex monomorphic adenoma of salivary glands. J Laryngol Otol 103:756–759, 1989.
188. Stanley MW, Horwitz CA, Rollins SD, et al. Basal cell (monomorphic) and minimally pleomorphic adenomas of the salivary glands: distinction from the solid (anaplastic) type of adenoid cystic carcinoma in fine-needle aspiration. Am J Clin Pathol 106:35–41, 1996.
189. López JI, Ballestin C. Fine-needle aspiration cytology of a membranous basal cell adenoma arising in an intraparotid lymph node. Diagn Cytopathol 9:668–672, 1993.
190. Atula T, Klemi P-J, Donath K, Happonen R-P, Joensuu H, Grenman R. Pathology in focus: basal cell adenocarcinoma of the parotid: a case report and review of the literature. J Laryngol Otol 107:862–864, 1993.
191. Lucarini JW, Sciubba JJ, Khettry U, Nasser I. Terminal duct carcinoma. Recognition of a low-grade salivary adenocarcinoma. Arch Otolaryngol Head Neck Surg 120(9):1010–1015, 1994.
192. Nishimura T, Furukawa M, Kawahara E, Miwa A. Differential diagnosis of pleomorphic adenoma by immunohistochemical means. J Laryngol Otol 105:1157–1160, 1991.
193. Simpson RH, Clarke TJ, Sarsfield PT, Gluckman PG, Babajews AV. Polymorphous low-grade adenocarcinoma of the salivary glands: a clinicopathological comparison with adenoid cystic carcinoma (see comments). Histopathology 19(2):121–129, 1991.
194. Stanley MW, Horwitz CA, Henry MJ, Burton LG, Lowhagen T. Basal-cell adenoma of the salivary gland: a benign adenoma that cystologically mimics adenoid cystic carcinoma. Diagn Cytopathol 4(4): 242–246, 1988.
195. Löwhagen T, Tani EM, Skoog L. Salivary glands and rare head and neck lesions. In Bibbo M (ed.): Comprehensive Cytopathology (pp. 627–634). Philadelphia: W.B. Saunders, 1991.
196. Pisharodi LR. Basal cell adenocarcinoma of the salivary gland:

diagnosis by fine-needle aspiration cytology. Am J Clin Pathol 103:603–608, 1995.

197. Williams SB, Ellis GL, Auclair PL. Immunohistochemical analysis of basal cell adenocarcinoma. Oral Surg Oral Med Oral Pathol 75:64–69, 1993.
198. Scher RL, Feldman PS, Levine PA. Small-cell carcinoma of the parotid gland with neuroendocrine features. Arch Otolaryngol Head Neck Surg 114:319–321, 1988.
199. Gnepp DR, Wick MR. Small cell carcinoma of the major salivary glands: an immunohistochemical study. Cancer 66:185–192, 1990.
200. Koss LG, Spiro RH, Hajdu S. Small cell (oat cell) carcinoma of the minor salivary glands. Cancer 30:737–741, 1972.
201. Mullins RK, Thompson SK, Coogan PS, Shurbaji MS. Paranuclear blue inclusions: an aid in the cytopathologic diagnosis of primary and metastatic pulmonary small-cell carcinoma. Diagn Cytopathol 10(4):332–335, 1994.
202. Currens HS, Sajjad SM, Lukeman JM. Aspiration cytology of oat-cell carcinoma metastatic to the parotid gland. Acta Cytol 26(4):566–567, 1982.
203. Thompson MB, Nestok BR, Gluckman JL. Fine needle aspiration cytology of lymphoepitheliomalike carcinoma of the parotid gland: a case report. Acta Cytol 38(5):782–786, 1994.
204. Günhan O, Celasun B, Safli M, et al. Fine needle aspiration cytology of malignant lymphoepithelial lesion of the salivary gland: a report of two cases. Acta Cytol 38:751–754, 1994.
205. Malberger E, Tillinger R, Lichtig C. Diagnosis of basal-cell carcinoma with aspiration cytology. Acta Cytol 28(3):301–304, 1984.
206. Gherardi G, Marveggio C, Stiglich F. Parotid metastasis of Merkel cell carcinoma in a young patient with ectodermal dysplasia: diagnosis by fine needle aspiration cytology and immunocytochemistry. Acta Cytol 34(6):831–836, 1990.
207. Domagala W, Lubinski J, Lasota J, Giryn I, Weber K, Osborn M. Neuroendocrine (Merkel-cell) carcinoma of the skin: cytology, intermediate filament typing and ultrastructure of tumor cells in fine needle aspirates. Acta Cytol 31(3):267–274, 1987.
208. Pettinato G, De Chiara A, Insabato L. Diagnostic significance of intermediate filament buttons in fine needle aspirates of neuroendocrine (Merkel cell) carcinoma of the skin. Acta Cytol 33:420–421, 1989.
209. Whitlatch SP. Psammoma bodies in fine-needle aspiration biopsies of acinic cell tumor. Diagn Cytopathol 2(3):268–269, 1986.
210. McCutcheon JM, Mancer K, Dardick I. Acinic cell tumour: a metastasis in the lung diagnosed by electron microscopy of aspirated material. Cytopathology 3(6):373–377, 1992.
211. Sherman ME, Magro C, Berry Y, Szyfelbein WM. Oncocytic nodule: an unusual case of a submaxillary gland mass in an elderly patient. Acta Cytol 34(6):827–829, 1990.
212. Laforga JB, Aranda FI. Oncocytic carcinoma of parotid gland: fine-needle aspiration and histologic findings. Diagn Cytopathol 11:376–379, 1994.
213. Austin MB, Frierson HF, Feldman PS. Oncocytoid adenocarcinoma of the parotid gland: cytologic, histologic and ultrastrutural findings. Acta Cytol 31(3):351–356, 1987.
214. Brandwein MS, Huvos AG. Oncocytic tumors of major salivary glands: a study of 68 cases with follow-up of 44 patients. Am J Surg Pathol 15(6):514–528, 1991.
215. Rajan PB, Wadehra V, Hemming JD, Hawkesford JE. Fine needle aspiration cytology of malignant oncocytoma of the parotid gland—a case report. Cytopathology 5(2):110–113, 1994.
216. Haba R, Kobayashi S, Miki H, et al. Polymorphous low-grade adenocarcinoma of submandibular gland origin. Acta Pathol Jpn 43(12):774–778, 1993.
217. Batsakis JG, El-Naggar AK. Terminal duct adenocarcinomas of salivary tissues. Ann Otol Rhinol Laryngol 100:251–253, 1991.
218. Dee S, Masood S, Isaacs JH, Hardy NM. Cytomorphologic features of salivary duct carcinoma on fine needle aspiration biopsy: a case report. Acta Cytol 37(4):539–542, 1993.
219. Fyrat P, Cramer H, Feczko JD, et al. Fine-needle aspiration biopsy of salivary duct carcinoma: report of five cases. Diagn Cytopathol 16:526–530, 1997.
220. Elsheikh TM, Bernacki EG, Pisharodi L. Fine-needle aspiration cytology of salivary duct carcinoma. Diagn Cytopathol 1:47–51, 1994.
221. Gutmann EJ. Lymphangioma presenting as a primary parotid neoplasm in an adult: report of a case with the diagnosis suggested by fine needle aspiration biopsy. Acta Cytol 38(5):747–750, 1994.
222. Stanley MW, Skoog L, Tani EM, Horwitz CA. Nodular fasciitis: spontaneous resolution following diagnosis by fine needle aspiration. Diagn Cytopathol 9:322–324, 1993.
223. Herold J, Nicholson AG. Fine needle aspiration cytology in the diagnosis of amyloid in the submandibular gland. Br J Oral Maxillofac Surg 30(6):393–394, 1992.
224. Franquemont DW, Mills SE. Plasmacytoid monomorphic adenoma of salivary glands: absence of myogenous differentiation and comparison of spindle cell myoepithelioma. Am J Surg Pathol 17(2):146–153, 1993.
225. Torlakovic E, Ames E, Manivel JC, Stanley MW. Benign and malignant neoplasms of myoepithelial cells: cytologic findings. Diagn Cytopathol 9:655–660, 1993.
226. Bondeson L, Andreasson L, Olsson M, Rausing A. Salivary gland anlage tumor: cytologic features in a case examined by fine-needle aspiration. Diagn Cytopathol 16:518–521, 1997.
227. Dehner LP, Valbuena L, Perez-Atayde A, Reddick RL, Askin FB, Rosai J. Salivary gland anlage tumor ("congenital pleomorphic adenoma"): a clinicopathologic, immunohistochemical and ultrastructural study of nine cases. Am J Surg Pathol 18:25–36, 1994.

CHAPTER 5

Bone

Fine needle aspiration (FNA) biopsy of bone is being performed with increasing frequency in the United States. It is, therefore, paramount that pathologists have more than a rudimentary understanding of the cytologic features of bone tumors. It is our intent to provide a practical overview of the utility of FNA biopsy in the diagnosis of bone tumors. In addition, accurate bone tumor diagnosis, whether related to surgical pathology curettage specimens or FNA biopsy cytopathology smears, requires a multidisciplinary approach. Therefore, an integrated assessment of clinical features by the orthopedic surgeon, radiographic interpretation by the radiologist, and cytologic and/or histologic confirmation by the pathologist is required.

IMPORTANCE OF RADIOLOGIC INTERPRETATION

One must not underestimate the value of radiographic interpretation in the diagnosis of bone tumors. It is often preferable to begin with radiographs for several reasons. First, accurate radiographic interpretation may allow recognition of certain benign tumors or tumor-like conditions that do not require further investigation or therapy. One such example is metaphyseal fibrous defect; unless occupying greater than 50% of the diameter of the bone or complicated by fracture, this could be left untreated, because most of these lesions completely resolve on their own. Second, when necessary, radiographs help the orthopedic surgeon choose an optimal site for biopsy. Third, for the pathologist, assessment of such features as location, size, shape, pattern of bone destruction, matrix, margins, periosteal reactions, and concomitant soft-tissue abnormalities usually correlates with the aggressiveness or biologic behavior of the lesion and, therefore, allows a more specific diagnosis. In this manner, radiographs serve as the "gross specimen" for the pathologist. Interestingly, most of the cytopathology literature regarding FNA biopsy of bone lesions has largely relied on cytomorphologic features alone, giving little emphasis to the importance of clinicoradiologic correlation. Seventy-five years ago, Ewing stated, "The gross anatomy (as revealed in radiographs) is often a safer guide to correct clinical conception of the disease than the variable and uncertain nature of a small piece of tissue."[1] However, as with all tests, limitations do exist regarding the interpretation of radiographs. In some cases, the radiographic appearance of a tumor may even be misleading. For example, an aneurysmal bone cyst may show very aggressive features radiologically; conversely, low-grade central osteosarcomas may appear innocuous. For these reasons, radiologic interpretations should not supplant close scrutiny of cytologic preparations and/or histologic specimens for establishing a bone tumor diagnosis.

FNAB CONSIDERATIONS

We, as well as others, define "fine needle" aspiration as using 22-gauge or smaller diameter needles for diagnostic purposes.[2] At our institution, we routinely use 23- and 25-gauge needles for FNA biopsy of bone tumors, usually in an outpatient setting. Some lesions that would be inaccessible to thin-needle penetration may be accessible to larger "intermediate"-diameter (18–20-g) needles with a cutting edge. Some authors have used the latter in conjunction with fluoroscopic guidance to evaluate bone tumors that would normally be inaccessible by FNA biopsy.[3,4]

Some bone lesions do not readily lend themselves to analysis by "strictly defined" FNA biopsy. Most of these lesions are completely intraosseous and are usually surrounded by an intact, noninvolved cortex. Not surprisingly, these features are usually indicative of benignancy and include entities such as metaphyseal fibrous defect (nonossifying fibroma), benign (atypical) fibrous histiocytoma, fibrous dysplasia, osteofibrous dysplasia, hemangioma, osteochondroma, and enchondroma. However, similar problems regarding FNA biopsy access are not restricted to benign bone lesions but may be encountered in low-grade (less aggressive) malignancies such as low-grade central osteosarcomas and, in our experience, most examples of low-grade chondrosarcoma. In addition to their frequent intraosseous location, osteoid osteoma and osteoblastoma are also not usually evaluated by FNA biopsy because of their densely ossified matrix. Table 5–1 summarizes bone lesions that are generally not accessible for FNA biopsy evaluation. In contrast, bone tumors or tumor-like lesions that are accessible to analysis by FNA biopsy generally form large lytic defects, have a sufficiently thinned or perforated cortex, may be associated with concomitant soft-tissue extension, and/or tend to be composed of a softer, less mineralized matrix. Such features are more commonly encountered in more locally aggressive and/or malignant bone tumors. Occasionally, relatively inaccessible, intraosseous lesions complicated by pathologic fracture may also be analyzed by FNA biopsy.

TABLE 5–1. Bone Lesions Generally Inaccessible for FNA Biopsy Analysis

Metaphyseal fibrous defect (nonossifying Fibroma)
Benign (atypical) fibrous histiocytoma
Fibrous dysplasia
Osteofibrous dysplasia
Hemangioma
Osteochondroma
Enchondroma
Osteoid osteoma
Osteoblastoma
Low-grade central osteosarcoma
Low-grade chondrosarcoma

PERTINENT BONE TERMINOLOGY

Before proceeding to a more detailed discussion of the pathology of bone tumors, one should be acquainted with basic bone terminology. Although it is not unusual for the terms osteoid, woven bone, and lamellar bone to be used interchangeably, these terms actually represent specific forms of bone matrix. By definition, osteoid is unmineralized (noncalcified) bone. Under normal circumstances, it is seen as a thin, pale pink-staining layer rimming active mature bone in hematoxylin and eosin-stained sections. The width of the layer is dependent on the rate of bone formation and that of mineralization. Not surprisingly, osteoid tends to be prominent in areas of healing fracture (callus) and absent in areas of inactivity. In bone tumors, especially osteosarcomas, osteoid tends to be produced by the neoplastic cells in a lacelike or pericellular pattern. Woven bone is a form of immature, mineralized bone usually organized in small trabeculae and spicules. Its collagen fibers are described as having a feltlike pattern, and osteocytes tend to be randomly distributed. Lamellar bone is mature mineralized bone organized concentrically within the cortex (compact bone) and longitudinally in trabeculae of the medullary cavity (cancellous or spongy bone). The collagen fibers within the latter areas are arranged in distinct parallel sheets and bundles surrounding uniformly distributed osteocytes, with their long axes parallel to the collagen lamellae. With the exception of a thin rim of osteoid lining an occasional active mature bony trabecula, the presence of osteoid or woven bone always represents a pathologic process. Further, as a general rule, only benign tumors or tumor-like processes and rare examples of low-grade malignant neoplasms (i.e., periosteal osteosarcoma, low-grade central osteosarcoma) are capable of producing mature lamellar bone. Hence, most osteosarcomas, which are usually of high grade, are not capable of producing mature lamellar bone; instead, they form predominately osteoid and/or woven bone. For this reason, the softer matrix of most examples of osteosarcoma is easily evaluated by FNA biopsy, whereas osteoid osteoma and most examples of osteoblastoma, which contain a much greater amount of mineralized matrix, are generally inaccessible.

Confusion may also be caused by the improper use of the terms chondroid and cartilage. In hematoxylin and eosin-stained sections, chondroid is a pink-brown, amorphous material that resembles osteoid but is generally produced by cartilaginous neoplasms, the prototype being chondroblastoma. As with its mature counterpart, hyaline cartilage, chondroid is deposited as sheets, plates, and nodules and lacks the lacelike pattern typically seen with osteoid-producing tumors. True "chondroid" is not usually seen in chondrosarcomas, which are generally associated with the production of mature hyaline cartilage. The latter represents a blue- to gray-staining material that, similar to chondroid, is also arranged in plates and nodules when produced by tumors and, in normal circumstances, is associated with the articular surfaces of bone and the growth plate. Hyaline cartilage is rich in proteoglycans, which are responsible for imparting a glistening blue-gray and somewhat mucinous appearance to the material. In contrast, chondroid lacks abundant proteoglycans and thus appears pink to brown.

Although distinguishing the variety of matrices on surgical specimens is usually easily accomplished, it is important for practitioners of FNA biopsy to understand the limitations of making cytologic distinctions. We believe bone lesions should be evaluated by both a modified rapid Wright-Giemsa stain (Diff-Quik stain) and a Papanicolaou stain. In our experience, matrix material is better visualized on air-dried, Diff-Quik-stained smears. With both Diff-Quik and Papanicolaou staining, osteoid, mineralized bone, chondroid, hyaline cartilage, and even dense collagen appear tinctorially similar, having a magenta to metachromatic appearance with the former stain and yellow-green to brown with the latter. Chondroid and hyaline cartilage tend to have a more fibrillar quality than osteoid and bone and, as previously mentioned, are more commonly arranged in nodules and plates. Nevertheless, extensive morphologic overlap exists between the different forms of matrix, making reliable distinction in most cases difficult or impossible. For this reason, close scrutiny of the individual cells (cytomorphology), degree of hypercellularity, and clinicoradiologic correlation are more important for rendering a specific bone tumor diagnosis. Even when FNA biopsy yields nodules of metachromatic material with distinct lacunae characteristic of hyaline cartilage, this does not allow distinction between a chondroblastic osteosarcoma and a chondrosarcoma. In addition, osteosarcomas may produce osteoid, mineralized bone, chondroid, and even hyaline cartilage; nevertheless, several examples of chondroblastic osteosarcomas have been initially misdiagnosed as chondrosarcomas on FNA biopsy.[5–7] When matrix material is associated with numerous pleomorphic tumor cells, an osteosarcoma is likely, regardless of the appearance of the matrix material. Chondrosarcomas, which are usually of low grade, generally have a more abundant amount of matrix material relative to the malignant cells. The latter usually show far less atypia than most osteosarcomas.[8] Nevertheless, although matrix material may not be easily subtyped by FNA biopsy, its presence is important to recognize and helps to classify bone tumors more accurately.

BONE CYSTS

Simple Bone Cyst

Simple bone cysts (solitary bone cysts, unicameral bone cysts) represent unilocular, intramedullary cysts typically seen within the first and second decades of life.[9] By definition, simple bone cysts involve the metaphysis, with absent to slight expansion of the cortex. The radiologic appearance may appear trabeculated and even multilobulated (Fig. 5–1). Aspiration biopsy generally yields a colorless to amber, virtually acellular fluid; subsequent curettage generates very little tissue, which usually consists only of a thin fibrous tissue lining, often associated with a fibrin or cementum-like material. In the absence of a confirmatory open curettage, if the fluid aspirated is colorless to amber and the clinical and radiologic features are consistent with a simple bone cyst, FNA biopsy may be considered diagnostic. However, the presence of bloody fluid mandates further investigation.[10]

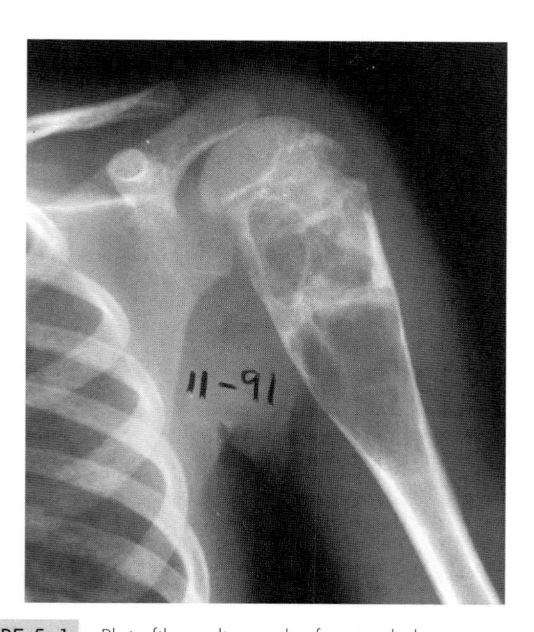

FIGURE 5–1. Plain film radiograph of a simple bone cyst in the proximal humeral metaphysis of a 6-year-old boy. The lesion appears sharply marginated, lytic, and trabeculated but shows little to no cortical expansion. This classic radiographic appearance, coupled with aspiration of clear to amber acellular fluid, may be considered diagnostic of simple bone cyst. The presence of bloody fluid mandates further investigation.

Aneurysmal Bone Cyst

Aneurysmal bone cysts are characteristically expansile, destructive, and multiloculated lesions, usually arising within the metaphyseal region of the long bones or dorsal elements of the vertebral spine (Fig. 5–2).[11] Most patients are in the first or second decade of life. FNA biopsy generally yields predominantly blood or blood-tinged fluid with few scattered stromal fibroblasts, histiocytes, and rare multinucleated osteoclast-type giant cells (Fig. 5–3). In our experience, surgical curettage is almost always necessary to exclude other neoplastic processes that may be engrafted onto the aneurysmal bone cyst (i.e., chondroblastoma or giant cell tumor), as well as, more ominously, telangiectatic osteosarcoma.[8,12] Histologically, aneurysmal bone cysts are characterized by multiloculated cavernous spaces filled with blood and lined by fibrous to fibromyxoid septa with scattered osteoclast-type giant cells and hemosiderin-laden macrophages. Osteoid formation is common, and occasionally chondroid is observed.[11] As a result, a few fragments of matrix material may be evident on FNA biopsy.

OSSEOUS TUMORS

Osteoid Osteoma

Osteoid osteomas represent sharply circumscribed osteoblastic proliferations, generally < 1 cm in greatest dimension, usually arising within the appendicular skeleton, especially the long bones of the extremities. Most examples occur in patients in the second decade of life. Patients frequently complain of pain, which is often described as worse at night and dramatically relieved by aspirin.[13] The radiographic appearance is fairly characteristic, with dense cortical sclerosis surrounding a small central lucency (nidus). Histologically, the nidus is composed of anastomosing strands and trabeculae of osteoid and woven bone within a highly vascularized fibrous stroma. Osteoblasts prominently rim the anastomosing trabeculae. Because FNA biopsy is not successful in sampling osteoid osteomas, we are unaware of any reports describing the diagnosis of osteoid osteoma by FNA biopsy.

Osteoblastoma

Osteoblastoma, like osteoid osteoma, is a benign osteoblastic neoplasm that histologically appears identical to osteoid osteoma but is generally larger (> 1 cm), lacks the characteristic radiographic and clinical findings of osteoid osteoma, and shows a predilection for the axial skeleton, in particular the skull and craniofacial region and the dorsal elements of the vertebral spine.[14] Osteoblastoma may be referred to as cementoblastoma when arising around the root of a tooth. Less commonly, osteoblastoma involves the appendicular skeleton. When osteoblastoma involves the long bones of the extremities, metaphyseal or diaphyseal localization is typical.[15] Radiographically, os-

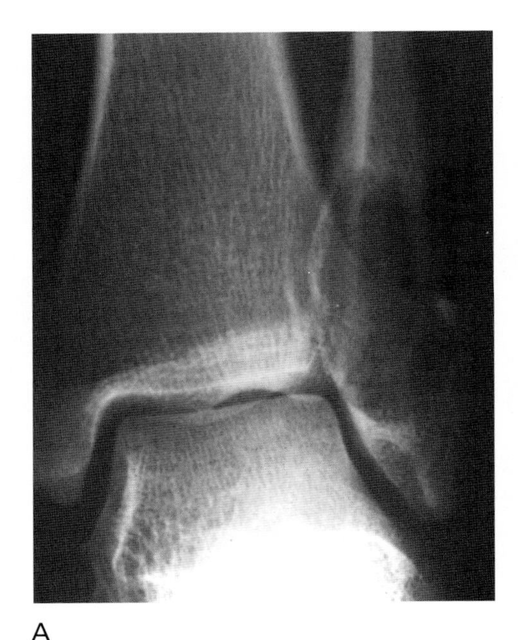

A

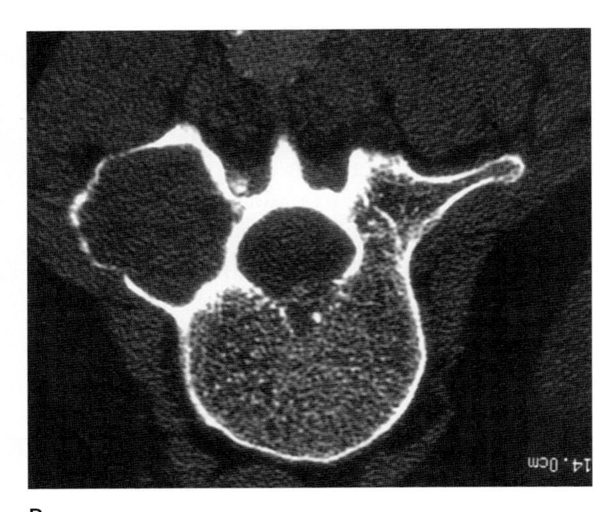

B

FIGURE 5–2. **(A)** Plain film radiograph of an aneurysmal bone cyst forming an expansile lytic lesion in the distal fibula metaphysis of a 17-year-old girl. **(B)** Aneurysmal bone cyst frequently involves the dorsal elements of the spine. This 60-year-old man had a lytic and expansile lesion involving the transverse process of the lumbar (L1) vertebra. Note the peripheral rim of ossification. (Computed tomography)

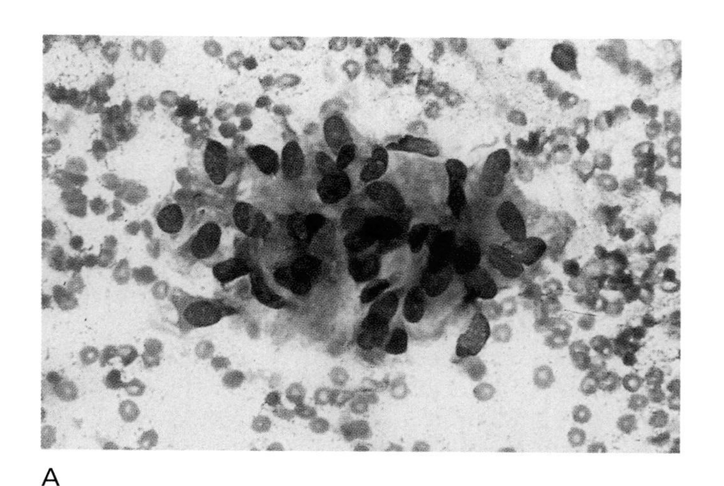

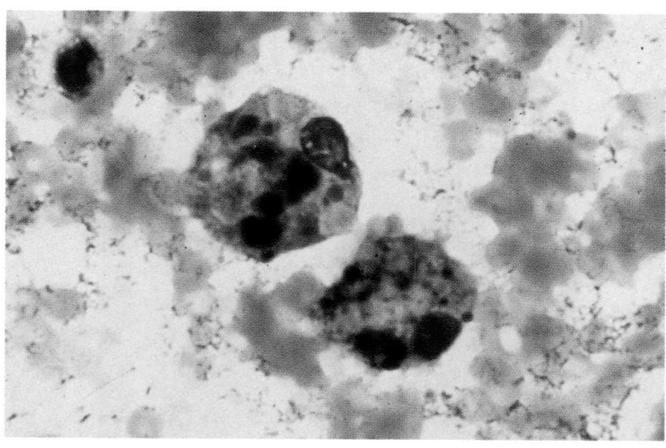

C

FIGURE 5–3. (**A**) FNA biopsy of aneurysmal bone cyst showing a rare aggregate of uniform-appearing stromal fibroblasts. (**B**) Pigmented histiocytes within a bloody aspirate are also a common finding. (**C**) Osteoclast-type giant cells are a nonspecific finding and may be seen in a variety of bone tumors and tumor-like lesions. (Diff-Quik)

teoblastomas are usually well circumscribed, expansile, and often radiolucent. Calcifications within the lesion are frequent.[15] Such radiographic features, although nonspecific, usually indicate a benign process; however, occasional examples, especially recurrent tumors, may appear quite aggressive. Most cases of osteoblastoma, although lacking the sclerotic cortex of osteoid osteoma, have a sufficiently ossified matrix to preclude adequate sampling by FNA biopsy. In those rare instances in which FNA biopsy has been performed, the cytologic evaluation has been nondiagnostic or nonspecific.[16] Other investigators have been more successful with FNA biopsy, showing evidence of benign-appearing osteoblasts intimately associated with matrix production.[3] The osteoblasts were described as usually mononucleated but occasionally binucleated, with abundant cytoplasm and a mostly round, eccentric nucleus. Mitotic figures were absent.[3] However, mitotic figures may appear in the stromal cells but are never present in the osteoblasts.[15] Further, these authors used larger "intermediate"-bore needles to obtain the samples and not "fine"-needles *per se.* Nevertheless, we would caution against attempting FNA biopsy diagnoses in suspected lesions of osteoblastoma, especially if the clinical and radiologic features are the least atypical. Osteoblastoma-like osteosarcomas may appear quite similar to osteoblastomas; the presence of clear-cut permeation of surrounding bone, a feature not generally observed with FNA biopsy, is suggestive of osteosarcoma. Further, the osteoblasts of osteoblastoma do not appear appreciably different from the reactive osteoblasts associated with a periosteal reaction or callus.

Osteosarcoma

Conventional intramedullary osteosarcoma represents a malignant neoplasm of bone that displays a variety of histologic appearances, linked by the production of osteoid and/or bone by the malignant tumor cells. Most osteosarcomas arise within the metaphyseal regions of the long bones of the extremities, especially around the knee.[17] Patients are usually diagnosed in the first or second decade of life. Extraskeletal osteosarcoma and examples of skeletal osteosarcoma arising in association with pre-existing conditions (e.g., bone infarcts, Paget's disease, prior irradiation) tend to occur in older age groups (fifth

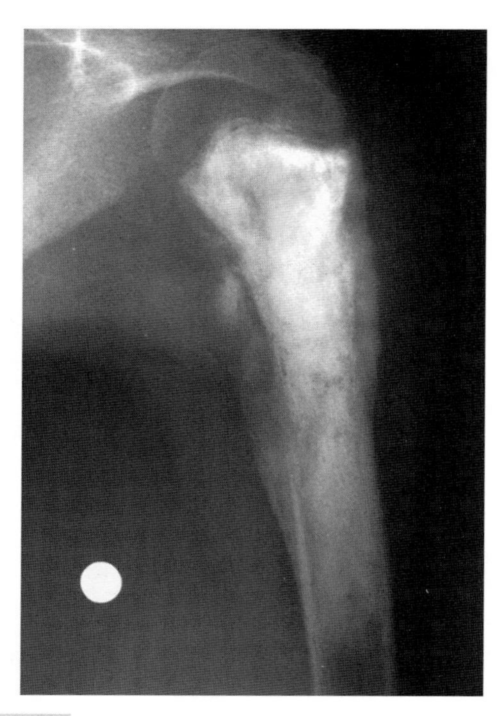

FIGURE 5–4. Plain film radiograph of high-grade intramedullary osteosarcoma involving the proximal humeral metaphysis in a 10-year-old boy. The ossified tumor shows aggressive radiologic features with evidence of permeation, periosteal reactions (Codman's angle), and soft-tissue extension.

decade or older).[18,19] Radiologically, a permeative and aggressive appearance is typical and characterized by ill-defined margins, periosteal reactions, and concomitant soft-tissue extension (Fig. 5–4). Osteosarcomas may appear lytic or sclerotic or show varying combinations of the two. FNA biopsy generally yields hypercellular smears composed of spindled to round, pleomorphic, obviously malignant tumor cells, often associated with and/or admixed with a prominent metachromatic matrix material (Fig. 5–5).[15] Although the tumor cells vary considerably in size and shape, most appear large and round to polygonal, but occasional spindled forms also are found. The nuclei are often eccentric, have irregular nuclear membranes with coarse chromatin, and contain one or more nucleoli. Binucleated and multinucleated forms are usually observed. The majority of the tumor cells have moderate amounts of dense cytoplasm. Subclassifying the matrix material (i.e., osteoid, dense collagen, or cartilage) is usually of little help in diagnosing the lesion as an osteosarcoma or chondrosarcoma.[8,17] Many cases of histologically confirmed high-grade osteosarcoma show no evidence of matrix production on FNA biopsy smears (Fig. 5–6).[17] Typical clinical and radiologic findings in conjunction with hypercellular smears of obviously pleomorphic tumor cells are usually diagnostic of high-grade osteosarcoma. At our institution, we render a definitive diagnosis of osteosarcoma by FNA biopsy only in cases where the clinical and radiologic features are characteristic and the cytologic smears show evidence of a pleomorphic sarcoma, with or without matrix production. Using this approach for a 3-year period, we have rendered on site diagnoses of high-grade osteosarcoma in thirteen primary tumors. We have had only one unsatisfactory example, but no misdiagnoses. Whether or not such rare primary bone lesions as fibrosarcoma or malignant fibrous histiocytoma can be reliably separated from high-grade fibroblastic osteosarcomas solely on FNA biopsy remains to be seen. At our institution, as well as others, primary fibrosarcomas and malignant fibrous histiocytomas of bone are treated as high-grade osteosarcomas (i.e., preoperative chemotherapy and, if feasible, limb salvage surgery).[17]

In contrast, there is considerably greater difficulty in making the diagnosis of low-grade osteosarcoma by FNA biopsy. Such cases may appear innocuous radiographically and are not usually accessible to FNA biopsy. For this diagnosis, we prefer open histologic biopsy, because the distinction between entities such as fibrous dysplasia and low-grade intraosseous osteosarcoma may depend, at least partially, on the presence of permeation of surrounding bone by the neoplastic cells.[20] Although our experience is limited, it is also probably worthwhile to perform open biopsy on all surface (juxtacortical) lesions, where the differential diagnosis includes periosteal osteosarcoma, juxtacortical chondrosarcoma, and juxtacortical myositis ossificans. The histologic appearance of parosteal osteosarcoma is virtually identical to that of low-grade intraosseous osteosarcoma, consisting of slender trabeculae of mature lamellar bone within a rather bland-appearing fibrous stroma. Periosteal osteosarcoma is essentially a juxtacortical chondroblastic osteosarcoma. Whether it can be reliably distinguished from juxtacortical chondrosarcoma by FNA biopsy is not certain.

Cytomorphologic Features of Osteosarcoma

- Moderately to markedly cellular smears with individually dispersed and small cohesive groups of cells
- Variably sized, mostly large cells ranging from round and polygonal to spindled
- Moderate to marked nuclear pleomorphism with coarse chromatin and one or more nucleoli; binucleated and multinucleated forms are frequent
- Variable amounts of delicate strands and amorphous fragments of matrix material

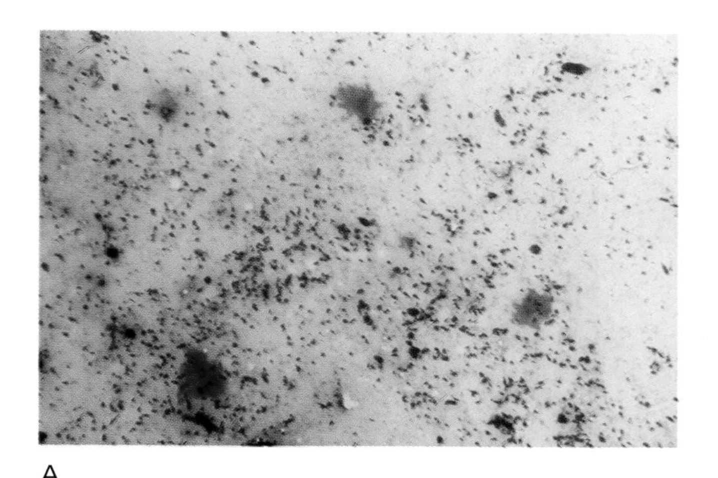

A

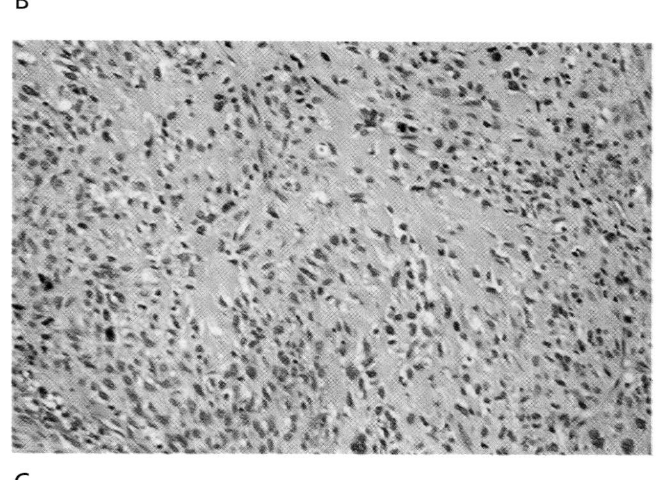

B

C

FIGURE 5–5. **(A)** FNA biopsy of osteosarcoma generally yields hypercellular smears composed of mostly solitary spindle-shaped tumor cells and fragments of matrix material. **(B)** High-power view of metachromatic matrix material and moderately pleomorphic, spindled to polygonal tumor cells. The matrix material is tinctorially similar to chondroid, hyaline cartilage, and dense collagen. Although reliable separation among the latter matrices may not be possible with cytologic examination, this morphologic appearance, coupled with the typical clinical and radiologic features, is diagnostic of osteosarcoma. (Diff-Quik) **(C)** Classic-appearing high-grade osteoblastic osteosarcoma with "lacelike" osteoid deposition. (Hematoxylin and eosin). (From Kirkpatrick SE, Pike EJ, Geisinger KR, Ward WG. Chondroblastoma of bone: use of fine needle aspiration biopsy and potential diagnostic pitfalls. Diagn Cytopathol 16:65, 1997. Copyright © 1997 by John Wiley & Sons, Inc. Reprinted by permission of Wiley-Liss, Inc., a division of John Wiley & Sons, Inc.)

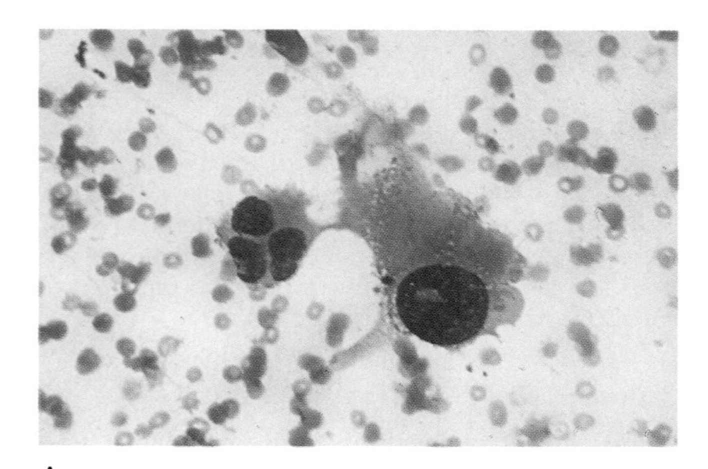

A

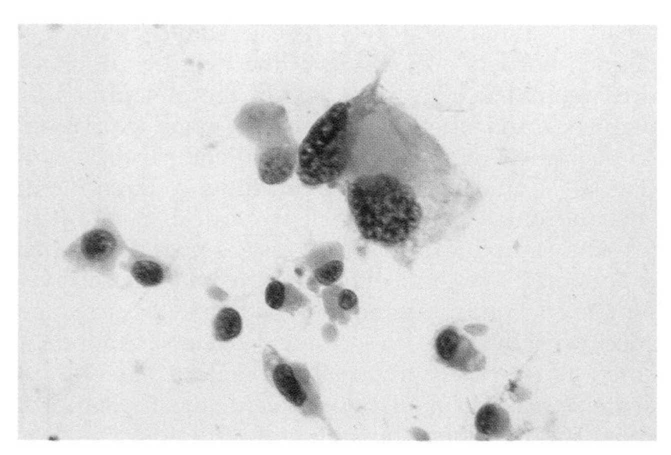

B

FIGURE 5–6. Many cases of osteosarcoma lack identifiable matrix production and are instead composed of mostly solitary and small loose clusters of anaplastic tumor cells. Such examples may be difficult if not impossible to distinguish from malignant fibrous histiocytoma of bone. (Diff-Quik and Papanicolaou)

CARTILAGINOUS TUMORS

Enchondroma

Enchondromas are benign, intramedullary neoplasms composed of lobules of hyaline cartilage. The majority arise within the appendicular skeleton, especially the small tubular bones of the hands and feet. Although occurring at any age, most patients are in the second or third decade of life at the time of diagnosis. Complaints of localized pain may be present, but many patients are asymptomatic. Radiologically, enchondromas show characteristic lytic lesions with benign features, including sharply-defined margins. Intramedullary collections of punctate to stippled and ringlike calcifications are frequently present. Because the cortex surrounding enchondromas is often intact, sampling by FNA biopsy is usually not possible. On Diff-Quik staining, reported accessible examples (usually involving "intermediate"-bore cutting needles with fluoroscopic guidance) of enchondroma show lobules of matrix and scattered small and uniform-appearing chondrocytes within distinct lacunae.[4] Binucleated cells and mitotic figures are generally absent. Correlation with radiographic findings is essential to exclude low-grade chondrosarcoma.[4,21] We do not analyze suspected cases of enchondroma by FNA biopsy.

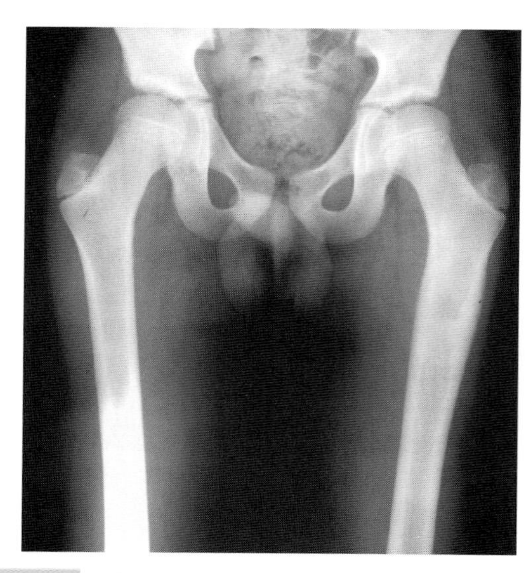

FIGURE 5–7. Chondrosarcoma arising within the diaphysis of the left proximal femur in an 11-year-old boy. Most chondrosarcomas occur in older patients (older than 40 years of age). However, the combination of cortical expansion and thickening coupled with intramedullary calcifications is characteristic of low-grade chondrosarcoma. No soft-tissue extension of periosteal reaction is evident. Such cases are inaccessible to usual FNAB sampling.

Chondrosarcoma

Chondrosarcomas represent malignant neoplasms of hyaline cartilage, typically arising within the pelvis and long tubular bones of the extremities in patients in the fifth to seventh decades of life.[22] Almost always, patients with chondrosarcoma complain of localized pain in the affected region. Radiologically, most examples appear uniformly calcified with sharp margination, although purely lytic examples with less defined borders occur. Within the long bones, chondrosarcomas usually arise within the metaphysis or diaphysis and often show areas of bone expansion, cortical thickening, and/or cortical endosteal erosion (Fig. 5–7). The latter feature precludes a diagnosis of enchondroma. When sufficient cortical thinning is present, sampling by FNA biopsy is possible. In our experience, low-grade chondrosarcomas closely resemble enchondromas not only in biopsy or curettage specimens but also in cytologic smears. Abundant amounts of mostly hypocellular matrix material are evident, accompanied by scattered, mostly uniform chondrocytes, often showing very little to no nuclear atypia (Fig. 5–8). Most tumor cells are uninucleated and larger than those seen in enchondroma and contain an oval to round, slightly hyperchromatic nucleus. Binucleated forms are occasionally observed. Nucleoli range from small to inconspicuous. Myxoid change or degeneration is also a helpful diagnostic feature in chondrosarcoma. On cytologic smears, myxoid change appears as a metachromatic granular background film or precipitate (Fig. 5–9). Nevertheless, clinicoradiologic correlation is of the utmost importance in establishing a diagnosis.[4,21] Higher-grade (grade 2 or 3) examples of chondrosarcoma are generally more cellular, exhibit more nuclear pleomorphism and atypia, and are much more easily recognized as malignant (Fig. 5–10).[4,7] The main differential diagnosis in the latter includes chondroblastic osteosarcoma. Layfield et al.[6] documented two examples of chondroblastic osteosarcomas initially misdiagnosed as chondrosarcomas. Nevertheless, we believe this diagnostic pitfall can usually be avoided if one pays attention to the older age of most patients with chondrosarcoma and the distinctive radiologic features. In limited histologic specimens (i.e., cell blocks), chondroblastic osteosarcomas tend to show increased cellularity at the periphery of the cartilaginous lobules and greater nuclear atypia compared to chondrosarcoma (Fig. 5–11).[23]

Cytomorphologic Features of Chondrosarcoma

- Abundant fragments of mostly hypocellular matrix, often accompanied by a background myxoid granular film
- Slightly to moderately cellular smears with mostly individually dispersed tumor cells, some present within distinct lacunae

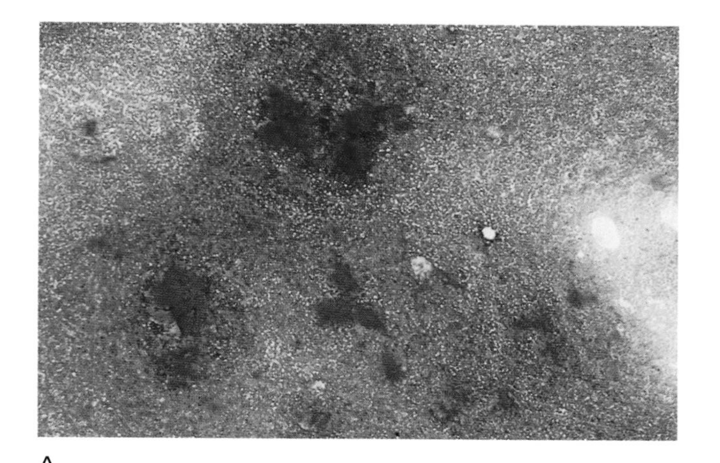

A

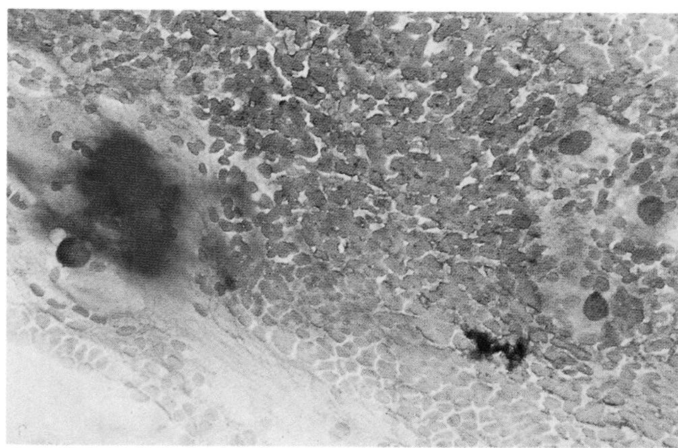

B

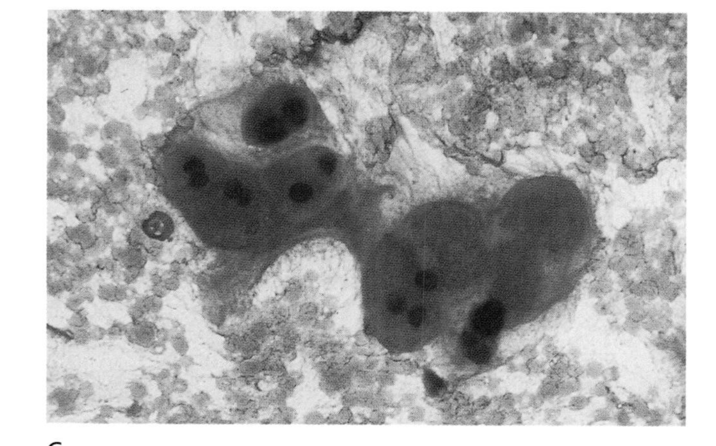

C

FIGURE 5–8. (**A**) FNA biopsy of low-grade chondrosarcoma generally reveals paucicellular smears and abundant fragments of matrix material. (**B**) Scattered, mostly uniform neoplastic chondrocytes show round to ovoid nuclei and clear cytoplasm. (**C**) Occasionally, the nonspecific matrix material is more easily identified as cartilage because of the presence of clearly visible lacunae. Nevertheless, this finding does not allow for distinction between enchondroma or chondroblastic osteosarcoma. (**A,B,C,** Diff-Quik)

- Mostly uniform but large, round tumor cells with distinct cell borders, round nucleus, and slightly coarse chromatin
- Binucleated cells sometimes observed

Approximately 10% of patients with chondrosarcoma develop dedifferentiated chondrosarcoma.[24] In general, the skeletal distribution of dedifferentiated chondrosarcoma corresponds to that of conventional chondrosarcoma. Radiographs usually reveal an intraosseous lytic lesion with an associated large soft-tissue mass; the latter is only rarely associated with punctate calcified densities. Focally, juxtaposed to the larger lytic mass is usually a smaller, more calcified "chondrosarcomatous" lesion. Although the vast majority appear to arise from more centrally located in-

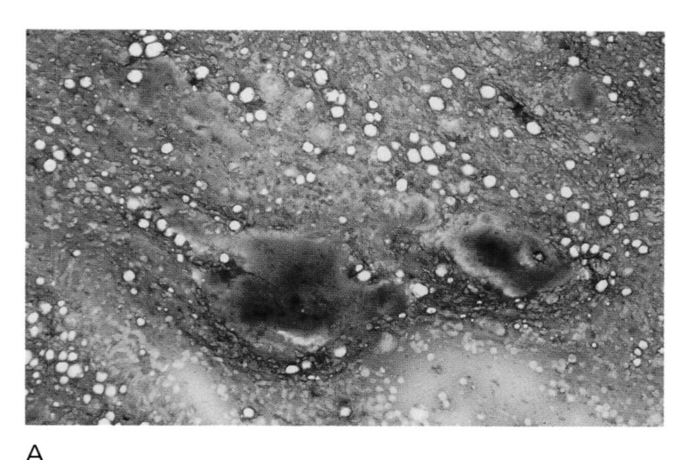

A

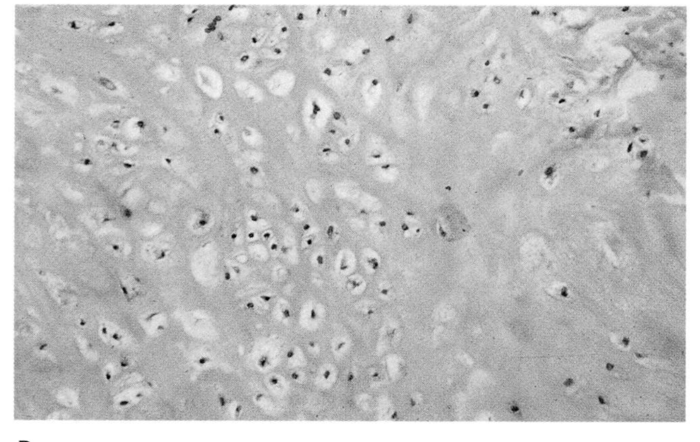

B

FIGURE 5–9. (**A**) In addition to matrix material, chondrosarcomas frequently demonstrate myxoid changes/degeneration, as evidenced by a metachromatic granular film or precipitate. (Diff-Quik) (**B**) Cell block of classic low-grade chondrosarcoma with a relatively hypocellular appearance and uniform distribution of neoplastic cells. Clinicoradiologic correlation is essential. (Hematoxylin and eosin)

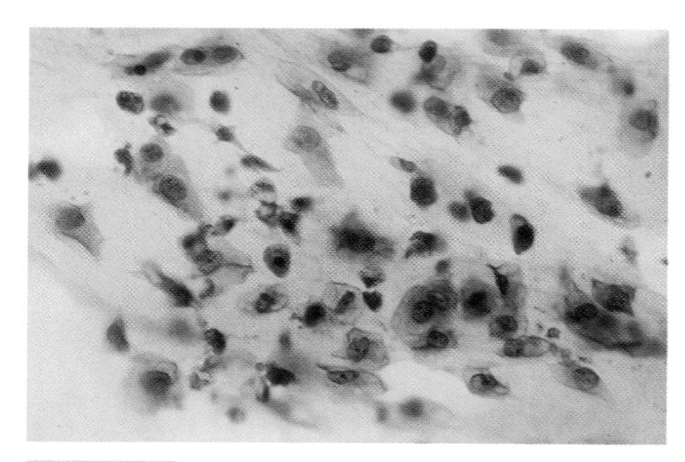

FIGURE 5–10. High-grade chondrosarcomas are generally more cellular, although the degree of nuclear atypia is usually less than that seen in osteosarcoma. (Papanicolaou)

tramedullary chondrosarcomas, some examples also occur in peripheral chondrosarcomas arising from osteochondromas.[25] FNA biopsy usually reveals "nonspecific" pleomorphic sarcoma; the low-grade chondrosarcomatous portion is rarely sampled (Fig. 5–12).[25,26] The surgical excision specimen generally demonstrates a low-grade chondrosarcoma sharply juxtaposed to a high-grade "nonchondrosarcomatous" pleomorphic sarcoma; the latter usually has features of malignant fibrous histiocytoma or high-grade osteosarcoma (Fig. 5–13).

FNA biopsy of mesenchymal chondrosarcoma has only rarely been reported.[4] The smears are typically hypercellular and composed of mostly uniform, rounded to slightly oval cells with solitary, hyperchromatic nuclei and high nuclear to cytoplasmic (N/C) ratios, morphologically resembling the malignant elements of Ewing's sarcoma. Less commonly, matrix material suggestive of cartilaginous differentiation is evident. Only if the primitive cellular component is identified along with a definite cartilaginous matrix can a diagnosis of mesenchymal chondrosarcoma be rendered by FNA biopsy. We are not convinced that mesenchymal chondrosarcoma can be reliably separated from small cell osteosarcoma and Ewing's sarcoma based solely on the cytologic features of the tumor cells. Further confounding the differential diagnosis of Ewing's sarcoma, CD99 immunopositivity is also a common finding in mesenchymal chondrosarcoma.[27]

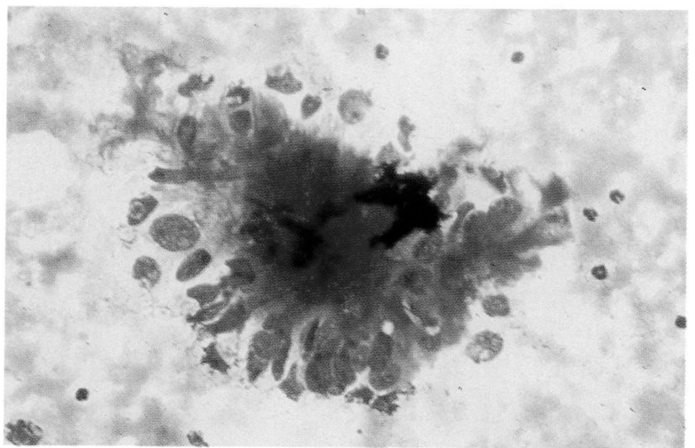

A

B

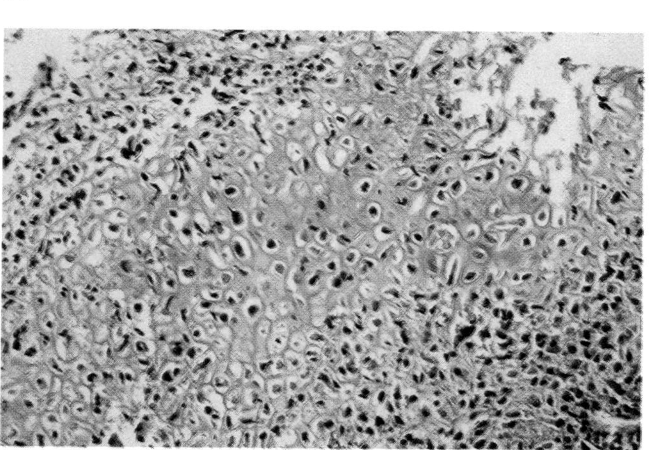

C

FIGURE 5–11. FNA biopsy of chondroblastic osteosarcoma. **(A)** Pleomorphic tumor cells are associated with a large fragment of matrix material tinctorially similar to osteoid. (Diff-Quik) **(B,C)** Cell block shows chondroid and hyaline cartilage differentiation. Note the marked nuclear atypia and variability of the neoplastic cells and the "spindling" and/or condensation of cells at the periphery, features helpful in distinguishing chondroblastic osteosarcoma from chondrosarcoma.

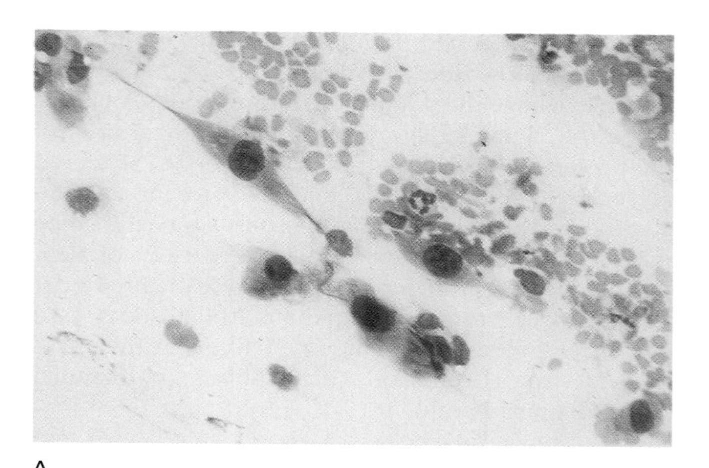

A

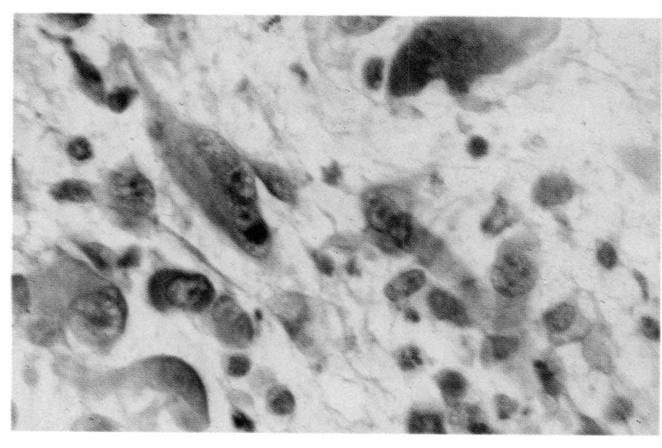

B

FIGURE 5–12. FNA biopsy of dedifferentiated chondrosarcoma generally samples only the high-grade component. In this instance the smears reveal a pleomorphic sarcoma reminiscent of malignant fibrous histiocytoma. (Diff-Quik, Papanicolaou)

FNA biopsy of clear cell chondrosarcomas generally yields small, rounded tumor cells and clusters of larger, epithelioid-like cells within a background of amorphous matrix material. The epithelioid tumor cells have abundant amounts of finely vacuolated cytoplasm, one or two round, hyperchromatic nuclei, and a central nucleolus.[4] As in most bone tumors, the diagnosis of clear cell chondrosarcoma by FNA biopsy requires clinicoradiologic correlation. The epiphyses of the long bones are characteristically involved, and most patients are in the third or fourth decade of life.

Myxoid chondrosarcoma, also known as chordoid sarcoma, is a rare malignancy. It usually arises within the extraskeletal deep soft tissues of the extremities but rarely occurs in bone.[28] Although FNA biopsy of extraskeletal myxoid chondrosarcoma has been reported,[29] to our knowledge no examples have been described in bone.

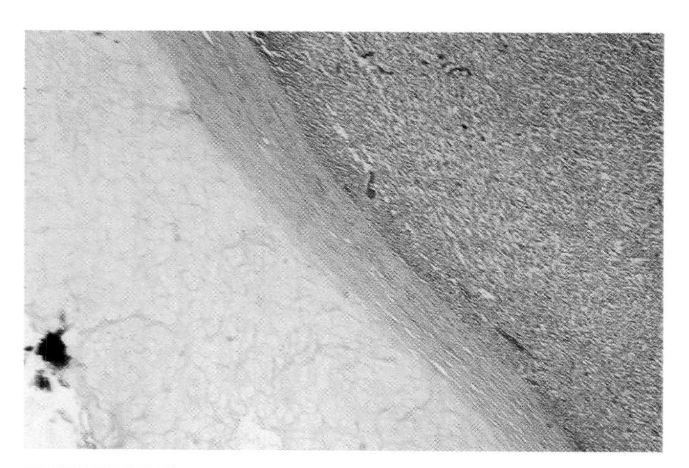

FIGURE 5–13. Dedifferentiated chondrosarcoma characterized by a low-grade chondrosarcoma (*left*) juxtaposed to a high-grade nonchondrosarcomatous sarcoma (*right*). (Hematoxylin and eosin)

Chondroblastoma

Chondroblastoma is a benign, chondroid-producing tumor characteristically arising within the epiphysis of a skeletally immature patient.[30] Most patients are younger than 20 years of age at diagnosis. Radiographically, chondroblastomas are sharply demarcated, lytic lesions, often with sclerotic margins, localized to the epiphysis or epimetaphysis of a long bone (Fig. 5–14).[30] FNA biopsy yields hypercellular smears

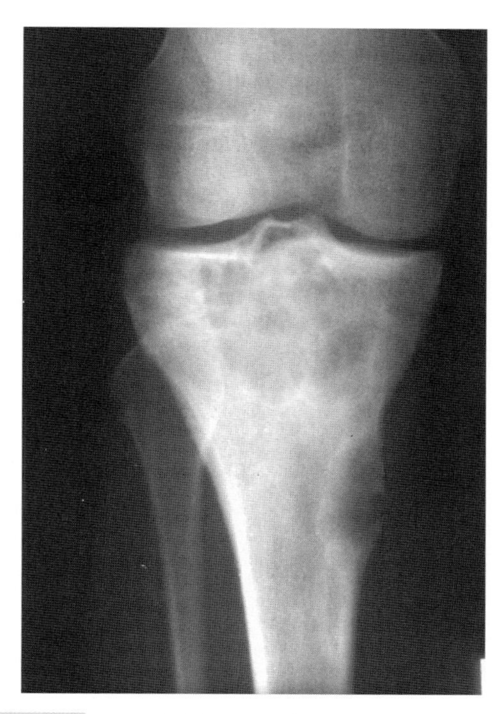

FIGURE 5–14. Chondroblastoma of bone forming a geographic lytic defect within the proximal tibial epiphysis. Note the incidental metaphyseal fibrous defect involving the tibial metaphysis eccentrically.

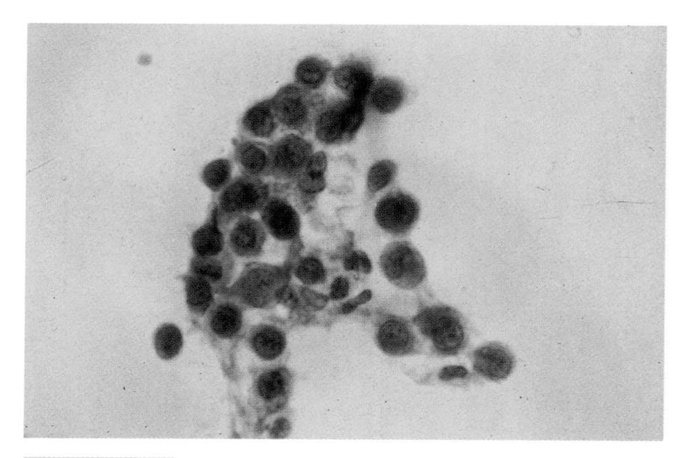

FIGURE 5–15. FNA biopsy of chondroblastoma revealing mostly uniform chondroblasts with round to ovoid, reniform-shaped nuclei, some of which contain nuclear grooves. (Papanicolaou)

composed of uniform-appearing, round to polygonal chondroblasts with central to eccentrically located nuclei, often exhibiting prominent nuclear grooves (Fig. 5–15).[8,31] The nuclei generally have a fine chromatin pattern and inconspicuous nucleoli. The cytologic features may closely resemble those of Langerhans' cell histiocytosis; however, the lack of inflammatory cells (i.e., eosinophils, neutrophils, and lymphocytes) and the presence of scattered fragments of matrix material (i.e., chondroid) help exclude the latter diagnosis. Further, when Langerhans' cell histiocytosis involves the long bones of the extremities, diaphyseal involvement is far more common than epiphyseal origin.[32] Cytologically, the matrix material of chondroid appears tinctorially similar to osteoid but tends to be arranged in more well-defined nodules and plates reminiscent of cartilaginous differentiation (Fig. 5–16).[8] Nevertheless, we believe the diagnostic finding on FNA biopsy to be the characteristic chondroblasts, regardless of the presence or absence of matrix material.[8] Scattered osteoclast-type giant cells may be present, and in some cases may be quite numerous. Osteoclast-type giant cells may be seen in a wide variety of neoplasms, however, including metaphyseal fibrous defect, giant cell tumor, Langerhans' cell histiocytosis, and osteosarcoma. Meticulous attention to the background stromal cells (i.e., chondroblasts) is of much grater diagnostic value.

Cytomorphologic Features of Chondroblastoma

- Moderate to markedly cellular smears composed of mostly individually dispersed tumor cells
- The neoplastic chondroblasts are mostly uniform, round to polygonal, with central to eccentric, round to oval nuclei, frequently containing nuclear grooves or folds
- Variable numbers of benign osteoclast-type giant cells
- Rare fragments of fibrillar to amorphous matrix material

Chondromyxoid Fibroma

Chondromyxoid fibromas are benign tumors usually arising within the metaphyseal regions of the long bones of the extremities and the pelvis (i.e., ilium).[33] Most patients are initially diagnosed in the second or third decade of life. Radiologically, most examples ap-

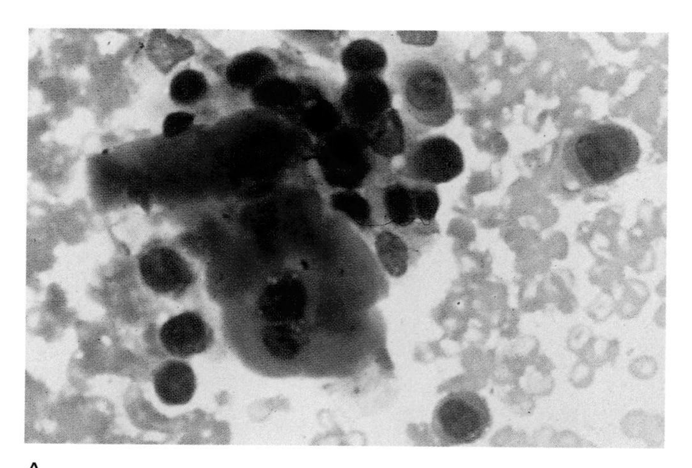

A

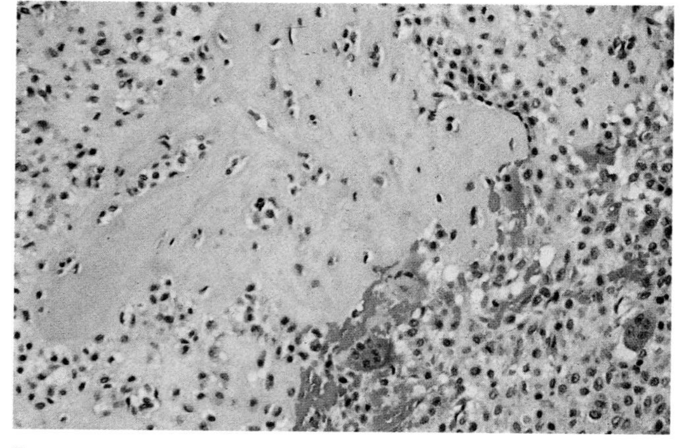

B

FIGURE 5–16. **(A)** Chondroblasts with associated matrix material consistent with chondroid. The latter cannot be reliably distinguished from hyaline cartilage in cytologic preparations. (Diff-Quik) **(B)** Cell block of chondroblastoma. Note the island of "chondroid" tinctorially similar to osteoid. Numerous chondroblasts and a few osteoclast-type giant cells are seen peripherally. (Hematoxylin and eosin). (From Kirkpatrick SE, Pike EJ, Geisinger KR, Ward WG. Chondroxblastoma of bone: use of fine needle aspiration biopsy and potential diagnostic pitfalls. Diagn Cytopathol 16:65, 1997. Copyright © 1997 by John Wiley & Sons, Inc. Reprinted by permission of Wiley-Liss, Inc., a division of John Wiley & Sons, Inc.)

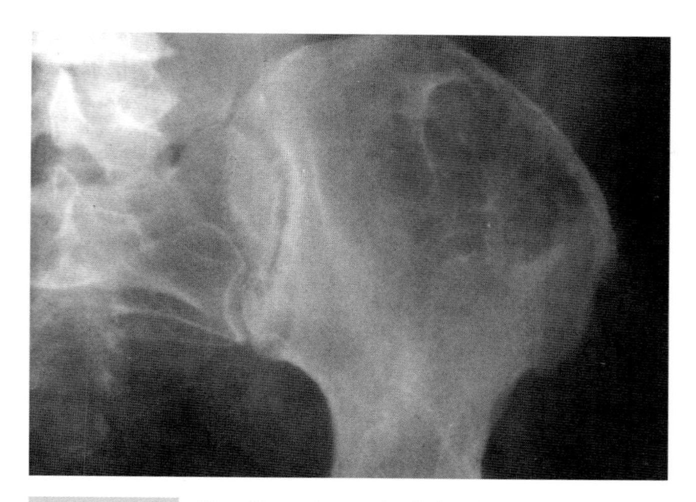

FIGURE 5–17. Plain film radiograph of chondromyxoid fibroma involving the left iliac wing. The lesion is lytic with geographic (well-defined), sclerotic borders, indicative of a benign process.

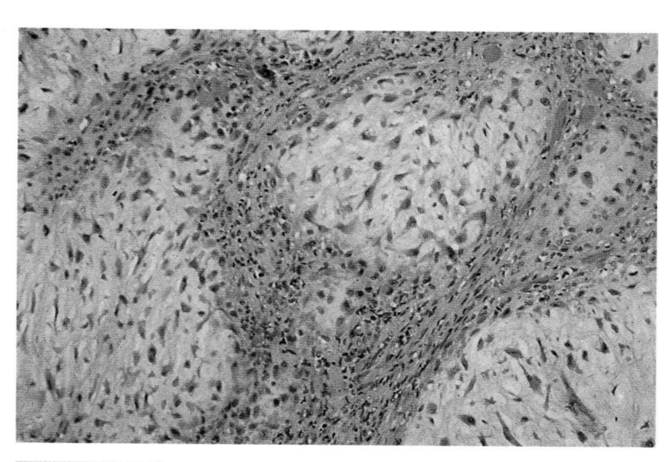

FIGURE 5–18. Classic histiologic appearance of chondromyxoid fibroma with myxomatous lobules containing spindled to stellate-appearing cells separated by variably cellular fibrous tissue septa. (Hematoxylin and eosin)

pear benign as eccentrically located, lytic lesions with sharply scalloped, frequently sclerotic margins. Unlike enchondroma or chondrosarcoma, calcifications within the lesion are rarely seen (Fig. 5–17). Chondromyxoid fibromas are characterized histologically by lobules of myxoid to chondroid-appearing stroma rimmed by hypercellular fibrous tissue septa with scattered multinucleated osteoclast-type giant cells. Within the matrix material are variably sized, spindled to stellate-appearing cells, which may be alarmingly atypical. Completely developed hyaline cartilage is rarely observed (Fig. 5–18). FNA biopsy specimens have been reported infrequently.[34,35] Smears are usually moderately cellular, with ill-defined fragments of background myxoid fibrillar matrix and mostly discohesive, variably sized, round to stellate-appearing cells (Fig. 5–19). Most appear uninucleated, but occasional bi- and multinucleated cells are also observed. The nuclear chromatin ranges from fine to "smudgy and condensed."[34] Likewise, the cytoplasm is variable, ranging from foamy to dense. As in histologic sections, well-developed cartilage with individual cells lying within distinct lacunae is a rare finding. The amount of atypia in the neoplastic cells of chondromyxoid fibroma may be enough to cause diagnostic confusion with more ominous entities, such as chondrosarcoma or osteosarcoma. Meticulous attention to the clinical features (e.g., age of patient) and radiologic findings will help avoid this diagnostic pitfall.

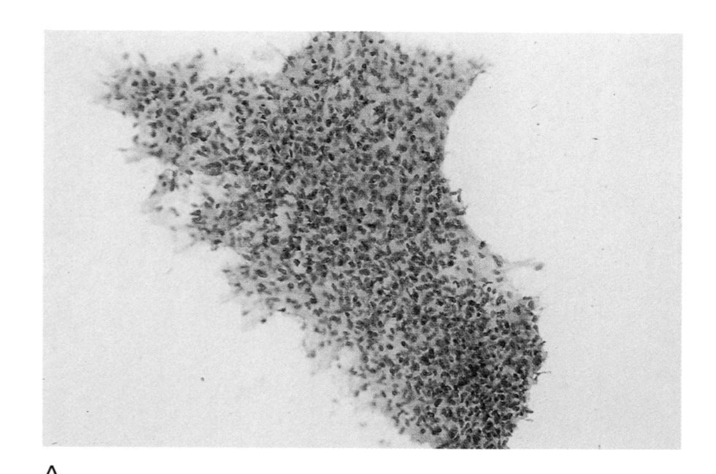

A

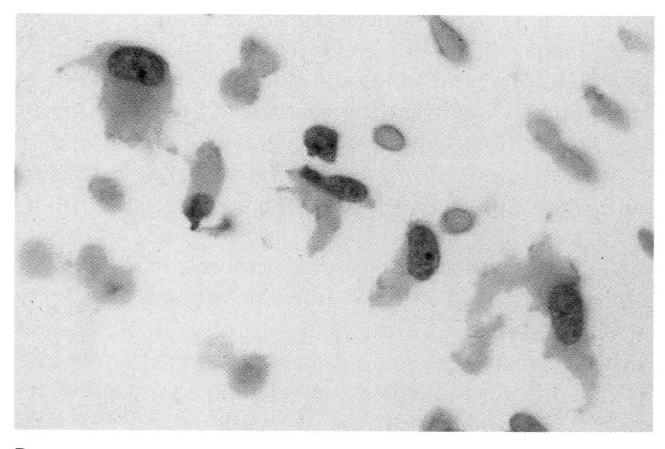

B

FIGURE 5–19. **(A)** FNA biopsy of chondromyxoid fibroma with a large hypercellular myxoid stromal fragment. **(B)** The neoplastic cells are usually spindled to stellate in configuration and may display nuclear atypia. (**A,B,** hematoxylin and eosin) (Courtesy of Dr. Joshua Z. Sickel, MD, Rochester, NY)

FIBROUS/FIBROHISTIOCYTIC TUMORS

Desmoplastic Fibroma

Desmoplastic fibroma is a locally aggressive, nonmetastasizing fibrous tumor most commonly occurring in patients in the second decade of life. The anatomic site most commonly affected is the mandible, followed by the metaphyseal regions of the long bones of the extremities.[36] Radiographically, most cases appear lytic and expansile and demonstrate a "soap bubble" appearance traversed by irregular trabeculations. Sharp margination is usually accompanied by a rind of sclerosis, indicating a benign process. Occasional examples show more aggressive radiologic features with concomitant soft-tissue extension. Histologically, desmoplastic fibroma represents the intraosseous counterpart of soft-tissue desmoid tumor. To the best of our knowledge, there have been no adequate cytologic descriptions of FNA biopsy of desmoplastic fibroma; however, we expect that cytologic specimens obtained from desmoplastic fibromas will be similar to what has been previously described in soft-tissue desmoid tumors.[37] In a large series of primary neoplasms of bone analyzed by FNA biopsy, Layfield et al.[6] reported a case of desmoplastic fibroma initially misdiagnosed as osteosarcoma, presumably related to misinterpretation of reactive osteoblasts, osteoid and banal-appearing spindle cell elements.

Fibrosarcoma

Although occurring in age groups and anatomic sites similar to those of osteosarcoma, fibrosarcomas are extremely rare primary bone tumors. High-grade fibrosarcomas appear radiologically aggressive with evidence of permeation, ill-defined margins, and concomitant soft-tissue extension. In contrast, low-grade fibrosarcomas may show some radiologic overlap with desmoplastic fibroma, appearing more benign.[38] The histologic and cytologic appearances usually recapitulate the radiographic features. In tissue sections, high-grade fibrosarcomas are generally characterized by closely packed, hyperchromatic, spindle-shaped uniform tumor cells often arranged in regular intersecting fascicles, the so-called herringbone pattern. Its low-grade counterpart contains more stromal collagen, shows less overall nuclear atypia, and may closely resemble desmoplastic fibroma, with the exceptions of greater anaplasia of the tumor cells and higher mitotic activity in the former. Only a few reports have described the cytology of primary fibrosarcoma of bone.[39,40] The smears are generally cellular and composed of mostly spindle-shaped cells with a single oval to elongated nucleus, a hyperchromatic chromatin pattern, and inconspicuous nucleoli. Unlike malignant fibrous histiocytoma, marked nuclear atypia is not a feature. We have no experience with FNA biopsy of this lesion. Whether or not fibrosarcomas can be reliably distinguished from high-grade fibroblastic osteosarcomas by FNA biopsy will require further analysis.

Malignant Fibrous Histiocytoma

As with fibrosarcoma of bone, malignant fibrous histiocytoma (MFH) of bone is a rare primary bone tumor usually involving the metaphyseal region of the long bones of the extremities. Although MFH occurs in patients over a wide age range, most patients are older (older than 30 years of age) than those with osteosarcoma.[41] The radiologic features of MFH are that of an aggressive, usually lytic neoplasm with ill-defined margination and soft-tissue extension. MFH cytologic smears are characteristically slightly to moderately cellular, containing numerous single and discohesive, markedly pleomorphic tumor cells, many of which appear stripped of their cytoplasm. Nuclei are variable both in size and contour, ranging from round to highly irregular and multilobulated. The chromatin pattern is darkly stained and coarsely granular. One or more prominent nucleoli are typical. Multinucleated tumor giant cells are frequent.[42] The overall appearance mirrors the more common MFH observed in soft tissue (Fig. 5–20). The presence of identifiable matrix (i.e., osteoid or chondroid) production by the tumor cells precludes a diagnosis of MFH. However, as emphasized earlier, reliably distinguishing between dense collagen and osteoid in cytologic preparations is not generally possible.[8,17] At our institution, the distinction between MFH of bone and a high-grade fibroblastic osteosarcoma is of no therapeutic importance because both are treated with the high-

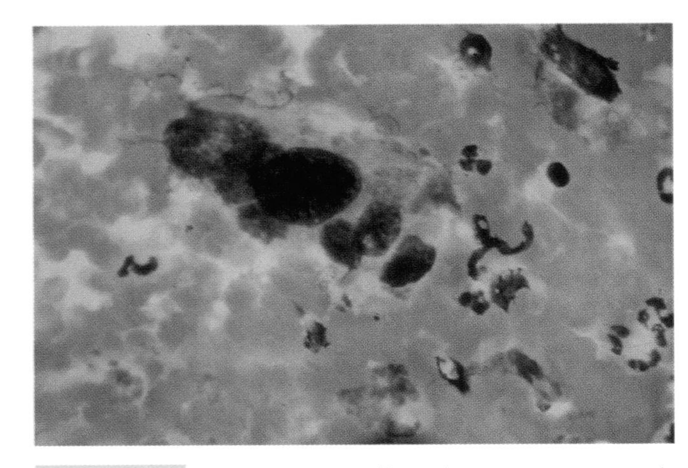

FIGURE 5–20. FNA of malignant fibrous histiocytoma composed of mostly single cells with markedly pleomorphic and anaplastic nuclei. Multinucleated giant cells are typical. At our institution, such lesions in bone are treated as high-grade osteosarcomas. (Diff-Quik)

grade osteosarcoma protocol. Although some cases of MFH of bone may be more amenable to radiation therapy, there appears to be no significant difference in survival rates between MFH of bone and osteosarcoma.[41]

HEMATOPOIETIC TUMORS

Multiple Myeloma

Multiple myeloma, or plasma cell myeloma, represents an intraosseous, solitary or multicentric, plasma cell proliferation arising most commonly in the vertebral spine, pelvis, ribs, and skull. Most patients are older than 50 years of age at diagnosis.[43] Sharply marginated, "punched out," lytic areas of bone destruction are usually evident (Fig. 5–21). Immunoelectrophoresis, in the vast majority of patients with multiple myeloma, reveals increased levels of monoclonal immunoglobulin in the serum and/or immunoglobulin light chains (Bence Jones protein) in the urine. FNA biopsy preparations show numerous, mostly solitary plasma cells and "plasmacytoid" cells, which may range from mature to immunoblastlike in appearance (Fig. 5–22).[44] The cytoplasm is usually deeply basophilic (Diff-Quik staining), and paranuclear clear zones are often evident, surrounding round, eccentrically situated nuclei. Bi- and multinucleated forms are frequent. The chromatin pattern varies from coarse to finely granular; nucleoli may be present, sometimes appearing quite large. Intranuclear and/or intracytoplasmic inclusions are occasionally observed. Some cases of multiple myeloma appear uniformly poorly differentiated and anaplastic, making distinction from metastatic anaplastic carcinoma, large cell lymphoma, or malignant melanoma difficult. In such cases, clinical and radiologic correlation and immunohistochemical analysis are necessary for definitive diagnosis.[45]

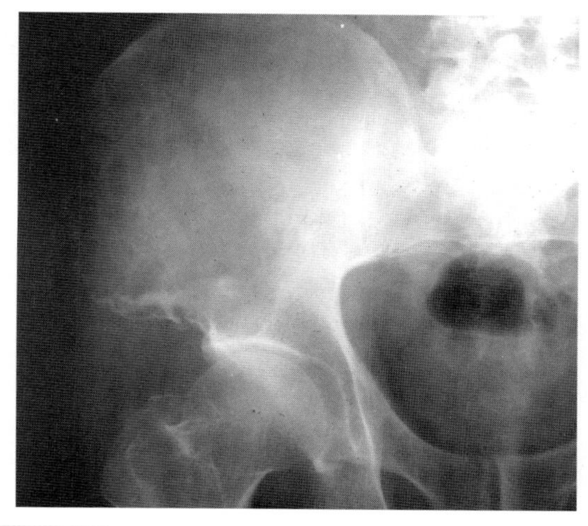

FIGURE 5–21. Plain film radiograph of multiple myeloma forming multiple lytic defects in the right ilium.

Malignant Lymphoma

Primary malignant lymphomas of bone are rare; more commonly, bone involvement by malignant lymphoma represents a secondary spread from a preexisting known primary. Primary intraosseous lymphoma most often involves the long bones of the extremities, especially the lower extremities, and the pelvis. Although patients at any age may be affected, it is distinctly rare in children.[46] Most primary lymphomas of bone appear radiologically malignant, but the radiographic features are relatively nonspecific. Blastic, lytic, and mixed blastic and lytic patterns may

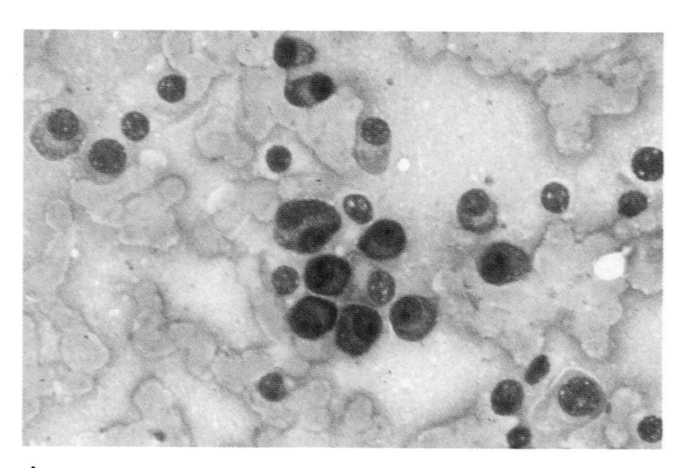

A

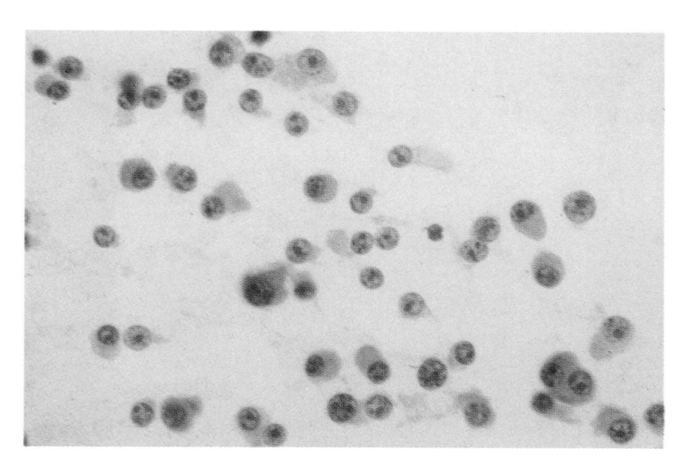

B

FIGURE 5–22. Numerous plasma cells with enlarged, atypical nuclei and occasional prominent nucleoli. Binucleated forms are also present. (Diff-Quik, Papanicolaou)

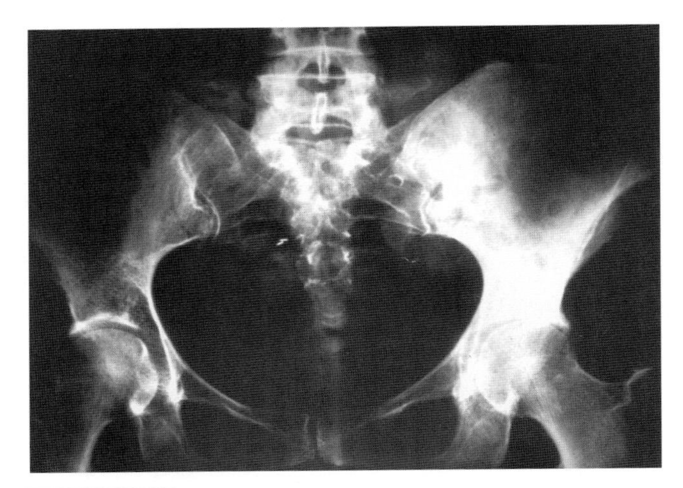

FIGURE 5–23. Plain film radiograph of malignant lymphoma involving the left pelvis of a 61-year-old woman. Malignant lymphomas may be osteoblastic but frequently show aggressive radiologic features with extensive involvement.

be seen (Fig. 5–23). Most primary bone lymphomas represent B-cell proliferations, usually of the large cell type. FNA biopsy smears are hypercellular and resemble those arising in lymph nodes, consisting predominantly of discohesive, solitary, large cells with high N/C ratios, surrounded by a thin rim of cytoplasm. The nuclei may be cleaved, folded, or polylobated, but virtually always have irregular contours (Fig. 5–24).[47] Small, mature-appearing lymphocytes are often admixed with the larger neoplastic lymphoid cells. Numerous fragments of lymphoid cell cytoplasm (lymphoglandular bodies) are a helpful diagnostic feature. However, because many intraosseous lymphomas are associated with a dense, fibroblastic background, FNA biopsy may yield low cellularity and even unsatisfactory specimens.[47] Extensive crush artifact may also further confound the diagnosis. If lymphoma is suspected with immediate interpretation, material can be obtained for immunophenotyping by flow cytometry or immunocytochemistry. In our experience, low-grade (small cell) lymphomas are generally more amenable to analysis by flow cytometry than high-grade (large cell) lymphomas.

Langerhans' Cell Histiocytosis (Histiocytosis X)

Langerhans' cell histiocytosis (LCH), or histiocytosis X, represents a clonal proliferation of the distinctive Langerhans' cell histiocytes, often accompanied by numerous inflammatory cells (eosinophils, neutrophils, lymphocytes, and plasma cells). Patients with osseous LCH may range in age from birth to 71 years; however, the majority of cases occur early within the first decade of life.[32] The most common anatomic site affected is the craniofacial region, especially the skull and posterior jaw. Involvement of the long bones is usually associated with diaphyseal localization; this may radiographically appear quite aggressive, having ill-defined marginization and periosteal reactions, mimicking Ewing's sarcoma (Fig. 5–25).[32] We are aware of at least one case report of Langerhans' cell histiocytosis involving the diaphysis of the right femur that was misdiagnosed by FNA biopsy as Ewing's sarcoma, mostly on the basis of the radiographic findings.[48] Extraosseous manifestations of LCH include diabetes insipidus, a seborrhea-like skin rash, interstitial pulmonary abnormalities, and lymphadenopathy. LCH of bone in aspiration cytologic preparations is characteristically hypercellular and composed of numerous, discohesive Langerhans' cell histiocytes with ovoid to reniform and/or convol-

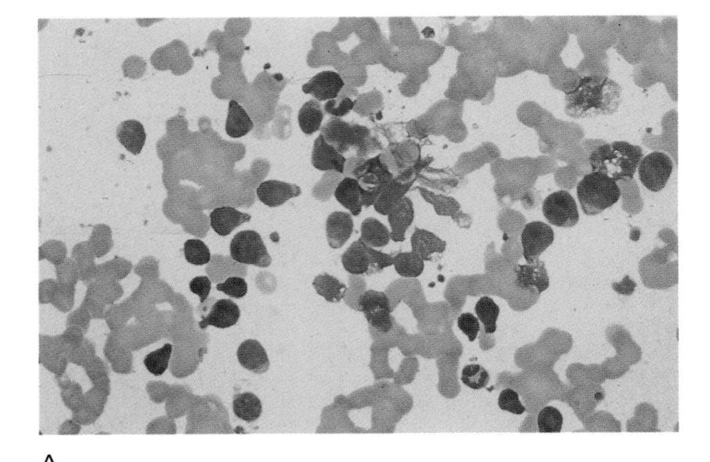

A

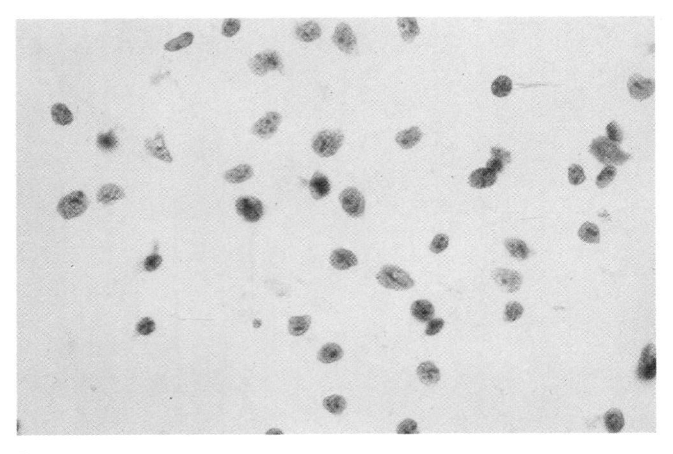

B

FIGURE 5–24. **(A)** FNA biopsy of malignant lymphomas of bone with numerous discohesive large lymphoid cells and scattered cytoplasmic blobs (lymphoglandular bodies). (Diff-Quik) **(B)** The nuclear contours are often more irregular, with polylobation, folds, and grooves. Nucleoli are often well developed. (Papanicolaou)

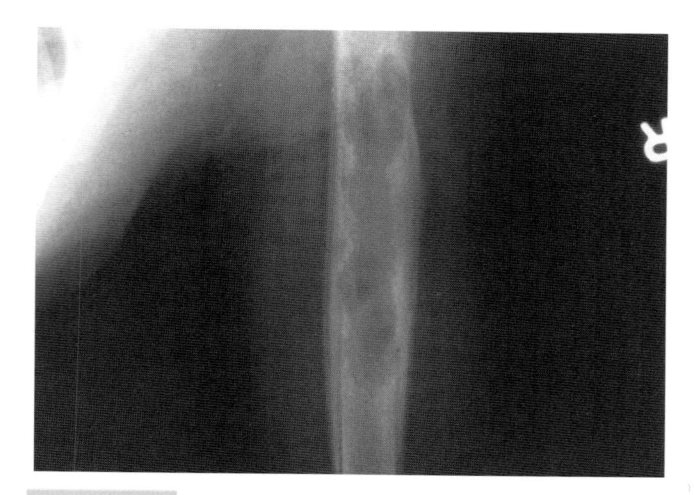

FIGURE 5–25. Plain film radiograph of Langerhans' cell histiocytosis of bone involving the shaft (diaphysis) of the proximal humerus in a 15-month-old boy. Such cases with prominent periosteal reactions may be radiologically confused with Ewing's sarcoma.

uted nuclei, often containing distinct longitudinal grooves.[49] Nucleoli are usually small and inconspicuous but may be prominent. Binucleated and multinucleated forms are frequently present. Accompanying the diagnostic histiocytes are variable numbers of eosinophils, lymphocytes, and even neutrophils, which may occasionally predominate (Fig. 5–26). Cytologically, LCH of bone can be distinguished from osteomyelitis by the absence of the distinctive Langerhans' cell histiocyte in the latter lesion. The lack of matrix material and the typical clinical and radiologic features help distinguish LCH of bone from chondroblastoma.[8] The histiocytes of Rosai-Dorfman disease (sinus histiocytosis with massive lymphadenopathy) are generally characterized by vesicular nuclei, lacking convolutions and longitudinal grooves, and are surrounded by an abundant amount of foamy cytoplasm. Leukophagocytosis by the histiocytes and absence of eosinophils are further helpful diagnostic features of Rosai-Dorfman disease.[50]

Whether or not ancillary studies (immunocytochemistry or electron microscopy) are needed for "definitive" diagnosis is debatable. Many, but not all, Langerhans' cell histiocytes coexpress S-100 protein and CD1a. In our opinion, although supportive data may be obtained from immunohistochemical staining and ultrastructural examination (Birbeck granules), a definitive diagnosis of LCH of bone may be rendered based on cytomorphology alone, especially if correlated with clinical and radiologic data.[32,51]

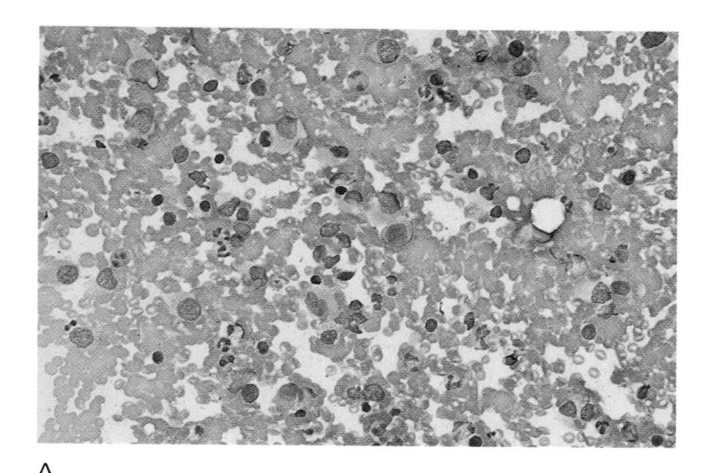

A

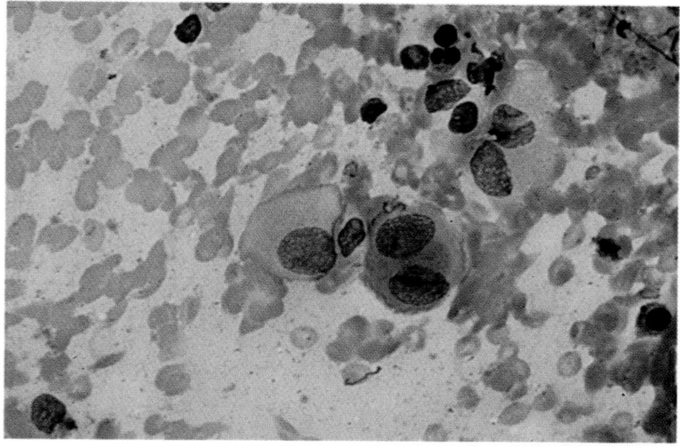

B

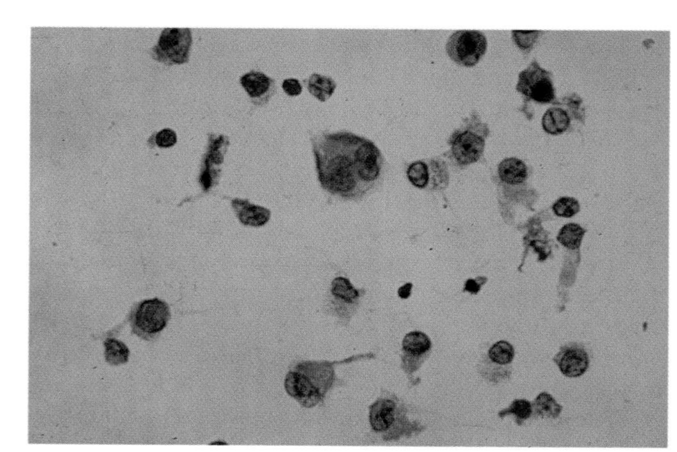

C

FIGURE 5–26. **(A)** FNA biopsy of Langerhans' cell histiocytosis reveals a hypercellular smear composed of diagnostic, discohesive histiocytes in a background of inflammatory cells. (Diff-Quik) **(B,C)** High-power view illustrating prominent nuclear grooves, binucleated forms, and occasional prominent nucleoli. (Diff-Quik and Papanicolaou)

Cytomorphologic Features of LCH of Bone

- Moderately to highly cellular smears composed of individually dispersed large histiocytes
- The diagnostic histiocytes have round to oval nuclei, finely granular chromatin, inconspicuous to prominent nucleoli, and distinct longitudinal grooves or folds
- Binucleated and multinucleated histiocytes frequently present
- Background inflammatory cells with variable numbers of eosinophils, neutrophils, lymphocytes, and plasma cells

VASCULAR TUMORS

Hemangioma

Intraosseous hemangiomas are relatively rare, often incidental findings that have a predilection for the axial skeleton, especially the skull and spine. Although they may arise in patients at any age, most are seen in older patients (fifth decade or older).[52] Radiologically, hemangiomas are wholly intramedullary and lytic, often containing vertical trabecular striations. Lesions arising within the long bones may also produce a lytic or "sunburst" appearance, with reactive bone appearing to radiate from the center. Most intraosseous hemangiomas are inaccessible to FNA biopsy sampling. Isolated case reports of accessible lesions have been nondiagnostic.[6,16] This is not surprising, because most cases of hemangioma of bone are composed of large cavernous vascular spaces lined by thin fibrous tissue septa and largely containing blood.

Hemangioendothelial Sarcoma

Malignant vascular tumors of bone have not been as clearly defined as their soft-tissue counterparts; thus, terminology is often inconsistent and confusing. Some authors prefer the term "hemangioendothelial sarcoma," which encompasses a spectrum of designations, including hemangioendothelioma and angiosarcoma.[53] The former has generally referred to low-grade hemangioendothelial sarcoma, the latter high-grade tumors. Most hemangioendothelial sarcomas involve the long bones of the extremity and occur in middle-aged to older patients.[53] The radiographic features vary depending on the grade (aggressiveness) of the lesion. Nevertheless, one characteristic feature of hemangioendothelial sarcomas of bone is their tendency for multicentricity, especially with low-grade epithelioid neoplasms (up to 50% of cases).[54] Most are characterized by multicentric radiologic lucencies that often have at least partially sclerotic margins, occupying a single bone or anatomic area (Fig. 5–27). In tissue sections, hemangioendothelial sarcomas range from obviously low-grade vasoformative or epithelioid lesions to high-grade anaplastic tumors with little to no vascular differentiation. Likewise, aspiration cytologic preparations may show great variability. Low-grade epithelioid hemangioendotheliomas of bone tend to be hypercellular and composed of varying amounts of spindle-shaped to epithelioid-appearing cells. Rare, sharply demarcated, intracytoplasmic vacuoles and intranuclear cytoplasmic pseudoinclusions are sometimes observed (Fig. 5–28).[55] Metachromatic stromal fragments probably representing chondromyxoid stroma may also be evident (Fig. 5–29). In contrast, high-grade angiosarcomas show relatively nonspecific cytologic features and thus may be difficult to distinguish from other pleomorphic sarcomas.[56] Along with clinical and radiologic correlation, immunohistochemistry, especially factor VIII and CD31, may be especially helpful in confirming vascular differentiation. Occasional epithelioid vascular tumors may also express cytokeratin, and this may further confound the differential diagnosis of metastatic carcinoma.[57]

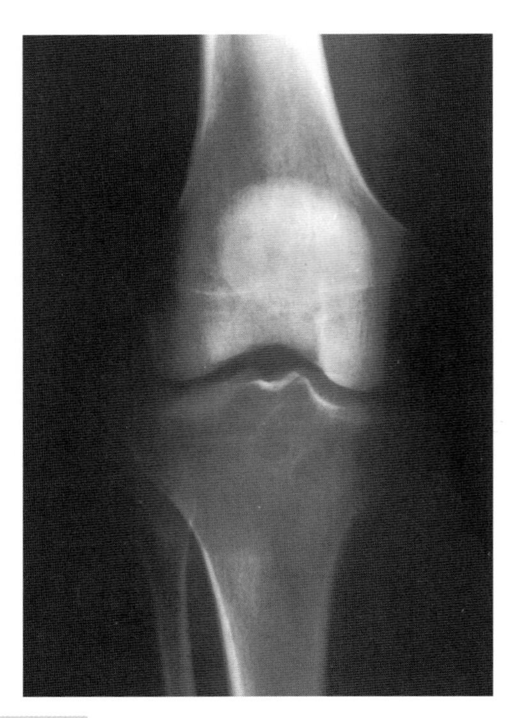

FIGURE 5–27. Plain film radiograph of epithelioid hemangioendothelioma of bone involving the proximal tibia in a 40-year-old man. Multicentric lucencies with rinds of sclerosis are observed.

TUMORS OF UNCERTAIN ORIGIN

Giant Cell Tumor

Giant cell tumor of bone is a benign but locally aggressive lesion usually arising within the epiphysis of a long bone in a skeletally mature patient. Most pa-

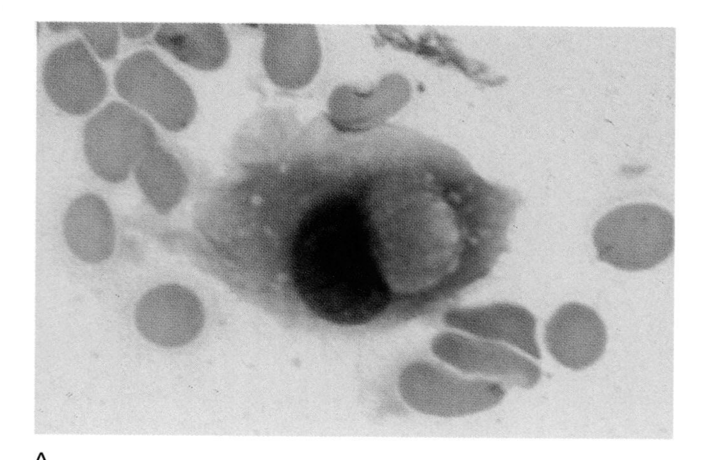

A

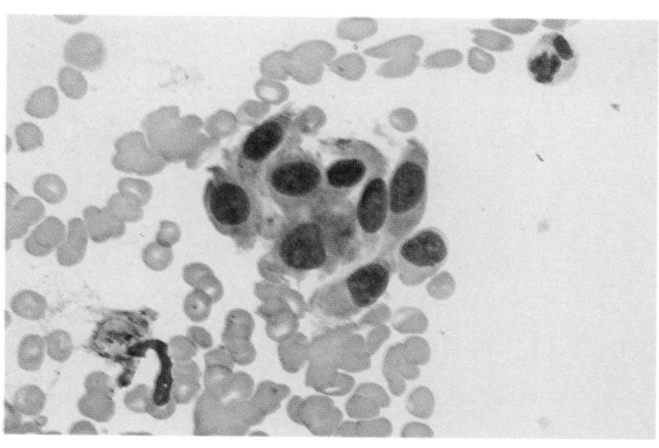

B

FIGURE 5–28. (**A**) FNA biopsy of epithelioid hemangioendothelioma showing a large, sharply demarcated intracytoplasmic vacuole. (**B**) The neoplastic cells may range from epithelioid to spindled-shaped. (**A,B,** Diff-Quik)

tients are in the third or fourth decade of life at presentation.[58] Radiographically, giant cell tumors are characteristically lytic, with sharply defined but nonsclerotic margins (Fig. 5–30). Soft-tissue extension occasionally occurs. FNA biopsy of giant cell tumor generally yields hypercellular smears composed of single and discohesive, spindled to ovoid-shaped uniform-appearing neoplastic stromal cells admixed with numerous osteoclast-type giant cells.[12] The spindled cells have a generally uniform, oval nucleus, a finely granular chromatin pattern, and inconspicuous to small nucleoli. Although mitotic figures may be observed, the stromal cells of giant cell tumor lack malignant features (Fig. 5–31). As a point of comparison, the nuclei of the stromal cells should approximate, in size and configuration, the nuclei within the osteoclast-type giant cells. In addition to pertinent clinicoradiologic findings, the cytologic diagnosis of giant cell tumor is based on the background stromal cells and not the presence or absence of osteoclast-type giant cells.[12]

Cytomorphologic Features of Giant Cell Tumor of Bone

- Moderately to highly cellular smears composed of variable amounts of osteoclast-type giant cells
- Oval to spindle-shaped, mostly uniform stromal cells arranged individually and in small clusters
- Nuclei of the stromal cells approximates, both in size and configuration, the nuclei of the osteoclast-type giant cells

Ewing's Sarcoma

Ewing's sarcoma of bone occurs mostly in the first and second decades of life but is distinctly uncommon in African-Americans.[59] The radiographic appearance is characteristically lytic and aggressive, with ill-defined margins and often a concomitant soft-tissue mass (Fig. 5–32). Although not specific, a distinctive radiologic feature is the characteristic "onion skin" periosteal reaction. Any bone may be affected, but most cases of

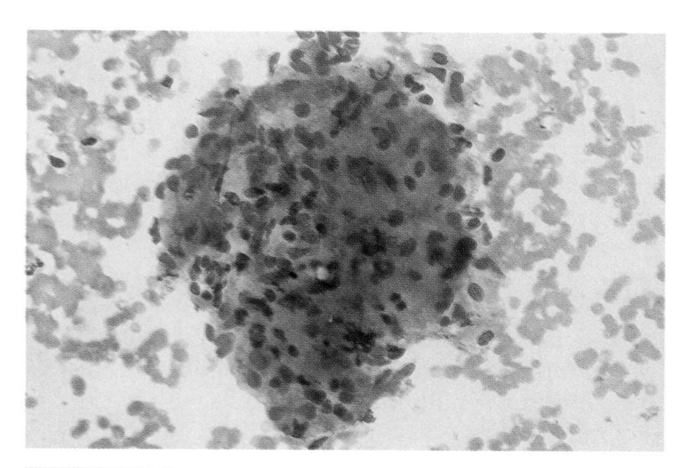

FIGURE 5–29. The chondromyxoid stroma of epithelioid hemangioendothelioma in cytologic preparations appears as stromal fragments. (Diff-Quik)

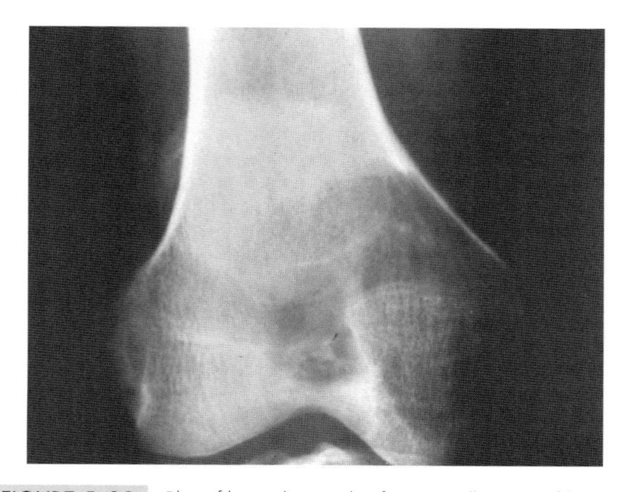

FIGURE 5–30. Plain film radiograph of giant cell tumor of bone involving the distal femoral epiphysis in a 64-year-old man. The lesion is geographic and purely lytic and lacks a sclerotic rim.

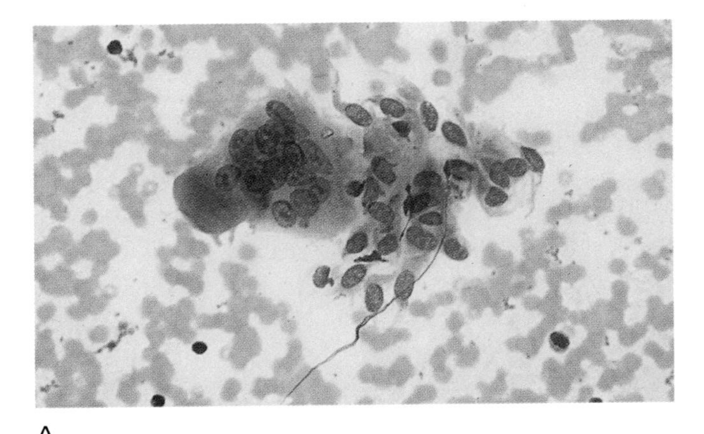

A

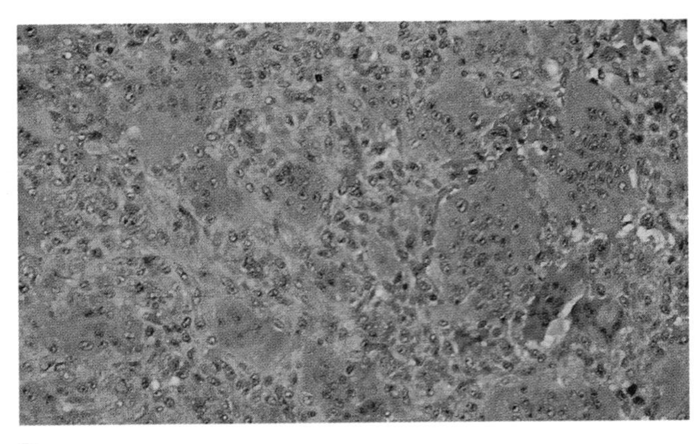

B

FIGURE 5–31. (**A**) Giant cell tumor of bone characterized by mostly spindle-shaped stromal cells and a benign-appearing osteoclast-type giant cell. The stromal cells have mostly uniform nuclei and approximate the nuclei of the giant cells. (Diff-Quik) (**B**) Classic histologic appearance of giant cell tumor. (Hematoxylin and eosin)

Ewing's sarcoma involve the long bones of the extremities, with a predilection for the diaphysis. On FNA biopsy, Ewing's sarcoma demonstrates hypercellular smears composed of both solitary uniform, round to ovoid tumor cells and small cohesive clusters with some nuclear molding (Fig. 5–33).[60] Occasional pseudorosette formation is observed. Evidence of matrix production by the tumor cells is distinctly absent. The nuclei are generally regular and round to ovoid, and lack the markedly irregular nuclear contours and convolutions typically observed in primary osseous lymphomas (Fig. 5–34). The chromatin is finely granular and uniformly distributed; nucleoli are generally inconspicuous. N/C ratios are exceedingly high. Rarely, atypical (large cell) variants of Ewing's sarcoma exhibit irregular, vesicular nuclei and prominent nucleoli, but marked nuclear pleomorphism is never a feature; if present, it should cause reassessment of the diagnosis (Fig. 5–35).[60] Although scattered cytoplasmic fragments (lymphoglandular bodies) are rarely observed, they are never as numerous as in lymphoid proliferations. The small amount of cytoplasm present sometimes contains minute vacuoles, presumably representing glycogen. However, intracytoplasmic glycogen is a relatively nonspecific feature and may be seen in a variety of small blue cell tumors (e.g., lymphoma, rhabdomyosarcoma). In difficult cases, immunocytochemistry may be helpful: Ewing's sarcomas typically express CD99 but are nonreactive with leukocyte common antigen.[61] Cytogenetic analysis in most cases demonstrates the t(11;22)(q24:q12) translocation. We do not believe that Ewing's sarcoma can be reliably distinguished from small cell osteosarcoma based on cytomorphology alone. Only the presence of matrix production by the malignant cells can establish the latter diagnosis. Further, some examples of small cell osteosarcoma also express CD99, limiting its ultimate usefulness.[61] If the diagnosis of Ewing's sarcoma is suspected at the time of rapid (immediate) on-site FNA biopsy interpretation, material should be obtained for cytogenetic analysis and/or a cell block for immunocytochemistry. Final interpretation of such techniques should always be correlated with cytomorphologic and clinicoradiologic features.

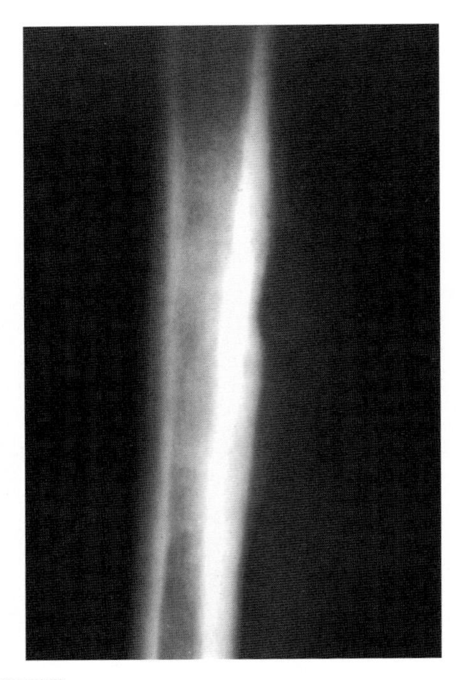

FIGURE 5–32. Plain film radiograph of Ewing's sarcoma involving the diaphysis of the femur. Note the ill-defined margins and laminated periosteal reaction.

Cytomorphologic Features of Ewing's Sarcoma

- Highly cellular smears with mostly discohesive, individual tumor cells and small cohesive clusters
- Uniform, small round tumor cells with high N/C ratios
- Round to oval nuclei with dark, finely granular chromatin and inconspicuous nucleoli
- Absence of matrix material

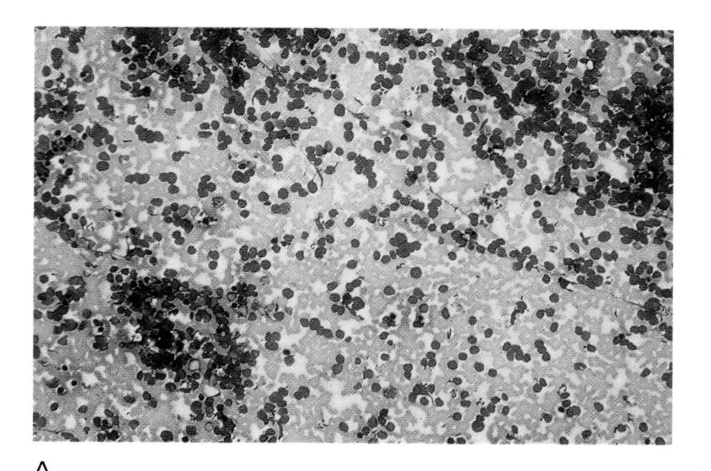

A

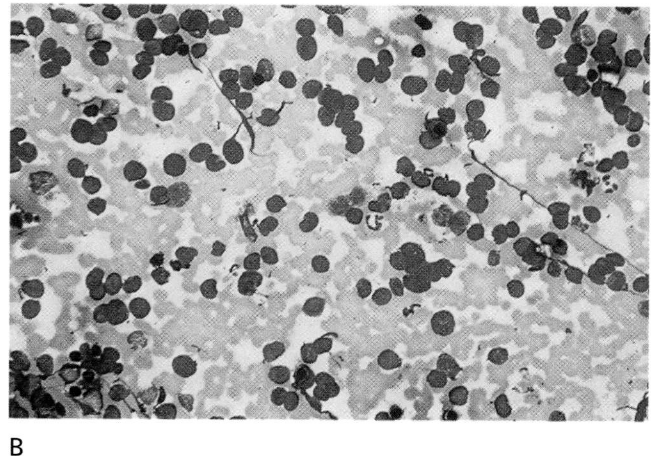

B

FIGURE 5–33. FNA biopsy of Ewing's sarcoma usually reveals hypercellular smears of mostly dissociated single cells and a few cohesive clusters. The nuclei are round to oval and uniform, and N/C ratios are very high. (Diff-Quik)

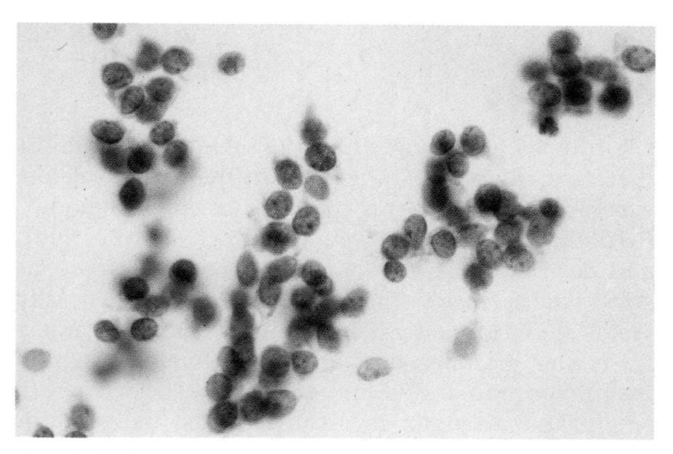

FIGURE 5–34. Ewing's sarcoma characterized by mostly uniform round nuclei with fine, even chromatin and inconspicuous to small nucleoli. Irregular nuclear contours are generally lacking. (Papanicolaou)

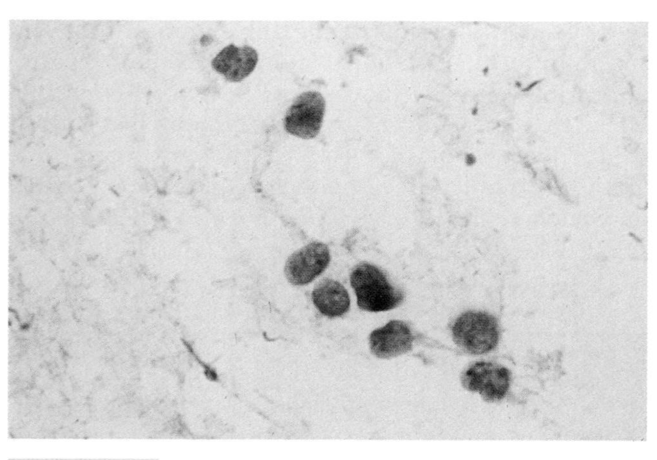

FIGURE 5–35. Rarely, atypical (large cell) forms of Ewing's sarcoma show considerably larger nuclei with irregular contours and folds. (Papanicolaou)

Adamantinoma

Adamantinoma is a low-grade malignancy of bone arising almost exclusively in the diaphysis and metadiaphysis of the tibia and/or fibula. The majority of cases occur in patients in the second or third decade of life.[62] Radiographically, adamantinomas are eccentrically located, lytic to multilobulated defects with sharply defined margins and are rarely multicentric (Fig. 5–36). Cytologic smears from adamantinomas

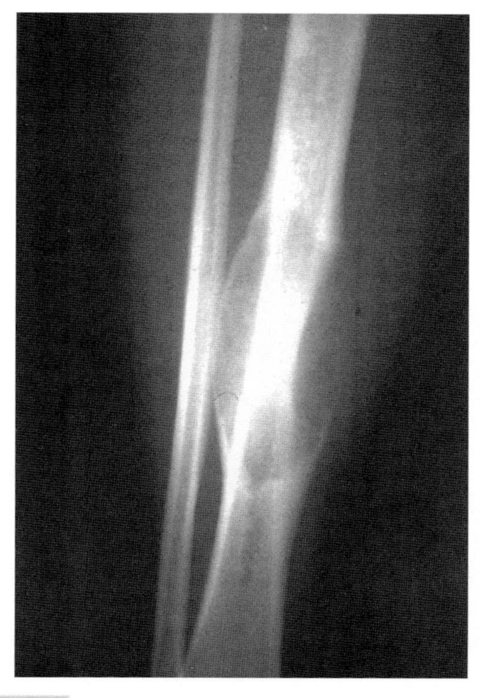

FIGURE 5–36. Plain film radiograph of adamantinoma of bone involving the tibial diaphysis in a 16-year-old girl. The lesion is lytic and expansile but overall well delineated from the surrounding parent bone.

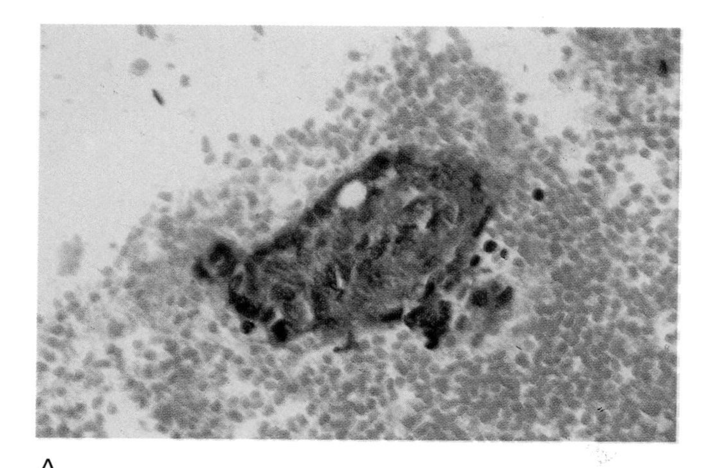

A

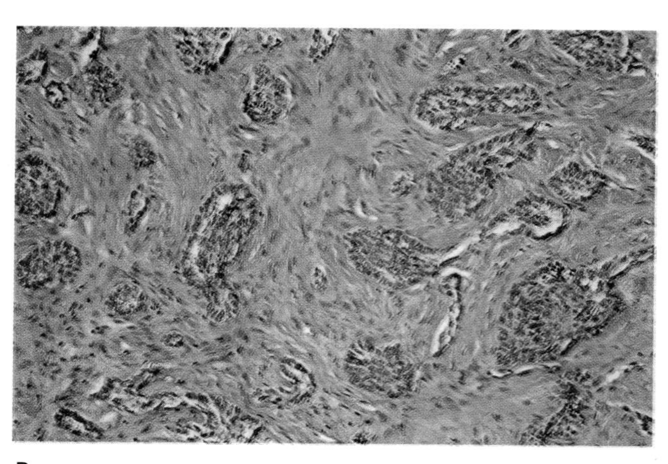

B

FIGURE 5–37. **(A)** FNA biopsy of adamantinoma yielding a cohesive cluster of mostly uniform epithelial tumor cells. **(B)** Classic histologic appearance of adamantinoma of bone characterized by islands of epithelium, reminiscent of basal cell carcinoma of skin, within a bland-appearing spindle cell fibrous stroma. (**A,B,** hematoxylin and eosin)

are most often hypercellular and contain varying proportions of benign-appearing spindle-shaped cells admixed with cohesive clusters of small, uniform, ovoid to round epithelial cells (Fig. 5–37).[63,64] The latter have oval to round, mostly uniform nuclei with a finely granular chromatin pattern and inconspicuous to small nucleoli. Correlation with clinical and radiographic findings usually prevents a misdiagnosis of metastatic carcinoma. Further, the epithelial cells of adamantinoma do not appear anaplastic. Occasional examples of adamantinoma are composed entirely of spindle-shaped cells and histologically resemble fibrosarcoma. The characteristic clinical and radiographic features, as well as the lack of anaplasia and significant mitotic activity, generally preclude the latter diagnosis.

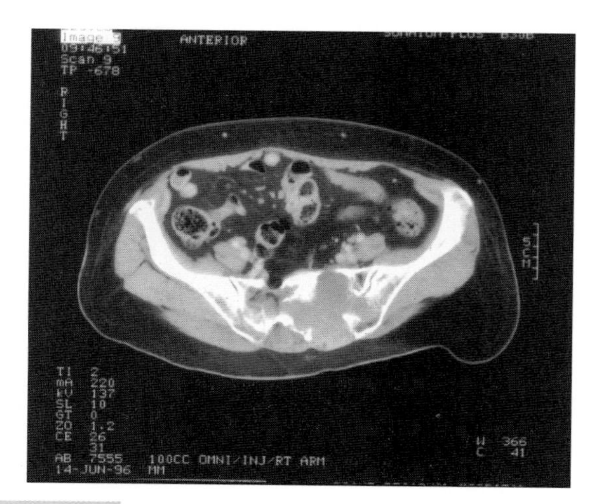

FIGURE 5–38. Computed tomographic scan of chondroma involving the sacrum in a 74-year-old woman. Such lesions are often very destructive and associated with extensive soft-tissue extension.

Chordoma

Chordomas are malignant but slow-growing tumors that virtually always arise within the axial skeleton, especially the sacral and sphenoid-occipital regions.[65] Radiographically, they are usually expansile, lytic, and destructive lesions associated with extraosseous soft-tissue extension (Fig. 5–38). FNA biopsy generally yields slight to moderately cellular smears and an extracellular granular film of myxoid material within the background. Individual neoplastic cells occur both singly and arranged in small clusters and strands. Most have round to ovoid nuclei surrounded by moderate amounts of dense to vacuolated cytoplasm (Fig. 5–39).[28,66] In some cases, large cytoplasmic vacuoles cause displacement and scalloping of the nucleus, forming the so-called physaliferous cell. Intranuclear inclusions may also be seen. Nucleoli range from small and inconspicuous to prominent. Although most examples of chordoma lack significant nuclear pleomorphism, occasional tumors exhibit bizarre-appearing anaplastic nuclei (Fig. 5–40).[67] This is especially true of recurrent tumors. Immunohistochemically, the neoplastic cells commonly coexpress epithelial markers (i.e., epithelial membrane antigen and cytokeratin) and S-100 protein.

METASTASES

The most common role for FNA biopsy in the diagnosis of intraosseous neoplasms is metastatic tumor. Most patients with metastases have a prior history of carcinoma, especially adenocarcinoma. Occasional examples of sarcoma also metastasize to bone. The majority of patients are older, usually over 50 years of

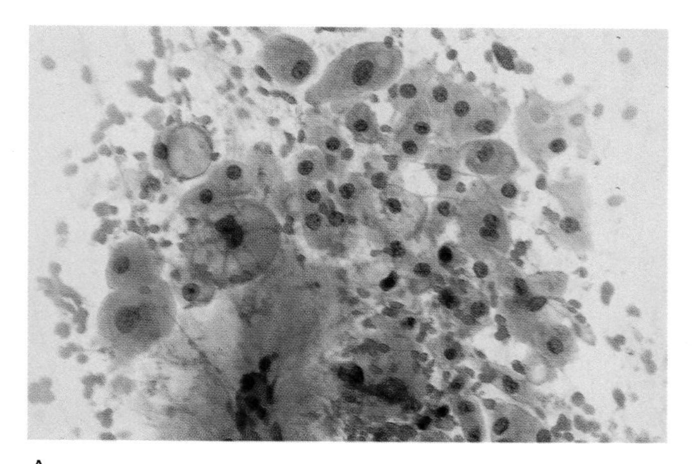

A

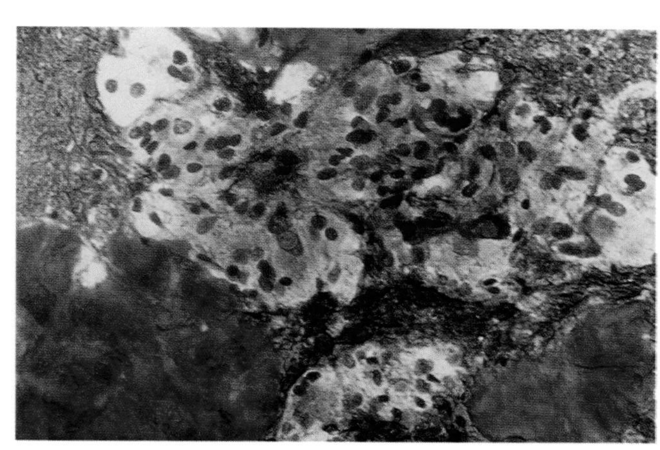

B

FIGURE 5–39. **(A)** FNA biopsy of chordoma generally yields hypercellular smears admixed with mucinous material. The neoplastic cells are round to polygonal with variably sized nuclei and abundant amounts of eosinophilic to vacuolated cytoplasm. (Hematoxylin and eosin). **(B)** In Diff-Quik preparations, the myxoid material is metachromatic, tinctorially similar to chondroid or hyaline cartilage. (Courtesy of Dr. Paul Wakely, Jr., Charlotte, NC)

age. In children, neuroblastoma is the most common nonlymphoreticular osseous metastasis. Radiographically, metastatic lesions usually appear aggressive but may range from purely lytic to osteoblastic (Fig. 5–41). The cytologic features of the more commonly associated metastatic carcinomas to bone, including breast, colon, lung, and renal adenocarcinomas, are described elsewhere in detail (Fig. 5–42).

CONCLUSION

FNA biopsy of bone tumors is a field in its infancy; we still have a great deal to learn. Nevertheless, it is clear that FNA biopsy is a reliable tool for establishing diagnoses in chondroblastoma, chondromyxoid fibroma, chordoma, giant cell tumor, Ewing's sarcoma, high-grade intramedullary osteosarcoma, multiple myeloma, Langerhans' cell histiocytosis, and adamantinoma, especially when correlated with the clinical and radiologic findings. Indeed, some cases of chondrosarcoma may be easily diagnosed by FNA biopsy. It is also extraordinarily useful for helping to exclude metastases, malignant lymphoma, infection, and even gouty tophus.[68] Whether or not FNA biopsy proves

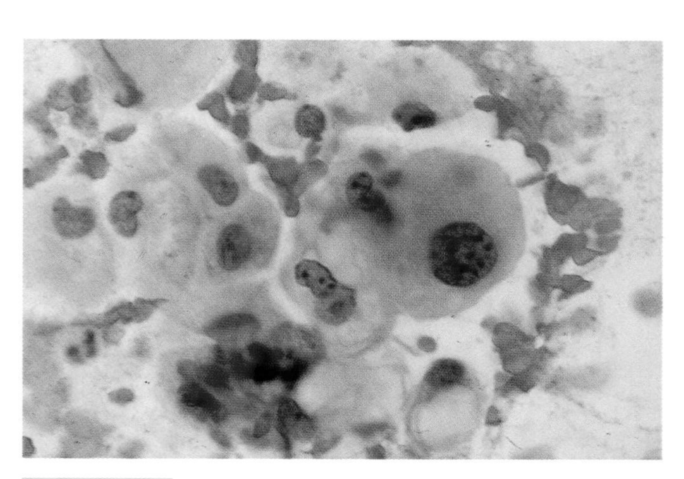

FIGURE 5–40. Occasional examples of chordoma show enlarged, more anaplastic-appearing nuclei. (Hematoxylin and eosin)

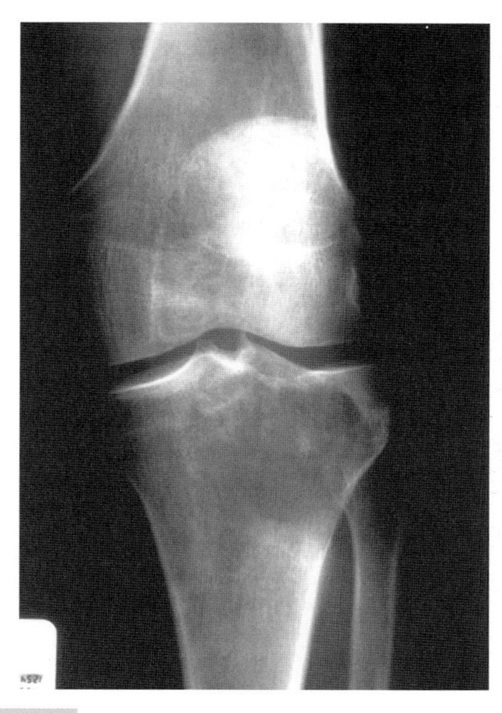

FIGURE 5–41. Plain film radiograph of metastatic renal cell carcinoma involving the proximal tibial epimetaphysis of a 64-year-old man. The lesion is purely lytic; the differential diagnosis includes giant cell tumor.

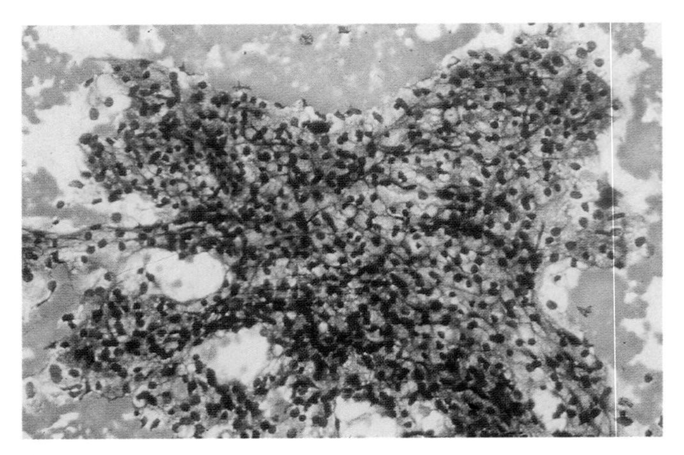

A

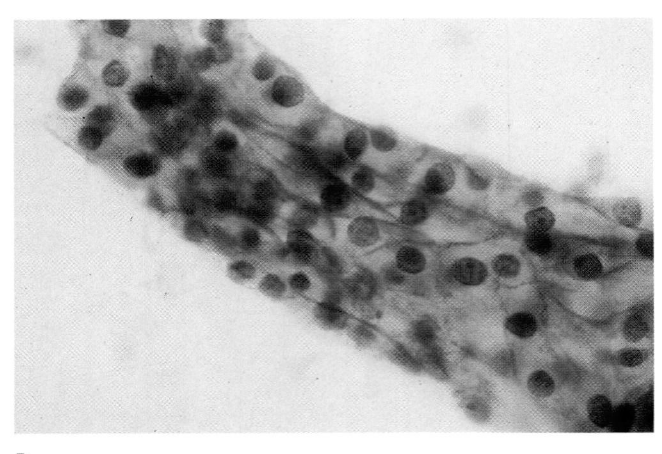

B

FIGURE 5–42. Metastatic renal cell carcinoma (same patient as Fig. 5–41) composed of cellular aggregates of mostly uniform cells with round nuclei and abundant amounts of clear cytoplasm. (Diff-Quik, Papanicolaou)

reliably useful for diagnosing such rare bone tumors as desmoplastic fibroma, hemangioendothelial sarcoma, fibrosarcoma, and infantile myofibromatosis awaits further investigation. Although immunohistochemistry and cytogenetic analysis may be supportive, the gold standard for accurate diagnosis remains with meticulous attention to cytomorphologic features and correlation with clinical and radiologic data.

REFERENCES

1. Ewing J. A review and classification of bone sarcomas. Arch Surg 4:485, 1922.
2. DeMay RM. The art and science of cytopathology: aspiration cytology. Chicago: ASCP Press, 1996.
3. Walaas L, Kindblom L-G. Light and electron microscopic examination of fine needle aspirates in the preoperative diagnosis of osteogenic tumors: a study of 21 osteosarcomas and two osteoblastomas. Diagn Cytopathol 6:27, 1990.
4. Walaas L, Kindblom L-G, Gunterberg B, Bergh P. Light and electron microscopic examination of fine-needle aspirates in the preoperative diagnosis of cartilaginous tumors. Diagn Cytopathol 6:396, 1990.
5. Layfield LJ, Glasgow BJ, Anders KH, Mirra JM. Fine needle aspiration cytology of primary bone lesions. Acta Cytol 31:177, 1987.
6. Layfield LJ, Armstrong K, Zaleski S, Eckardt J. Diagnostic accuracy and clinical utility of fine-needle aspiration cytology in the diagnosis of clinically primary bone lesions. Diagn Cytopathol 9:168, 1993.
7. Tunc M, Ekinci C. Chondrosarcoma diagnosed by fine needle aspiration cytology. Acta Cytol 40:283, 1996.
8. Kilpatrick SE, Pike EJ, Geisinger KR, Ward WG. Chondroblastoma of bone: use of fine needle aspiration biopsy and potential diagnostic pitfalls. Diagn Cytopathol 16:65, 1997.
9. Campanacci M, Capanna R, Picci P. Unicameral and aneurysmal bone cysts. Clin Orthop 204:25, 1986.
10. Mirra JM, Picci P, Gold RH. Bone tumors: clinical, radiologic, and pathologic correlations. Philadelphia: Lea & Febiger, 1989.
11. Vergel DE, Dios AM, Bond JR, et al. Aneurysmal bone cyst: a clinicopathologic study of 238 cases. Cancer 69:2921,1992.
12. Sneige N, Ayala AG, Carrasco CH, et al. Giant cell tumor of bone: a cytologic study of 24 cases. Diagn Cytopathol 1:111, 1985.
13. Klein MH, Shankman S. Osteoid osteoma: a radiologic and pathologic correlation. Skeletal Radiol 21:23, 1992.
14. Lucas D, Unni KK, McLeod RA, et al. Osteoblastoma: a clinicopathologic study of 306 cases. Hum Pathol 25:117, 1994.
15. Rocca CD, Huvos AG. Osteoblastoma: varied histological presentations with a benign clinical course: an analysis of 55 cases. Am J Surg Pathol 20:841, 1996.
16. Mondal A, Misra DK. CT-guided needle aspiration cytology (FNAC) of 112 vertebral lesions. Indian J Pathol Microbiol 37:255, 1994.
17. Nicol KK, Ward WG, Savage PD, Kilpatrick SE. Fine needle aspiration biopsy of skeletal versus extraskeletal osteosarcoma. Cancer (Cancer Cytopathol) 84:176, 1998.
18. Lee JSY, Fetsch JF, Wasdhal DA, et al. A review of 40 patients with extraskeletal osteosarcoma. Cancer 76:2253, 1995.
19. Mirra JM, Bullough P, Marcove RC. Malignant fibrous histiocytoma and osteosarcoma in association with bone infarcts. J Bone Joint Surg [Am] 56:932, 1974.
20. Kurt A-M, Unni KK, McLeod RA, Pritchard DJ. Low grade intraosseous osteosarcoma. Cancer 65:1418, 1990.
21. Mirra JM, Gold R, Downs J, Eckardt JJ. A new histologic approach to the differentiation of enchondroma and chondrosarcoma of the bones. A clinicopathologic analysis of 51 cases. Clin Orthop 201:214, 1985.
22. Henderson ED, Dahlin DC. Chondrosarcoma of bone—a study of two hundred and eighty-eight cases. J Bone Joint Surg [Am] 45:1450, 1963.
23. Unni KK. Dahlin's bone tumors: general aspects and data on 11,087 cases, 5th Ed. Philadelphia: Lippincott-Raven, 1996.
24. Dahlin D, Beabout J. Dedifferentiation of low grade chondrosarcomas. Cancer 28:461, 1971.
25. Kilpatrick SE, Pike EJ, Ward WG, Pope TL. Dedifferentiated chondrosarcoma in patients with multiple osteochondromatosis: report of a case and review of the literature. Skeletal Radiol 26:370, 1997.
26. Dee S, Meneses M, Ostrowski ML, et al. Pleomorphic ("dedifferentiated") chondrosarcoma. Report of a case initially examined by fine needle aspiration biopsy. Acta Cytol 35:467, 1991.
27. Granter SR, Renshaw AA, Fletcher CD, Bhan AK, Rosenberg AE. CD99 reactivity in mesenchymal chondrosarcoma. Hum Pathol 27:1273, 1996.
28. Kilpatrick SE, Inwards CY, Fletcher CDM, et al. Myxoid chondrosarcoma (chordoid sarcoma) of bone: a report of two cases and a review of the literature. Cancer 79:1903, 1997.
29. Wakely PE Jr, Geisinger KR, Cappellari JO, et al. Fine needle aspiration cytopathology of soft tissue: chondromyxoid and myxoid lesions. Diagn Cytopathol 12:101, 1995.

30. Turcotte RE, Kurt A-M, Sim GH, et al. Chondroblastoma. Hum Pathol 24:944, 1993.
31. Fanning CV, Sneige NS, Carrasco CH, et al. Fine needle aspiration cytology of chondroblastoma of bone. Cancer 65:1847, 1990.
32. Kilpatrick SE, Wenger DE, Gilchrist GS, et al. Langerhans' cell histiocytosis (histiocytosis X) of bone: a clinicopathologic analysis of 263 pediatric and adult cases. Cancer 76:2471, 1995.
33. Rahimi A, Beabout JW, Ivins JC, Dahlin DC. Chondromyxoid fibroma: a clinicopathologic study of 76 cases. Cancer 30:726, 1972.
34. Gupta S, Dev G, Marya S. Chondromyxoid fibroma: a fine needle aspiration diagnosis. Diagn Cytopathol 9:63, 1993.
35. Layfield LJ, Ferreiro JA. Fine-needle aspiration cytology of chondromyxoid fibroma: a case report. Diagn Cytopathol 4:148, 1988.
36. Inwards CY, Unni KK, Beabout JW, Sim FH. Desmoplastic fibroma of bone. Cancer 68:1978, 1991.
37. Raab SS, Silverman JF, McLeod DL, et al. Fine needle aspiration biopsy of fibromatoses. Acta Cytol 37:323, 1993.
38. Bertoni F, Capanna R, Calderoni P, et al. Primary central (medullary) fibrosarcoma of bone. Semin Diagn Pathol 1:185, 1984.
39. Sanerkin NG, Jeffree GM. Cytology of bone tumors. A color atlas with text. Philadelphia: JB Lippincott, 1980.
40. Hajdu SI, Hajdu EO. Cytopathology of soft tissue and bone tumors. In: Wied GL (ed): Monographs in clinical cytology. Basel: Kager, 1989.
41. Nishida J, Sim GH, Wenger DE, Unni KK. Malignant fibrous histiocytoma of bone: a clinicopathologic study of 81 patients. Cancer 79:482, 1997.
42. Kannan V, von Ruden D. Malignant fibrous histiocytoma of bone: initial diagnosis by aspiration biopsy cytology. Diagn Cytopathol 4:262, 1988.
43. Kyle RA. Multiple myeloma: review of 869 cases. Mayo Clin Proc 50:29, 1975.
44. Karmakar T, Dey P. Fine needle aspiration of plasma cell disorders: special emphasis on plasma cell subtype. Diagn Cytopathol 11;119, 1994.
45. Powers CN, Wakely PE Jr, Silverman JF, et al. Fine needle aspiration biopsy of extramedullary plasma cell tumors. Mod Pathol 3:648, 1990.
46. Ostrowski ML, Unni KK, Banks PM, et al. Malignant lymphoma of bone. Cancer 58:2646, 1986.
47. Htwe HM, Lucas DR, Bedrossian CWM, Ryan JR. Fine needle aspirate of primary lymphoma of bone. Diagn Cytopathol 15:421, 1996.
48. Biswal BM, Lai P, Uppal R, Mallik S. Unifocal Langerhans' cell histiocytosis (eosinophilic granuloma) resembling Ewing's sarcoma. Australasian Radiol 38:313, 1994.
49. Elsheikh T, Silverman JF, Wakely PE, Jr et al. Fine needle aspiration cytology of Langerhans' cell histiocytosis (eosinophilic granuloma) of bone in children. Diagn Cytopathol 7:261, 1991.
50. Rosai J, Dorfman RF. Sinus histiocytosis with massive lymphadenopathy: a pseudolymphomatous benign disorder. Cancer 30:1174, 1972.
51. Kilpatrick SE. Fine needle aspiration biopsy of Langerhans' cell histiocytosis of bone: are ancillary studies necessary for "definitive" diagnosis? Acta Cytol 42:820, 1998.
52. Unni KK, Ivins JC, Beabout J, Dahlin DC. Hemangioma, hemangiopericytoma, and hemangioendothelioma (angiosarcoma) of bone. Cancer 27:1403, 1971.
53. Wold LE, Unni KK, Beabout JW, et al. Hemangioendothelial sarcoma of bone. Am J Surg Pathol 6:59, 1982.
54. Kleer CG, Unni KK, McLeod RA. Epithelioid hemangioendothelioma of bone. Am J Surg Pathol 20:1301, 1996.
55. Kilpatrick SE, Koplyay PD, Ward WG, Richards F. Epithelioid hemangioendothelioma of bone and soft tissue: a fine-needle aspiration biopsy study with histologic and immunohistochemical confirmation. Diagn Cytopathol 19:38, 1998.
56. Khiyami A, Green LK, Gyorkey F, Landon G. Primary angiosarcoma of the cuboidal bone: a case report. Diagn Cytopathol 7:520, 1991.
57. Gray MH, Rosenberg AE, Dickerson GR, Bhan AK. Cytokeratin expression in epithelioid vascular neoplasms. Hum Pathol 21:212, 1990.
58. Hutter RVP, Worcester JN Jr, Francis KS, et al. Benign and malignant giant cell tumors of bone: a clinicopathological analysis of the natural history of the disease. Cancer 15:653, 1972.
59. Kissane JM, Askin FB, Fokulkes M, et al. Ewing's sarcoma of bone: clinicopathologic aspects of 303 cases from the intergroup Ewing's sarcoma study. Hum Pathol 14:773, 1983.
60. Renshaw AA, Perez-Atayo AR, Fletcher JA, Granter SR. Cytology of typical and atypical Ewing's sarcoma/PNET. Am J Clin Pathol 106:620, 1996.
61. Stevenson AJ, Chatten J, Bertoni F, Miettinen M. CD99 (p 30/32 Mic2) neuroectodermal/Ewing's sarcoma antigen as an immunohistochemical marker: review of more than 600 tumors and the literature experience. Appl Immunohistochem 2:231, 1994.
62. Campanacci M, Giunti A, Bertoni F, et al. Adamantinoma of the long bones: the experience at the Instituto Orthopedico Rizzoli. Am J Surg Pathol 5:533, 1981.
63. Tabei SZ, Abdollahi B, Nili F. Diagnosis of metastatic adamantinoma of the tibia by pulmonary brushing cytology. Acta Cytol 32:579, 1988.
64. Galero-Davidson H, Fernandez-Rodriguez A, Torres-Olivera FJ, et al. Cytologic diagnosis of a case of recurrent adamantinoma. Acta Cytol 33:635, 1989.
65. Kaiser TE, Pritchard DJ, Unni KK. Clinicopathologic study of sacrococcygeal chordoma. Cancer 53:2574, 1984.
66. Nijhawan VS, Rajwanshi A, Das A, et al. Fine needle aspiration cytology of sacrococcygeal chordoma. Diagn Cytopathol 5:404, 1989.
67. Ronc R, Ramzy I, Duncan D. Anaplastic saccrococcygeal chordoma: fine needle aspiration cytologic findings and embryologic considerations. Acta Cytol 30:183, 1986.
68. Nicol KK, Ward WG, Geisinger KR, Cappellari JO, Kilpatrick SE. Fine needle aspiration biopsy of gouty tophi: lessons in cost-effective patient management. Diagn Cytopathol 17:30, 1997.

CHAPTER 6

Skin and Superficial Soft Tissue

Clinically apparent skin lesions and "lumps and bumps" in the underlying soft tissue are one of the most common reasons patients present to physicians. Most of these lesions are benign and are often treated without pathologic confirmation. However, skin cancer also constitutes the most common organ-specific malignancy, with more than 600,000 cases diagnosed annually in the United States; most are squamous cell and basal cell carcinomas.[1] In addition, many primary skin cancers go unnoticed by patients and may not always have significant clinical consequences. Likewise, many soft-tissue "lumps and bumps" are benign and require no surgical or medical intervention. However, approximately 8000 new cases of sarcomas are diagnosed in the United States annually.[1] A large proportion produce superficial masses that are palpable. Unlike most skin cancers (except melanoma), sarcomas produce significant morbidity. Therefore, early intervention is indicated for all clinically suspicious soft-tissue masses.

The punch biopsy is the primary technique for morphologically evaluating cutaneous lesions.[1] However, there is still a role for the cytologic evaluation of some epidermal skin lesions, which may facilitate therapy. In addition to the Tzanck smear for herpes, cytologic preparations with potassium hydroxide (KOH) are routinely used for epidermal fungal infections (e.g., the "spaghetti and meatballs" of tinea versicolor) and arthropod infections.[2–4] Exfoliative cytologic techniques with material obtained by scraping a lesion with a scalpel or curet and spreading the specimen on a slide can be used to diagnose a variety of epidermal skin lesions, with accuracy reportedly ranging from 88–100% for basal cell carcinoma.[5–7] Likewise, the cytologic examination of unroofed vesicular lesions can often rapidly distinguish life-threatening staphylococcal scalded skin syndrome and toxic epidermal necrolysis.[4] Direct imprints of lesions, and especially the base of vesicles, can also be performed with satisfactory results.[4] The practical utility of fine needle aspiration (FNA) for skin lesions is primarily confined to neoplasms that form nodules large enough for aspiration, such as basal cell carcinoma, squamous cell carcinoma, adnexal tumors, cysts, and recurrent or metastatic melanoma (Table 6–1).[8] However, except for the above-mentioned scenarios, skin lesions are best diagnosed by punch biopsy, especially suspected primary melanocytic lesions (melanoma) and inflammatory skin disease.[1]

FNA is an important technique in the evaluation of dermal and superficial soft-tissue pathologic processes (see Table 6–1).[8–13] The accuracy of FNA of superficial soft-tissue masses has been reported to range from 64–100% and is directly related to the experience of the aspirator and diagnostician.[8,10,13] The sensitivity of FNA of soft-tissue masses is believed to be higher

TABLE 6–1. Clinical Situations in Which Skin and Superficial Soft-Tissue FNA Biopsy is Recommended

- Evaluation of a superficially palpable mass
- Diagnosis of primary, recurrent, and metastatic malignancies
- Primary melanocytic lesions should **NEVER** be evaluated by FNA
- Acquisition of material for microbiologic culture
- Acquisition of material for immunophenotypic (flow cytometry)

than the specificity; FNA is more accurate in diagnosing malignant lesions than benign lesions.[8,13] The false-positive (malignant) rate for FNA of soft-tissue sarcomas has been reported to range from 2–4%, with a false-negative (benign) rate of 0.2%.[1,10] The most common cause of an unsatisfactory soft-tissue FNA biopsy is failure to aspirate the suspected lesion.[13] Other scenarios that may lead to an interpretation of an inadequate sample are difficulty in aspirating a sufficient number of cells, especially from benign lesions, and the reluctance of the pathologist to make a definite benign diagnosis in a sample that contains a small number of normal cells for fear that the lesion in question was not adequately sampled. An inherent difficulty with soft-tissue tumors may be the ability to subtype a particular sarcoma by FNA.[11] Less often, grading may pose a challenge, largely related to insufficient sampling of a large mass.[11] However, the use of immunohistochemistry and/or electron microscopy on direct smears or cell block specimens is often helpful.

SKIN

Infections and Inflammation

Infectious agents that may involve the skin and superficial soft tissues include viruses, bacteria, fungi, parasites, and arthropods.[4,9,14] Some agents are specific for a particular cellular host, such as molluscum contagiosum for keratinocytes, whereas others, such as bacteria, are more generalized and induce pathologic consequences at various body sites and with a variety of cell types. Geographic location, occupation, age, lifestyle, and immune status are important clinical features in evaluating skin and superficial soft-tissue infections.[14] Both FNA and exfoliative scrape cytology are important techniques in examining a lesion of suspected infectious etiology.[4,9,14] An attempt to triage obtained material for appropriate microbiologic culture as well as for cytologic examination should be made routinely.[9,14]

Viruses

Viral infections are regularly diagnosed by exfoliative scrape cytology but not generally by FNA biopsy. However, familiarity with common viral cytopathic effects may be beneficial in certain instances. Herpesvirus (herpes simplex, herpes zoster, and varicella) induces multinucleation, nuclear molding, margination of chromatin (ground glass), and nuclear inclusions surrounded by halos (Fig. 6–1).[2,15] Warts (human papillomavirus) are common infections of the skin. Cytologic studies are often nonspecific and may demonstrate increased parakeratotic cells and occasionally intranuclear inclusions. The koilocytic changes in cervical lesions are not always appreciated with skin warts. Molluscum contagiosum, caused by a poxvirus, clinically presents as a papillary lesion with a collarette from which curdlike material may be expressed. The cytologic finding of large oval intracytoplasmic inclusions up to 35 μ (Henderson-Patterson bodies) adjacent to a compressed pyknotic nucleus is characteristic. Other pox-type viral diseases such as smallpox, orf, and milker's nodules also produce homogeneous cytoplasmic inclusions (Guarnieri body) with occasional intranuclear inclusions. In addition, atypical bullous measles may show Warthin-Finkeldey giant cells, which are irregularly shaped histiocytic cells with multiple nuclei.

Bacteria and Mycobacteria

Bacterial infections of the skin such as furuncles and caruncles are common but are not regularly subjected to cytologic examination. However, bacterial and mycobacterial infections of the dermis and superficial soft tissue are often diagnosed by FNA biopsy.[9,14] Bacterial infestation of the dermis and superficial soft tissue generally results in cellulitis, fasciitis, and ultimate abscess formation if the patient is not immunosuppressed or treated. The cytologic appearance of cellulitis and fasciitis generally includes abundant inflammatory cells, including neutrophils and mononuclear inflammatory cells as well as fibroblasts and histiocytes, which are often vacuolated.[14] Reactive atypia, especially in fibroblasts, is common and must be distinguished from a malignant process. Reparative features may also be observed in adjacent or involved epithelium or muscle. Necrosis is seen in advanced cases and with abscess formation. Bacteria can often be better observed with the Diff-Quik stain or a Gram's stain.

Mycobacteria (e.g., tuberculosis leprosy) typically induce a granulomatous response with epithelioid histiocytes, multinucleated giant cells, lymphocytes, and necrosis. In leprosy, globi, which are globules of lipid, are characteristically observed both in histiocytes and extracellularly.[16] Negative images of the

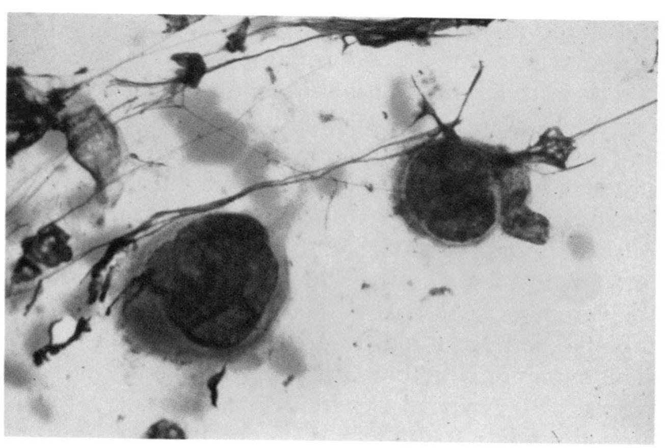

A

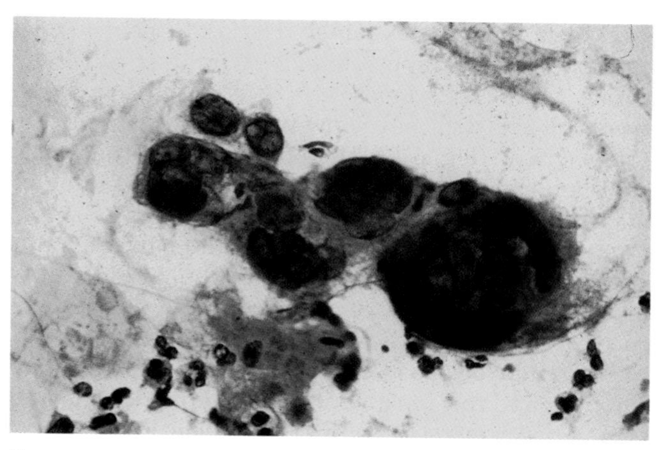

B

FIGURE 6–1. Herpesvirus infection. Exfoliative scrape cytology of the base of a vesicular skin lesion demonstrates multinucleation, nuclear molding, and margination of the chromatin. (**A,** Diff-Quik; **B,** Papanicolaou)

mycobacterium may be appreciated with the Diff-Quik stain and can be confirmed with acid-fast or Fite stain.[16] Other granulomatous entities include fungal infection, sarcoidosis, foreign-body material, rheumatoid nodule, gout, and pseudogout (see noninfectious granulomatous diseases).

Fungi

Superficial fungal infection of the skin, including dermatophytes, tinea versicolor (*Malassezia furfur*), and candidiasis, may be diagnosed by exfoliative scrape cytology.[4] In contrast, subcutaneous mycoses can be sampled by FNA biopsy and generally display a granulomatous cytologic appearance. Fungal pseudohyphae and yeast may be visualized as negative images with the Diff-Quik stain.[14] Methenamine silver, mucicarmine, and Fontana-Masson stain can be used to delineate the specific organism.[14,16] The following fungal organisms are often diagnosed by FNA: *Aspergillus, Candida, Blastomyces, Cryptococcus, Histoplasma, Coccidioides,* and *Zygomycetes* (Fig. 6–2).[14,16–18]

Other Infections

Other infectious agents of the skin and superficial soft tissue that can be diagnosed cytologically by FNA biopsy or exfoliative scrape cytology include protozoa (cutaneous leishmaniasis and cutaneous amebiasis), parasites (cysticercosis, pinworm), and arthropods (*Demodex folliculorum,* scabies mite).[19] Cutaneous leishmaniasis is a zoonotic disease endemic to the Mediterranean area. In cytologic preparations, organisms with small round nuclei and associated kinetoplasts are seen within histiocytes or extracellularly in a background of chronic inflammation.[20] Cysticercosis is a worldwide parasite infection of muscle. FNA biopsy of a suspected cysticercosis lesion should be performed with caution because of reports of anaphylaxis and possible dissemination of parasites.[21,22] Cytologic examination reveals various parts of the organism, such as hooklets, scolex, spiral wall fragments, sucker, or calcareous corpuscles, within a background of inflammation in which eosinophils predominate.[21,22]

Basal Cell Carcinoma

Basal cell carcinoma is the most common skin cancer. It typically presents in middle-aged to elderly patients in sun-damaged skin as nodules, often with central ulceration (rodent ulcer). Although often locally aggres-

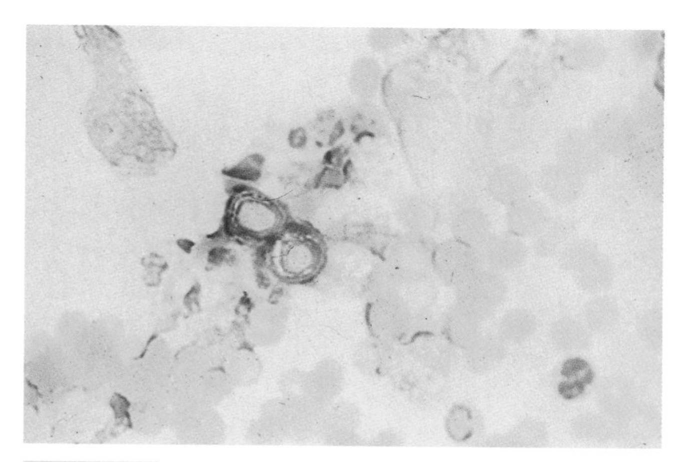

FIGURE 6–2. Blastomycosis. Aspirated lesions that contain blastomycosis display spherical (8–20 microns in diameter) organisms with thick refractile walls. Broad-based budding is occasionally observed. (Diff-Quik)

sive, metastases are extremely rare. Both FNA biopsy and exfoliative scrape cytologic techniques allow accurate cytologic diagnoses.[3,4,7,8,23–28] Tightly crowded sheets of cells that may show nuclear overlapping are characteristic (Fig. 6–3). The borders of the fragments are often smooth and may demonstrate palisading of nuclei without molding.[4,23–28] The cells are small, round to oval, and hyperchromatic and have high nuclear to cytoplasmic (N/C) ratios.[4,26,28] Spindling of the basal cells may be seen in cystic tumors. The presence of mitotic figures and nucleoli is variable.[28] Melanin may be observed in pigmented basal cell carcinomas, and foci of squamous differentiation may be seen in basosquamous tumors.[25] Metachromatic, pink-staining, amorphous material is usually present in the background.[27] The differential diagnosis includes Merkel cell carcinoma, metastatic small cell carcinoma, and adnexal tumors. Merkel cell carcinoma and metastatic small cell carcinoma are generally positive for neuron-specific enolase and chromogranin and demonstrate perinuclear dot immunocytochemically positive for low molecular keratin. Basal cell carcinomas, on the other hand, are usually diffusely keratin-positive and negative for neuron-specific enolase and chromogranin.[28] However, rare cases of basal cell carcinoma exhibit neuroendocrine differentiation.[29]

Squamous Lesions

Squamous cell carcinoma is the second most common human malignancy. It generally appears in sun-exposed skin in middle-aged and elderly persons. Lesions may evolve from actinic (solar) keratosis to carcinoma *in situ* to squamous cell carcinoma. Keratoacanthomas are a specific squamous proliferation with a characteristic clinical appearance of a central crater; many investigators consider them to represent a well-differentiated squamous carcinoma. The cytologic differentiation of squamous lesions is difficult and requires correlation of the clinical appearance and history, as well as often subsequent biopsy for primary lesions. Actinic keratosis, carcinoma *in situ*, keratoacanthoma, and well-differentiated squamous cell carcinoma are generally flat-surfaced lesions, often with ulceration or scaly crust formation. Thus, they are not usually aspirated but are accessible to exfoliative scrape cytology.[4–7] However, nodular invasive squamous lesions can be evaluated by FNA biopsy.[8,30,31] Atypical hyperchromatic squamous cells are observed with well-differentiated keratinized squamous cell carcinoma (Fig. 6–4). Numerous mitotic figures and atypical vesicular nuclei with prominent nucleoli are often seen with poorly differentiated nonkeratinized squamous cell carcinoma. Dense cytoplasm and distinct cell borders are characteristic findings of most squamous lesions. Neoplastic cells may be observed singly (especially with keratinized squamous cell carcinoma) or in tight sheets. Within the latter, the cells may appear spindled.[30] The differential diagnosis of squamous cell carcinoma includes pilomatrixoma and epidermoid cysts, which do not generally demonstrate atypical features. However, pilomatrixomas may display slight atypia, with syncytial clusters of "basaloid" cells having high N/C ratios; this can be a potential pitfall for incorrectly diagnosing a high-grade malignancy (see Pilomatrixoma).

Paget's Disease

Paget's disease of the nipple or vulva is rarely diagnosed by exfoliative scrape cytology in ulcerated lesions. Usually, only benign superficial squamous cells

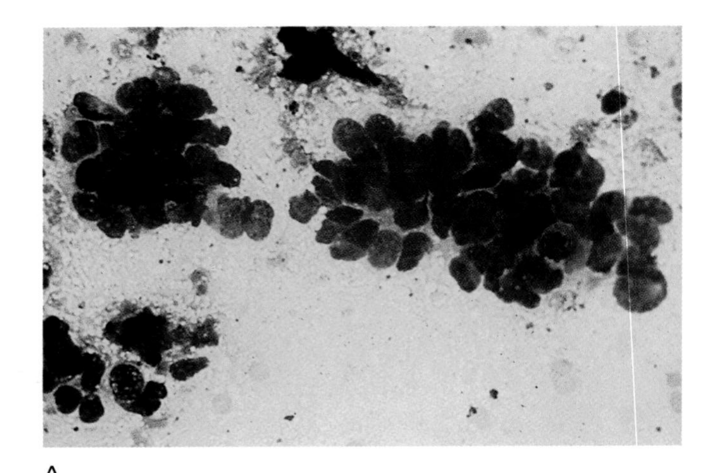

A

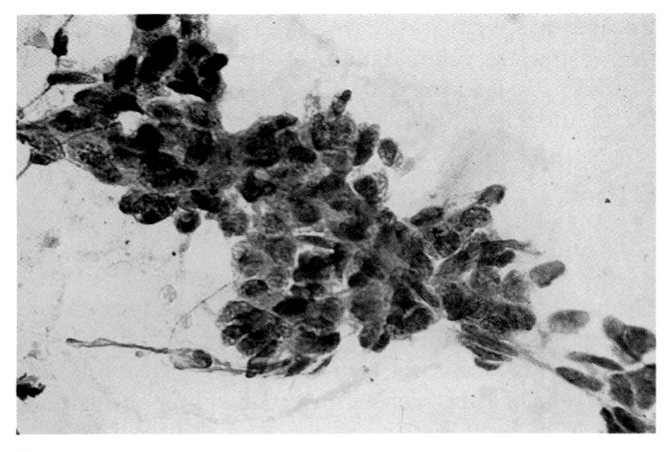

B

FIGURE 6–3. FNA biopsy of basal cell carcinoma of the skin displays tightly crowded sheets with nuclear overlapping. Nuclear palisading along the edge is present focally. Cells are small and round to oval, with an increased N/C ratio. (**A,** Diff-Quik; **B,** Papanicolaou)

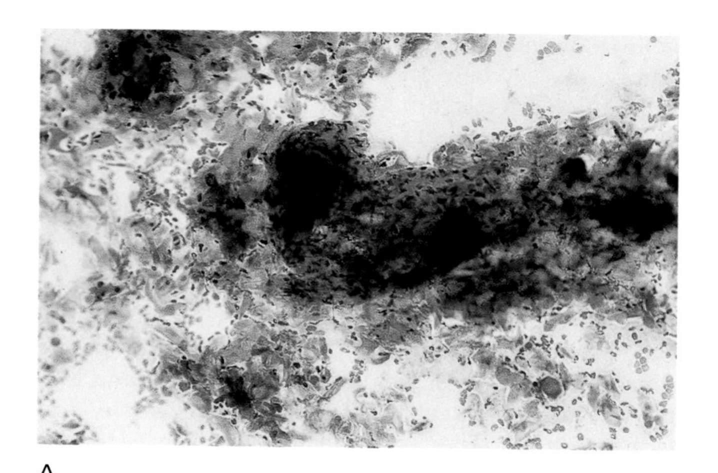

A

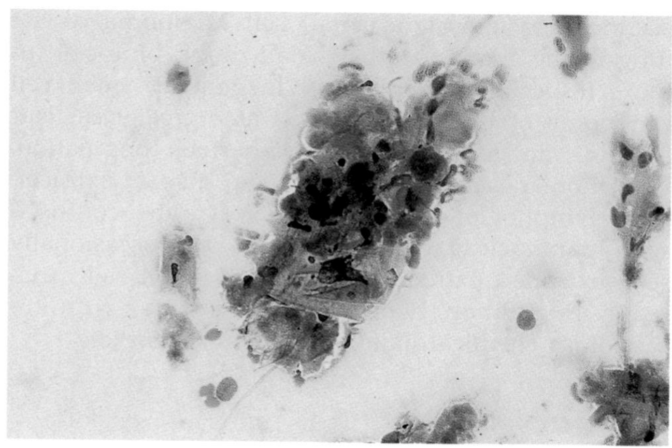

B

FIGURE 6–4. FNA biopsy of squamous cell carcinoma displays sheets and atypical single cells. The nuclei of well-differentiated lesions are hyperchromatic and irregular; the cytoplasm is "hard," with evidence of keratinization. (Papanicolaou)

are observed, and the individually scattered pleomorphic cells that are infiltrating the epidermis in histologic sections are not sampled. FNA biopsy has no clinical role in the diagnosis of Paget's disease.

Merkel Cell Carcinoma

Merkel cell carcinoma is a rare neuroendocrine tumor of the skin that most commonly presents as a subcutaneous mass in the head and neck of elderly patients. Metastases are relatively common compared to most other skin cancers. FNA biopsy demonstrates abundant small round blue cells that tend to be single or loosely cohesive (Fig. 6–5).[32–40] Rare rosette-like formations may be observed. Slight molding may be seen, and there is usually some crush artifact (nuclear streaking).[34] The nucleus is round and demonstrates the classic "salt and pepper" pattern of neuroendocrine tumors with the Papanicolaou stain.[32–40] Nuclear pleomorphism is generally slight, and binucleated cells are occasionally present. Paranuclear globules are seen in some cases.[33,35,36] The N/C ratio is quite high, often with only a thin rim of finely granular cytoplasm. Naked nuclei, numerous mitoses, and individual cell necrosis may all be present.[32,37] Immunohistochemically, the cells are positive for cytokeratin (perinuclear dotlike staining) and neuron-specific enolase, and usually for S-100.[35–37,39] Variable immunohistochemical staining for epithelial membrane antigen (EMA) has been reported.[37] The cells are negative for vimentin, calcitonin, serotonin, and leukocyte common antigen.[36] The differential diagno-

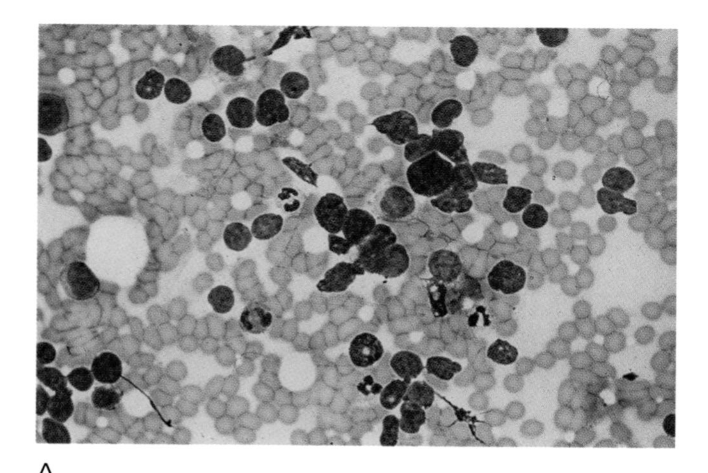

A

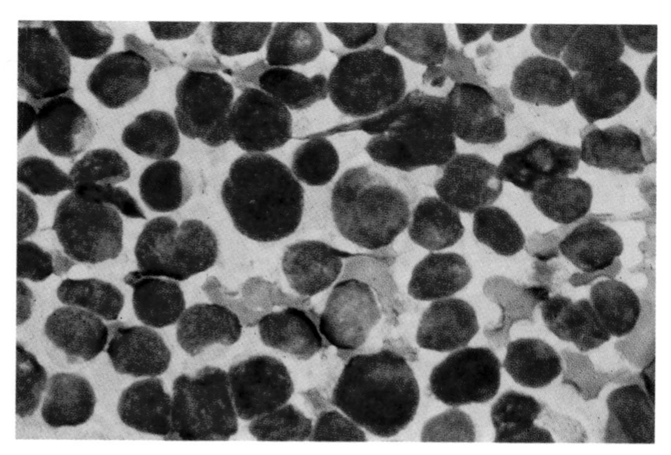

B

FIGURE 6–5. FNA biopsy of Merkel cell carcinoma shows loosely cohesive and single cells with slight nuclear molding. The cells demonstrate round to oval nuclei, with "salt and pepper" chromatin and increased N/C ratios. Note the absence of tight clusters and lymphoglandular bodies, as seen with basal cell carcinoma and lymphoma, respectively. (Diff-Quik) sis includes metastatic small cell carcinoma of the lung to the skin and other small round blue cell tumors (especially lymphoma). In general, small cell carcinoma of the lung demonstrates greater nuclear molding and a more pleomorphic cell population. Lymphomas lack true cohesion and display extracellular lymphoglandular bodies, not to be confused with paranuclear globules, which are occasionally seen extruded extracellularly with Merkel cell carcinoma. In addition, lymphomas are positive with immunohistochemical stains for lymphoid markers.

Pilomatrixoma

Pilomatrixoma, or calcifying epithelioma of Malherbe, is a benign tumor of the hair follicle that usually occurs in the head, neck, and upper extremities of young people. The characteristic histologic features are a tumor consisting of basaloid and nucleated squamous cells merging with anucleate "ghost" or "shadow" cells surrounded by fibrous stroma containing multinucleated giant cells. Calcification of this neoplasm is common. By FNA, three characteristic cell types should be identified: basaloid cells, nucleated squamous cells, and anucleate ghost cells (Fig. 6–6).[8,41–43] The basaloid and nucleated squamous cells are present in sheets or singly and may demonstrate considerable variability, including degenerative changes. Clusters of basaloid cells in syncytial groups may be misdiagnosed as a high-grade malignancy if the clinical findings and other cytologic features, such as the presence of ghost cells, are not appreciated.[42,43] Anucleate ghost cells are seen singly and in clumps. With the Papanicolaou stain, the ghost cells display light-yellow, abundant cytoplasm with a centrally located empty space where the nucleus previously resided. Inflammatory cells, multinucleated giant cells, debris, and calcium fragments may be observed in the background.[41,42]

Sebaceous Carcinoma

Sebaceous carcinoma is an uncommon adnexal tumor that most commonly occurs on the eyelid, often associated with the meibomian glands. The FNA biopsy of sebaceous carcinoma yields numerous single and lobule-like clusters of cells.[8] The neoplastic cells are very large, with abundant vacuolated cytoplasm. The nucleus is vesicular, with prominent nucleoli and frequent mitotic figures. Squamous and basaloid cells may be admixed with the neoplastic cells. The background often contains lipid and granular debris. The anaplastic appearance of the vacuolated neoplastic cells should be differentiated from the smaller, benign-appearing foam cells (histiocytes) identified in chalazion, which occurs solely in the eyelid.[8] Also within the differential is fat necrosis and metastatic adenocarcinoma.

Apocrine Carcinoma

Apocrine papillary adenocarcinoma is a rare malignant adnexal tumor that most frequently involves the axilla. An FNA biopsy case report of a lymph node metastasis from an adjacent primary scalp lesion demonstrated monomorphic cells with abundant, finely granular cytoplasm with central nuclei arranged in clusters and papillary formations.[44]

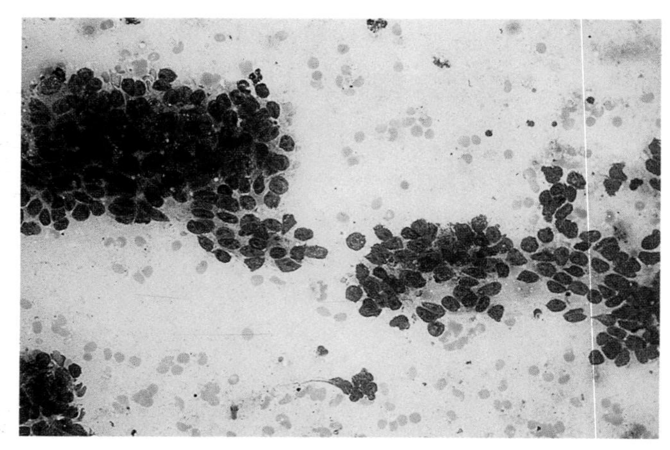
A

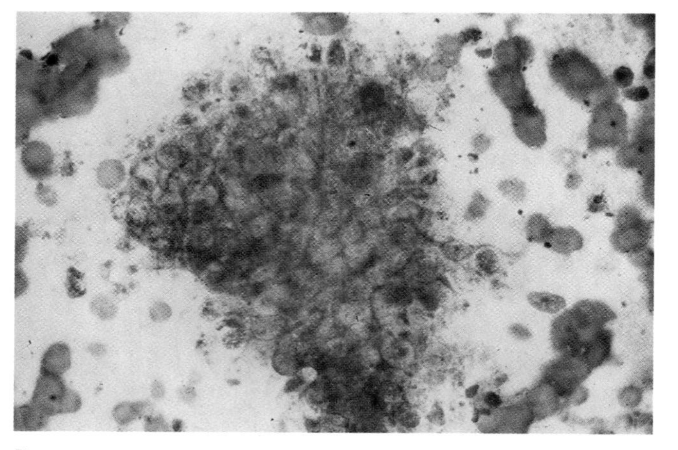
B

FIGURE 6–6. FNA biopsy of pilomatrixoma should demonstrate three cell types: basaloid cells, nucleated squamous cells, and anucleated ghost or shadow cells. The basaloid cells often are arranged in clusters with increased N/C ratios. The ghost or shadow cells are an important diagnostic clue, dependably identifying pilomatrixoma from other skin and superficial soft-tissue lesions. (Diff-Quik)

Hidradenoma

Hidradenoma is an uncommon adnexal tumor that occurs predominantly in the female perineum. The aspirate smears generally display cellular specimens with a monomorphic cell population, with many of the cells arranged in sheets, some with a papillary appearance.[45] The neoplastic cells have a moderate amount of cytoplasm and bland oval nuclei without nucleoli. Metachromatic hyaline globules may be observed. A reported case of a rare malignant hidradenoma demonstrated mild atypia and occasional tubule formation, as well as mitotic figures and necrosis.[46]

Syringocystadenoma Papilliferum

Syringocystadenoma papilliferum occurs most commonly on the scalp or face, is usually first noted at birth or in childhood, and often increases in size during puberty. The FNA biopsy appearance is that of discohesive papillary-like clusters of round to oval monomorphic cells with central bland nuclei.[47] The background contains a mixed inflammatory infiltrate with many plasma cells.

Cylindroma

Cylindromas occur in early adulthood, most commonly on the head and face. These tumors may be inherited and when multiple and on the scalp have been referred to as "turban tumors." They are often associated with trichoepitheliomas and eccrine spiradenomas. FNA biopsy of cylindromas reveals clusters of tightly packed monomorphic cells with a thin rim of cytoplasm or naked nuclei.[48] The nuclei are round to oval with smooth nuclear contours, finely granular chromatin, and inconspicuous nucleoli. The Diff-Quik stain highlights metachromatic pink to red hyaline globules, which may be encircled by the neoplastic cells. Alternatively, this hyaline substance may be observed investing clusters of cells. The cytologic appearance of a benign dermal cylindroma is virtually identical to the more malignant adenoid cystic carcinoma of the salivary gland. The identification of hyaline globules with basaloid cells from a skin FNA is consistent with cylindroma; however, rare basal cell carcinomas also demonstrate hyaline globules, as do metastatic adenoid cystic carcinomas.[48]

Eccrine Spiradenoma

Eccrine spiradenomas usually arise during early adulthood as a solitary tender nodule with no characteristic location. Aspirates are cellular, with bland basaloid cells arranged in an irregular tubular or clustered pattern associated with amorphous basement membrane-like material. A published FNA case report described a rare malignant example that contained cells with atypical nuclear features (hyperchromasia, irregular shapes, and prominent irregular nucleoli) and mitoses.[49]

Chondroid Syringoma

Chondroid syringomas occur most commonly on the head and neck. Most are benign. These tumors are cytologically similar to pleomorphic adenoma of the salivary gland. Benign-appearing round to oval epithelioid cells with monomorphic nuclei and a moderate amount of cytoplasm are present singly and in tubules.[47,50,51] Admixed with the epithelioid cells are occasional spindled mesenchymal cells. Typically, the tumor cells are embedded in a myxoid matrix that demonstrates a characteristic bright-magenta color with the Diff-Quik stain.

Melanoma

Although melanoma is the third most common skin cancer, it is most responsible for mortality associated with cutaneous malignancies. It has a high rate of metastasis and can occur in any age group. As with other skin cancers, its prevalence increases with sun exposure. In general, melanoma presents as a pigmented lesion with asymmetry, irregular borders, variegated color, and a diameter greater than 6 mm.[1] Suspected primary melanomas should **NEVER** be subjected to primary diagnosis by FNA (see Table 6–1).[1] Full-thickness histologic biopsy for the evaluation of depth of invasion is required. However, FNA biopsy is commonly used to diagnose recurrent or metastatic melanoma.[1]

Malignant melanoma has been termed the "great mimicker" because of its diverse cytologic appearances. In general, the FNA biopsy obtains a moderate to highly cellular specimen, except in the occasional case of a desmoplastic melanoma (Fig. 6–7).[4,8,31,52–57] The neoplastic cells are usually singly dispersed. In general, the cells are large and polygonal, with round, eccentric nuclei. Binucleation is common. The nuclear contours are often smooth and the chromatin is finely granular. Macronucleoli and intracytoplasmic inclusions are characteristic. In 30–60% of the cases, the cells contain melanin pigment characterized as a finely granular, brown-black cytoplasmic pigment with the Papanicolaou stain and blue-black pigment with the Diff-Quik stain (Fig. 6–8). Melanin is usually not present in all cells. In some cases, prominent melanin pigment is grossly seen in the smears, and

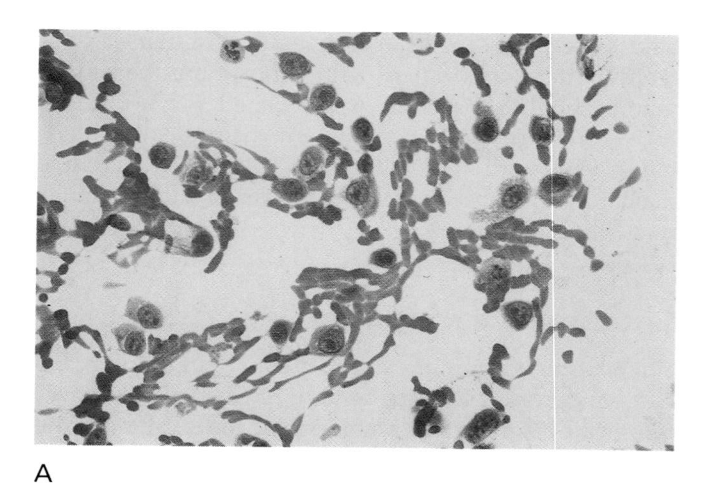

A

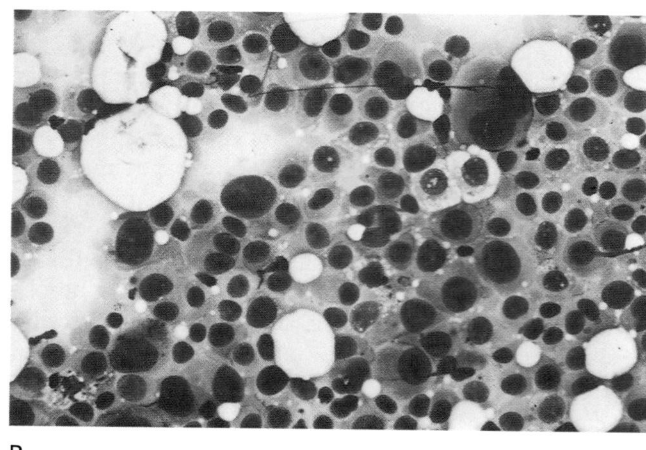

B

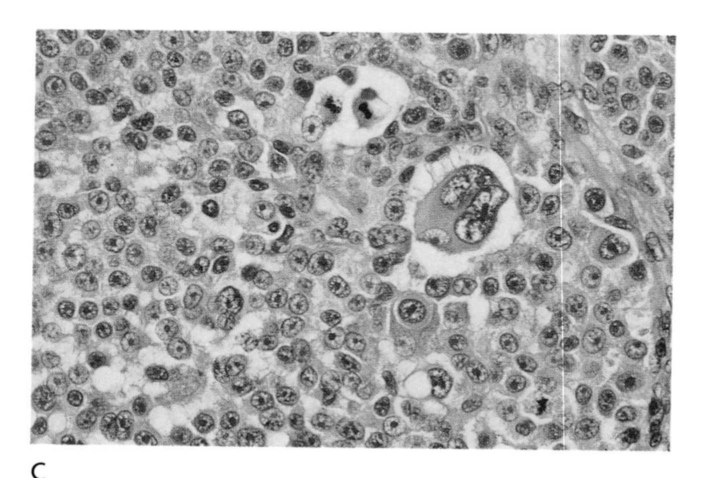

C

FIGURE 6–7. Melanoma. Cellular specimens with anisonucleosis and prominent nuclei are characteristic of melanoma. The nuclei are often eccentric with prominent nucleoli, imparting a rhabdoid appearance. Binucleation is common. Cytoplasm may be scant or abundant. (**A,** Papanicolaou; **B,** Diff-Quik; **C,** cell block)

the melanin bleach technique with hydrogen peroxide may be required to view cellular detail adequately.[58] In cases that do not contain melanin, fine cytoplasmic vacuolization is common. The cytologist should always be aware of the potential of this "great mimicker" to have a variety of cytologic appearances, including rhabdoid, giant cell, clear cell, signet-ring cell, desmoplastic (spindled), myxomatous, and amelanotic variants (Fig. 6–9).[52–57,59–61] Thus, the differential diagnosis of malignant melanoma includes carcinomas, sarcomas, and lymphomas. The immunohistochemical profile of malignant melanoma is usually positive for S-100, HMB-45, and vimentin.[54–57,62] Electron microscopy is especially helpful in identifying premelanosomes within amelanotic melanomas. However, the most useful diagnostic clue, in addition to the cytologic appearance, is the clinical history, especially in instances of suspected recurrent or metastatic malignant melanoma.

Cysts

Skin and superficial soft-tissue cysts are extremely common, with 90% of the cases representing epidermoid (infundibular) cysts. FNA biopsy of an epidermoid cyst generally yields abundant yellow-tan, pastelike, and often malodorous material (Fig. 6–10).[63] Benign nucleated and anucleated squamous cells, often with degenerative features and keratinous debris, are characteristic.[8,30,31,63] Inflammatory cells, including multinucleated giant cells, may be seen, especially if rupture has occurred. The cytologic features of epidermoid cysts are indistinguishable from epidermal inclusion cysts (traumatic epidermoid cysts) and milia. It is also difficult cytologically to distinguish an epidermoid cyst from a trichoepithelioma (pilar or isthmus catagen) cyst; the only distinguishing feature is the lack of parakeratotic cells with the latter. Identification of basaloid cells helps to distin-

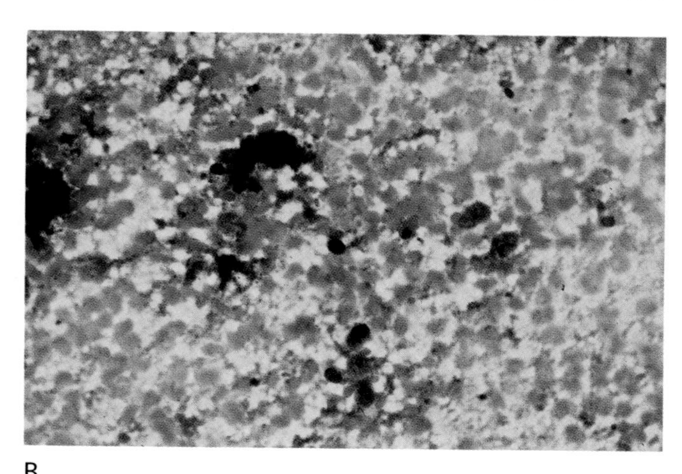

C

FIGURE 6–8. Melanoma. (**A**) Melanin pigment is occasionally observed grossly on direct smears. Microscopically, melanin pigment is granular and brown-black with Papanicolaou (**B**) and blue-black with the Diff-Quik stain (**C**).

guish a pilomatrixoma from an epidermoid cyst. Careful attention should be directed toward the nuclear features in the squamous cells of all cystic structures so as not to overlook a degenerating cystic squamous cell carcinoma.[30]

Ganglion (synovial) cysts occur most frequently in the region of the wrist, foot, or knee. They are often associated with a joint and may become painful from compression of adjacent structures. FNA biopsy of a ganglion cyst typically obtains thick, semitransparent gelatinous myxoid fluid containing a variable number of vacuolated histiocytes (Fig. 6–11).[64,65]

A number of cysts located in the head and neck region (bronchogenic, branchial cleft, median raphe, thyroglossal, and thymic) are discussed in detail in the salivary gland chapter.[30,63] Other benign superficial cystic structures of the skin and superficial soft tissue include steatocystoma, cutaneous ciliated cysts, digital mucous cyst, mucocele, and endometriosis

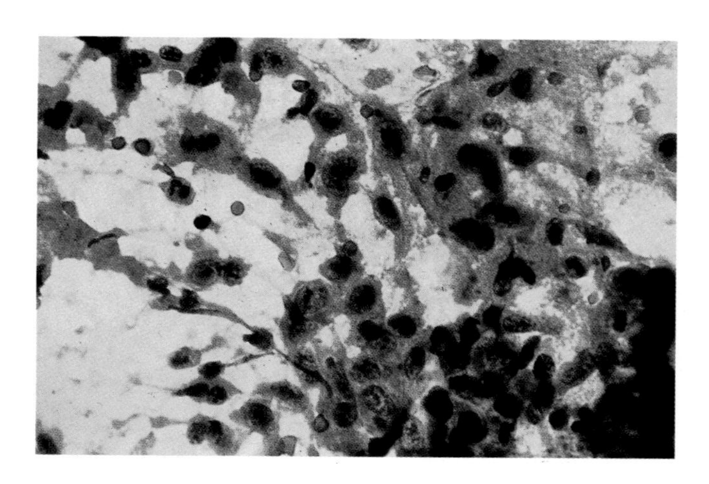

FIGURE 6–9. FNA biopsy of desmoplastic melanoma demonstrates atypical spindle cells with irregular vesicular nuclei and prominent nucleoli. Differentiation from other spindle cell tumors often requires immunocytochemical expression of HMB-45 or electron microscopic identification of melanosomes. (Papanicolaou)

A

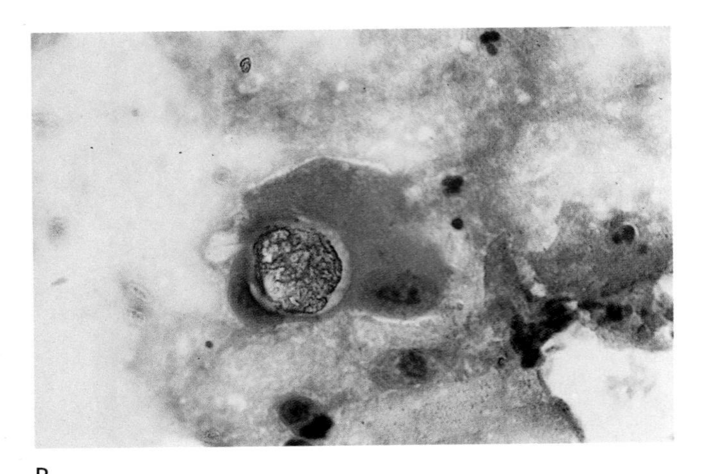

B

FIGURE 6–10. FNA biopsy of epidermal inclusion cyst demonstrates nucleated and anucleated squamous cells, inflammatory cells, multinucleated giant cells, and keratinous debris, which may be refractile. (Papanicolaou)

(discussed later in the chapter), as well as other entities.[30,31,63] Cystic degeneration of a primary or metastatic neoplasm should always be considered when evaluating a cystic structure.[30]

SOFT TISSUE

Mesenchymal Reparative and Reactive Processes

Reparative and/or reactive mesenchymal processes involving the superficial soft tissues may induce a nodular mass suspicious for a neoplasm. Examples include nodular fasciitis, proliferative fasciitis/myositis, and myositis ossificans.[66–72] These lesions are more common in young adults and demonstrate rapid growth. Local trauma or inflammation is often cited as the etiologic agent. The FNA cytologic appearance of nodular fasciitis is dependent on the age of the growth because it procedes through histologic developmental stages.[69] Early lesions yield more cellular smears than later lesions and show a wide diversity of cell types (Fig. 6–12). In the early phase, reactive spindled fibroblasts and myofibroblasts are common; they possess round to oval nuclei, with evenly dispersed chromatin and small to large nucleoli.[67,69] Mitotic figures may be plentiful. Lymphocytes, histiocytes, and an occasional neutrophil may be seen. Some of the histiocytes may contain cytoplasmic lipid (lipophages) or granular hemosiderin (siderophages). Occasional multinucleated giant cells and fragments of blood vessels may also be seen. As the lesion evolves and becomes less active, spindled fibroblasts with less cytoplasm and inconspicuous nucleoli become more prominent. The background matrix changes from a loose myxoid tissue to collagenous in nature. In reparative/reactive masses in which ganglion-like giant cells are present in addition to the typical fibroblasts, proliferative fasciitis/myositis is diagnosed (Fig. 6–13).[66,72] Ganglion-like giant cells display one or two round to oval, eccentrically placed vesicular nuclei with one or multiple prominent nucleoli. The cytoplasm is abundant and may contain vacuoles or be indistinct. The cytologic features of myositis ossificans are also similar to nodular fasciitis, except that foci of cartilaginous or osseous material are present.[68,70,71] Regenerating muscle fibers are also typically seen in myositis ossificans.

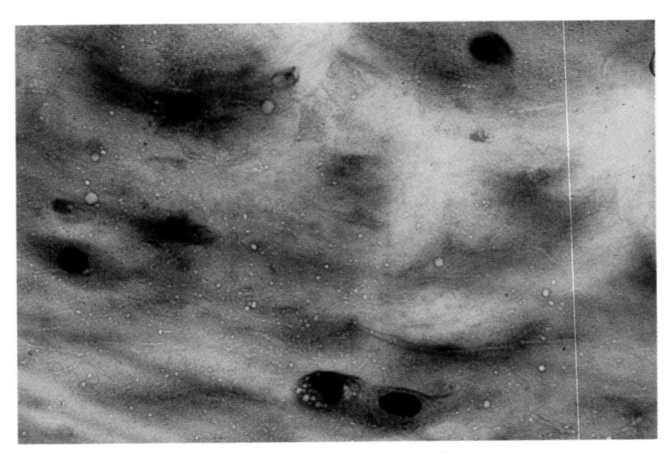

FIGURE 6–11. FNA biopsy of a ganglion (synovial) cyst typically displays thick semitransparent gelatinous myxoid fluid and scattered vacuolated histocytes. (Diff-Quik)

The cytologic features of the active or early phase of reactive/reparative mesenchymal processes may be easily confused with sarcomas because of abundant mitosis and active-appearing fibroblasts and myofibroblasts with prominent nucleoli.[67] In addition, the myxoid background may suggest a true myxoid neoplasm. However, correlation of the clinical findings,

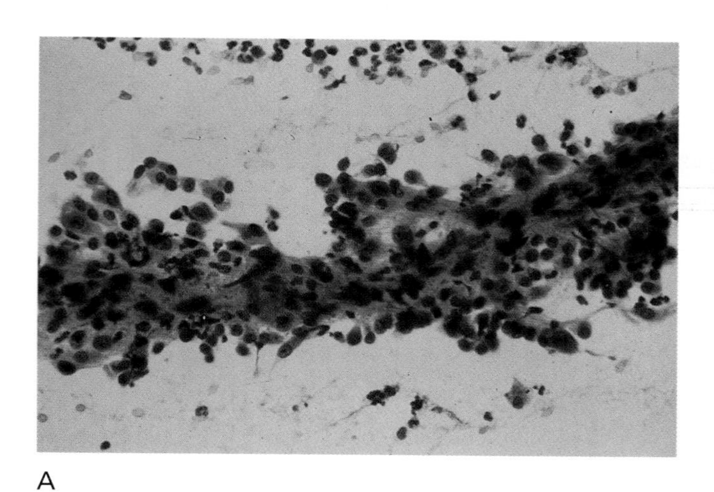

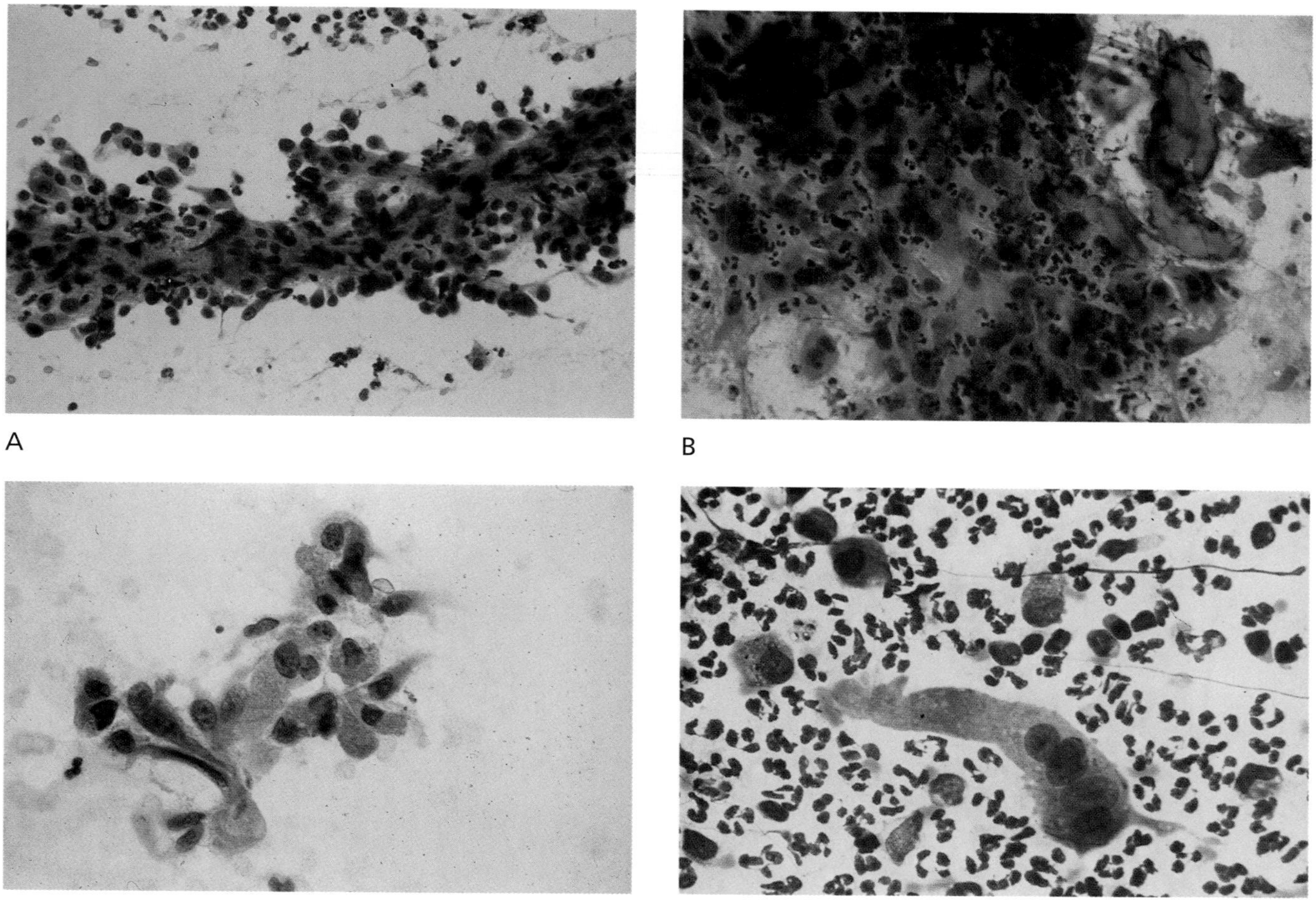

A

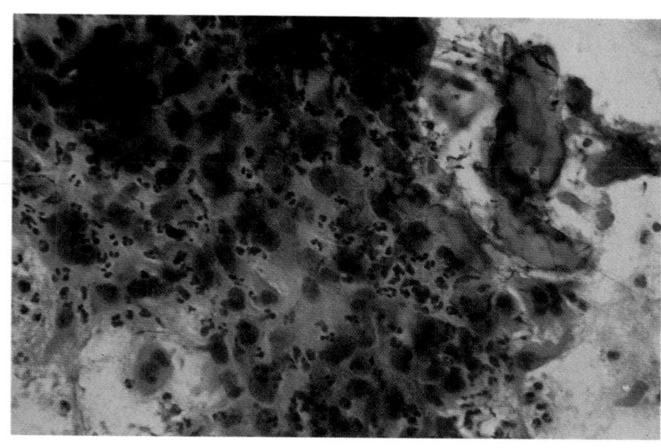

B

C

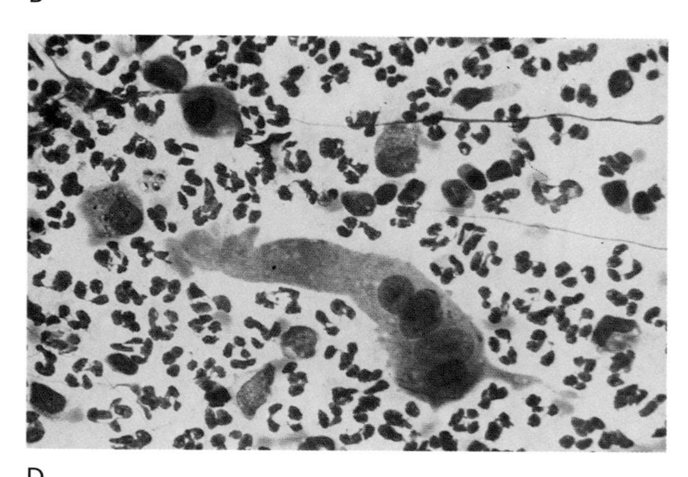

D

FIGURE 6–12. FNA biopsy of mesenchymal repair displays spindled "active" proliferating fibroblasts in clusters and as single cells. A "streaming" pattern may be observed. Active fibroblasts may show a variety of cell shapes. Prominent nucleoli are often present, and the chromatin is evenly distributed, with smooth nuclear contours. N/C ratios are generally not increased. Numerous mitoses and multinucleation may be present. The observation of the inflammatory cells in the background and the clinical history are helpful to avoid a potential diagnosis of sarcoma. (**A,B,C,** Papanicolaou; **D,** Diff-Quik)

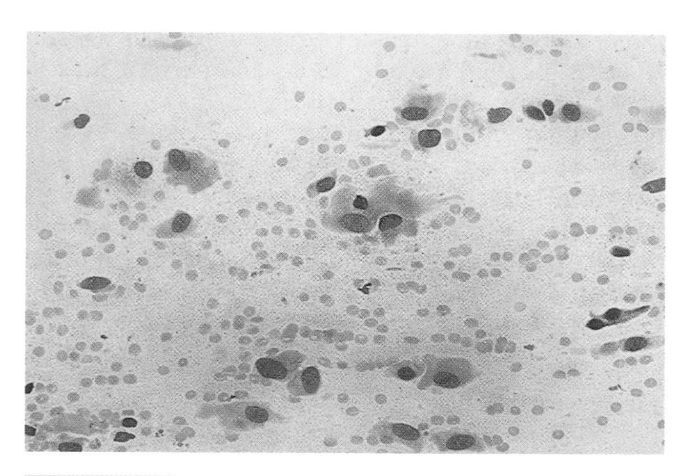

FIGURE 6–13. Aspirates from proliferative fasciitis demonstrate ganglion-like cells with abundant basophilic cytoplasm and eccentrically placed nuclei. Inflammatory cells, histiocytes, fibroblasts, and myofibroblasts are also commonly present. (Diff-Quik)

especially a history of a rapidly enlarging mass in a young person that is less than 3 cm in diameter, along with the characteristic cytologic appearance should preclude a misdiagnosis.

Fibromatosis

The fibromatoses are benign, but locally aggressive proliferations of fibroblasts that are classified with regard to specific location and the gender and age of the patient. Relatively common variants include Dupuytren's contracture (wrist), Peyronie's disease (penis), desmoid (abdominal), and fibromatosis coli (neck of infants). Central to these lesions is the fibroblast. These lesions usually present in the superficial soft tissue as a firm mass. The FNA biopsy generally yields low to moderate cellularity, depending on the degree of collagenization (Fig. 6–14).[73–78] The cells

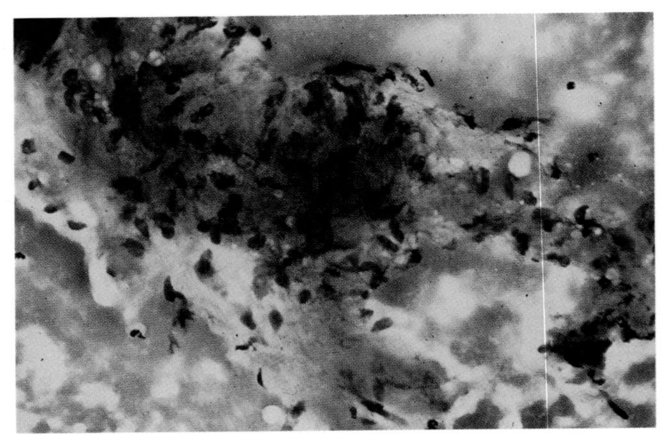

A

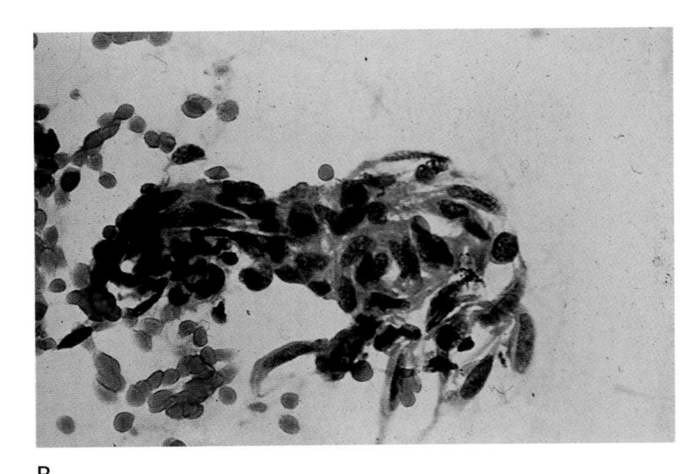

B

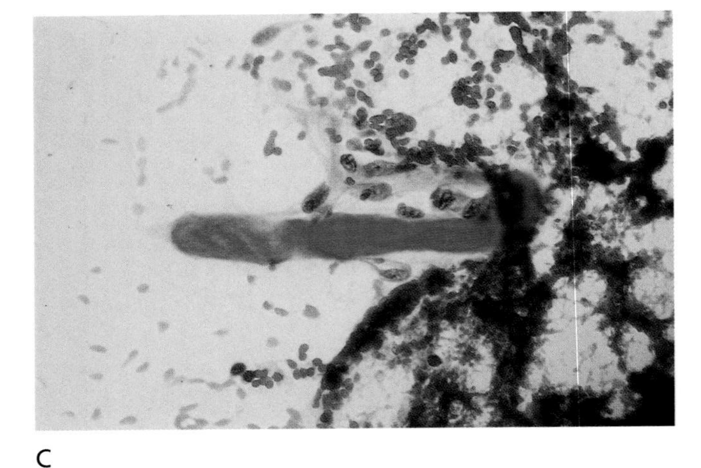

C

FIGURE 6–14. FNA biopsy of fibromatosis reveals low to moderate cellularity composed of single and loosely clustered spindle-shaped fibroblasts. The amount of collagenization is relative to the age of the lesion, with other processes demonstrating hyalanized bands of keloidal collagen. (**A,** Diff-Quik; **B,C,** Papanicolaou)

may be single or in loose clusters, often associated with collagen, which may be hyalinized. The spindle-shaped fibroblasts have centrally placed round to oval, vesicular nuclei with evenly dispersed chromatin. The nucleolus may be prominent in actively growing lesions. Cytoplasm is moderate in volume and may show blebs or tags.[76] In some specimens, usually in the early phase of the lesion, metachromatic myxoid material is present in the background. In older lesions, hyalinized bands of keloid collagen can be seen (see Fig. 6–14).[76] Rare inflammatory cells may also be seen but are not prominent.[73,76] Calcification as well as osseous or chondroid metaplasia may rarely occur. In lesions associated with skeletal muscle, large pleomorphic multinucleated giant cells may be present in the smears.[74,75] They represent atrophic or regenerative muscle.[78] The main differential diagnosis is fibrosarcoma, which generally demonstrates greater cellularity, with more tightly cohesive groups of cells having greater nuclear atypia.[79] However, this distinction can be difficult, and histologic confirmation is often recommended to rule out well-differentiated fibrosarcoma as well as other spindle cell neoplasms.[76]

Elastofibroma

Elastofibroma is a slow-growing tumor that occurs most commonly in the shoulder of elderly patients. FNA biopsy reveals poorly cellular specimens consisting of spindled, benign-appearing fibroblasts as well as occasional mature adipocytes and skeletal muscle fibers.[80] The most characteristic finding is the presence of elastin in the background characterized by homogenous, waxy matrix material with interspersed aggregates of serrated globular material.[80] This material is best appreciated by lowering the substage condenser (non-Koehler illumination) and has been described as fernlike or braid-like. Special stains confirm the presence of elastin in these globular structures.[80]

Fibrosarcoma

Fibrosarcomas may be congenital or occur at any age but are most common in middle-aged adults. Women are affected more often than men. Superficial fibrosarcomas are rare, with palpable lesions usually involving the deep tissues of the extremities or trunk. The

TABLE 6–2. Spindled Cells of the Skin and Superficial Soft Tissues

Cell	Nucleus	Bi-nucleate	Chromatin	Nucleolus	Intranuclear Inclusion	Longitudinal Nuclear Groove	Cytoplasm	Asociated with Matrix	Special Stains	Immunohisto-chemistry	Electron Microscopy
Fibroblast	Spindled, oval	Occasionally	Even	+–+++	⊖	⊖	Abundant, pale, occasional arborizations	0–+++		Vimentin ⊕	Pinocytotic vesicles, RER
Schwann cell	Spindled (wavy), oval	⊖	Hyper-chromatic	0–+	Occasionally	⊖	delicate, pale wispy	+–+++		S-100 ⊕, EMA ⊖	Interdig-tating cell mem-branes, myelin
Perineural cell	Spindled, oval	⊖	Even	0–++	⊖	⊖	Moderate, pale	+–+++		S-100 ⊖, EMA ⊕	
Endothelium	Spindled, polygonal	Rarely	Even	0–+	⊖	Occasionally	Abundant, pale, rare intra-cyto-plasmic hemo-siderin	0–+	Ulex Europas	F VIII, CD 31 and 34 ⊕	Weibel-Palade bodies, pinocy-totic vesi-cles, tight junctions
Smooth muscle	Spindled, rarely polygonal	⊖	Even, rare chromo-centers	0–++	⊖	Rarely	Abudant, pale, perinuclear vacuoles	0–+		Actin, desmin ⊕	Thin fila-ments focal densities, interme-diate junctions

tumors are known for recurrences occurring many years after primary resection. Rarely, these neoplasms are associated with radiation exposure and burns. The FNA biopsy may show single, loosely clustered, or tightly packed fascicles of spindle cells. Collagen fragments are generally scarce. The cells demonstrate oval nuclei with finely granular, hyperchromatic chromatin and indistinct nucleoli. The nuclear membranes may be irregular and mitosis may be present. The cytoplasm lacks distinguishing vacuoles, globules, pigment, or inclusions. Necrosis may be present. The differential diagnosis of fibrosarcoma includes benign fibromatosis, which generally demonstrates blander nuclei, lower cellularity, and more single cells and collagen fragments. Other spindle cell soft-tissue neoplasms enter into the differential diagnosis (Table 6–2).[76,79] The presence of nuclear palisading (neural), myxoid background (e.g., malignant fibrous histiocytoma [MFH]), painful lesion (neural, synovial sarcoma), biphasic pattern (synovial sarcoma), giant cells, bizarre and pleomorphic cells (MFH), cytoplasmic differentiation, osteoid, osseous metaplasia, cartilaginous metaplasia, and calcification suggest tumors other than fibrosarcoma. Fibrosarcomas express only vimentin by immunocytochemistry and are generally a diagnosis of exclusion (Fig. 6–15).[81]

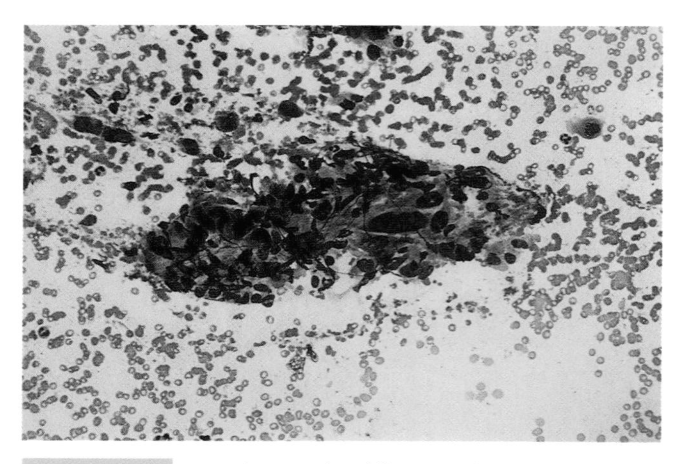

FIGURE 6–15. FNA biopsy of undifferentiated sarcoma with highly atypical spindle nuclei in a background of stromal material. Ancillary techniques did not identify histogenesis, and the lesion was cytologically diagnosed as a spindle cell neoplasm, which includes fibrosarcoma as a diagnosis of exclusion. (Diff-Quik)

Noninfectious Granulomatous Diseases

The hallmark cytologic feature of granulomatous tissue is histiocytes, which are often epithelioid in appearance with associated lymphocytes (Fig. 6–16). The cytopathologist should attempt to elucidate the responsible etiologic agent or cause of the granulomatous inflammation. As discussed previously, mycobacteria and fungi are the usual infectious agents that generally induce caseating (necrotic) granulomatous inflammation.[14] Cytologic examination with special stains and obtaining additional material for microbiologic culture are within the cytopathologist's armamentarium in the workup of a possible infectious etiology.[9,14] Other conditions that frequently induce a

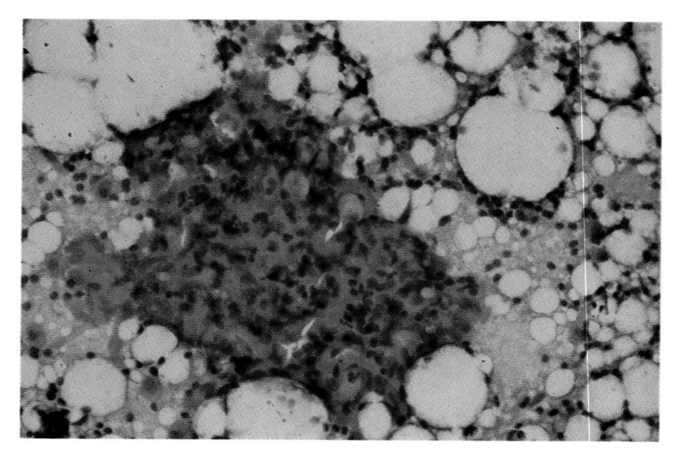

A

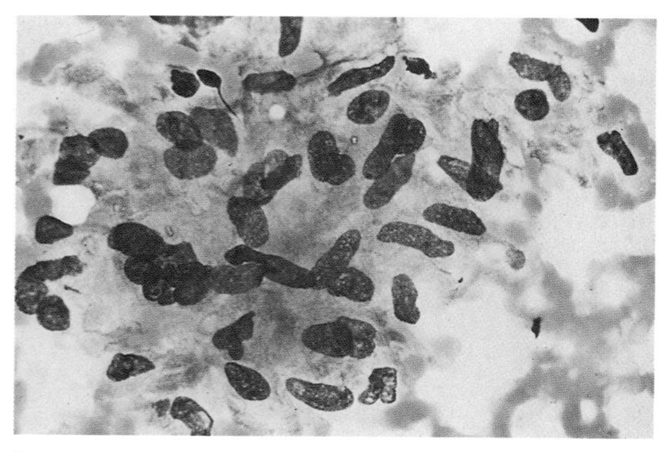

B

FIGURE 6–16. FNA biopsy of granuloma displays clusters of epithelioid histiocytes, often with "bean-shaped" or "snowshoe-shaped" nuclei and abundant cytoplasm. Lymphocytes may be seen in the background as well as in the cytoplasm of the epithelioid aggregates. (**A,** Papanicolaou; **B,** Diff-Quik) granulomatous tissue reaction are sarcoidosis, rheumatoid arthritis, metabolic deposition disorders, and the implantation of foreign material.[82–85] In addition, chalazion is a site-specific granulomatous response in the eyelid.[8]

Sarcoidosis is a systemic disease of unknown origin that classically involves the lungs. However, granulomas may occur in the dermis and subcutaneous fat, causing plaques, papules, and nodules. There is no specific cytologic feature to identify sarcoidosis; it often is a diagnosis of exclusion when correlated with clinical history. Rheumatoid nodules occur in approximately 20% of patients with rheumatoid arthritis. These nodules frequently localize on extensor surfaces of the forearm but may also be seen on the hands, knees, and scalp. As with sarcoid, there is no specific cytologic finding. Correlating clinical information and serology is helpful.

Gout and calcium pyrophosphate deposition disease may produce small to large masses amenable to FNA biopsy. Gout crystals are needle-shaped; calcium pyrophosphate crystals are rhomboidal (Fig. 6–17).[84] The crystals may be seen within histiocytes or extracellularly. With Diff-Quik-stained air-dried smears, monosodium urate crystals (gout) demonstrate negative needle-shaped images, whereas calcium pyrophosphate crystals stain yellowish-orange.[84] Monosodium urate crystals are strongly birefringent and are yellow when parallel to the compensator with polarization microscopy. Calcium pyrophosphate crystals are weakly birefringent and blue when examined with polarization microscopy.[83] Polarization microscopy is also helpful for the evaluation of other inorganic materials that may induce a granulomatous response, such as various metals, glass, silicates, talc, tumoral calcinosis, and suture (Figs. 6–18 and 6–19). Implantable biomaterials may produce a granulomatous response related to the wear-and-tear shedding of biomaterials into surrounding soft tissue.[82,85] Silicone globules from breast implants are best observed with non-Koehler illumination.[82,85]

Dermatofibromas (Benign Histiocytoma)

Dermatofibromas are slow-growing subcutaneous lesions that present as slightly raised, often pigmented nodules in the lower extremities of women in the third and fourth decades. The lesions may grow during pregnancy. Simple excision is generally curative. The FNA biopsy of dermatofibroma consists of loosely grouped, benign-appearing, spindle-shaped fibroblasts and histiocytes that may manifest a storiform arrangement. The histiocytes may contain hemosiderin (siderophages) or lipid (lipophages) and may be multinucleated.[86]

Dermatofibrosarcoma Protuberans

Dermatofibrosarcomas protuberans (DFSP) are slow-growing infiltrative tumors of the dermis and subcutaneum. They are more common in middle-aged men, with a predilection for the trunk and proximal extremities. Metastases are uncommon, but the neoplasm recurs in more than 50% of patients. The biologic behavior of DFSP is believed to be intermediate between benign dermatofibroma and MFH.

The FNA biopsy of DFSP includes moderate to high cellularity, with predominantly spindled, benign-appearing fibroblasts.[86,87] Atypical nuclear features may be slight and mitoses are rare. The cells may be

A

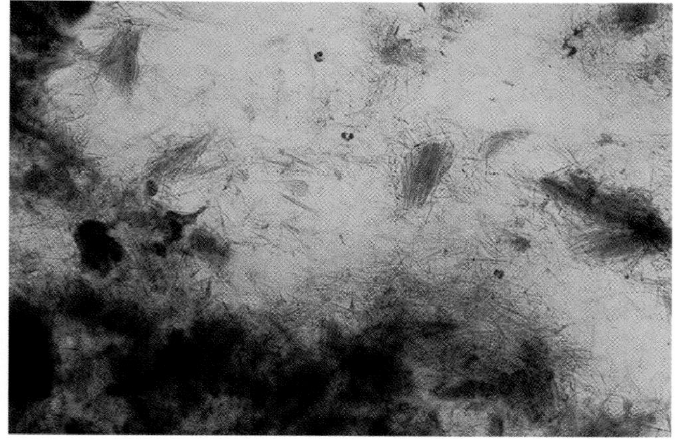

B

FIGURE 6–17. Gout. Aspirates of tophi demonstrate needle-shaped crystals, necrosis, inflammatory cells, and multinucleated giant cells. The crystals are polarizable and may be intracellular or extracellular. (**A,** Diff-Quik; **B,** Papanicolaou)

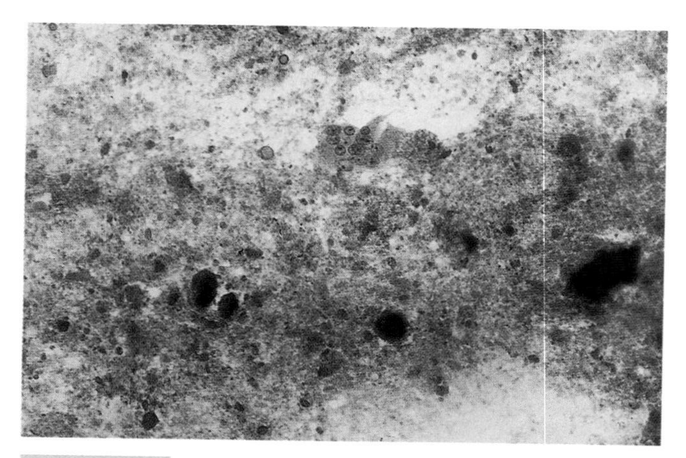

FIGURE 6–18. FNA biopsy of tumoral calcinosis demonstrates abundant finely particulate material as well as variably staining, often refractile, calcium fragments. Foreign-body giant cells may also be observed. (Diff-Quik)

arranged in a storiform or cartwheel-like pattern.[86] Scattered histiocytes may be identified. Aspiration of tissue fragments may reveal entrapped adipocytes and fibroblasts in a finely fibrillar metachromatic matrix. Some cases may contain melanin pigment or reveal a myxoid background.[87] The main differential diagnoses include dermatofibroma, nodular fasciitis, and MFH.[86] Dermatofibromas generally show a more prominent two-cell population of benign spindled cells and histiocytes, without the entrapped adipocytes of the more infiltrative DFSP.[86] However, histologic biopsy may be indicated to distinguish dermatofibroma from DFSP definitively. MFH typically demonstrates frank atypia and bizarre pleomorphic cells and thus is not generally confused with DFSP.

Malignant Fibrous Histiocytoma (MFH)

MFH is a sarcoma of uncertain histogenesis that generally occurs in older adults, with a male predominance. MFH grows rapidly, often metastasizes, and recurs frequently after radical surgical resection. Various histologic types of MFH have been described, including classic, myxoid, giant cell, inflammatory, and angiomatoid. The myxoid and angiomatoid variants are more common than the other three types in the superficial soft tissues. Many investigators believe that the atypical fibrous xanthoma of the dermis represents a biologically benign variant of the classic MFH of soft tissue.

The FNA cytology of the classic variant of MFH demonstrates a highly cellular smear, with cells arranged in single fashion or loosely cohesive clusters (Fig. 6–20).[88–90] The cells are pleomorphic and bizarre with irregular nuclear membranes, irregularly distributed coarse dark chromatin, and prominent nucleoli. Multinucleation is common, as are mitoses, which may be atypical.[88–90] Cytoplasm is variable but often is vacuolated.[89] Indentation of the nucleus from cytoplasmic vacuolization may be observed, but the vacuoles contain mesenchymal mucin and are thus not indicative of lipoblasts.[89] Similarly, elongated straplike cells may be seen with longitudinal striations but do not demonstrate cross-striations of rhabdomyoblasts. Cytoplasmic hyaline globules may be present. Other cells such as inflammatory cells, multinucleated giant cells, and osseous and chondroid differentiation may be observed. A myxoid variant is distinctive in that the cells are not as pleomorphic and are set in a myxoid ground substance.[88] The angiomatoid variant demonstrates spindle (endothelial) cells and atypical-appearing histiocytes with irregular nuclear mem-

A

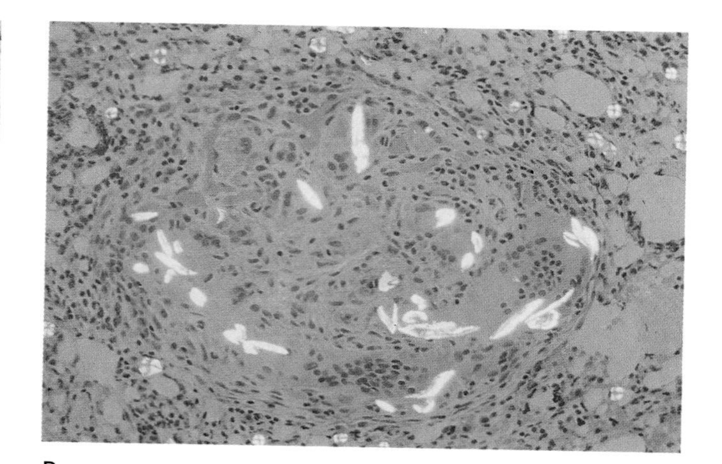

B

FIGURE 6–19. Suture granuloma. (**A**) Cell block preparation demonstrating granuloma with multinucleated giant cells. (**B**) Polarization of granulomas is often helpful to classify the lesion. In this instance, polarization enhances the suture material. (Hematoxylin and eosin)

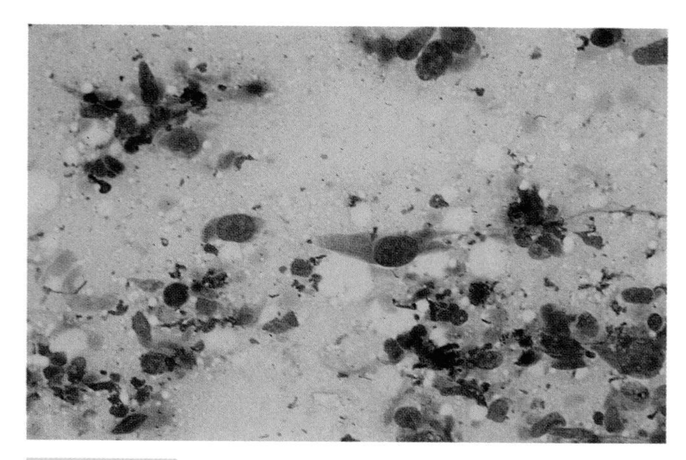

FIGURE 6–20. The FNA appearance of MFH is highly variable. It often displays single cells and discohesive groups of pleomorphic cells. (Diff-Quik)

branes, prominent nucleoli, and ill-defined cell borders. The giant cell MFH demonstrates, in addition to classic MFH features, numerous osteoclast-like giant cells as well as numerous monomorphic histiocytic cells. Areas of osteoid may be present. Inflammatory MFH displays a diffuse inflammatory infiltrate as well as foamy histiocytic cells, which often phagocytize the inflammatory cells. As noted above, the cytologic appearance of MFH is extremely varied, and the diagnosis should always be suspected when strikingly malignant cells are aspirated from a superficial soft-tissue mass. The tumor cells of MFH are characteristically positive for vimentin and histiocytic markers by immunocytochemistry. Within the broad differential diagnosis of MFH are other soft-tissue malignancies, Hodgkin's and non-Hodgkin's lymphomas, melanoma, metastatic poorly differentiated sarcomatoid (spindle) carcinomas, and reactive/reparative mesenchymal processes.

Fat Necrosis

Fat necrosis is a common lesion most often seen in the breast. It is believed to be related to trauma and eventually heals in most instances without clinical intervention. However, because of a mass effect and the occasional induration of the overlying skin, fat necrosis may be evaluated by FNA biopsy. The cytologic appearance includes numerous vacuolated benign-appearing histiocytes with reniform nuclei, reactive lipocytes, lipocytes without nuclei (ghost lipocytes), acute and chronic inflammation, spindle-shaped mesenchymal (fibroblastic) repair cells, and granular necrotic debris (Fig. 6–21).[67] Multinucleated giant cells are present in many instances and may contain needle-shaped negative-staining crystals in their cytoplasm.[91] The multinucleated histiocytes and reactive lipocytes of fat necrosis may be confused with the lipoblasts of liposarcomas and atypical lipomas as well as metastatic renal cell carcinoma, clear cell carcinoma, and other adenocarcinomas.[67] Attention to benign nuclear features, the lack of nuclear scalloping, and the presence of inflammatory cells and ghost lipocytes, together with the clinical history, directs the cytologist to the correct diagnosis.

Hibernoma

Hibernomas are uncommon benign neoplasms of brown fat that present as subcutaneous masses, most commonly in the neck, shoulder, and back. FNA biopsy yields moderately cellular smears of large

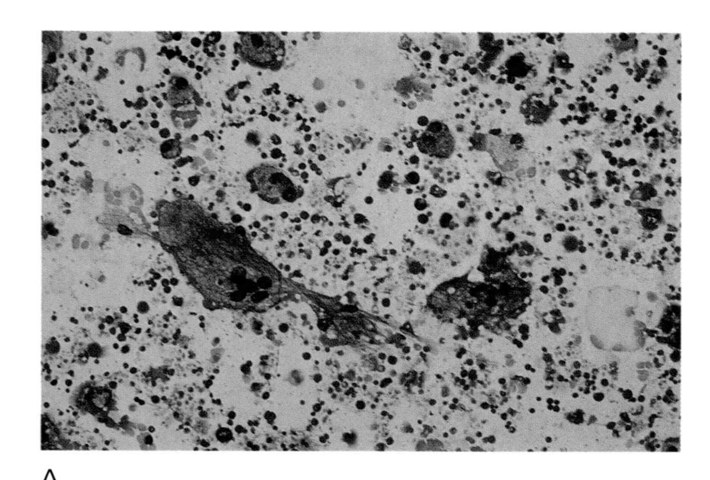

A

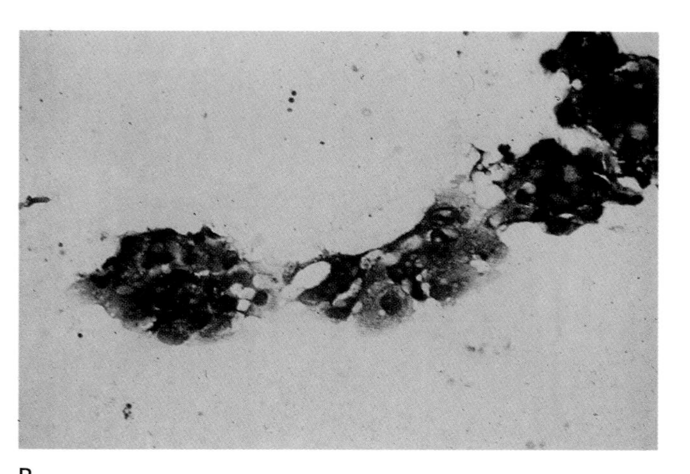

B

FIGURE 6–21. FNA biopsy of fat necrosis displays a variety of cytologic findings, including numerous vacuolated benign histiocytes; lipocytes, some without nuclei (ghost lipocytes); inflammation; mesenchymal (fibroblastic) repair cells; multinucleated giant cells; and granular necrotic debris. (Diff-Quik) round to oval cells and lobules in loose clusters.[92,93] The cells display low N/C ratios with oval, centrally or eccentrically placed, nuclei with evenly dispersed chromatin and a small uniform nucleolus. The cytoplasm shows characteristic microvacuolization that does not indent the nucleus, as observed with the lipoblasts of liposarcoma or atypical lipoma.[67,92]

Lipoma

Lipomas are the most common superficial soft-tissue neoplasm. They are slow-growing, soft, benign lesions and are most common in women older than 40. These lesions vary in size and are located throughout the body but rarely occur in head, hands, or feet. The FNA biopsy of a lipoma yields lobulated tissue fragments and oily metachromatic lipid micelles from disruption of the adipocytes (Fig. 6–22).[93,94] Within the cellular lobules reside adipocytes with their large globoid cytoplasmic fat vacuoles and their spindled, bland, eccentrically located nucleus. Capillaries and scattered fibroblasts may also be observed. Differentiation of lipoma from surrounding normal adipose tissue is not possible by cytologic features alone and requires clinical confidence of needle localization within the suspected lesion.

There are a number of variants of lipoma that may be suggested or diagnosed based on cytologic and clinical features. Fibrolipoma, spindle cell lipoma, and pleomorphic lipomas all show mature adipocytes and fibroblasts with associated collagen. The fibroblasts of spindle cell lipoma are slightly more numerous than those associated with the typical lipoma and may show mild nuclear atypia.[95] Pleomorphic lipomas also demonstrate bizarre pleomorphic giant cells, some of which have a peripheral wreath of nuclei (floret-like cells). Both spindle cell and pleomorphic lipomas most commonly occur in men after the sixth decade in the shoulder or neck region. Angiolipoma and angiomyolipoma both demonstrate vascularized fragments of mature adiopocytes. Spindled smooth muscle cells are also present in angiomyolipoma. Angiolipoma characteristically presents as a subcutaneous painful mass.[93] Angiomyolipomas are rare in the superficial soft tissues and are more commonly found in the kidney.[96] Atypical lipoma is unique among the lipoma variants in that it contains lipoblasts and is generally considered a well-differentiated liposarcoma of the superficial soft tissues that clinically behaves in a fairly benign fashion.[93]

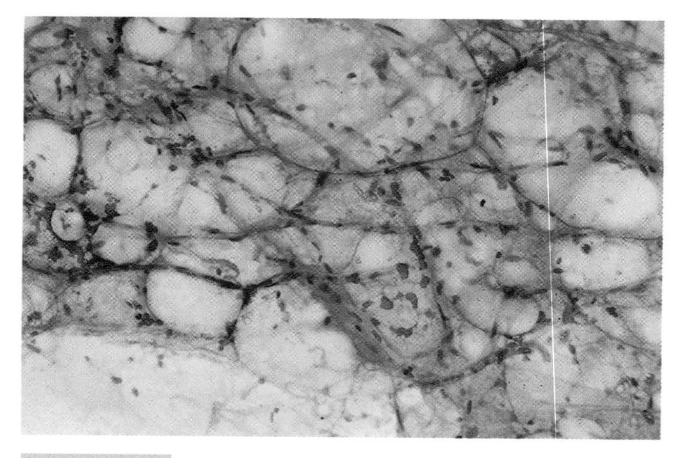

FIGURE 6–22. Lipoma. Vascularized fragment of adipose tissue. Clinical suspicion of a superficial mass lesion is necessary to suggest the diagnosis of lipoma. (Papanicolaou)

Liposarcoma

Liposarcomas may occur in the superficial soft tissues but are more commonly seen in deep sites in adults. Men are more commonly affected than women. There are four major histologic types, directly related to prognosis. Myxoid and well-differentiated liposarcomas portend a good prognosis; round cell and pleomorphic liposarcomas are high-grade neoplasms with frequent local recurrences and metastases. Central to the four types of liposarcoma is the lipoblast.[93,94,97,98] Lipoblasts demonstrate a range of cytologic appearances, reflecting their degree of differentiation toward the mature lipocyte. The early-stage lipoblast is small with round to oval, bland nuclei and one or two small cytoplasmic vacuoles. The midstage lipoblast is the classic diagnostic lipoblast and demonstrates a variably sized, hyperchromatic, sharply scalloped nucleus with multiple cytoplasmic vacuoles (Fig. 6–23). The late-stage lipoblast displays a large globoid vacuole similar to mature adipocytes, plus a few small perinuclear vacuoles. The presence of early, mid, or late lipoblasts is necessary for a confident diagnosis of liposarcoma. The presence of myxoid matrix with a prominent plexiform delicate capillary network defines myxoid liposarcoma (see Fig. 6–23).[93,94] A well-differentiated liposarcoma may be difficult to differentiate from a lipoma: both may demonstrate lobules of mature vascularized fat with occasionally interspersed spindle cells. Therefore, careful scrutiny for lipoblasts is necessary to diagnose a well-differentiated liposarcoma.[93,94,97] Within the superficial soft tissues and especially the immediate subcutaneous tissue, well-differentiated liposarcomas are called atypical lipomas because they have benign biologic behavior.[93] Round cell liposarcoma displays round, variably vacuolated small cells that may be single or in tight clusters, in addition to lipoblasts.[93,94] Pleomorphic liposarcoma demonstrates a spectrum of cell types, including lipoblasts and bizarre multinucleated tumor giant cells.[93,94] Numerous mitotic figures are usually present, and intracytoplasmic hyaline globules may be seen.

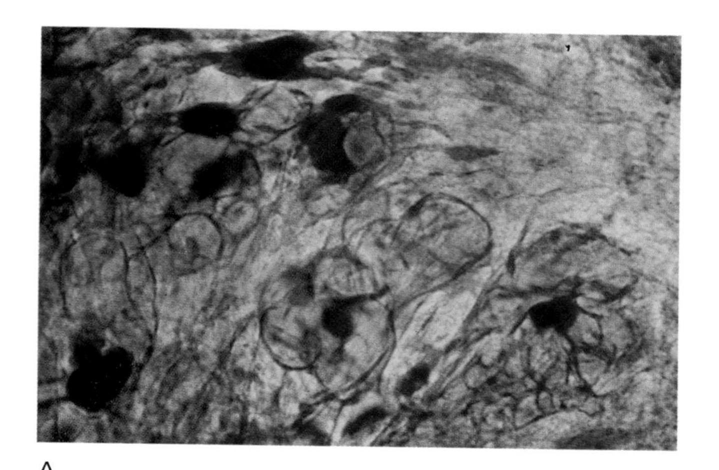

A

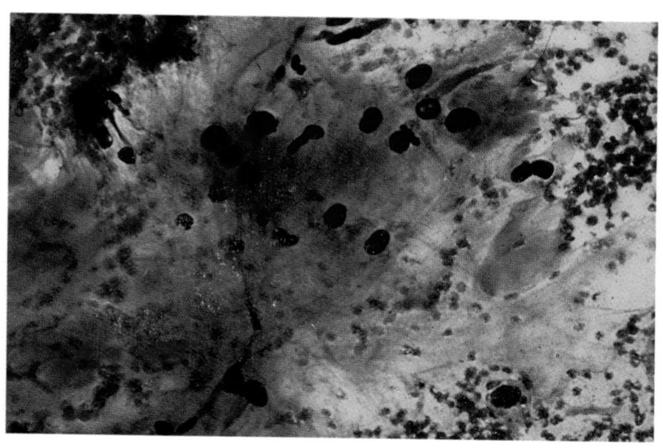

B

FIGURE 6–23. (**A**) The characteristic identifying cell within the aspirate of a liposarcoma is the lipoblast, which is often multinucleated, with indented scalloped nuclei. (Papanicolaou) (**B**) Myxoid material that stains magenta with the Diff-Quik stain is present with myxoid liposarcoma.

Leiomyoma

Although leiomyomas most commonly occur in the uterus and the gastrointestinal tract, they may also occur in the superficial soft tissues. Angioleiomyoma is a specific superficial soft-tissue leiomyoma that characteristically is painful. FNA biopsy of leiomyomas demonstrates low to moderate cellularity with spindled bland-appearing cells, often in cohesive sheets.[99] Single tumor cells are rare. Elongated nuclei with blunt or round ends and evenly dispersed chromatin and occasional nucleoli are a helpful distinguishing feature of smooth muscle cells, as is the occasional presence of perinuclear vacuoles. Tumor cells stain immunocytochemically with desmin and actin, which differentiates spindled smooth muscle cells from other spindled cells, such as Schwann cells, fibroblasts, and endothelium (see Table 6–2).

Leiomyosarcoma

Leiomyosarcomas are uncommon malignancies of the superficial soft tissues, usually located in the extremities or trunk. In FNA biopsy, well-differentiated low-grade leiomyosarcomas are cytologically nearly identical to benign leiomyomas, except for the possible presence of slightly atypical nuclei (irregular nuclear membranes, chromatin coarseness) and increased mitoses.[99] Distinction requires histologic confirmation of a smooth muscle tumor with at least one mitosis per 10 high-power fields, or evidence of metastases. High-grade leiomyosarcomas yield more cellular smears, with neoplastic clusters and single cells (Fig. 6–24). The tumor cells may be highly pleomorphic with spindle cell features, multinucleation, irregular nuclei, coarse chromatin, and prominent nucleoli.[99,100] Epithelioid variants also exist (Fig. 6–24).[99,100] Necrosis and mitotic figures are usually present. The differential diagnosis of a high-grade leiomyosarcoma includes other pleomorphic malignancies, especially MFH. Immunocytochemical demonstration of actin and desmin is often required.

Rhabdomyoma

Rhabdomyomas are rare, benign tumors that most often arise in the head and neck region, associated with small muscles. There is a predilection for males. The adult type usually occurs in persons older than 40, the fetal type in children younger than 3. The FNA biopsy demonstrates large round to polygonal myoblasts, although occasional elongated cells are seen.[101] Nuclei may be multiple and distributed randomly within the cells. The nuclei are round and vesicular and display prominent nucleoli and occasional intranuclear cytoplasmic inclusions. Cytoplasm is abundant and may demonstrate cross-striations, vacuoles, granules, or rod-shaped crystals.[101] A particular type of myoblastic cell termed the "spider cell" is characterized by thin cytoplasmic processes. The cytologic appearance of rhabdomyoma is very similar to that of regenerative muscle.[67]

Rhabdomyosarcoma

Rhabdomyosarcomas are rare tumors that occur most commonly in children and adolescents, with boys being affected more often than girls. The tumor has a predilection for the head and neck region (orbit, sinuses, oral cavity, neck), retroperitoneum, genitourinary tract, and extremities. Three types of rhabdomyosarcomas (embryonal, alveolar, and pleomorphic) may be responsible for a superficially palpable mass. Early rhabdomyoblasts are small round blue

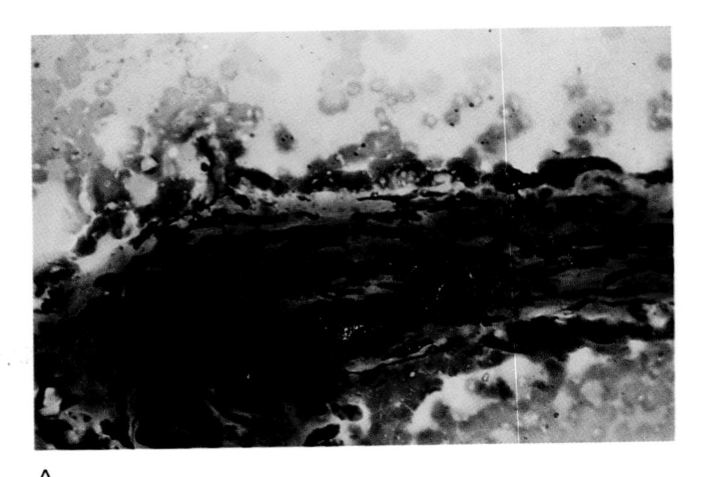

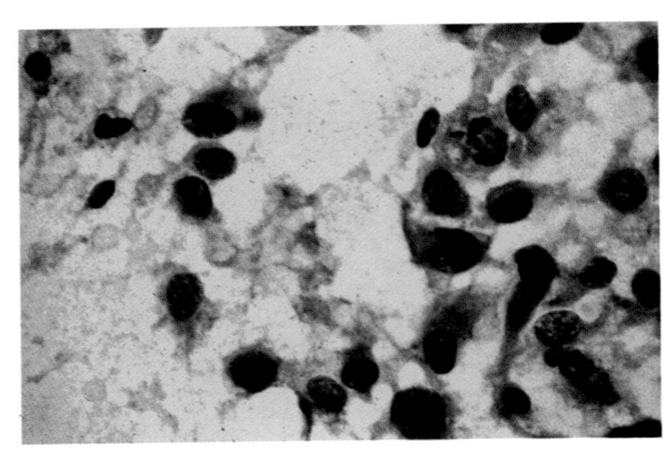

FIGURE 6–24. (**A**) FNA biopsy of a typical leiomyosarcoma demonstrates aggregates of spindled cells that have hyperchromatic chromatin, irregular nuclear membranes, and increased mitoses. (**B**) Epithelioid leiomyosarcomas display pleomorphic to epithelioid cell shapes, as well as atypical nuclei. Epithelioid leiomyosarcomas may be confused with malignant fibrous histiocytoma or carcinomas. Immunohistochemical demonstration of actin and/or desmin is often necessary for definitive classification. (**A,** Diff-Quik; **B,** Papanicolaou)

cells with scant cytoplasm. Occasional cells are binucleated. Intermediate rhabdomyoblasts demonstrate more abundant pale to vacuolated cytoplasm with radially arranged fibrillar material. Late rhabdomyoblasts display the classic strap cell appearance; these elongated to triangular cells have abundant, sometimes vacuolated, cytoplasm that in addition to intracytoplasmic fibrillar material may display more organized cross-striations. Multinucleation is common, and intranuclear cytoplasmic invaginations may be observed.[67]

FNA biopsy of embryonal rhabdomyosarcoma yields cellular smears with many round and spindled cells.[102–105] The nuclear features are uniform, with fine chromatin and conspicuous nucleoli. Extracellular myxoid matrix material may appear "tigroid" in the background, similar to that seen in seminomas.[104] FNA biopsy of alveolar rhabdomyosarcoma is similar, with the presence of small round blue cells having irregular nuclei with prominent nucleoli (Fig. 6–25).[104] Spindle cells are not generally seen, whereas giant multinucleated myoblasts and distinctive floret cells (multiple nuclei arranged in a wreathlike configuration) may be observed. The background can be fibrotic or necrotic. FNA biopsy of pleomorphic rhabdomyosarcoma presents as bizarre tumor giant cells that may display myoblastic features such as cross-striations or evidence of skeletal muscle differentiation by electron microscopy or immunocytochemistry.

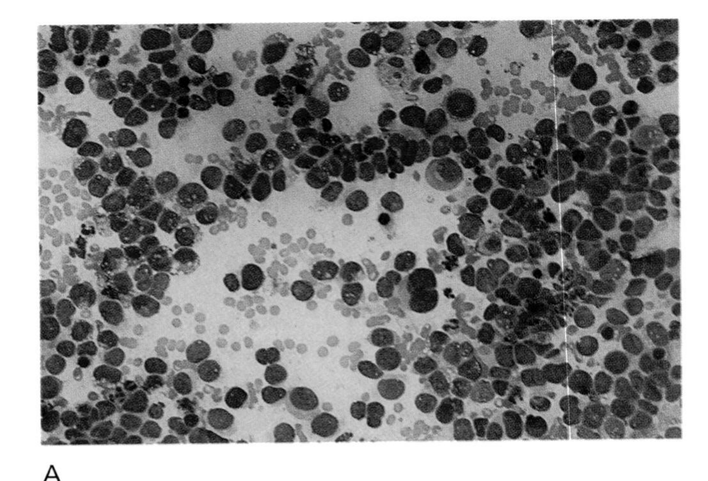

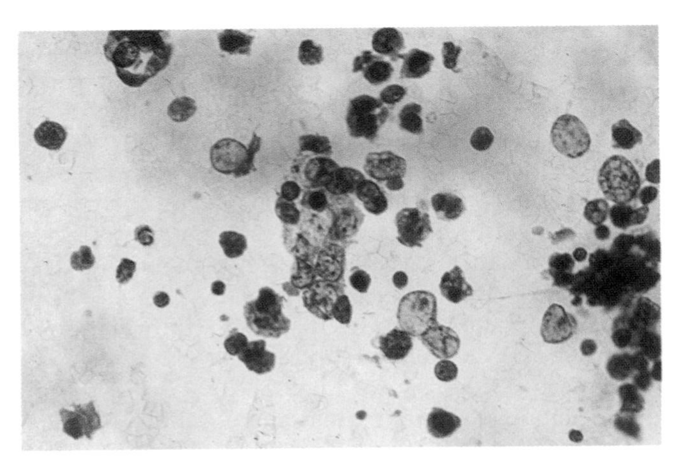

FIGURE 6–25. FNA biopsy of alveolar rhabdomyosarcoma displays abundant small round blue cells with irregular nuclei and often prominent nucleoli. (**A,** Diff-Quik; **B,** Papanicolaou)

The differential diagnosis of embryonal and alveolar rhabdomyosarcomas include other tumors considered as small round blue cell tumors (Ewing's sarcoma, peripheral primitive neuroectodermal tumor, lymphoma). Identification of cytologic myoblastic features such as cytoplasmic fibrillar material and cross-striations, multinucleation, and the expression of desmin and myoglobin by immunocytochemistry and the identification of thick and thin filaments by electron microscopy are helpful diagnostic features.[102] Pleomorphic rhabdomyosarcomas resemble other pleomorphic sarcomas; as with the embryonal and alveolar variants, demonstration of myofilaments is required to make the correct diagnosis.

Traumatic Neuroma

Traumatic neuromas represent a disordered proliferation of neural and fibrous elements at the point of traumatic nerve damage from an injury, surgery, or amputation. The lesions are typically tender and become exquisitely painful during FNA biopsy. The FNA biopsy demonstrates a specimen with low cellularity consisting of Schwann cells, myelin, axons, and interspersed fibroblasts. The cytologic features are identical to the FNA biopsy of normal nerve; thus, the diagnosis may be suggested only by correlating the clinical history and presentation.

Benign Nerve Sheath Tumors (Schwannoma, Neurofibroma, and Perineurona)

Schwannomas (neurilemmoma) and neurofibromas are common benign lesions of the nerve sheath that occur throughout the superficial soft tissues. Multiple neurofibromas are associated with von Recklinghausen's disease. The FNA biopsy is often painful. Perineuromas are relatively uncommon benign proliferations of perineural cells that surround the peripheral nerves.

The FNA cytomorphology of schwannoma and neurofibroma is similar, with single cells, small clusters, and thick fragments (Fig. 6–26). Spindle cells predominate and are characteristically squiggly or comma-shaped.[106–112] Intranuclear inclusions may be present, and the chromatin is evenly distributed, without a prominent nucleolus.[110] Cytoplasm is fragile, and single naked nuclei may be common. Tumor cell nuclei may also be embedded in finely fibrillar metachromatic ground substance.[111] Nuclear palisading and Verocay bodies (palisaded cells around a fibrillar core) are generally specific for schwannomas.[106,110,111] In addition, Antoni A (ordered hypercellular areas) and Antoni B (hypocellular, loosely and haphazardly arranged cells) may be observed with schwannomas. Myxoid areas may be seen in the Antoni B areas of schwannomas and in neurofibromas.

Neurofibromas also typically demonstrate scattered spindled fibroblasts and mast cells. Foamy histiocytes may be seen with degenerating or cystic schwannomas, and occasionally melanin pigment is observed within the spindled Schwann cells.[107] With degeneration or cystic changes, both schwannomas and neurofibromas may display cytologic atypia with multilobulation and irregular hyperchromatic nuclei with coarse chromatin; however, mitoses are not seen.[107,109] The presence of mitoses suggests a malignant nerve sheath tumor.[114] Perineuromas similarly present cytologically as spindled cells within a myxoid background. In addition, the spindled cells may have long cytoplasmic extensions and pseudo-signet-ring cells may be present.[113] Differentiation of perineuroma from schwannoma and neurofibroma can be

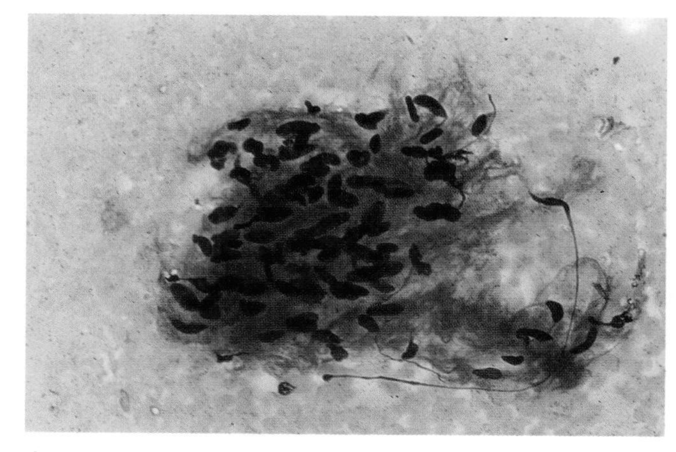

A

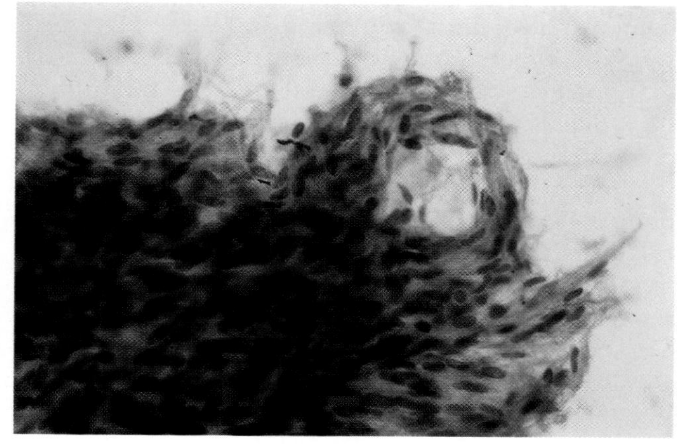

B

FIGURE 6–26. Compressed hyperchromatic wavy nuclei with evenly dispersed chromatin are observed in schwannomas. The nuclei are embedded in a finely fibrillar metachromatic ground substance. (**A,** Diff-Quik; **B,** Papanicolaou)

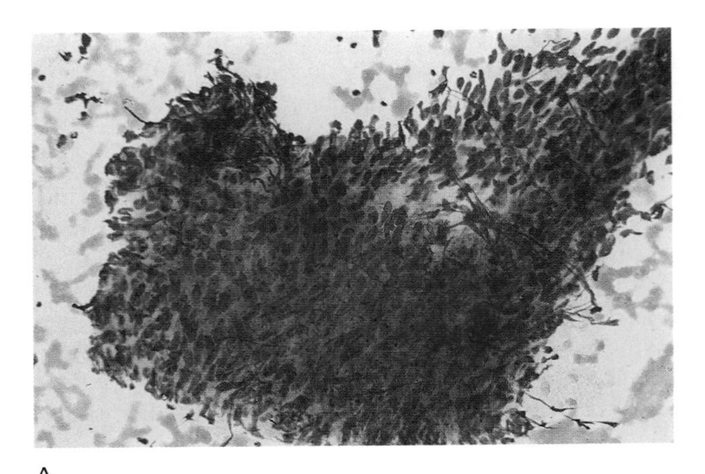

A

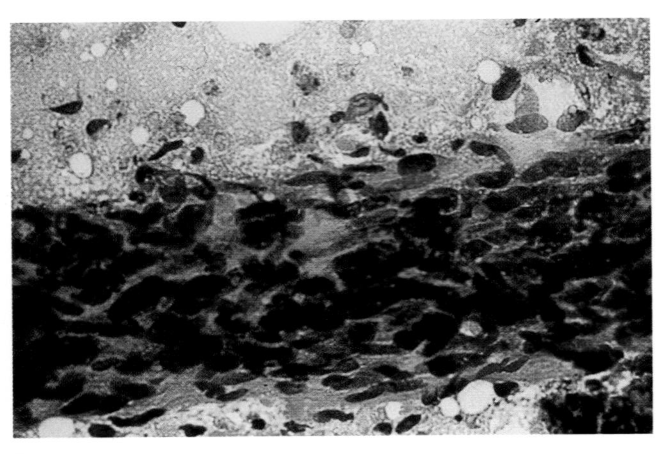

B

FIGURE 6–27. FNA biopsy of malignant nerve sheath tumor demonstrates numerous spindled cells within a delicate fibrillar matrix. The nuclei are hyperchromatic, with a serpentine appearance. Slight to severe nuclear atypia may be present. Immunohistochemical demonstration of S-100 protein is a helpful delineating feature. (**A,** Diff-Quik; **B,** Papanicolaou)

achieved by immunocytochemistry (perineural cells are S-100-negative and EMA-positive; Schwann cells are S-100-positive and EMA-negative).[108,112,113]

Malignant Nerve Sheath Tumors

Malignant nerve sheath tumors (malignant schwannoma) account for approximately 10% of all soft-tissue sarcomas. They typically present as rapidly enlarging, often painful masses, occasionally with associated paresthesias. Middle-aged men are most frequently affected. The neck, extremities, and buttocks are often involved. These tumors may arise *de novo* or from malignant degeneration of a neurofibroma (but not schwannoma). Malignant nerve sheath tumors are often seen in von Recklinghausen's disease.

The FNA biopsy is usually cellular, with numerous spindled cells (Fig. 6–27).[114] The appearance of the spindled cells is variable; they may or may not demonstrate atypical cytologic features.[108,114,115] Wavy, squiggly hyperchromatic nuclei suggest a neural origin; however, the spindled cells may be more fibroblast-like.[114] Occasionally, a biphasic pattern of spindled and larger epithelioid cells may be seen (malignant epithelioid schwannoma). The cells are often cohesive or interspersed within a delicately fibrillar matrix.[114,115] Mitoses are typically frequent.[114] Evidence of rhabdomyoblasts together with malignant neural cells defines a triton tumor.[108]

Important diagnostic clues to malignant nerve sheath tumors are spindle-shaped cells with numerous mitoses, loss of polarity, nuclear pleomorphism, and an appropriate clinical history.[108,114,115] Immunocytochemical positive staining for S-100 antigen is a helpful feature to distinguish malignant nerve sheath tumors from fibrosarcoma and other S-100-negative spindled cell neoplasms (see Table 6–2).[108,112,113] However, S-100 staining can often be weak or focal. Malignant nerve sheath tumors that demonstrate a biphasic pattern may be confused with synovial sarcoma but do not display immunocytochemical expression of both mesenchymal and epithelial components.

Peripheral Primitive Neuroectodermal Tumor/Ewing's Sarcoma

Peripheral primitive neuroectodermal tumors (PNETs) often involve the superficial soft tissues of the trunk, pelvis, or extremities. PNETs occur in adolescents and carry a poor prognosis. Although PNETs are of neural origin, they are not directly associated with nerves in every case. FNA biopsy of PNET is moderate to highly cellular, consisting of small round blue cells with scant cytoplasm, often with vacuolization, and round nuclei, which may exhibit molding.[116] The tumor cells may be seen individually scattered or in clusters.[116] Ewing's sarcoma is also a small round blue cell tumor that may be present in soft tissue (Fig. 6–28). However, it is more common in the bone and is described in detail in Chapter 5. Features helpful in the differentiation of PNET from other small round blue cell tumors of childhood include t11;22 translocation, CD99 and neuron-specific enolase positivity, and negative staining for leukocyte common antigen (LCA), actin, myoglobin, and desmin.

Hemangioma

Hemangiomas are benign tumors of proliferating blood vessels that commonly present within the skin and superficial soft tissue of children and adolescents, especially girls. Often they spontaneously regress. The

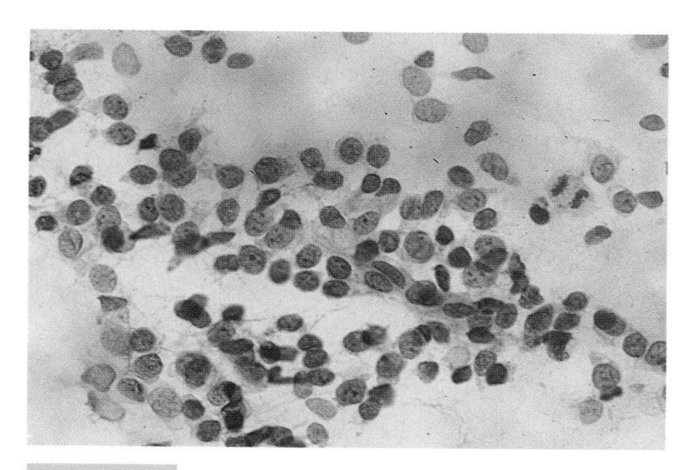

FIGURE 6–28. FNA biopsy of Ewing's sarcoma displays small round cells, often in small groups or sheets as well as singly. The nuclei are round to oval with fine chromatin. Nuclear grooves and small nucleoli are present. Cytoplasm is pale and scant. (Papanicolaou)

clinical appearance is usually that of a flat or polypoid, often pigmented, lesion that may blanch with externally applied pressure. Three major histologic types are recognized: capillary, cavernous, and mixed. Other variants include pyogenic granuloma (increased amount of granulation tissue and inflammatory cells), sclerosing hemangioma, and intramuscular hemangioma (usually young adults). FNA biopsy is usually poorly cellular, with mostly blood.[8] Scattered spindled or polygonal endothelial cells may be seen singularly or in strips or sheets. Their nuclei have finely dispersed chromatin, inconspicuous nucleoli, and occasional longitudinal nuclear grooves.[8] Rare cells may be bi- or multinucleated. Intracytoplasmic hemosiderin may be present in endothelial cells, especially with thrombosed vessels. Hemosiderin-laden histiocytes and fibroblasts may also be seen. The differential diagnosis includes other spindle cell neoplasms and pigmented lesions (see Table 6–2). Dermatofibromas are also grossly pigmented lesions with spindled and histiocytic cells that may be confused with a thrombosed hemangioma. The spindled cells of dermatofibroma are not positive immunocytochemically for endothelial markers (factor VIII, CD31, CD34). Melanocytic lesions may also demonstrate spindled cells, but melanin granules, as well as S-100 and HMB-45 positivity, are generally present.

Hemangiopericytoma

Hemangiopericytomas are rare soft-tissue vascular neoplasms of intermediate malignant potential that occur predominantly in adults. These tumors occur throughout the body but are more common within the soft tissues, especially in the lower extremities. FNA biopsy is moderate to highly cellular, consisting of tissue fragments, loose cellular aggregates, and individual tumor cells.[117,118] Many of the fragments are tightly packed and demonstrate vascular slits and channels coursing through the fragments.[117,118] Tumor cells are variable in shape and often have high N/C ratios, one or two distinct nucleoli, evenly dispersed chromatin, and scant, fragile, vacuolated or granular cytoplasm.[117,118] Nuclear notches and invaginations may be present (Fig. 6–29). Mitoses may be numerous; this has been suggested to reflect malignant potential.[117] A central core of metachromatic basement membrane-like material may separate radiating tumor cells from the more spindled endothelial cells (see Fig. 6–29).[118] Knoblike formations of neoplastic cells around capillaries have been described.[118] The differential diagnosis includes a variety of mesenchymal tumors, especially those that tend to display cellular cohesion and pleomorphism such as synovial sarcoma, MFH, and leiomyosarcoma. The tumor cells of hemangiopericytoma are actin- and vimentin-positive (variably positive for S-100) and negative for desmin, myoglobin, factor VIII, and epithelial markers by immunohistochemistry. A specific diagnosis may not always be achieved with FNA biopsy.

Epithelioid Hemangioendothelioma

Epithelioid hemangioendotheliomas are rare neoplasms of intermediate malignant potential that may occur in the superficial soft tissues. In the lung, they were previously referred to as intravascular bronchioloalveolar tumors. FNA biopsy displays moderate cellularity with isolated cells, sheets, and rosette-like and acinar-like formations.[119–123] The tumor cells are variable in size and shape but display characteristic abundant cytoplasm with prominent vacuolization

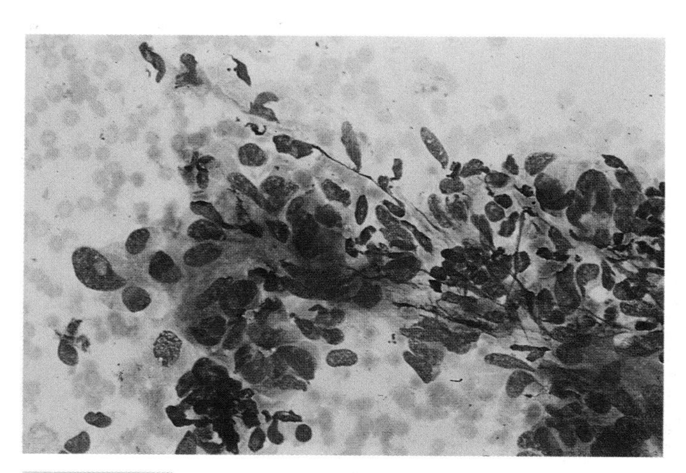

FIGURE 6–29. FNA biopsy of hemangiopericytoma demonstrates variably sized tumor cells, nuclear notches and involutions, and scant fragile cytoplasm. Metachromatic basement membrane-like material is often associated with the tumor cells with the Diff-Quik stain.

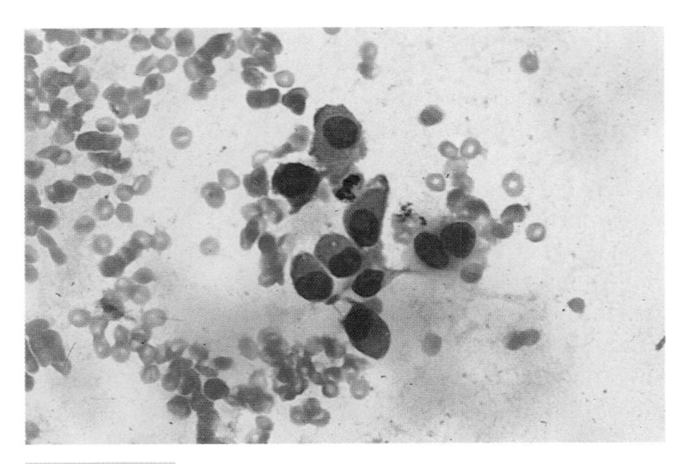

FIGURE 6–30. FNA biopsy of epithelioid hemangioendothelioma shows cells with abundant cytoplasm and occasional nucleoli, which may be misdiagnosed as carcinoma. (Diff-Quik)

(Fig. 6–30).[122] The nuclei may demonstrate irregular membranes and intranuclear cytoplasmic inclusions.[120,121,123] One or more nucleoli are generally present. The tumor cells of epithelioid hemangioendothelioma could be misdiagnosed as adenocarcinoma because of the prominent cytoplasmic vacuolization and occasional gland-like cellular configurations. Carcinomas are cytokeratin-positive and factor VIII-negative; epithelioid hemangioendotheliomas demonstrate the reverse pattern.[120,122,123]

Angiosarcoma

Angiosarcomas are rare malignant vascular tumors that often occur in the skin and superficial soft tissues. They are often associated with prior radiation therapy, lymphedema, or the use of Thortrast. FNA biopsy cellularity depends on the composition of the tumor—for instance, more solid tumors with less vascular spaces result in more cellular specimens.[124–126] The neoplastic cells may be arranged in a variety of clustered patterns, including papillary, pseudoacinar, or rosette, or in single fashion.[124] The clusters and sheets may be prominently arranged around anastomosing vascular channels (Fig. 6–31).[124–126] The tumor cells vary in shape; however, the nuclei are most often large and irregular, with elongated to spindled shapes and round to pointed edges (see Fig. 6–31).[8,126] The chromatin varies from evenly and finely distributed to dense.[8,126] Longitudinal nuclear grooves are present in some cases.[125] Nucleoli may be inconspicuous to prominent.[124] Cytoplasm is usually moderate, pale, and amphophilic; it occasionally is vacuolated.[126] Erythrophagocytosis may be seen, and numerous red blood cells are present in the background.[125] The differential diagnosis includes other sarcomas, especially those displaying a prominent vascular pattern.

Myxoma

Myxomas are benign neoplasms of unknown etiology, usually occurring in adults but occasionally in younger persons. Women are affected more often than men. The lesion presents as a palpable mass, usually in the muscles of the extremities, shoulder, or buttocks. Myxomas are usually solitary, but multiple lesions may occur in association with fibrous dysplasia of the bone. FNA biopsy generally yields a smear consisting of poorly cellular viscous material.[127–129] Neoplastic cells vary in size and shape; most are elongated, with spindle-shaped nuclei and evenly dis-

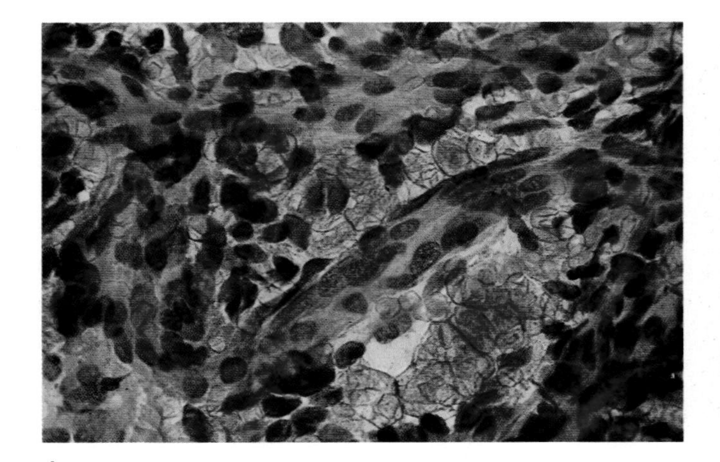

A

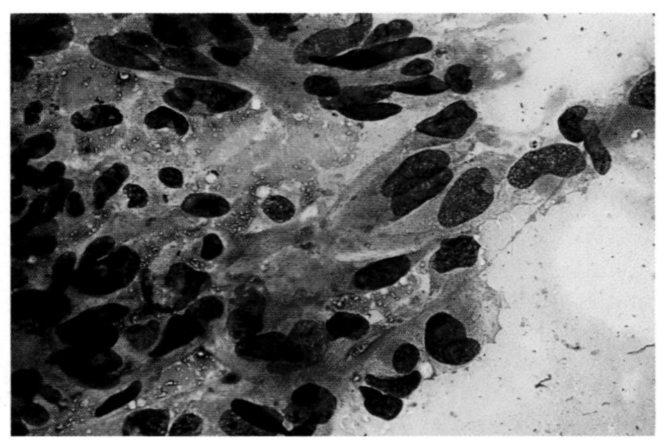

B

FIGURE 6–31. FNA biopsy of angiosarcoma shows atypical cells arranged as single cells or sheets, often with prominent anastomosing vascular channels. Although the tumor cells vary in size and shape, many show a spindled morphology. A moderate amount of pale to amphophilic cytoplasm is present, which may demonstrate vacuolization or erythrophagocytosis. (**A,** Papanicolaou; **B,** Diff-Quik) persed chromatin and small nucleoli.[127–129] Occasional cells are binucleated.[127] The cells often demonstrate abundant vacuoles or occasionally one large vacuole.[127] Finely vacuolated histiocytes and nonbranching capillaries may be observed. A finely granular to fibrillar background myxoid material is basophilic with the Papanicolaou stain and metachromatic with the Diff-Quik stain.[127] Vacuolization of the background myxoid material may be observed. The differential diagnosis includes a variety of neoplasms with prominent myxoid change such as liposarcoma, chondrosarcoma, MFH, metastatic colloid carcinoma, mucocele, and neural tumors. The low cellularity, lack of cytologic atypia, and nonbranching vascular pattern are useful in diagnosing myxoma.

Granular Cell Tumor

Granular cell tumors (granular cell myoblastomas) occur within the skin, superficial soft tissues, and oral cavity, where they present as well-circumscribed, raised, nodular lesions that may be tender or pruritic. FNA biopsy reveals monomorphic cells with oval to slightly spindle-shaped nuclei, many of which may be naked because of stripping of the abundant, delicate, granular cytoplasm (Fig. 6–32).[8,130–132] Atypical nuclear features are generally not appreciated, and the cells are S-100-positive.[132]

Alveolar Soft-Part Sarcoma

Alveolar soft-part sarcomas are rare malignant tumors affecting adolescents and young adults, with a male predominance. The tumor most often occurs in the lower extremities. FNA biopsy is generally cellular, consisting of single or discohesive aggregates of monomorphic cells with large, often naked nuclei, fine chromatin, distinctive macronucleoli, and occasional binucleation (Fig. 6–33).[133–138] The finely granular to vacuolated cytoplasm is delicate and abundant. Glycogen (periodic acid-Schiff [PAS]-positive) and crystals (periodic acid-Schiff with predigestion with diastase [PAS-D]-positive) are generally present. The differential diagnosis includes renal cell carcinoma (no PAS-D crystals), paraganglioma (primary never occurring in the extremities), granular cell tumor (no PAS-D crystals), and malignant melanoma.

Rhabdoid Tumor

Malignant rhabdoid tumors are much more prevalent in the kidney; however, cases have been reported in the soft tissues of the trunk and paravertebral area. Children and young adults are more frequently affected. As with renal rhabdoid tumors, FNA biopsy demonstrates relatively monomorphic cells in tight and loose clusters as well as singly.[139] Tumor cells display characteristic rounded or bean-shaped, eccentrically located vesicular nuclei with a central prominent nucleolus and occasionally distinctive globular hyaline PAS-positive cytoplasmic inclusions, which are most often perinuclear and indent the nucleus (Fig. 6–34).[139]

Epithelioid Sarcoma

Epithelioid sarcoma is a soft-tissue tumor that affects adolescents and young adults, with a male predominance. It tends to occur in the extremities, especially the upper extremities. The neoplasms occur in the subcutis and deeper tissues, are often multinodular, and grow along neurovascular bundles. Tumor nod-

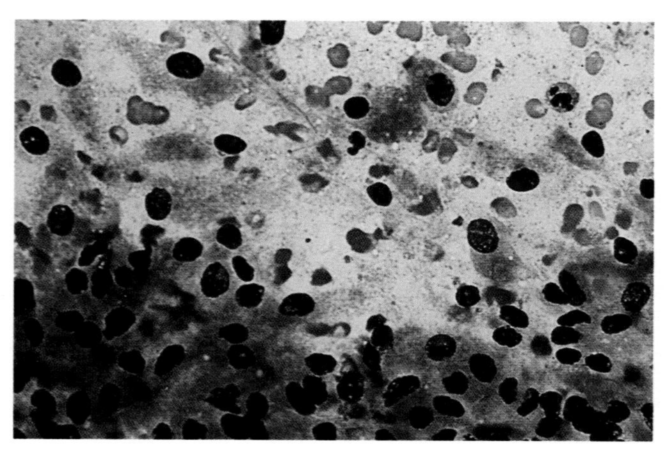

A

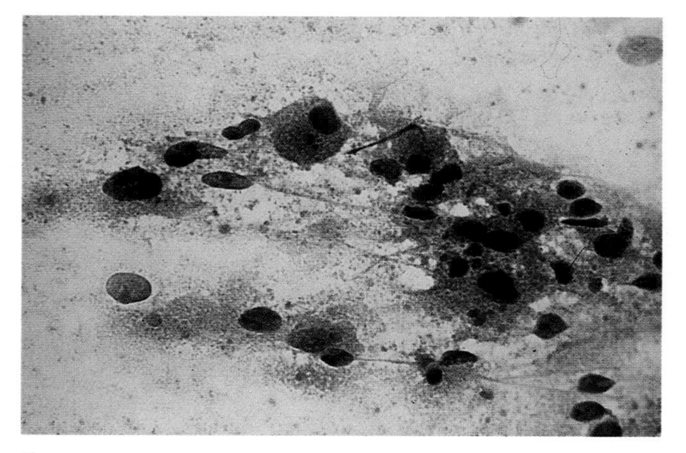

B

FIGURE 6–32. Granular cell tumor smears reveal monomorphic oval to slightly spindled nuclei with stripped to a moderate amount of granular, delicate cytoplasm. (**A,** Diff-Quik; **B,** Papanicolaou)

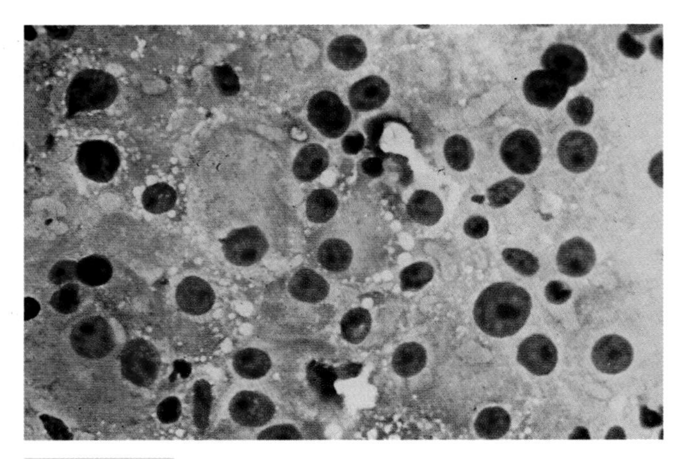

FIGURE 6–33. FNA biopsy of alveolar soft-part sarcoma is generally cellular, with single or discohesive large monomorphic cells with prominent macronucleoli. The cytoplasm is abundant, delicate, and granular to vacuolated. (Diff-Quik)

ules within the dermis often rapidly ulcerate the overlying epidermis. There is a high recurrence rate despite wide surgical resection. Metastases are common to the lung and lymph nodes.

FNA biopsy is generally cellular, with cells arranged in cohesive three-dimensional groups, sheets, noncohesive clusters, and single cells. Most of the cells are polygonal and may display a variable N/C ratio.[140–142] A minority have spindled contours.[140,141] Nuclei vary from smooth to irregular nuclear membranes, inconspicuous to prominent nucleoli, variable chromatin distribution, and occasional multilobulation and/or binucleation.[141,142] Mitoses may be plentiful. A perinuclear condensation or clearing (halo-like) has been described.[142] The cytoplasm has a variable appearance and may appear keratinized.[140] In the background, necrotic debris, inflammatory cells, and fibrous stroma are usually observed.[142] The cells are positive for epithelioid markers (keratin and EMA) as well as vimentin, but negative for other mesenchymal antigens.[141,142] The differential diagnosis includes metastatic carcinoma and granulomas.

Synovial Sarcoma

Synovial sarcoma affects the extremities of young adults. It may or may not be associated with the synovium of tendon sheaths, bursae, and joints. It is more common in males and often presents as a painful lesion. It metastasizes to lymph nodes more frequently than most other sarcomas. Typically, FNA biopsy is markedly cellular, with a biphasic pattern of spindled and epithelioid cells (Fig. 6–35).[143] The spindled cells are arranged in aggregates and dissociated single cells. Nuclei are hyperchromatic and oval to elongated, occasionally with irregular longitudinal folded outlines. Nucleoli are inconspicuous to multiple. There is a delicate amount of tapered cytoplasm, often stripped from the nucleus. Spindled cells predominate in the smears over epithelioid cells, which may be arranged in nests or glands. The epithelioid cells contain regular, centrally placed, round nuclei with even chromatin and sometimes multiple nucleoli. Mitoses have been variably reported. Calcification, myxoid change, squamous metaplasia or keratin pearls, cartilage, bone, histiocytes, and multinucleated giant cells may be seen. Monophasic variants of synovial sarcoma exist in which only the spindled or epithelioid component (exceedingly rare) is present.

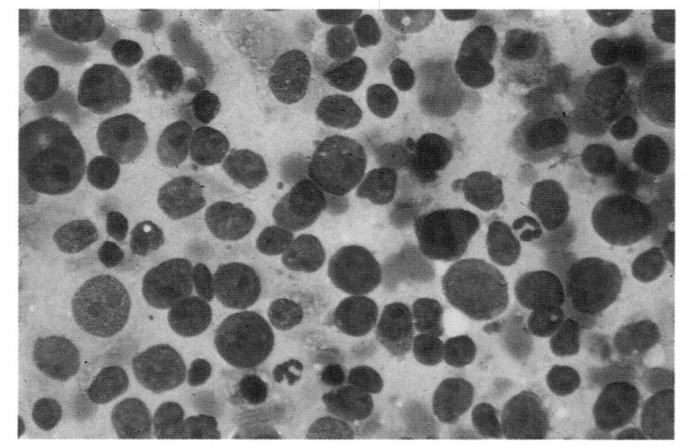

A

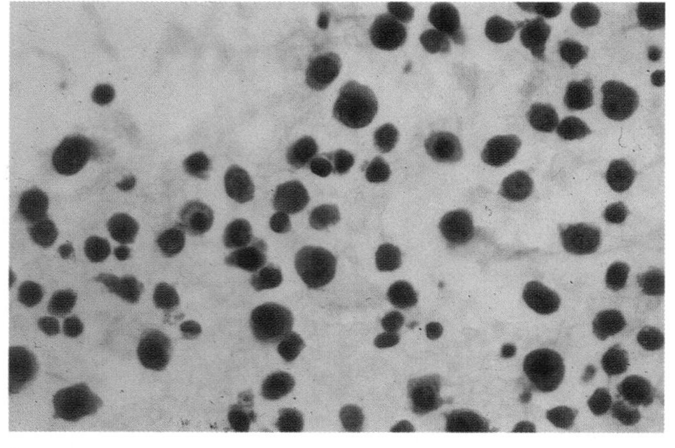

B

FIGURE 6–34. FNA biopsy of extrarenal rhabdoid tumor displays tumor cells with round to bean-shaped (reniform), eccentrically placed nuclei with prominent nucleoli. Perinuclear globules are responsible for the indentation of some of the nuclei. (**A,** Diff-Quik; **B,** Papanicolaou)

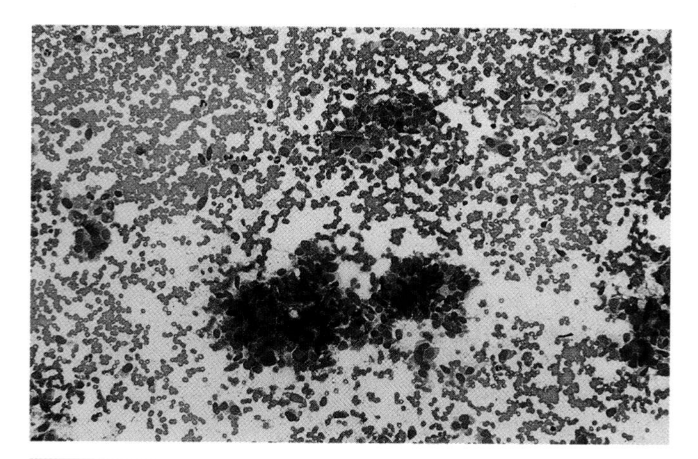

FIGURE 6–35. Cellular smear from biphasic synovial sarcoma demonstrates a biphasic pattern of epithelioid and spindled cells. Aggregates and single cells are present, with many of the single cells being stripped of cytoplasm. (Diff-Quik)

The differential diagnosis of synovial sarcoma is extensive. Synovial sarcomas demonstrate epithelioid cells positive for PAS, alcian blue, and mucicarmine, but only alcian blue-positive spindle cells. Epithelioid cells immunohistochemically express keratin and EMA, with occasional spindled cells showing positive staining.[143] The spindled cells usually express only vimentin.[143] Biphasic synovial sarcomas are more readily diagnosed than the monophasic variants. Aspirated material may be submitted for karyotypic analysis and may manifest a specific translocation, t (x,18). The monophasic spindled cell synovial sarcoma demonstrates a cytologic appearance similar to fibrosarcoma, hemangiopericytoma, and other spindle cell neoplasms.[143] Epithelioid variant monophasic synovial sarcomas have a differential diagnosis of metastatic carcinoma, malignant melanoma, epithelioid schwannoma, and epithelioid sarcoma.

Cutaneous Endometriosis

Cutaneous endometriosis typically occurs in cesarean section sites or other surgical scars in premenopausal women. It presents as blue nodules that may enlarge and become more painful during menses. Three different cell types should be observed for a definitive diagnosis: endometrial stromal cells, endometrial glandular cells, and hemosiderin-laden histiocytes (Fig. 6–36).[31,141–147] Often observed individually and in loose aggregates, endometrial stromal cells are elongated, with oval to spindle-shaped nuclei, finely granular chromatin, and indistinct cytoplasm. Endometrial glandular cells may be seen in honeycomb sheets of closely packed round monotonous columnar cells with evenly dispersed chromatin. Nuclear palisading and occasional nuclear overlap may be observed. Histiocytes typically demonstrate hemosiderin within abundant vacuolated cytoplasm. Mucinous, squamous, and tubal metaplasia may occur within cutaneous endometriosis.[145] Because of the proliferative nature of the epithelial cells, a misdiagnosis of metastatic adenocarcinoma is possible.

Granulocytic Sarcoma

Granulocytic sarcoma (extramedullary leukemia) or chloroma can occur in the superficial soft tissues and skin in addition to palpable lymph nodes and deep body sites. The cytologic features reflect characteristics specific for leukemic involvement.[148–150] Romanowsky-type stains (Diff-Quik and Giemsa) on

A

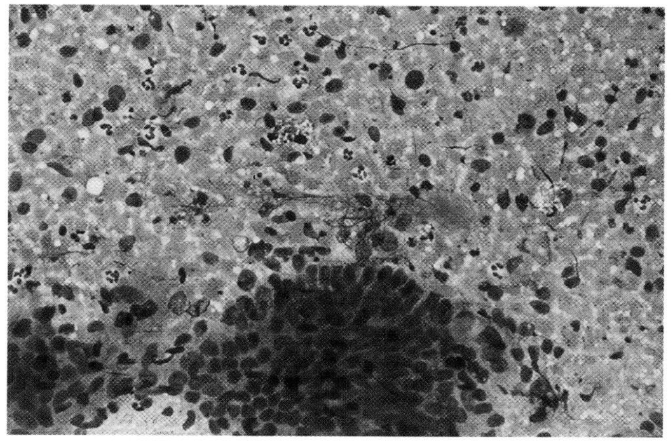

B

FIGURE 6–36. FNA smear from cutaneous endometriosis complicating a cesarean section. Present are tightly clumped glandular cells, single spindled endometrial stromal cells, and histocytes. (**A,** Papanicolauou; **B,** Diff-Quik)

air-dried smears are preferred and are especially useful in those with hematopathology abilities.[151] The smears are generally hypercellular and should demonstrate single cells showing a range of hematopoietic maturation with a predominance of cells specific for the type of leukemia.[148–150] Myeloid lineage leukemias may demonstrate azurophilic granules and often Auer rods.[148] Special stains such as Sudan black, choroacetate esterase, PAS, and myeloperoxidase are helpful in morphologically differentiating cell type.[148–150] Triaging the specimen for immunophenotypic analysis, gene rearrangement studies, and cytogenetics is also recommended.[149,150] Blood smear, bone marrow biopsy and aspirate, and the clinical history are additional important diagnostic clues.[150]

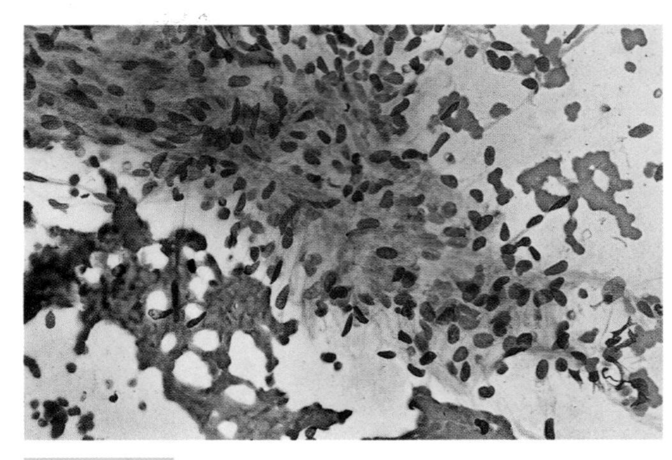

FIGURE 6–37. FNA biopsy of Kaposi's sarcoma displays clusters of bland, overlapping spindled cells with evenly dispersed chromatin, inconspicuous nucleoli, and indistinct cell borders. (Diff-Quik)

Metastasis and Recurrences

Cutaneous and superficial soft-tissue metastases occur in 0.7–4.4% of patients dying from a malignancy, often representing the terminal manifestation.[152,153] These metastases increase proportionally with age and are most common after the fifth decade.[152] Cutaneous and superficial soft-tissue metastases are available to direct palpation, inspection, and FNA biopsy. The most common sites for cutaneous and superficial soft-tissue metastases are the back, upper extremities, and scalp.[154] The most common metastatic malignancies to the superficial skin and superficial soft tissue include lung carcinoma, malignant melanoma, head and neck squamous cell carcinoma, breast carcinoma, and gastrointestinal tumors.[153,155,156] Neuroblastoma has a propensity to metastasize to the skin of the shoulder and produce the so-called "blueberry muffin baby."[154] The differential diagnosis of suspected metastatic or recurrent malignancies is extensive and includes adnexal tumors, subcutaneous glands (e.g., Bartholin's gland), heterotopic tissue (ectopic breast) pleomorphic sarcomas, lymphomas, granulomas, and other entities.

Other

A variety of other lesions and entities may present on FNA biopsy of the skin and soft tissue. Ectopic breast tissue may be present in many sites in the trunk but is more common in the axilla; it often presents as an enlarging mass during pregnancy. FNA biopsy demonstrates honeycomb clusters of benign ductal epithelium and associated fat.[157] A case of invasive ductal carcinoma with frankly malignant cytologic features has been reported in an ectopic breast.[158] Superficial soft-tissue FNA biopsies that demonstrate adnexal structures or subcutaneous glands (e.g., Bartholin's) as well as heterotopic tissue may be misinterpreted as a metastatic neoplasm.[8,31,159] Inflammatory breast carcinoma of the skin does not present as a mass lesion but may be diagnosed by multiple "blind" dermal FNA biopsies that reveal small clusters of malignant ductal cells.[160] A variety of soft-tissue sarcomas generally present as deep masses but occasionally present in superficial soft tissues. These have been described cytologically and include malignant melanoma of soft parts (clear cell sarcoma), giant cell tumor of tendon sheath, cartilaginous tumors, and osseous tumors.[161–169] In addition, the cytologic characteristics of cutaneous lymphomas and plasmacytomas, postradiation sarcomas, Kaposi's sarcoma, and bacillary angiomatosis have been reported (Fig. 6–37).[3,8,31,170–172]

REFERENCES

1. Arca MJ, Biermann JS, Johnson TM, Chang BE. Biopsy techniques for skin, soft tissue and bone neoplasms. Surg Oncol Clin North Am 4:157–174, 1995.
2. Tzanck A. Le cytodiagnostic immediat en dermatologie. Bull Soc Fr Dermatol Syph 7:68–70, 1947.
3. Goldman L, McCabe RM, Sawyer F. The importance of cytology technique for the dermatologist in office practice. AMA Arch Dermatol 81:359–368, 1960.
4. Barr RJ. Cutaneous cytology. J Am Acad Dermatol 10:163–180, 1984.
5. Hitch JM, Wilson TB, Scoggin A. Evaluation of a rapid method of cytologic diagnosis in suspected skin cancer. South Med J 44:407–414, 1951.
6. Graham J, Bingul O, Urbach F, et al. Papanicolaou smears and frozen sections on selected cutaneous neoplasms. JAMA 178:106–385, 1961.
7. Selbach G, Heisel E. The cytologic approach to skin disease. Acta Cytol 6:439–442, 1962.
8. Layfield LJ, Glasgow BJ. Aspiration biopsy cytology of primary cutaneous tumors. Acta Cytol 37:679–688, 1993.
9. Lee PC, Turnidge J, McDonald PJ. Fine needle aspiration biopsy in diagnosis of soft tissue infections. J Clin Microbiol 22:80–83, 1985.
10. Barth RD, Merino MJ, Solomon D, et al. A prospective study of the value of core needle biopsy and fine needle aspiration in the diagnosis of soft tissue masses. Surgery 112:536–543, 1992.

11. Bennert KW, Abdul-Karim FW. Fine needle aspiration vs. needle core biopsy of soft tissue tumors—a comparison. Acta Cytol 38:381–384, 1994.
12. Ayala AG, Ro JY, Fanning CV, et al. Core needle biopsy and fine needle aspiration in the diagnosis of bone and soft-tissue lesions. Hemato/Oncol Clin North Am 9:633–651, 1995.
13. Akerman M. The cytology of soft tissue tumors. Acta Orthop Scand (Suppl 273) 68:54–59, 1997.
14. Silverman JF, Gay RM. Fine-needle aspiration and surgical pathology of infectious lesions. Morphologic features and the role of the clinical microbiology laboratory for rapid diagnosis. Clin Lab Med 15:251–278, 1995.
15. Blank H, Burgoon CF, Baldridge GD, McCarthy PL, Urbach F. Cytologic smears in diagnosis of herpes simplex, herpes zoster and varicella. JAMA 146:1410–1412, 1951.
16. Singh N, Arora VK, Ramam M. Nodular lepromatous leprosy: report of a case diagnosed by FNA. Diagn Cytopathol 11:373–375, 1994.
17. Hicks MJ, Green LK, Clarridge J. Primary diagnosis of disseminated coccidioidomycosis by fine needle aspiration of a neck mass. A case report. Acta Cytol 38:422–426, 1994.
18. Stong GC, Raval HB, Martin JW, Kazragis RJ, Enghardt MH. Nodular subcutaneous histoplasmosis. A case report with diagnosis by fine needle aspiration biopsy. Acta Cytol 38:777–781, 1994.
19. Fujita WH, Barr RJ, Gottschalk HR. Cutaneous amebiasis. Arch Dermatol 117:309–310, 1981.
20. Al-Jitawi SA, Farraj SE, Ramahi SA. Conventional scraping versus fine needle aspiration cytology in the diagnosis of cutaneous leishmaniasis. Acta Cytol 39:82–84, 1995.
21. Arora VK, Gupta K, Singh N, Bhatia A. Cytomorphologic panorama of cysticercosis on fine needle aspiration. A review of 298 cases. Acta Cytol 38:377–380, 1994.
22. Yue XH. Fine needle aspiration biopsy diagnosis of cysticercosis. A case report. Acta Cytol 38:90–92, 1994.
23. Brown CL, Klaber MR, Robertson MG. Rapid cytological diagnosis of basal cell carcinoma of the skin. J Clin Pathol 32:361–367, 1979.
24. Gordon LA, Orell SR. Evaluation of cytodiagnosis of cutaneous basal cell carcinoma. J Am Acad Dermatol 11:1082–1086, 1984.
25. Malberger E, Tillinger R, Lichtig C. Diagnosis of basal-cell carcinoma with aspiration cytology. Acta Cytol 28:301–304, 1984.
26. Bocking A, Schunck K, Auffermann W. Exfoliative-cytologic diagnosis of basal-cell carcinoma, with the use of DNA image cytometry as a diagnostic aid. Acta Cytol 31:143–149, 1987.
27. Derrick EK, Smith R, Melcher DH, Morrison EA, Kirkham N, Darley CR. The use of cytology in the diagnosis of basal cell carcinoma. Br J Dermatol 130:561–563, 1994.
28. Haddad MG, Silverman JF. Fine-needle aspiration cytology of metastatic basal cell carcinoma of the skin to the lung. Diagn Cytopathol 10:15–19, 1994.
29. Aneiros-Cachaza J, Caracuel MD, Camara M, Alonso J. Neuroendocrine changes in basal cell carcinoma. Acta Cytol 32:431, 1988.
30. Ramzy I, Rone R, Schantz D. Squamous cells in needle aspirates of subcutaneous lesions: a diagnostic problem. Am J Clin Pathol 85:319–324, 1986.
31. Nadji M, Defortuna S, Sevin BU, Ganjei P. Fine-needle aspiration cytology of palpable lesions of the lower female genital tract. Int J Gynecol Pathol 13:54–61, 1994.
32. Mellblom L, Akerman M, Carlen B. Aspiration cytology of neuroendocrine (Merkel cell) carcinoma of the skin. Report of a case. Acta Cytol 28:297–300, 1984.
33. Pettinato G, De Chiara A, Insabato L, Iaffaioli V. Neuroendocrine (Merkel cell) carcinoma of the skin. Fine needle aspiration cytology and clinicopathologic study of a case. Acta Cytol 28:283–289, 1984.
34. Szpak CA, Bossen EH, Linder J, Johnston WW. Cytomorphology of primary small-cell (Merkel cell) carcinoma of the skin in fine needle aspirates. Acta Cytol 28:290–296, 1984.
35. Domagala W, Lubinski J, Lasota J, Giryn I, Weber K, Osborn M. Neuroendocrine (Merkel cell) carcinoma of the skin. Cytology, intermediate filament typing and ultrastructure of tumor cells in fine needle aspirates. Acta Cytol 31:267–275, 1987.
36. Gherardi G, Marveggio C, Stiglich F. Parotid metastasis of Merkel cell carcinoma in a young patient with ectodermal dysplasia. Diagnosis by fine needle aspiration cytology and immunocytochemistry. Acta Cytol 34:831–836, 1990.
37. Skoog L, Schmitt FC, Tani E. Neuroendocrine (Merkel-cell) carcinoma of the skin: immunocytochemical and cytomorphologic analysis on fine-needle aspirates. Diagn Cytopathol 6:53–57, 1990.
38. Al-Kaisi NK. Fine-needle aspiration cytology of a metastatic Merkel-cell carcinoma. Diagn Cytopathol 7:184–188, 1991.
39. Perez-Guillermo M, Sola-Perez J, Abad-Montano C, Quirante P, Romero M. Merkel cell tumor of the eyelid and the cytologic aspect in fine needle aspirates: report of a case. Diagn Cytopathol 10:146–151, 1994.
40. Gottschalk-Sabag S, Ne'eman Zvi, Glick T. Merkel cell carcinoma diagnosed by fine-needle aspiration. Am J Dermatopathol 18:269–272, 1996.
41. Gomez-Aracil V, Azua J, Pedro CS, Romero J. Fine needle aspiration cytologic findings in four cases of pilomatrixoma (calcifying epithelioma of Malherbe). Acta Cytol 34:842–846, 1990.
42. Ma KF, Tsui MS, Chan SK. Fine needle aspiration diagnosis of pilomatrixoma. A monomorphic population of basaloid cells with squamous differentiation not to be mistaken for carcinoma. Acta Cytol 35:570–574, 1991.
43. Kinsey W, Coghill SB. A case of pilomatrixoma misdiagnosed as squamous cell carcinoma. Cytopathology 4:167–171, 1993.
44. Ray R, Dey P, Dutta BN. Fine needle aspiration cytology of metastatic apocrine carcinoma. Acta Cytol 38:283–285, 1994.
45. Rollins SD. Fine needle aspiration diagnosis of a vulvar papillary hidradenoma: a case report. Diagn Cytopathol 10:60–61, 1994.
46. Ray R, Dey P. Fine needle aspiration cytology of malignant hidradenoma. Acta Cytol 37:842–843, 1993.
47. Srinivasan R, Rajwanshi A, Padmanabhan V, Dey P. Fine needle aspiration cytology of chondroid syringoma and syringocystadenoma papilliferum. A report of two cases. Acta Cytol 37:535–538, 1993.
48. Bondeson L, Lindholm K, Thorstenson S. Benign dermal eccrine cylindroma. A pitfall in the cytologic diagnosis of adenoid cystic carcinoma. Acta Cytol 27:326–328, 1983.
49. Varsa EW, Jordan SW. Fine needle aspiration cytology of malignant spiradenoma arising in congenital eccrine spiradenoma. Acta Cytol 34:275–277, 1990.
50. Gottschalk-Sabag S, Glick T. Chondroid syringoma diagnosed by fine-needle aspiration: a case report. Diagn Cytopathol 10:152–155, 1994.
51. Sliwa-Hahnle K, Obers V, Lakhoo M, Saadia R. Chondroid syringoma of the abdominal wall. SAJS 34:46–48, 1996.
52. Woyke S, Domagala W, Czerniak B, Strokowska M. Fine needle aspiration cytology of malignant melanoma of the skin. Acta Cytol 24:529–538, 1980.
53. Gupta SK, Rajwanshi AK, Das DK. Fine needle aspiration cytology smear patterns of malignant melanoma. Acta Cytol 29:983–988, 1985.
54. Perry MD, Gore M, Seigler HF, Johnston WW. Fine needle aspiration biopsy of metastatic melanoma. A morphologic analysis of 174 cases. Acta Cytol 30:385–395, 1986.
55. Layfield LJ, Ostrzega N. Fine needle aspirate smear morphology in metastatic melanoma. Acta Cytol 33:606–612, 1989.
56. Vance KV, Park HK, Silverman JF. Fine needle aspiration cytology of desmoplastic malignant melanoma. A case report. Acta Cytol 35:765–769, 1991.
57. Wakely PE, Frable WJ, Geisinger KR. Aspiration cytopathology of malignant melanoma in children. A morphologic spectrum. Am J Clin Pathol 103:231–234, 1995.

58. Frangioni G, Borgioli G. One-hour bleach for melanin. Stain Tech 63:325–331, 1988.
59. Lindholm K, de la Torre M. Fine needle aspiration cytology of myxoid metastatic malignant melanoma. Acta Cytol 32:719–720, 1988.
60. Chang ES, Wick MR, Swanson PE, Dehner LP. Metastatic malignant melanoma with "rhabdoid" features. Am J Clin Pathol 102:426–431, 1994.
61. Chong SM, Nilsson BS, Quah TC, Wee A. Malignant melanoma: an uncommon cause of small round cell malignancy in childhood. Acta Cytol 41:609–610, 1997.
62. Banks ER, Jansen JF, Oberle E, Davey DD. Cytokeratin positivity in fine-needle aspirates of melanomas and sarcomas. Diagn Cytopathol 12:230–233, 1995.
63. Roy M, Bhattacharyya A, Sanyal S, Dasgupta S. Study of benign superficial cysts by fine needle aspiration cytology. J Indian Med Assoc 93(1):8, 1995.
64. Oertel YC, Beckner ME, Engler WF. Cytologic diagnosis and ultrastructure of fine-needle aspirates of ganglion cysts. Arch Pathol Lab Med 110:938–942, 1986.
65. Dodd LG, Layfield LJ. Fine-needle aspiration cytology of ganglion cysts. Diagn Cytopathol 15:377–381, 1996.
66. Anglo-Henry MR, Seaquist MB, Marsh WL. Fine needle aspiration of proliferative fasciitis. A case report. Acta Cytol 29:882–886, 1985
67. James LP. Cytopathology of mesenchymal repair. Diagn Cytopathol 1:91–104, 1985.
68. Popok SM, Naib ZM. Fine needle aspiration cytology of myositis ossificans. Diagn Cytopathol 1:236–240, 1985.
69. Stanley MW, Skoog L, Tani EM, Horwitz CA. Nodular fasciitis: spontaneous resolution following diagnosis by fine-needle aspiration. Diagn Cytopathol 9:322–324, 1993.
70. de Almeida MM, Abecassis N, Almeida MO, Mendonca ME. Fine-needle aspiration cytology of myositis ossificans: a case report. Diagn Cytopathol 10:41–43, 1994.
71. Wakely PE, Almeida M, Frable WJ. Fine-needle aspiration biopsy cytology of myositis ossificans. Mod Pathol 7:23–25, 1994.
72. Jacobs JC. Aspiration cytology of proliferative myositis. A case report. Acta Cytol 39:535–538, 1995.
73. El-Naggar A, Abdul-Karim FW, Marshalleck JJ, Sorensen K. Fine-needle aspiration of fibromatosis of the breast. Diagn Cytopathol 3:320–322, 1987.
74. Wakely PE, Price WG, Frable WJ. Sternomastoid tumor of infancy (fibromatosis colli): diagnosis by aspiration cytology. Mod Pathol 2:378–381, 1989.
75. Zaharopoulos P, Wong JY. Fine-needle aspiration cytology in fibromatoses. Diagn Cytopathol 8:73–78, 1992.
76. Raab SS, Silverman JF, McLeod DL, Benning TL, Geisinger KR. Fine needle aspiration biopsy of fibromatoses. Acta Cytol 37:323–328, 1993.
77. Carabias E, Dhimes P, de Agustin P, Gutierrez E. Nodular cervical induration in a violinist. Report of a case with fine needle aspiration cytologic findings. Acta Cytol 40:1301–1303, 1996.
78. Schwartz RA, Powers CN, Wakely PE, Kellman RM. Fibromatosis colli: the utility of fine-needle aspiration in diagnosis. Arch Otolaryngol Head Neck Surg 123:301–304, 1997.
79. Silverman JF, Geisinger KR, Frable WJ. Fine-needle aspiration cytology of mesenchymal tumors of the breast. Diag Cytopathol 4:50–58, 1988.
80. Pisharodi LR, Cary D, Bernacki EG. Elastofibroma dorsi: diagnostic problems and pitfalls. Diagn Cytopathol 10:242–244, 1994.
81. Wick MR, Swenson PF, Manivel JC. Immunohistochemical analysis of soft tissue sarcomas: comparisons with electron microscopy. Appl Pathol 6:169–196, 1988.
82. Dodd LG, Sneige N, Reece GP, Fornage B. Fine-needle aspiration cytology of silicone granulomas in the augmented breast. Diagn Cytopathol 9:498–502, 1993.
83. Allen EA, Ali SZ, Erozan YS. Tumoral calcium pyrophosphate dihydrate deposition disease: cytopathologic findings on fine-needle aspiration. Diagn Cytopathol 15:349–351, 1996.
84. Liu K, Moffat EJ, Hudson ER, Layfield LJ. Gouty tophus presenting as a soft-tissue mass diagnosed by fine-needle aspiration: a case report. Diagn Cytopathol 15:246–249, 1996.
85. Raso DS, Greene WB. Silicone breast implants—Pathology. Ultrastructural Pathol 21:263–271, 1997.
86. Powers CN, Hurt MA, Frable WJ. Fine-needle aspiration biopsy: dermatofibrosarcoma protuberans. Diagn Cytopathol 9:145–150, 1993.
87. Perry MD, Furlong JW, Johnston WW. Fine needle aspiration cytology of metastatic dermatofibrosarcoma protuberans. A case report. Acta Cytol 30:507–512, 1986.
88. Merck C, Hagmar B. Myxofibrosarcoma. A correlative cytologic and histologic study of 13 cases examined by fine needle aspiration cytology. Acta Cytol 24:137–144, 1980.
89. Kim K, Goldblatt PJ. Malignant fibrous histiocytoma. Cytologic, light microscopic and ultrastructural studies. Acta Cytol 26:507–511, 1982.
90. Luzzatto R, Grossmann S, Scholl JG, Recktenvald M. Postradiation pleomorphic malignant fibrous histocytoma of the breast. Acta Cytol 30:48–50, 1986.
91. Walker WP, Smith RJH, Cohen MB. Fine-needle aspiration biopsy of subcutaneous fat necrosis of the newborn. Diagn Cytopathol 9:329–332, 1993.
92. Hashimoto CH, Cobb CJ. Cytodiagnosis of hibernoma: a case report. Diagn Cytopathol 3:326–329, 1987.
93. Walaas L, Kindblom LG. Lipomatous tumors: a correlative cytologic and histologic study of 27 tumors examined by fine needle aspiration cytology. Hum Pathol 16:6–18, 1985.
94. Akerman M, Rydholm A. Aspiration cytology of lipomatous tumors: a 10-year experience at an orthopedic oncology center. Diag Cytopathol 3:295–302, 1987.
95. Lew WYC. Spindle cell lipoma of the breast: a case report and literature review. Diagn Cytopathol 9:434–437, 1993.
96. Wadih GE, Raab SS, Silverman JF. Fine needle aspiration cytology of renal and retroperitoneal angiomyolipoma. Report of two cases with cytologic findings and clinicopathologic pitfalls in diagnosis. Acta Cytol 39:945–950, 1995.
97. Shattuck MC, Victor TA. Cytologic features of well-differentiated sclerosing liposarcoma in aspirated samples. Acta Cytol 32:896–901, 1988.
98. Foust RL, Berry AD, Moinuddin SM. Fine needle aspiration cytology of liposarcoma of the breast. A case report. Acta Cytol 38:957–960, 1994.
99. Tao LC, Davidson DD. Aspiration biopsy cytology of smooth muscle tumors. A cytologic approach to the differentiation between leiomyosarcoma and leiomyoma. Acta Cytol 37:300–308, 1993.
100. Smith MB, Silverman JF, Raab SS, Towell BD, Geisinger KR. Fine-needle aspiration cytology of hepatic leiomyosarcoma. Diagn Cytopathol 11:321–327, 1994.
101. Yu GH, Kussmaul WG, DiSesa VJ, Lodato RF, Brooks JSJ. Adult intracardiac rhabdomyoma resembling the extracardiac variant. Hum Pathol 24:448–451, 1993.
102. Seidal T, Walaas L, Kindblom LG, Angervall L. Cytology of embryonal rhabdomyosarcoma: a cytologic, light microscopic, electron microscopic, and immunohistochemical study of seven cases. Diagn Cytopathol 4:292–299, 1988.
103. Moriarty AT, Nelson WA, McGahey B. Fine needle aspiration of rhabdomyosarcoma of the heart. Light and electron microscopic findings and histologic correlation. Acta Cytol 34:74–78, 1990.
104. de Almeida M, Stastny JF, Wakely PE, Frable WJ. Fine-needle aspiration biopsy of childhood rhabdomyosarcoma: reevaluation of the cytologic criteria for diagnosis. Diagn Cytopathol 11:231–236, 1994.
105. Agarwal PK, Srivastava M, Mathur N, Pant MC, Agarwal S. Fine needle aspiration of poorly differentiated rhabdomyosarcoma presenting with quadriparesis. A case report. Acta Cytol 40:985–988, 1996.
106. Ramzy I. Benign schwannoma: demonstration of Verocay bodies using fine needle aspiration. Acta Cytol 21:316–319, 1977.

107. Dahl I, Hagmar B, Idvall I. Benign solitary neurilemoma (schwannoma). A correlative cytological and histological study of 28 cases. Acta Path Microbiol Immunol Scand Sect 92:91–101, 1984.
108. Hood IC, Qizilbash AH, Young JEM, Archibald SD. Needle aspiration cytology of a benign and a malignant schwannoma. Acta Cytol 28:157–164, 1984.
109. Ryd W, Mugal S, Ayyash K. Ancient neurilemoma: a pitfall in the cytologic diagnosis of soft-tissue tumors. Diagn Cytopathol 2:244–247, 1986.
110. Vendraminelli R, Fiorentino R, Collazzo R, Filippo LD, Delendi N. Cervical schwannoma with intranuclear vacuoles by fine-needle sampling without aspiration. Diagn Cytopathol 4:335–338, 1988.
111. Mair S, Leiman G. Benign neurilemoma (schwannoma) masquerading as a pleomorphic adenoma of the submandibular salivary gland. Acta Cytol 33:907–910, 1989.
112. Bernardello F, Caneva A, Bresaola E, et al. Breast solitary schwannoma: fine-needle aspiration biopsy and immunocytochemical analysis. Diagn Cytopathol 10:221–223, 1994.
113. Housini I, Dabbs DJ. Fine needle aspiration cytology of perineurioma: report of a case with histologic, immunohistochemical and ultrastructural studies. Acta Cytol 34:420–424, 1990.
114. Silverman JF, Weaver MD, Gardner N, Larkin EW, Park HK. Aspiration biopsy cytology of malignant schwannoma metastatic to the lung. Acta Cytol 29:15–18, 1985.
115. Schwartz JG, Dowd DC. Fine needle aspiration cytology of metastatic malignant schwannoma. A case report. Acta Cytol 33:377–380, 1989.
116. Auger M, Bedard YC, Keating S. Diagnosis of a peripheral neuroectodermal tumour by fine needle aspiration. Cytopathology 1:243–249, 1990.
117. Geisinger KR, Silverman JF, Cappellari JO, Dabbs DJ. Fine-needle aspiration cytology of malignant hemangiopericytomas with ultrastructural and flow cytometric analyses. Arch Pathol Lab Med 114:705–710, 1990.
118. Kumar ND, Misra K. Aspiration cytology of hemangiopericytoma: a report of two cases. Diagn Cytopathol 6:341–344, 1990.
119. Jayaram G. Cytology of hemangioendothelioma. Acta Cytol 28:153–156, 1984.
120. Pettinato G, Insabato L, De Chiara A, Forestieri P, Manco A. Epithelioid hemangioendothelioma of soft tissue. Fine needle aspiration cytology, histology, electron microscopy and immunohistochemistry of a case. Acta Cytol 30:194–200, 1986.
121. Pettinato G, Insabato L, De Chiara A, Manco A, Forestieri P. Intranuclear cytoplasmic inclusions in epithelioid hemangioendothelioma. Acta Cytol 32:604–605, 1988.
122. Gambacorta M, Bonacina E. Epithelioid hemangioendothelioma: report of a case diagnosed by fine-needle aspiration. Diagn Cytopathol 5:207–210, 1989.
123. Nowels KW, Burford-Foggs A, Benson AB III, Hidvegi DF. Epithelioid hemangioendothelioma: cytomorphology and histological features of a case. Diagn Cytopathol 5:75–78, 1989.
124. De Gaetani CF, Trentini GP. Gastric angioendothelioma: a cytologic evaluation. Acta Cytol 21:306–309, 1977.
125. Abele JS, Miller T. Cytology of well-differentiated and poorly differentiated hemangiosarcoma in fine needle aspirates. Acta Cytol 26:341–348, 1982.
126. Silverman JF, Lannin DL, Larkin EW, Feldman P, Frable WJ. Fine-needle aspiration cytology of postirradiation sarcomas, including angiosarcoma, with immunocytochemical confirmation. Diagn Cytopathol 5:275–281, 1989.
127. Akerman M, Rydholm A. Aspiration cytology of intramuscular myxoma. A comparative clinical and histologic study of ten cases. Acta Cytol 27:505–510, 1983.
128. Mockli GC, Ljung BM, Goldman RL. Fine needle aspiration of intramuscular myxoma of the tongue. A case report. Acta Cytol 37:226–228, 1993.
129. Fornage BD, Romsdahl MM. Intramuscular myxoma: sonographic appearance and sonographically guided needle biopsy. J Ultrasound Med 13:91–94, 1994.
130. Franzen S, Stenkvist B. Diagnosis of granular cell myoblastoma by fine-needle aspiration biopsy. Acta Path Microbiol Scand 72:391–395, 1968.
131. Geisinger KR, Kawamoto EH, Marshall RB, Ahl ET, Cooper MR. Aspiration and exfoliative cytology, including ultrastructure, of a malignant granular-cell tumor. Acta Cytol 29:593–597, 1985.
132. Strobel SL, Shah NT, Lucas JG, Tuttle SE. Granular-cell tumor of the breast. A cytologic, immunohistochemical and ultrastructural study of two cases. Acta Cytol 29:598–601, 1985.
133. Kapila K, Chopra P, Verma K. Fine needle aspiration cytology of alveolar soft-part sarcoma. A case report. Acta Cytol 29:559–561, 1985.
134. Persson S, Willems JS, Kindblom LG, Angervall L. Alveolar soft-part sarcoma. Virchows Archiv A Pathol Anat Histopathol 412:499–513, 1988.
135. Drachenberg CB, Papadimitriou JC. Alveolar soft-part sarcoma. A case report with correlation of fine needle aspiration and ultrastructural cytologic features. Acta Cytol 35:746–752, 1991.
136. Shabb N, Sneige N, Fanning CV, Dekmezian R. Fine-needle aspiration cytology of alveolar soft-part sarcoma. Diagn Cytopathol 7:293–298, 1991.
137. Husain M, Nguyen GK. Alveolar soft-part sarcoma. Report of a case diagnosed by needle aspiration cytology and electron microscopy. Acta Cytol 39:951–954, 1995.
138. Liu TT, Chou YH, Lai CR, et al. Breast mass due to alveolar soft-part sarcoma of the pectoris major muscle. Eur J Radiol 24:57–59, 1997.
139. Perez JS, Perez-Guillermo M, Bernal AB, Lopez TM. Malignant rhabdoid tumor of soft tissues: a cytopathological and immunohistochemical study. Diagn Cytopathol 8:369–373, 1992.
140. Ahmed MN, Feldman M, Seemayer TA. Cytology of epithelioid sarcoma. Acta Cytol 18:459–461, 1974.
141. Goswitz JJ, Kappel T, Klingaman K. Fine-needle aspiration of epithelioid sarcoma. Diagn Cytopathol 9:677–681, 1993.
142. Hernandez-Ortiz MJ, Valenzuela-Ruiz P, Gonsalez-Estecha A, Santana-Acosta A, Ruiz-Villaespesa A. Fine needle aspiration cytology of primary epithelioid sarcoma of the vulva. A case report. Acta Cytol 39:100–103, 1995.
143. Kilpatrick SE, Teot LA, Stanley MW, Ward WG, Savage PD, Geisinger KR. Fine needle aspiration biopsy of synovial sarcoma. A cytomorphologic analysis of primary, recurrent and metastatic tumors. Am J Clin Pathol 106:769–775, 1996.
144. Griffin JB, Betsill WL. Subcutaneous endometriosis diagnosed by fine needle aspiration cytology. Acta Cytol 29:584–588, 1985.
145. Leiman G, Naylor G: Mucinous metaplasia in scar endometriosis. Diagnosis by aspiration cytology. Diagn Cytopathol 1:153–156, 1985.
146. Ashfaq R, Molberg KH, Vuitch F. Cutaneous endometriosis as a diagnostic pitfall of fine needle aspiration biopsy. A report of three cases. Acta Cytol 38:577–581, 1994.
147. Gupta RK, Naran S. Aspiration cytodiagnosis of endometriosis in an abdominal scar after cesarean section. Acta Cytol 39:603–604, 1995.
148. Kumar PV. Soft tissue chloroma diagnosed by fine needle aspiration cytology. A case report. Acta Cytol 38:83–86, 1994.
149. Mourad WA, Sneige N, Huh YO, et al. Fine needle aspiration cytology of extramedullary chronic myelogenous leukemia. Acta Cytol 39:706–712, 1995.
150. Dey P, Varma S, Varma N. Fine needle aspiration biopsy of extramedullary leukemia. Acta Cytol 40:252–256, 1996.
151. Berman JJ, McNeill RE. Cytologic evaluation of Papanicolaou-stained bone marrow aspirates. Diagn Cytopathol 5:383–387, 1989.
152. Reyes CV, Jensen J, Eng AM. Fine needle aspiration cytology of cutaneous metastases. Acta Cytol 37:142–148, 1993.
153. Srinivasan R, Ray R, Nijhawan R. Metastatic cutaneous and subcutaneous deposits from internal carcinoma. An analysis of cases diagnosed by fine needle aspiration. Acta Cytol 37:894–898, 1993.

154. Pak HY, Foster BA, Yokota SB. The significance of cutaneous metastasis from visceral tumors diagnosed by fine-needle aspiration biopsy. Diagn Cytopathol 3:24–29, 1987.
155. Gattuso P, Castelli MJ, Reyes CV, Reddy V. Cutaneous and subcutaneous masses of the chest wall: a fine-needle aspiration study. Diagn Cytopathol 15:374–376, 1996.
156. Dey P, Karmakar T. Fine needle aspiration cytology in operative scar nodules in breast carcinoma. Acta Cytol 39:137–139, 1995.
157. Velanonich V. Fine-needle aspiration cytology in the diagnosis and management of ectopic breast tissue. Am Surg 61:277–278, 1995.
158. Vargas J, Nevado M, Rodriquez-Peralto JL, De Agustin PP. Fine needle aspiration diagnosis of carcinoma arising in an ectopic breast. A case report. Acta Cytol 39:941–944, 1995.
159. Aisner SC, Burke KC, Resnik CS. Aspiration cytology of heterotopic ossification. A case report. Acta Cytol 36:159–162, 1992.
160. Dodd LG, Layfield LS. Fine needle aspiration of inflammatory carcinoma of the breast. Diag Cytopathol 15:363–366, 1996.
161. Caraway NP, Fanning CV, Wojcik EM, Staerkel GA, Benjamin RS, Ordonez NG. Cytology of malignant melanoma of soft parts: fine needle aspirates and exfoliative specimens. Diagn Cytopathol 9:632–638, 1993.
162. Almeida MM, Nunes AM, Frable WJ. Malignant melanoma of soft tissue. A report of three cases with diagnosis by fine needle aspiration cytology. Acta Cytol 38:241–246, 1994.
163. Husain M, Nguyen GK. Malignant melanoma of soft parts diagnosed by needle aspiration cytology and electron microscopy. Diagn Cytopathol 13:89–92, 1995.
164. Wakely PE, Frable WJ. Fine-needle aspiration biopsy cytology of giant-cell tumor of tendon sheath. Am J Clin Pathol 102:87–90, 1994.
165. Gonzalez-Campora R, Herrero ES, Otal-Salaverri C, et al. Diffuse tenosynovial giant cell tumor of soft tissues. Report of a case with cytologic and cytogenetic findings. Acta Cytol 39:770–776, 1995.
166. Agarwal PK, Gupta M, Srivastava A, Agarwal S. Cytomorphology of giant cell tumor of tendon sheath. A report of two cases. Acta Cytol 41:587–589, 1997.
167. Gordon MD, Vetto J, Meshul CK, Schmidt WA. FNA of extraskeletal myxoid chondrosarcoma: cytomorphologic, EM, and x-ray microanalysis features. Diagn Cytopathol 10:352–356, 1994.
168. Hazarika D, Kumar RV, Rao CR, Mukherjee G, Pattabhiraman V, Shekar MC. Fine needle aspiration cytology of chondroblastoma and chondromyxoid fibroma. A report of two cases. Acta Cytol 38:592–596, 1994.
169. Niemann TH, Bottles K, Cohen MB. Extraskeletal myxoid chondrosarcoma: fine-needle aspiration biopsy findings. Diagn Cytopathol 11:363–366, 1994.
170. Silverman JF, Lannin DL, Larkin EW, Feldman P, Frable WJ. Fine needle aspiration cytology of postirradiation sarcomas, including angiosarcoma, with immunocytochemical confirmation. Diagn Cytopathol 5:275–281, 1989.
171. Pai RR, Raghuveer CV. Extramedullary plasmacytoma diagnosed by fine needle aspiration cytology. A report of four cases. Acta Cytol 40:963–966, 1996.
172. Sanchez MA, Rorat E. Fine needle aspiration diagnosis of intramuscular bacillary angiomatosis. A case report. Acta Cytol 40:751–755, 1996.

Index

Note: Page numbers in italics indicate figures; those followed by t indicate tables.

C

D

E

R

S

T

V

W

ISBN 0-443-07963-3